Ilya Narsky and Frank C. Porter

**Statistical Analysis Techniques
in Particle Physics**

Related Titles

Quarks and Leptons
An Introductory Course in Modern Particle Physics

1984
ISBN: 978-0-471-88741-6

Barlow, R.J.

Statistics
A Guide to the use of StatisticMethods in the Physical Science

1989
ISBN: 978-0-471-92295-7, e-publication available

Talman, R.

Accelerator X-Ray Sources

2006
ISBN: 978-3-527-40590-9, e-publication available, ISBN: 978-3-527-61030-3

Griffiths, D.

Introduction to Elementary Particles

2008
ISBN: 978-3-527-40601-2

Wangler, T.P.

RF Linear Accelerators

2008
ISBN: 978-3-527-40680-7, e-publication available

Reiser, M.

Theory and Design of Charged Particle Beams

2008
ISBN: 978-3-527-40741-5, -publication available

Russenschuck, S.

Field Computation for Accelerator Magnets
Analytical and Numerical Methods for Electromagnetic Design and Optimization

2010
ISBN: 978-3-527-40769-9, e-publication available

Padamsee, H., Knobloch, J., Hays, T.

RF Superconductivity for Accelerators

2008
ISBN: 978-3-527-40842-9

Brock, I., Schörner-Sadenius, T. (eds.)

Physics at the Terascale

2011
ISBN: 978-3-527-41001-9, e-publication available

Behnke, O., Kröninger, K., Schott, G., Schörner-Sadenius, T. (eds.)

Data Analysis in High Energy Physics
A Practical Guide to Statistical Methods

2013
ISBN: 978-3-527-41058-3, e-publication available

Ilya Narsky and Frank C. Porter

Statistical Analysis Techniques in Particle Physics

Fits, Density Estimation and Supervised Learning

WILEY-VCH

Authors

Dr. Ilya Narsky
MathWorks
Natick
United States of America
inarsky@yahoo.com

Prof. Frank C. Porter
California Inst. of Technology
Physics Department
Pasadena
United States of America
fcp@hep.caltech.edu

Cover
CMS Experiment at LHC, CERN
Data recorded Mon Nov 8, 11:30:43 2010 CEST
Run/Event 150431/630470
Lumi section 173

■ All books published by Wiley-VCH are carefully produced. Nevertheless, authors, editors, and publisher do not warrant the information contained in these books, including this book, to be free of errors. Readers are advised to keep in mind that statements, data, illustrations, procedural details or other items may inadvertently be inaccurate.

Library of Congress Card No.:
applied for

British Library Cataloguing-in-Publication Data:
A catalogue record for this book is available from the British Library.

Bibliographic information published by the Deutsche Nationalbibliothek
The Deutsche Nationalbibliothek lists this publication in the Deutsche Nationalbibliografie; detailed bibliographic data are available on the Internet at http://dnb.d-nb.de.

© 2014 WILEY-VCH Verlag GmbH & Co. KGaA, Boschstr. 12, 69469 Weinheim, Germany

All rights reserved (including those of translation into other languages). No part of this book may be reproduced in any form – by photoprinting, microfilm, or any other means – nor transmitted or translated into a machine language without written permission from the publishers. Registered names, trademarks, etc. used in this book, even when not specifically marked as such, are not to be considered unprotected by law.

Print ISBN 978-3-527-41086-6
ePDF ISBN 978-3-527-67731-3
ePub ISBN 978-3-527-67729-0
Mobi ISBN 978-3-527-67730-6
oBook ISBN 978-3-527-67732-0

Cover Design Grafik-Design Schulz
Typesetting le-tex publishing services GmbH, Leipzig
Printing and Binding Markono Print Media Pte Ltd, Singapore

Printed on acid-free paper

Contents

Acknowledgements *XIII*

Notation and Vocabulary *XV*

1 **Why We Wrote This Book and How You Should Read It** *1*

2 **Parametric Likelihood Fits** *5*
2.1 Preliminaries *5*
2.1.1 Example: CP Violation via Mixing *7*
2.1.2 The Exponential Family *9*
2.1.3 Confidence Intervals *10*
2.1.4 Hypothesis Tests *11*
2.2 Parametric Likelihood Fits *12*
2.2.1 Nuisance Parameters *16*
2.2.2 Confidence Intervals from Pivotal Quantities *17*
2.2.3 Asymptotic Inference *19*
2.2.4 Profile Likelihood *20*
2.2.5 Conditional Likelihood *20*
2.3 Fits for Small Statistics *21*
2.3.1 Sample Study of Coverage at Small Statistics *22*
2.3.2 When the pdf Goes Negative *25*
2.4 Results Near the Boundary of a Physical Region *26*
2.5 Likelihood Ratio Test for Presence of Signal *28*
2.6 ✂ sPlots *31*
2.7 Exercises *35*
References *37*

3 **Goodness of Fit** *39*
3.1 Binned Goodness of Fit Tests *41*
3.2 Statistics Converging to Chi-Square *46*
3.3 Univariate Unbinned Goodness of Fit Tests *49*
3.3.1 Kolmogorov–Smirnov *49*
3.3.2 Anderson–Darling *50*
3.3.3 Watson *51*
3.3.4 Neyman Smooth *51*

3.4	Multivariate Tests 52
3.4.1	Energy Tests 53
3.4.2	Transformations to a Uniform Distribution 54
3.4.3	Local Density Tests 55
3.4.4	Kernel-based Tests 56
3.4.5	Mixed Sample Tests 57
3.4.6	Using a Classifier 58
3.5	Exercises 59
	References 61
4	**Resampling Techniques** 63
4.1	Permutation Sampling 63
4.2	Bootstrap 65
4.2.1	Bootstrap Confidence Intervals 68
4.2.2	Smoothed Bootstrap 70
4.2.3	Parametric Bootstrap 70
4.3	Jackknife 70
4.4	BC_a Confidence Intervals 76
4.5	Cross-Validation 78
4.6	✄ Resampling Weighted Observations 82
4.7	Exercises 86
	References 86
5	**Density Estimation** 89
5.1	Empirical Density Estimate 90
5.2	Histograms 90
5.3	Kernel Estimation 92
5.3.1	Multivariate Kernel Estimation 92
5.4	Ideogram 93
5.5	Parametric vs. Nonparametric Density Estimation 93
5.6	Optimization 94
5.6.1	Choosing Histogram Binning 97
5.7	Estimating Errors 100
5.8	The Curse of Dimensionality 102
5.9	Adaptive Kernel Estimation 103
5.10	Naive Bayes Classification 105
5.11	Multivariate Kernel Estimation 106
5.12	Estimation Using Orthogonal Series 108
5.13	Using Monte Carlo Models 111
5.14	Unfolding 112
5.14.1	Unfolding: Regularization 116
5.15	Exercises 120
	References 120
6	**Basic Concepts and Definitions of Machine Learning** 121
6.1	Supervised, Unsupervised, and Semi-Supervised 121

6.2	Tall and Wide Data *123*
6.3	Batch and Online Learning *124*
6.4	Parallel Learning *125*
6.5	Classification and Regression *127*
	References *128*

7	**Data Preprocessing** *129*
7.1	Categorical Variables *129*
7.2	Missing Values *132*
7.2.1	Likelihood Optimization *134*
7.2.2	Deletion *135*
7.2.3	Augmentation *137*
7.2.4	Imputation *137*
7.2.5	Other Methods *139*
7.3	Outliers *139*
7.4	Exercises *141*
	References *142*

8	**Linear Transformations and Dimensionality Reduction** *145*
8.1	Centering, Scaling, Reflection and Rotation *145*
8.2	Rotation and Dimensionality Reduction *146*
8.3	Principal Component Analysis (PCA) *147*
8.3.1	Theory *148*
8.3.2	Numerical Implementation *149*
8.3.3	Weighted Data *150*
8.3.4	How Many Principal Components Are Enough? *151*
8.3.5	Example: Apply PCA and Choose the Optimal Number of Components *154*
8.4	Independent Component Analysis (ICA) *158*
8.4.1	Theory *158*
8.4.2	Numerical implementation *161*
8.4.3	Properties *162*
8.5	Exercises *163*
	References *163*

9	**Introduction to Classification** *165*
9.1	Loss Functions: Hard Labels and Soft Scores *165*
9.2	Bias, Variance, and Noise *168*
9.3	Training, Validating and Testing: The Optimal Splitting Rule *173*
9.4	Resampling Techniques: Cross-Validation and Bootstrap *177*
9.4.1	Cross-Validation *177*
9.4.2	Bootstrap *179*
9.4.3	Sampling with Stratification *181*
9.5	Data with Unbalanced Classes *182*
9.5.1	Adjusting Prior Probabilities *183*
9.5.2	✂ Undersampling the Majority Class *184*

9.5.3	✂ Oversampling the Minority Class	*185*
9.5.4	Example: Classification of Forest Cover Type Data	*186*
9.6	Learning with Cost	*190*
9.7	Exercises	*191*
	References	*192*

10   Assessing Classifier Performance   *195*
- 10.1 Classification Error and Other Measures of Predictive Power   *195*
- 10.2 Receiver Operating Characteristic (ROC) and Other Curves   *196*
- 10.2.1 Empirical ROC curve   *196*
- 10.2.2 Other Performance Measures   *198*
- 10.2.3 Optimal Operating Point   *198*
- 10.2.4 Area Under Curve   *200*
- 10.2.5 Smooth ROC Curves   *200*
- 10.2.6 Confidence Bounds for ROC Curves   *205*
- 10.3 Testing Equivalence of Two Classification Models   *210*
- 10.4 Comparing Several Classifiers   *215*
- 10.5 Exercises   *217*
- References   *218*

11   Linear and Quadratic Discriminant Analysis, Logistic Regression, and Partial Least Squares Regression   *221*
- 11.1 Discriminant Analysis   *221*
- 11.1.1 Estimating the Covariance Matrix   *223*
- 11.1.2 Verifying Discriminant Analysis Assumptions   *225*
- 11.1.3 Applying LDA When LDA Assumptions Are Invalid   *226*
- 11.1.4 Numerical Implementation   *228*
- 11.1.5 Regularized Discriminant Analysis   *228*
- 11.1.6 LDA for Variable Transformation   *229*
- 11.2 Logistic Regression   *231*
- 11.2.1 Binomial Logistic Regression: Theory and Numerical Implementation   *231*
- 11.2.2 Properties of the Binomial Model   *233*
- 11.2.3 Verifying Model Assumptions   *233*
- 11.2.4 Logistic Regression with Multiple Classes   *234*
- 11.3 Classification by Linear Regression   *235*
- 11.4 ✂ Partial Least Squares Regression   *236*
- 11.5 Example: Linear Models for MAGIC Telescope Data   *239*
- 11.6 Choosing a Linear Classifier for Your Analysis   *247*
- 11.7 Exercises   *247*
- References   *248*

12   Neural Networks   *251*
- 12.1 Perceptrons   *251*
- 12.2 The Feed-Forward Neural Network   *254*
- 12.3 Backpropagation   *256*

12.4	Bayes Neural Networks	*260*
12.5	Genetic Algorithms	*262*
12.6	Exercises	*263*
	References	*263*

13 Local Learning and Kernel Expansion *265*

13.1	From Input Variables to the Feature Space	*266*
13.1.1	Kernel Regression	*269*
13.2	Regularization	*270*
13.2.1	Kernel Ridge Regression	*274*
13.3	Making and Choosing Kernels	*278*
13.4	Radial Basis Functions	*279*
13.4.1	Example: RBF Classification for the MAGIC Telescope Data	*280*
13.5	Support Vector Machines (SVM)	*283*
13.5.1	SVM with Weighted Data	*286*
13.5.2	SVM with Probabilistic Outputs	*288*
13.5.3	✄ Numerical Implementation	*288*
13.5.4	✄ Multiclass Extensions	*293*
13.6	Empirical Local Methods	*293*
13.6.1	Classification by Probability Density Estimation	*294*
13.6.2	Locally Weighted Regression	*295*
13.6.3	Nearest Neighbors and Fuzzy Rules	*298*
13.7	Kernel Methods: The Good, the Bad and the Curse of Dimensionality	*302*
13.8	Exercises	*303*
	References	*304*

14 Decision Trees *307*

14.1	Growing Trees	*308*
14.2	Predicting by Decision Trees	*312*
14.3	Stopping Rules	*312*
14.4	Pruning Trees	*313*
14.4.1	Example: Pruning a Classification Tree	*317*
14.5	Trees for Multiple Classes	*319*
14.6	✄ Splits on Categorical Variables	*320*
14.7	Surrogate Splits	*321*
14.8	✄ Missing Values	*323*
14.9	Variable importance	*324*
14.10	Why Are Decision Trees Good (or Bad)?	*327*
14.11	Exercises	*328*
	References	*329*

15 Ensemble Learning *331*

15.1	Boosting	*332*
15.1.1	Early Boosting	*332*
15.1.2	AdaBoost for Two Classes	*333*

15.1.3	Minimizing Convex Loss by Stagewise Additive Modeling *336*
15.1.4	Maximizing the Minimal Margin *343*
15.1.5	Nonconvex Loss and Robust Boosting *351*
15.1.6	Boosting for Multiple Classes *357*
15.2	Diversifying the Weak Learner: Bagging, Random Subspace and Random Forest *358*
15.2.1	Measures of Diversity *359*
15.2.2	Bagging and Random Forest *361*
15.2.3	Random Subspace *363*
15.2.4	Example: K/π Separation for BaBar PID *364*
15.3	Choosing an Ensemble for Your Analysis *365*
15.4	Exercises *367*
	References *367*

16 Reducing Multiclass to Binary *371*

16.1	Encoding *372*
16.2	Decoding *375*
16.3	Summary: Choosing the Right Design *378*
	References *379*

17 How to Choose the Right Classifier for Your Analysis and Apply It Correctly *381*

17.1	Predictive Performance and Interpretability *381*
17.2	Matching Classifiers and Variables *382*
17.3	Using Classifier Predictions *382*
17.4	Optimizing Accuracy *383*
17.5	CPU and Memory Requirements *383*

18 Methods for Variable Ranking and Selection *385*

18.1	Definitions *386*
18.1.1	Variable Ranking and Selection *386*
18.1.2	Strong and Weak Relevance *386*
18.2	Variable Ranking *389*
18.2.1	Filters: Correlation and Mutual Information *390*
18.2.2	Wrappers: Sequential Forward Selection (SFS), Sequential Backward Elimination (SBE), and Feature-based Sensitivity of Posterior Probabilities (FSPP) *394*
18.2.3	Embedded Methods: Estimation of Variable Importance by Decision Trees, Neural Networks, Nearest Neighbors, and Linear Models *400*
18.3	Variable Selection *401*
18.3.1	Optimal-Set Search Strategies *401*
18.3.2	Multiple Testing: Backward Elimination by Change in Margin (BECM) *403*
18.3.3	Estimation of the Reference Distribution by Permutations: Artificial Contrasts with Ensembles (ACE) Algorithm *410*
18.4	Exercises *413*

References *414*

19 Bump Hunting in Multivariate Data *417*
19.1 Voronoi Tessellation and SLEUTH Algorithm *418*
19.2 Identifying Box Regions by PRIM and Other Algorithms *420*
19.3 Bump Hunting Through Supervised Learning *422*
References *423*

20 Software Packages for Machine Learning *425*
20.1 Tools Developed in HEP *425*
20.2 R *426*
20.3 MATLAB *427*
20.4 Tools for Java and Python *428*
20.5 What Software Tool Is Right for You? *429*
References *430*

Appendix A: Optimization Algorithms *431*
A.1 Line Search *431*
A.2 Linear Programming (LP) *432*

Index *435*

Acknowledgements

We are grateful to our colleagues in high energy physics, astrophysics, and at Math-Works (I.N.) for many stimulating discussions and interesting problems. These interactions are ultimately the motivation for this book. While writing the book, we solicited and received comments from several colleagues who were willing to slog through unpolished material. We are certain that their suggestions have made this a better exposition. Specifically, we are indebted to physicists Roger Barlow, Tingting Cao, Bertrand Echenard, Louis Lyons, Piti Ongmongkolkul, and Steve Sekula; and statisticians Tom Lane, Crystal Linkletter, Gautam Pendse, and Ting Su.

We would like to thank Lars Sonnenschein for encouraging us to submit the book proposal to Wiley and coordinating the submission. We are grateful to Vera Palmer, Ulrike Werner, and Wiley for their technical guidance and support.

We thank Caltech High Energy Physics for hosting the web site for the book.

Notation and Vocabulary

We adopt several conventions in this book, which are gathered here for reference.

Random variables (RVs) are denoted with upper case Roman letters, for example, X. A value of such a random variable is denoted with the corresponding lower case letter, for example, x. Parameters needed to completely specify a distribution are usually denoted with Greek letters, for example, θ. Estimators for parameters consist of the same symbol, with an added "hat", for example, $\hat{\theta}$. Estimators are also RVs, but not distinguished with upper case from their values. Where desired for clarity, an argument is added, which provides this information, that is, $\hat{\theta}(X)$ is an RV and $\hat{\theta}(x)$ is a value of this RV at x.

Distributions are usually specified with an upper case Roman letter, such as F, for a cdf (cumulative distribution function), and the corresponding lower case letter, f, for the corresponding pdf (probability distribution function). Sometimes we use P to specify a pdf, where P stands for "probability". An estimator for a distribution is denoted with the hat notation, for example, $\hat{F}$. The notation $F_X(x;\theta)$ refers to the cdf to sample a value x from distribution F for random variable X, where the distribution depends on the value of parameter θ. Similarly, $f_X(x;\theta)$ refers to the respective pdf. We will generally use various simplifications of this notation where the meaning should be clear, for example, $F(x;\theta)$.

A joint distribution of two or more RVs separates arguments by commas, for example, $f_{X,Y}(X,Y)$. A conditional distribution uses the "|" notation. For example, the pdf of Y conditioned on X would be denoted by $f_{Y|X}(Y|X)$, and its value at specific x and y would be denoted by $f_{Y|X}(y|x)$. For brevity, we often drop the subscript $Y|X$.

Integrals are to be interpreted in the Lebesgue sense. In particular, $dF(x)$ and $f(x)dx$ are to be interpreted with the appropriate Dirac measure for discrete distributions. In other words, these are pmfs (probability mass functions) in the case of discrete distributions, with the integral sign becoming summation.

Vector RVs are set in bold Roman letters, for example, $\mathbf{X}$, and vector parameters are set in bold Greek, $\boldsymbol{\theta}$. A vector of observed values drawn from a vector RV is set in bold lowercase: $\mathbf{x}$. A vector consisting of i.i.d. (identical independently distributed) observations of a scalar RV is set in bold lowercase with a distinguishing font: **y**. A matrix of observed values is set in upper case bold with the same distinguishing font: **X**. Note that in this case we abandon the case distinction between a RV and an observation, in preference to emphasizing the matrix aspect.

We use the "E" notation to denote an expectation value. Thus $E(X)$ is the mean of the sampling distribution for RV X. Where necessary to make the sampling distribution clear, a subscript is added: $E_F(X)$ is the mean of X in distribution F. A subscript can be added to indicate what RV is used for taking an expectation of a function of several RVs. For example, if $G(X, Y)$ is a function of two RVs, $E_X(G)$ is used to denote a function of Y averaged over X. The variance of a RV is denoted by $\text{Var}(X)$. If the RV is multivariate, then $\text{Var}(\boldsymbol{X})$ is the full covariance matrix.

Response of a specific classification or regression model at point $\boldsymbol{x}$ is usually denoted by $f(\boldsymbol{x})$, and the respective RV is denoted by $F(\boldsymbol{x})$. We thus overload the letters f and F using them both for probability distributions and model response. It should be easy to avoid confusion because we never mix the notation in the same chapter. If f and F are used to specify response, P is used to specify a distribution.

In particle physics, an "event" represents a set of coincidental measurements. Adopting a vocabulary from the statistics literature, we use the word "observation" instead. An event or an observation is represented by a vector of recorded values, for example, $\boldsymbol{x}$. Unlike events, observations do not need to be ordered in time.

We use a capital letter to denote the size of a set with the respective lowercase letter used for indexing. For example, a dataset can have N observations indexed by $n = 1, \ldots, N$. The curly brackets are used to encapsulate a set, for example, $\{\boldsymbol{x}_n\}_{n=1}^N$ denotes a set of N observations represented by multivariate vectors $\boldsymbol{x}_n$. Sometimes we use a capital letter to specify a set, for example, $\mathcal{X}$; in that case, we use $|\mathcal{X}|$ to specify the set size.

We use $\|\boldsymbol{x}\|_q$ to denote the q-norm, $(\sum_{d=1}^D x_d^q)^{1/q}$, of vector $\boldsymbol{x}$ with D elements. The Euclidean distance between two vectors, $\boldsymbol{a}$ and $\boldsymbol{b}$, is the 2-norm of their difference, $\|\boldsymbol{a} - \boldsymbol{b}\|_2$. If the subscript q is omitted, the 2-norm $q = 2$ is assumed.

Elementwise (Hadamard) multiplication of two matrices of equal size is specified using $\circ$, for example, $\mathbf{A} = \mathbf{B} \circ \mathbf{C}$. This implies $a_{nd} = b_{nd} c_{nd}$ for $1 \leq n \leq N$ and $1 \leq d \leq D$, where N is the number of rows and D is the number of columns.

The $g : x \mapsto \mathcal{Y}$ notation specifies function g mapping argument x onto domain or set $\mathcal{Y}$.

The notation $\liminf_N$ refers to the limit of the greatest lower bound as $N \to \infty$, that is, $\liminf_N f(N) \equiv \lim_{N \to \infty} [\inf_{M \geq N} f(M)]$.

Various special symbols are summarized in Table 1.

Table 1 Special symbols used in the book.

Dimension	dim
Dirac delta function	$\delta(x)$
If and only if	iff
Gradient, or nabla operator	∇
k-dimensional real coordinate space	$\mathcal{R}^k$
Kronecker symbol	$\delta_{ij} = \begin{cases} 0 & \text{if } i \neq j \\ 1 & \text{if } i = j \end{cases}$
Laplace operator	∇^2
Normal (Gaussian) distribution	N

1
Why We Wrote This Book and How You Should Read It

The advent of high-speed computing has enabled a transformation in practical statistical methodology. We have entered the age of "machine learning". Roughly, this means that we replace assumptions and approximations by computing power in order to derive statements about characteristics of a dataset. Statisticians and computer scientists have produced an enormous literature on this subject. The jargon includes bootstrap, cross-validation, learning curves, receiver operating characteristic, decision trees, neural nets, boosting, bagging, and so on. These algorithms are becoming increasingly popular in analysis of particle and astrophysics data. The idea of this book is to provide an introduction to these tools and methods in language and context appropriate to the physicist.

Machine learning may be divided into two broad types, "supervised learning" and "unsupervised learning", with due caution toward oversimplifying. Supervised learning can be thought of as fitting a function in a multivariate domain over a set of measured values. Unsupervised learning is exploratory analysis, used when you want to discover interesting features of a dataset. This book is focused on the supervised learning side.

Supervised learning comes in two flavors: classification and regression. Classification aims at separating observations of different kinds such as signal and background. The fitted function in this case takes categorical values. Fitting a function with continuous values is addressed by regression. Fitting a scalar function of a scalar argument by least squares is a well-known regression tool. In case of a vector argument, classification appears to be used more often in modern physics analysis. This is why we focus on classification.

This book is not an introductory probability and statistics text. We assume our readers have been exposed to basic probability theory and to basic methods in parameter estimation such as maximum likelihood. We do not ignore these basic tools, but aim our discussion past the elementary development. Solid understanding of linear algebra, familiarity with multivariate calculus and some exposure to set theory are required as well.

Chapter 2 reviews techniques for parametric likelihood, a subject familiar to most physicists. We include discussion of practical issues such as fits for small statistics and fits near the boundary of a physical region, as well as advanced topics such as sPlots. Goodness of fit measures for univariate and multivariate data are reviewed in Chapter 3. These measures can be applied to distribution fitting by parametric likelihood described in Chapter 2 and nonparametric density estimation described in Chapter 5. Chapter 4 introduces resampling techniques such as the bootstrap, in the context of parameter estimation. Chapter 5 overviews techniques for nonparametric density estimation by histograms and kernel smoothing. The subjects reviewed in these chapters are part of the traditional statistician's toolkit.

In Chapter 6 we turn attention to topics in machine learning. Chapter 6 introduces basic concepts and gives a cursory survey of the material that largely remains beyond the scope of this book. Before learning can begin, data need to be cleaned up and organized in a suitable way. These basic processing steps, in particular treatment of categorical variables, missing values, and outliers, are discussed in Chapter 7. Other important steps, optionally taken before supervised learning starts, are standardizing variable distributions and reducing the data dimensionality. Chapter 8 reviews simple techniques for univariate transformations and advanced techniques such as principal and independent component analysis.

In Chapter 9 we shift the focus to classification. Chapters 9 and 10 are essential for understanding the material in Chapters 11–18. Chapter 9 formalizes the problem of classification and lays out the common workflow for solving this problem. Resampling techniques are revisited here as well, with an emphasis on their application to supervised learning. Chapter 10 explains how the quality of a classification model can be judged and how two (or more) classification models can be compared by a formal statistical test. Some topics in these chapters can be skipped at first reading. In particular, analysis of data with class imbalance, although an important practical issue, is not required to enjoy the rest of the book.

Chapters 11–15 review specific classification techniques. Although many of them can be used for two-class (binary) and multiclass learning, binary classification has been studied more actively than multiclass algorithms. Chapter 16 describes a framework for reducing multiclass learning to a set of binary problems.

Chapter 17 provides a summary of the material learned in Chapters 11–16. Summaries oversimplify and should be interpreted with caution. Bearing this in mind, use this chapter as a practical guide for choosing a classifier appropriate for your analysis.

Chapter 18 reviews methods for selecting the most important variables from all inputs in the data. The importance of a variable is measured by its effect on the predictive power of a classification model.

Bump hunting in multivariate data may be an important component of searches for new physics processes at the Large Hadron Collider, as well as other experiments. We discuss appropriate techniques in Chapter 19. This discussion is focused on multivariate nonparametric searches, in a setting more complex than univariate likelihood fits.

Throughout the book, we illustrate the application of various algorithms to data, either simulated or borrowed from a real physics analysis, using examples of MATLAB code. These examples could be coded in another language. We have chosen MATLAB for two reasons. First, one of the authors, employed by MathWorks, has been involved in design and implementation of the MATLAB utilities supporting various algorithms described in this book. We thus have intimate knowledge of how these utilities work. Second, MATLAB is a good scripting language. If we used tools developed in the particle physics community, these code snippets would be considerably longer and less transparent.

There are many software suites that provide algorithms described here besides MATLAB. In Chapter 20 we review several software toolkits, whether developed by physicists or not.

We hope that this book can serve both pedagogically and as a reference. If your goal is to learn the broad arsenal of statistical tools for physics analysis, read Chapters 2–9. If you are interested primarily in classification, read Chapters 9 and 10; then choose one or more chapters from Chapters 11–15 for in-depth reading. If you wish to learn a specific classification technique, read Chapters 9 and 10 before digging into the respective chapter or section.

Sections labeled with ✀ present advanced material and can be cut at first reading.

In various places throughout the book we use datasets, either simulated or measured. Some of these datasets can be downloaded from the Machine Learning Repository maintained by University of California in Irvine, http://www.ics.uci.edu/~mlearn.

A fraction of this book is posted at the Caltech High Energy Physics site, http://www.hep.caltech.edu/~NarskyPorter. The posted material includes code for all MATLAB examples, without comments or graphs. Sections not included in the book are posted there as well.

2
Parametric Likelihood Fits

The likelihood function is a pervasive object in physics analyses. In this chapter we review several basic concepts involving the likelihood function, with some simple applications to confidence intervals and hypothesis tests in the context of parametric statistics. We end with the technique of sPlots, providing an optimized method for background subtraction in multivariate settings.

2.1
Preliminaries

Given a collection of data, $\mathbf{X} = \{x_1, \ldots, x_N\}$, corresponding to a sampling of random variables (possibly vectors, hence the vector notation) from pdf $f(\mathbf{X}; \theta)$, where $\boldsymbol{\theta} = \{\theta_1, \ldots, \theta_R\}$ is a vector of unknown parameters, we define the *likelihood function* as a function of $\boldsymbol{\theta}$, for the sampled values $\mathbf{x}$ according to:

$$L(\boldsymbol{\theta}; \mathbf{X}) = f(\mathbf{X}; \boldsymbol{\theta}). \tag{2.1}$$

The normalization of the likelihood function (of $\boldsymbol{\theta}$) is typically unimportant in practice, in contrast to the pdf $f(\mathbf{X}; \boldsymbol{\theta})$, which must be normalized to one when integrated over the sample space. The likelihood function is a widely-used tool in making inferences concerning $\boldsymbol{\theta}$. It is of fundamental importance in Bayesian statistics. In frequency statistics, the likelihood function has no profound interpretation,[1] but nonetheless provides an important concept leading to algorithms with many useful properties.

There are some further notions that will be useful in discussing the use of the likelihood function and related topics. First, we shall typically suppose that the X_n are independent and identically distributed (i.i.d.) random variables. The dimension N is then called the *sample size*.

The likelihood function may be used to formulate a statistical measure for information, relevant to parameters of interest:

---

[1] This is controversial. The interested reader is referred to Cox and Mayo (2011); Mayo (2010), and references therein.

2 Parametric Likelihood Fits

Definition 2.1. *If $L(\theta; \mathbf{X})$ is a likelihood function depending on parameter θ, the Fisher information number (or Fisher information, or simply information), corresponding to θ, is:*

$$I(\theta) \equiv E\left\{\left[\frac{\partial \log L(\theta; \mathbf{X})}{\partial \theta}\right]^2\right\}. \tag{2.2}$$

This generalizes to the $R \times R$ *Fisher information matrix* in the case of a multidimensional parameter space:

$$I(\boldsymbol{\theta}) = E\left\{\frac{\partial \log L(\boldsymbol{\theta}; \mathbf{X})}{\partial \boldsymbol{\theta}}\left[\frac{\partial \log L(\boldsymbol{\theta}; \mathbf{X})}{\partial \boldsymbol{\theta}}\right]^\mathsf{T}\right\}. \tag{2.3}$$

We assume this matrix is positive definite for any $\boldsymbol{\theta}$ in the parameter space.

An intuitive view is that if L varies rapidly with θ, the experimental sampling distribution will be very sensitive to θ. Hence, a measurement will contain a lot of "information" relevant to θ. It is handy to note that (exercise for the reader):

$$E\left[\left(\frac{\partial \log L}{\partial \theta}\right)^2\right] = -E\left(\frac{\partial^2 \log L}{\partial \theta^2}\right). \tag{2.4}$$

In a multidimensional parameter space, this becomes the negative expectation value of the Hessian for $\log L$.

The quantity $\partial_\theta \log L$ is known as the *score function*:

$$S(\theta; \mathbf{X}) \equiv \frac{\partial \log L(\theta; \mathbf{X})}{\partial \theta}. \tag{2.5}$$

A property of the score function is that its expectation is zero:

$$E[S(\theta; \mathbf{X})] = 0. \tag{2.6}$$

Certain conditions must be satisfied for this to hold, notably that the sample space of $\mathbf{X}$ should be independent of θ.

When we estimate a population parameter given some sample $\mathbf{X}$, we strive for an estimate which is as close as possible to the true parameter value. That is, we wish to find an estimate with a small error. There are many ways one could measure error, but perhaps the most common measure is the *mean squared error*, or MSE. Given an estimator (random variable) $\hat{\theta}(\mathbf{X})$ for population parameter θ, the MSE is defined by:

$$\begin{aligned}\mathrm{MSE}(\hat{\theta}; \theta) &\equiv E[(\hat{\theta}(\mathbf{X}) - \theta)^2] \\ &= \mathrm{Var}[\hat{\theta}(\mathbf{X})] + \{E[\hat{\theta}(\mathbf{X})] - \theta\}^2.\end{aligned} \tag{2.7}$$

We recognize the second term in the second line as the square of the bias of the estimator. For an unbiased estimator, the MSE is simply the variance of the estimator. There is an obvious generalization of this definition to the multivariate case.

Using the information number, we may derive a lower bound on the variance of a parameter estimate for a given bias. Suppose that we have an estimator $\hat{\theta} = \hat{\theta}(\mathbf{X})$ for a parameter θ, with a *bias function* $b(\theta) \equiv E(\hat{\theta} - \theta)$. We have the theorem:

Theorem 2.2. *Rao–Cramér–Frechet (RCF)*
Assume:

1. The sample space of $\mathbf{X}$ is independent of θ.
2. The variance of $\hat{\theta}$ is finite, for any θ.
3. $\partial_\theta \int_{-\infty}^{\infty} g(\mathbf{X}) L(\theta; \mathbf{X}) d\mathbf{X} = \int_{-\infty}^{\infty} g(\mathbf{X}) \partial_\theta L(\theta; \mathbf{X}) d\mathbf{X}$, where $g(\mathbf{X})$ is any statistic of finite variance.

Then the variance, $\sigma_{\hat{\theta}}^2$ of estimator $\hat{\theta}$ obeys the inequality:

$$\sigma_{\hat{\theta}}^2 \geq \frac{[1 + \partial_\theta b(\theta)]^2}{I(\theta)}. \tag{2.8}$$

The proof is left as an exercise, but here is a sketch. First, show that

$$I(\theta) = \mathrm{Var}(\partial_\theta \log L). \tag{2.9}$$

Next, find the linear correlation parameter:

$$\rho \equiv \frac{E\{[S(\theta; \mathbf{X}) - E(S(\theta; \mathbf{X}))][\hat{\theta} - E(\hat{\theta})]\}}{\sqrt{\mathrm{Var}(\hat{\theta})\,\mathrm{Var}(S(\theta; \mathbf{X}))}}, \tag{2.10}$$

between the score function and $\hat{\theta}$. Finally, note that $\rho^2 \leq 1$.

In particular, for an *unbiased estimator* ($b(\theta) = 0, \forall \theta$), the minimum variance is

$$\sigma_{\hat{\theta}}^2 \geq \frac{1}{I(\theta)}. \tag{2.11}$$

An *efficient* unbiased estimator is one that achieves this bound, which we shall refer to as the RCF bound. The RCF bound also provides a handy tool to estimate uncertainties in planning an experiment, under the assumption that one can get close to the bound with a good estimator (and enough statistics).

2.1.1
Example: CP Violation via Mixing

For a practical illustration, consider the measurement of CP violation via $B^0 \bar{B}^0$ mixing at the $\Upsilon(4S)$. This measurement involves measuring the time difference, t, between the two B meson decays in an $\Upsilon \to B^0 \bar{B}^0$ event. The sign of t is determined relative to the decay that is the flavor "tag" B, for example a B decaying semileptonically. The pdf for this random variable may be written:

$$f(t; A) = \frac{1}{2} e^{-|t|}(1 + A \sin rt), \tag{2.12}$$

where $t \in (-\infty, \infty)$, $r = \Delta m/\Gamma$ is a known quantity, and A is the CP asymmetry parameter of interest. In the early days, when the B factory experiments to do this measurement were being proposed, there was some controversy concerning the efficiency of a simple estimation method.

The simplified analysis is to count the number of times $t < 0$, N_-, and the number of times $t > 0$, N_+. The expectation value of the difference between these, for a total sample size $N = N_- + N_+$, is

$$E(N_+ - N_-) = N \frac{rA}{1+r^2}. \tag{2.13}$$

In the *substitution method* we replace the expectation value by the observed difference, and invert to obtain an estimator for the asymmetry parameter:

$$\hat{A}_{\text{subs}} = d^{-1} \frac{N_+ - N_-}{n}, \tag{2.14}$$

where $d = r/(1+r^2)$ is known as the "dilution factor". We note that $\hat{A}$ is by definition an unbiased estimator for A. The question is, how efficient is it? In particular, we are throwing away detailed time information – does that matter very much, assuming our time resolution isn't too bad?

First, what is the variance of $\hat{A}$? For a given N, we may treat the sampling of $N_\pm$ as a binomial process, giving

$$\text{Var}(\hat{A}_{\text{subs}}) = (1 - d^2 A^2)/Nd^2. \tag{2.15}$$

Second, how well can we do, at least in principle, if we do our best? Let us use the RCF bound to estimate this (and argue that, at least asymptotically, we can achieve this bound, for example, with maximum likelihood estimation, described in Section 2.2):

For N-independent time samplings, the RCF bound on the variance of any unbiased estimator for A is

$$\text{Var}(\hat{A})_{\text{RCF}} \geq \frac{1}{E\left\{\left[\frac{\partial}{\partial A} \sum_{i=1}^{N} \log f(t_i; A)\right]^2\right\}}$$

$$\geq \frac{1}{NE\left[\left(\frac{\sin rt}{1 + A \sin rt}\right)^2\right]}. \tag{2.16}$$

Performing the integral gives:

$$\text{Var}(\hat{A})_{\text{RCF}} = \frac{1}{N}\left\{\sum_{k=1}^{\infty} A^{2(k-1)} \frac{r^{2k}(2k)!}{[1+(2r)^2][1+(4r)^2]\cdots[1+(2kr)^2]}\right\}^{-1}. \tag{2.17}$$

Figure 2.1 shows the comparison of this bound with the variance from the substitution method. The value $r = 0.7$ is used here, reflecting the state of knowledge

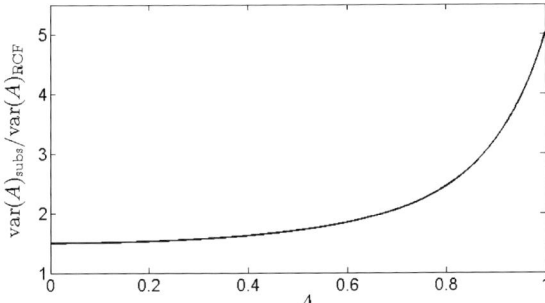

Figure 2.1 The variance according to the substitution method divided by the RCF bound on the variance in the asymmetry parameter estimates.

when the experiments were being proposed. We may conclude that, especially for large asymmetries, significant gains are obtained by using the detailed time information.

2.1.2
The Exponential Family

This leads to an interesting question: Under what (if any) circumstances can the minimum variance bound be achieved? To answer this, we consider an important class of distributions:

Definition 2.3. *Let $F_{\boldsymbol{\theta}}$ be a probability distribution (cdf) specified by parameters $\boldsymbol{\theta}$ (hence, a parametric distribution) on a sample space Ω with measure μ. The set of possible distributions $\{F_{\boldsymbol{\theta}} : \boldsymbol{\theta} \in \Theta\}$, where Θ is the parameter space, forms a collection of distributions which we call a family. The family $\{F_{\boldsymbol{\theta}}\}$ is called an* exponential family *iff*

$$\frac{dF_{\boldsymbol{\theta}}}{d\mu}(X) = h(X)e^{q(\boldsymbol{\theta})^{\top}g(X) - r(\boldsymbol{\theta})}, \quad X \in \Omega. \tag{2.18}$$

Here, $\mathbf{g}$ is a random vector mapping from X to $\mathcal{R}^k$ and $\mathbf{q}$ is a mapping from Θ to $\mathcal{R}^k$. Functions h and $\mathbf{g}$ depend only on X, while $\mathbf{q}$ and r depend only on $\boldsymbol{\theta}$.

We have defined the exponential family with respect to a measure μ on the sample space so that it applies to either discrete or continuous distributions, depending on the choice of measure. Many of our common distributions belong to this family. For example, the binomial distribution is an example of a discrete exponential family:

$$P_\theta(x; n) = \sum_{i=0}^{x} \binom{n}{i} \theta^i (1-\theta)^{n-i}, \quad \theta \in (0, 1). \tag{2.19}$$

The relevant measure is $\delta(x - j)$, where $j = 0, 1, \ldots, n$, hence the pdf is

$$f(x;n,\theta) = \binom{n}{x} \theta^x (1-\theta)^{n-x} \qquad (2.20)$$

$$= \binom{n}{x} \exp\left[x \log \frac{\theta}{(1-\theta)} + n \log(1-\theta) \right], \qquad (2.21)$$

where we have put it in the exponential form in the second line.

On the other hand, the Cauchy distribution centered at zero with full width at half maximum (FWHM) parameter Γ, has pdf:

$$f(x;\Gamma) = \frac{\Gamma}{2\pi} \frac{1}{x^2 + (\Gamma/2)^2}, \qquad (2.22)$$

This is not in the exponential family.

We have the following:

Theorem 2.4. *An efficient (perhaps biased) estimator for a parameter θ exists iff*

$$\frac{\partial \log L(\theta; \mathbf{X})}{\partial \theta} = [g(x) - s(\theta)] t(\theta). \qquad (2.23)$$

An unbiased efficient estimator exists iff we further have

$$s(\theta) = \theta. \qquad (2.24)$$

That is, an efficient estimator exists for members of the exponential family. The proof is again left to an exercise, but here is a hint: The RCF bound made use of the linear correlation coefficient, in which equality holds iff there is a linear relation:

$$\partial_\theta \log L(\theta; \mathbf{X}) = a(\theta) \hat{\theta} + b(\theta). \qquad (2.25)$$

2.1.3
Confidence Intervals

We have been discussing information and variance because they provide measures for how well we can estimate the value of a population parameter. Closely related is the concept of a *confidence interval*. More generally we could use the term *confidence set*, covering the possibilities of multidimensional parameter spaces and disconnected sets. We will treat these terms as synonymous. Unless explicitly discussing the Bayesian context, we work with frequentist statistics, constructing intervals in the Neyman sense (Neyman, 1937).

The basic idea behind the construction of a confidence set is as follows. For any possible $\boldsymbol{\theta}$ (possibly including "nonphysical" values), we construct a set of observations x for which that value of $\boldsymbol{\theta}$ will be included in the $1-\alpha$ confidence region. We call this set $S_\alpha(\boldsymbol{\theta})$. There are many ways we could imagine constructing such a set. We often look for as small a set as possible. In this case, the set $S_\alpha(\boldsymbol{\theta})$ is defined as the smallest set for which

$$\int_{S_\alpha(\boldsymbol{\theta})} f(\boldsymbol{x};\boldsymbol{\theta})\mu(dS) \geq 1 - \alpha \, , \qquad (2.26)$$

where $\boldsymbol{x} \in S_\alpha(\boldsymbol{\theta})$ for all $\boldsymbol{x}$ such that $f(\boldsymbol{x};\boldsymbol{\theta}) \geq \min_{\boldsymbol{x} \in S_\alpha(\boldsymbol{\theta})} f(\boldsymbol{x};\boldsymbol{\theta})$. That is, the set is constructed by ordering the probabilities, and including those $\boldsymbol{x}$ values for which the probabilities are greatest, for the given $\boldsymbol{\theta}$. Given an observation $\boldsymbol{x}$, the confidence interval $C_\alpha(\boldsymbol{x})$ for $\boldsymbol{\theta}$ at the $1 - \alpha$ *level of significance* or *confidence level*, is just the set of all values of $\boldsymbol{\theta}$ for which $\boldsymbol{x} \in S_\alpha(\boldsymbol{\theta})$. By construction, this set has a probability $1 - \alpha$ (or more) of including the true value of $\boldsymbol{\theta}$.

There are two technical remarks in order:

1. To be general, the measure $\mu(dS)$ is used; this may be defined as appropriate in the cases of a continuous or discrete distribution. The presence of the inequality in (2.26) is intended to handle the situation in which there is discreteness in the sampling space. In this case, the algorithm may err on the side of over-coverage.
2. It is possible that there will be degeneracies on sets of nonzero measure such that there is more than one solution for $S_\alpha(\boldsymbol{\theta})$. In this case, some other criteria will be needed to select a unique set.

2.1.4
Hypothesis Tests

Very closely related to the construction of confidence intervals is the subject of hypothesis tests. We consider the problem of testing between two hypotheses, H_0, called the *null hypothesis*, and H_1, called the *alternative hypothesis*. A *simple hypothesis* is one in which the both the null and alternative hypotheses are completely specified. For example, a test for $H_0: \theta = 0$ against $H_1: \theta = 1$ is simple, assuming everything else about the distribution is known. A *composite hypothesis* is one that is not simple. For example, a test for $H_0: \theta = 0$ against $H_1: \theta > 0$ is composite, because the alternative is not completely specified.

We set up the test by saying we are going to "reject" (i.e., consider unlikely) H_0 in favor of H_1 if the observation $\boldsymbol{x}$ lies in some region R of the sample space. This is known as the *critical region* for the test. When we reject one hypothesis in favor of another hypothesis, there are two types of error we could make:

Type I error: Reject H_0 when H_0 is true.
Type II error: Accept H_0 when H_1 is true.

The probability of making a Type I error is

$$\alpha = P(X \in R | H_0) \qquad (2.27)$$

$$= \int_R f(\boldsymbol{x}; H_0) d\boldsymbol{x} \, . \qquad (2.28)$$

The probability α is called the *significance level* of the test.

The probability of making a Type II error is

$$\beta = P(X \in \bar{R} | H_1) \tag{2.29}$$

$$= 1 - \int_R f(x; H_1) dx , \tag{2.30}$$

where $\bar{R}$ denotes the complement of R in the sample space. The quantity $1 - \beta$ is called the *power* of the test; it is the probability that H_0 is correctly rejected, that is, if H_1 is true. A "good" test is one which has high power for any given significance level. A test is called *Uniformly Most Powerful* (UMP) if, for specified significance it is at least as powerful as any other test, for all elements of H_1.

Another view of hypothesis testing is encapsulated in the *p-value*. The *p*-value for an observation is the probability, under the null hypothesis, that a fluctuation as large or larger than that observed, away from H_0, towards H_1, will occur. We will have opportunity to use both paradigms. In Chapter 3 we will see how hypothesis tests may be used to compute confidence intervals.

2.2
Parametric Likelihood Fits

The maximum likelihood method for parameter estimation consists of finding those values of $\boldsymbol{\theta}$ for which the likelihood is maximized. Call these values (estimators) $\hat{\boldsymbol{\theta}}$. They are determined by solving:

$$L(\hat{\boldsymbol{\theta}}; \mathbf{X}) = \max_{\boldsymbol{\theta}} L(\boldsymbol{\theta}; \mathbf{X}) . \tag{2.31}$$

Analytically, one solves the *likelihood equations*:

$$\left. \frac{\partial L(\boldsymbol{\theta}; x)}{\partial \theta_i} \right|_{\boldsymbol{\theta} = \hat{\boldsymbol{\theta}}} = 0; \quad i = 1, \ldots, R . \tag{2.32}$$

Oftentimes, this is intractable, and a numerical search for the maximum is necessary. It is also possible that the likelihood equation has no solution within the domain of $\boldsymbol{\theta}$. In this case, we find the $\boldsymbol{\theta}$ for which the likelihood achieves its maximum within the domain of definition (or, rather, its closure). The value of $\boldsymbol{\theta}$ for which the likelihood is maximized is called the *maximum likelihood estimator* (MLE) for $\boldsymbol{\theta}$. Since $\hat{\boldsymbol{\theta}} = \hat{\boldsymbol{\theta}}(\mathbf{X})$, the MLE is a random variable. There may be multiple solutions to the likelihood equation; these are called *roots of the likelihood equation* (RLE). Typically, we are interested in the RLE for which the maximum is achieved, that is, the MLE. However, if multiple roots have similar likelihood values, you should include these in the analysis, either as alternative solutions or by providing a possibly disconnected confidence set. In addition, it is usually more convenient to work with the logarithm of the likelihood, especially in numerical work where the likelihood itself may be an extremely tiny number.

A connection can be made with the least squares method of parameter estimation, using the logarithm of the likelihood. If f is a (multivariate) normal distribution, the likelihood function for a single observation is of the form:

$$L(\boldsymbol{\theta}; x) = \frac{1}{\sqrt{(2\pi)^D |\boldsymbol{\Sigma}|}} \exp\left\{-\frac{1}{2}[x - g(\boldsymbol{\theta})]^T \boldsymbol{\Sigma}^{-1}[x - g(\boldsymbol{\theta})]\right\}, \qquad (2.33)$$

where D is the dimension of X and $\boldsymbol{\Sigma} = \text{Cov}(X)$. Taking the logarithm and dropping the constant (independent of $\boldsymbol{\theta}$) terms yields

$$\log L(\boldsymbol{\theta}; x) = -\frac{1}{2}[x - g(\boldsymbol{\theta})]^T \boldsymbol{\Sigma}^{-1}[x - g(\boldsymbol{\theta})]. \qquad (2.34)$$

Thus, $-2 \log L$ is precisely the χ^2 expression in the least squares method. This connection is well-known. However, the assumption of normal sampling is often forgotten. For nonnormal distributions, the maximum likelihood and least squares procedures are distinct methods of parameter estimation, yielding different estimators.

The popularity of the maximum likelihood method is based on several nice properties, as well as reasonably simple computation. In particular:

1. The maximum likelihood estimator is *asymptotically* (i.e., as sample size $N \to \infty$) unbiased (hence *consistent*) and efficient.
2. If a *sufficient* statistic exists, the MLE (if a unique MLE exists) will be a function of the sufficient statistic. A sufficient statistic, $t(X)$, for $\boldsymbol{\theta}$ is a statistic such that the sampling distribution conditioned on t is independent of $\boldsymbol{\theta}$. Intuitively, a sufficient statistic is based on all of the available information relevant to $\boldsymbol{\theta}$. For example, the sample mean is a sufficient statistic for the mean of a normal distribution, while the sample median is not.
3. If an efficient unbiased estimator exists, the maximum likelihood algorithm will find it (readily seen from (2.23) and (2.24)).

The reader is referred to standard statistics textbooks for the fine print and proof of these statements, for example Shao (2003).

The nice asymptotic properties of the MLE do not necessarily carry over into small statistics cases. For example, with $(x_1, \ldots, x_N)$ an i.i.d. sampling from an $N(\mu, \sigma^2)$ distribution, the maximum likelihood estimators for μ and σ^2 are

$$\hat{\mu} = \bar{x} \equiv \frac{1}{N} \sum_{n=1}^{N} x_n$$

$$\hat{\sigma}^2 = \frac{1}{N} \sum_{n=1}^{N} (x_n - \bar{x})^2. \qquad (2.35)$$

In the first case, $\hat{\mu}$ is an unbiased estimator for μ for all values of $N > 0$. However, $\hat{\sigma}^2$ has a bias $b(\sigma^2) \equiv \langle \hat{\sigma}^2 \rangle - \sigma^2 = -\frac{1}{N}\sigma^2$. For small N, this bias can be very large. Fortunately, in this case it is easy to correct for, to obtain the familiar unbiased estimator $\frac{N}{N-1}\hat{\sigma}^2$.

A popular method for estimating errors in the maximum likelihood methodology is to look for parameter values $\boldsymbol{\theta}_\pm$, for which

$$-2\Delta \log L \equiv -2[\log L(\boldsymbol{\theta}_\pm; x) - \log L(\hat{\boldsymbol{\theta}}; x)] = 1 . \tag{2.36}$$

This method yields 68% confidence intervals (possibly disconnected) on the individual parameters as long as f_X is normal. This can be seen as follows. Suppose the parameters of interest are the g's in (2.33). In this case, it is straightforward to check that the region defined by $g_\pm$ according to $-2\Delta \log L(g; x) = 1$ provides a 68% confidence interval for each of the g's. To be clear, note that, if we are interested in g_1, the 68% confidence interval is obtained by including all values of g_1 within the multidimensional region given by the condition $-2\Delta \log L(g; x) = 1$ for the given x. This can be expressed with the notion of a profile likelihood introduced in Section 2.2.4. For arbitrary parameterization $\boldsymbol{\theta}$, where $g = g(\boldsymbol{\theta})$, we will obtain the same probability content in terms of frequency if we simply restate our region in g as a region in $\boldsymbol{\theta}$ according to the inverse mapping $g \to \boldsymbol{\theta}$. This is true even if, as is often the case, $\boldsymbol{\theta}$ is of lower dimension than g (i.e., there are relations among the components of g).

This property of the error estimates obtained with $-2\Delta \log L = 1$ is widely known. However, it is not so widely appreciated that it only holds for sampling from a normal distribution. If the sampling is not normal, then the probability content will be different, and must be determined for the correct sampling distribution. Often this can not be done analytically and must be done with Monte Carlo or other methods.

More generally, for example, when the distribution may not be normal, we return to our general Neyman construction, (2.26). Thus, let $h(\hat{\boldsymbol{\theta}}; \boldsymbol{\theta})$ be the probability distribution for maximum likelihood estimator $\hat{\boldsymbol{\theta}}$. Let now $S_\alpha(\boldsymbol{\theta})$ be the smallest set for which

$$\int_{S_\alpha(\boldsymbol{\theta})} h(\hat{\boldsymbol{\theta}}; \boldsymbol{\theta}) \mu(dS) \geq 1 - \alpha , \tag{2.37}$$

where $\hat{\boldsymbol{\theta}} \in S_\alpha(\boldsymbol{\theta})$ if $h(\hat{\boldsymbol{\theta}}; \boldsymbol{\theta}) \geq \min_{\hat{\boldsymbol{\theta}} \in S_\alpha(\boldsymbol{\theta})} h(\hat{\boldsymbol{\theta}}; \boldsymbol{\theta})$. Given an observation $\hat{\boldsymbol{\theta}}$, the confidence interval $C_\alpha(\hat{\boldsymbol{\theta}})$ for $\boldsymbol{\theta}$ at the $1 - \alpha$ confidence level, is just the set of all values of $\boldsymbol{\theta}$ for which $\hat{\boldsymbol{\theta}} \in S_\alpha(\boldsymbol{\theta})$. By construction, this set has a probability $1 - \alpha$ (or more) of including the true value of $\boldsymbol{\theta}$.

Consider now the *likelihood ratio*:

$$\lambda(\boldsymbol{\theta}, \hat{\boldsymbol{\theta}}) = \frac{L(\boldsymbol{\theta}; \hat{\boldsymbol{\theta}})}{L(\hat{\boldsymbol{\theta}}; \hat{\boldsymbol{\theta}})} . \tag{2.38}$$

The denominator is the maximum of the likelihood for the observation (MLE) $\hat{\boldsymbol{\theta}}$. We have $0 \leq \lambda(\boldsymbol{\theta}, \hat{\boldsymbol{\theta}}) \leq 1$. For any value of $\boldsymbol{\theta}$, we may make a table of possible results $\hat{\boldsymbol{\theta}}$ for which we will accept the value $\boldsymbol{\theta}$ with confidence level $1 - \alpha$. Consider the set:

$$A_\alpha(\boldsymbol{\theta}) \equiv \{\hat{\boldsymbol{\theta}} | \lambda(\boldsymbol{\theta}, \hat{\boldsymbol{\theta}}) \geq \lambda_\alpha(\boldsymbol{\theta})\} , \tag{2.39}$$

where $\lambda_\alpha(\boldsymbol{\theta})$ is chosen such that $A_\alpha(\boldsymbol{\theta})$ contains a probability fraction $1-\alpha$ of the sample space for $\{\hat{\boldsymbol{\theta}}\}$. That is,

$$P[\lambda(\boldsymbol{\theta},\hat{\boldsymbol{\theta}}) \geq \lambda_\alpha(\boldsymbol{\theta}); \boldsymbol{\theta}] \geq 1-\alpha. \tag{2.40}$$

Notice that we are ordering on likelihood ratio $\lambda(\boldsymbol{\theta},\hat{\boldsymbol{\theta}})$. Sometimes $\lambda_\alpha(\boldsymbol{\theta})$ is independent of $\boldsymbol{\theta}$.

We then use this table to construct confidence regions as follows: Suppose we observe a result $\hat{\boldsymbol{\theta}}$. We go through our table of sets $A_\alpha(\boldsymbol{\theta})$ looking for $\hat{\boldsymbol{\theta}}$. Every time we find it, we include that value of $\boldsymbol{\theta}$ in our confidence region. This gives a confidence region for $\boldsymbol{\theta}$ at the $1-\alpha$ confidence level. That is, the true value of $\boldsymbol{\theta}$ will be included in the interval with probability $1-\alpha$. For clarity, we will repeat this procedure more explicitly in algorithmic form.

The algorithm is the following:

1. Find $\hat{\boldsymbol{\theta}}$, the value of $\boldsymbol{\theta}$ for which the likelihood is maximized.
2. For any point $\boldsymbol{\theta}^*$ in parameter space, form the statistic

$$\lambda(\boldsymbol{\theta}^*,\hat{\boldsymbol{\theta}}) \equiv \frac{L(\boldsymbol{\theta}^*;\hat{\boldsymbol{\theta}})}{L(\hat{\boldsymbol{\theta}};\hat{\boldsymbol{\theta}})}. \tag{2.41}$$

3. Evaluate the probability distribution for λ (considering all possible experimental outcomes), under the hypothesis that $\boldsymbol{\theta} = \boldsymbol{\theta}^*$. Using this distribution, determine critical value $\lambda_\alpha(\boldsymbol{\theta}^*)$.
4. Ask whether $\hat{\boldsymbol{\theta}} \in A_\alpha(\boldsymbol{\theta}^*)$. It will be if the observed value of $\lambda(\boldsymbol{\theta}^*,\hat{\boldsymbol{\theta}})$ is larger than (or equal to) $\lambda_\alpha(\boldsymbol{\theta}^*)$. If this condition is satisfied, then $\boldsymbol{\theta}^*$ is inside the confidence region; otherwise it is outside.
5. Consider all possible $\boldsymbol{\theta}^*$ to determine the entire confidence region.

In general, the analytic evaluation of the probability in step (3) is intractable. We may employ the Monte Carlo method to compute this probability. In this case, steps (3)–(5) are replaced by the specific procedure:

3. Simulate many experiments with $\boldsymbol{\theta}^*$ taken as the true value(s) of the parameter(s), obtaining for each experiment the result $\boldsymbol{x}_{MC}$ and maximum likelihood estimator $\hat{\boldsymbol{\theta}}_{MC}$.
4. For each Monte Carlo simulated experiment, form the statistic

$$\lambda_{MC} \equiv \frac{L(\boldsymbol{\theta}^*;\hat{\boldsymbol{\theta}}_{MC})}{L(\hat{\boldsymbol{\theta}}_{MC};\hat{\boldsymbol{\theta}}_{MC})}. \tag{2.42}$$

The critical value $\lambda_\alpha(\boldsymbol{\theta}^*)$ is estimated as the value for which a fraction α of the simulated experiments have a larger value of λ_{MC}.

5. If $\lambda(\boldsymbol{\theta}^*,\hat{\boldsymbol{\theta}}) \geq \lambda_\alpha(\boldsymbol{\theta}^*)$, then $\boldsymbol{\theta}^*$ is inside the confidence region; otherwise it is outside. In other words, if $\lambda(\boldsymbol{\theta}^*,\hat{\boldsymbol{\theta}})$ is larger than at least a fraction α of the simulated experiments, then $\boldsymbol{\theta}^*$ is inside the confidence region.
6. This procedure is repeated for many choices of $\boldsymbol{\theta}^*$ in order to map out the confidence region.

2.2.1
Nuisance Parameters

It may happen that our sampling distribution depends on unknown parameters that are additional to the physically interesting parameters that we are trying to estimate. These additional parameters get in the way of making statements (e.g., confidence intervals) about the interesting physics; they are thus called *nuisance parameters*. Let $f(x;\boldsymbol{\theta})$ be a probability distribution with an R-dimensional parameter space. Divide this parameter space into subspaces $\mu_1 \equiv \theta_1, \ldots, \mu_k \equiv \theta_k$ and $\eta_1 \equiv \theta_{k+1}, \ldots, \eta_{r-k} \equiv \theta_r$, where $1 \leq k \leq R$. Let x be a sampling from f. We wish to obtain a confidence interval for $\boldsymbol{\mu}$, at the $1-\alpha$ confidence level. That is, for any observation x, we wish to have a prescription for finding sets $R_\alpha(x)$ such that

$$1 - \alpha = P[\boldsymbol{\mu} \in R_\alpha(x)]. \tag{2.43}$$

The $\boldsymbol{\eta}$ parameters are not of interest here, and are the nuisance parameters. This problem, unfortunately, does not have a nontrivial general exact solution, for $k < R$. We can see this as follows:

We use the same sets $S_\alpha(\boldsymbol{\theta})$ as constructed in (2.26). Given an observation x, the confidence interval $R_\alpha(x)$ for $\boldsymbol{\mu}$ at the $1-\alpha$ confidence level is just the set of all values of $\boldsymbol{\mu}$ for which $x \in S_\alpha[\boldsymbol{\theta} = (\boldsymbol{\mu}, \boldsymbol{\eta})]$. We take here the union over all values of the nuisance parameters in $S_\alpha[\boldsymbol{\theta} = (\boldsymbol{\mu}, \boldsymbol{\eta})]$ since, at most, one of those sets is from the true $\boldsymbol{\theta}$, and we want to make sure that we include this set if present. By construction, this set has a probability of at least $1-\alpha$ of including the true value of $\boldsymbol{\mu}$. The region may, however, substantially over-cover, depending on the problem.

In a Bayesian analysis, it is relatively easy to eliminate the nuisance parameters. They are simply integrated out of the likelihood function, giving the *marginal likelihood*:

$$L_M(\boldsymbol{\mu}) = \int L(\boldsymbol{\mu}, \boldsymbol{\eta}) d\boldsymbol{\eta}. \tag{2.44}$$

A prior distribution in the nuisance parameters must be included in a truly Bayesian analysis, but this is often taken to be uniform, as implicitly done in (2.44). However, consideration should be given in each case whether this approximation to prior knowledge is satisfactory. The marginal likelihood may also be useful in frequency estimation, for example with location parameters of a multivariate normal, but the properties must in general be studied.

We survey several techniques for interval estimation and elimination of nuisance parameters in the following sections. The reader is referred to textbooks such as Shao (2003) and the paper by Reid (2003) for further discussion including improvements in asymptotic methods. Chapter 4 introduces some further techniques using resampling methods.

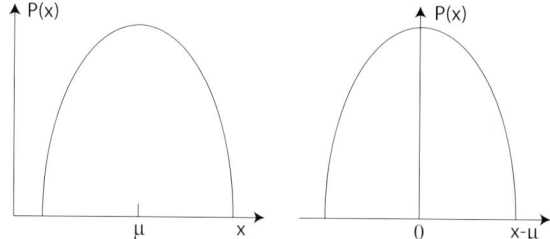

Figure 2.2 Example of a distribution with location parameter μ.

2.2.2
Confidence Intervals from Pivotal Quantities

A powerful tool, if available, in the construction of intervals and elimination of nuisance parameters is the method of pivotal quantities:

Definition 2.5. *Pivotal Quantity:* Consider a sample $\mathbf{X} = (X_1, X_2, \ldots, X_N)$ from population P, governed by parameters $\boldsymbol{\theta}$. A function $R(\mathbf{X}, \boldsymbol{\theta})$ is called pivotal iff the distribution of R does not depend on $\boldsymbol{\theta}$.

The notion of a pivotal quantity is a generalization of the feature of a *location parameter*: If $\boldsymbol{\mu}$ is a location parameter for X, then the distribution of $X - \boldsymbol{\mu}$ is independent of $\boldsymbol{\mu}$ (see Figure 2.2). Thus, $X - \boldsymbol{\mu}$ is a pivotal quantity. However, not all pivotal quantities involve location parameters.

If a suitable pivotal quantity is known, it may be used in the calculation of confidence intervals as follows (for now, consider a confidence interval for a single parameter θ): let $R(\mathbf{X}, \theta)$ be a pivotal quantity, and $1 - \alpha$ be a desired confidence level. Find c_1, c_2 such that

$$P[c_1 \leq R(\mathbf{X}, \theta) \leq c_2] \geq 1 - \alpha. \tag{2.45}$$

For simplicity, we will use "$= 1 - \alpha$" henceforth, presuming a continuous distribution. The key point is that, since R is pivotal, c_1 and c_2 are constants, independent of θ.

Now define:

$$C(\mathbf{X}) \equiv \{\theta : c_1 \leq R(\mathbf{X}, \theta) \leq c_2\}. \tag{2.46}$$

$C(\mathbf{X})$ is a confidence interval with $1 - \alpha$ confidence level, since

$$P[\theta \in C(\mathbf{X})] = P[c_1 \leq R(\mathbf{X}, \theta) \leq c_2] = 1 - \alpha. \tag{2.47}$$

Figure 2.3 illustrates the idea.

The present discussion, with constants c_1 and c_2 is framed as a one-parameter problem. However, it applies to a multidimensional parameter space as well: if $R(\mathbf{X}, \boldsymbol{\theta})$ is independent of $\boldsymbol{\theta}$, we look for a boundary B, enclosing region V, independent of $\boldsymbol{\theta}$, such that

$$P(R(\mathbf{X}, \boldsymbol{\theta}) \in V) \geq 1 - \alpha. \tag{2.48}$$

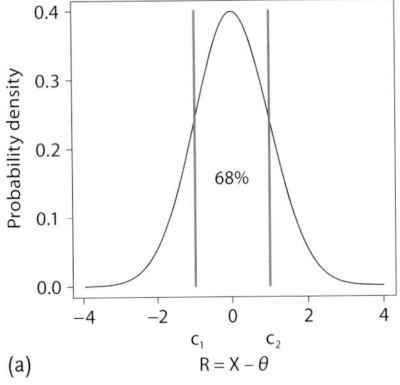
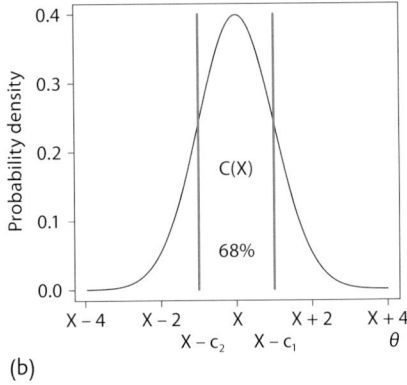

(a) R = X − θ

(b)

Figure 2.3 The difference between the random variable X and the mean is a pivotal quantity for a normal distribution. (a) Graph of $f(X - \theta)$ as a function of $X - \theta$, showing the constants c_1 and c_2 that mark a region with probability 68%. (b) The 68% confidence interval for parameter θ.

Then,

$$C(X) = \{\theta : R(X, \theta) \in V\} \tag{2.49}$$

provides confidence set $C(X)$ at the $1 - \alpha$ confidence level. If we can find a pivotal quantity,

$$R(X, \theta) = R(X, (\mu, \eta)) = R(X, (\mu)), \tag{2.50}$$

not depending on the nuisance parameters, this provides a means to eliminate the nuisance parameters. The following example illustrates this.

2.2.2.1 Pivotal Quantities: Example

Consider i.i.d. sampling $\mathbf{x} = X_1, \ldots, X_N$ from a pdf with location and scale parameters of the form:

$$P(x) = \frac{1}{\sigma} f\left(\frac{x - \mu}{\sigma}\right). \tag{2.51}$$

The pivotal quantity method may be applied to obtaining confidence intervals for different cases, according to:

- Case I: Parameter μ is unknown and σ is known. Then $X_i - \mu$, for any i, is pivotal. Also, the quantity $m - \mu$ is pivotal, where m is the sample mean, $m \equiv \frac{1}{N} \sum_{i=1}^{N} X_i$. As a sufficient statistic, m is a better choice for forming a confidence set for μ.
- Case II: Both μ and σ are unknown. Let S^2 be the sample variance:

$$S^2 \equiv \frac{1}{N} \sum_{i=1}^{N} (X_i - m)^2. \tag{2.52}$$

S/σ is a pivotal quantity (exercise), and can be used to derive a confidence interval for σ (since μ, considered a nuisance parameter here, does not appear). Another pivotal quantity is

$$t(\mathbf{x}) \equiv \frac{m - \mu}{(S/\sqrt{N})} . \tag{2.53}$$

This permits confidence intervals for μ:

$$\{\mu : c_1 \leq t(\mathbf{x}) \leq c_2\} = \left(m - c_2 s/\sqrt{N}, m - c_1 s/\sqrt{N} \right) \tag{2.54}$$

at the $1 - \alpha$ confidence level, where

$$P(c_1 \leq t(\mathbf{x}) \leq c_2) = 1 - \alpha . \tag{2.55}$$

Remark: $t(\mathbf{x})$ is often called a "Studentized[2] statistic" (though it isn't a statistic, since it depends also on unknown μ). In the case of normal sampling, the distribution of t is Student's t_{N-1}.

Suppose we are interested in some parameters $\boldsymbol{\mu} \subset \boldsymbol{\theta}$, where $\dim(\boldsymbol{\mu}) < \dim(\boldsymbol{\theta})$. Let $\boldsymbol{\eta} \subset \boldsymbol{\theta}$ stand for the remaining "nuisance" parameters. If you can find pivotal quantities, as above, then the problem is solved. Unfortunately, this is not always possible. The approach of test acceptance regions discussed in Section 3.1 is also problematic: H_0 becomes composite. since nuisance parameters are unspecified. In general, we don't know how to construct the acceptance region with specified significance level for

$$H_0: \boldsymbol{\mu} = \boldsymbol{\mu}_0, \quad \boldsymbol{\eta} \text{ unspecified} . \tag{2.56}$$

Instead, we may resort to approximate methods.

2.2.3
Asymptotic Inference

When it is not possible, or practical, to find an exact solution, it may be possible to base an approximate treatment on asymptotic criteria. We define some of the relevant concepts in this section.

Definition 2.6. *Consider sample* $\mathbf{X} = (X_1, \ldots, X_N)$ *from population* $P \in \mathcal{P}$, *where* $\mathcal{P}$ *is the space of possible populations. Let* $T_N(\mathbf{X})$ *be a test at the* α_{T_N} *significance level for*

$$H_0: P \in \mathcal{P}_0 \quad \text{against} \tag{2.57}$$

$$H_1: P \in \mathcal{P}_1 \tag{2.58}$$

---

[2] Student is a pseudonym for William Gosset, used because his employer did not permit employees to publish.

where $\mathcal{P}_0$ and $\mathcal{P}_1$ are disjoint subsets of $\mathcal{P}$. If

$$\lim_{N \to \infty} \sup \alpha_{T_N}(P) \leq \alpha \tag{2.59}$$

for any $P \in \mathcal{P}_0$, then α is called an asymptotic significance level of T_N.

Definition 2.7. Consider sample $\mathbf{X} = (X_1, \ldots, X_N)$ from population $P \in \mathcal{P}$. Let θ be a parameter vector for P, and let $C(\mathbf{X})$ be a confidence set for θ. If $\liminf_{N \to \infty} P[\theta \in C(\mathbf{X})] \geq 1 - \alpha$ for any $P \in \mathcal{P}$, then $1 - \alpha$ is called an asymptotic confidence level of $C(\mathbf{X})$.

Definition 2.8. If $\lim_{N \to \infty} P[\theta \in C(\mathbf{X})] = 1 - \alpha$ for any $P \in \mathcal{P}$, then $C(\mathbf{X})$ is a $1 - \alpha$ asymptotically correct *confidence set*.

There are many possible approaches, for example, one can look for "Asymptotically Pivotal" quantities; or invert acceptance regions of "Asymptotic Tests".

2.2.4
Profile Likelihood

We may compute approximate confidence intervals, in the sense of coverage, using the "profile likelihood". Consider likelihood $L(\boldsymbol{\mu}, \boldsymbol{\eta})$, based on observation $\mathbf{X} = \mathbf{x}$. Let

$$L_P(\boldsymbol{\mu}) = \sup_{\boldsymbol{\eta}} L(\boldsymbol{\mu}, \boldsymbol{\eta}). \tag{2.60}$$

$L_P(\boldsymbol{\mu}) = L(\boldsymbol{\mu}, \boldsymbol{\eta}(\boldsymbol{\mu}))$ is called the *Profile Likelihood* for $\boldsymbol{\mu}$. This provides a lower bound on coverage. Users of the popular fitting package MINUIT (James and Roos, 1975) will recognize that the MINOS interval uses the idea of the profile likelihood. We remind the reader that, for Gaussian sampling, intervals obtained with the profile likelihood have exact coverage (Section 2.2).

The profile likelihood has good asymptotic behavior: let $\dim(\boldsymbol{\mu}) = k$. Consider the likelihood ratio:

$$\lambda(\boldsymbol{\mu}) = \frac{L_P(\boldsymbol{\mu})}{\max_{\theta'} L(\theta')}, \tag{2.61}$$

where $\theta = (\boldsymbol{\mu}, \boldsymbol{\eta})$. The set

$$C(\mathbf{X}) = \{\boldsymbol{\mu} : -2 \log \lambda(\boldsymbol{\mu}) \leq c_\alpha\}, \tag{2.62}$$

where c_α is the χ^2 corresponding to the $1 - \alpha$ probability point of a χ^2 with k degrees of freedom, is an $1 - \alpha$ asymptotically correct confidence set. It may however not provide accurate coverage for small samples. Corrections to the profile likelihood exist that improve the behavior for finite samples (Reid, 2003).

2.2.5
Conditional Likelihood

Consider likelihood $L(\boldsymbol{\mu}, \boldsymbol{\eta})$. Suppose $T_\eta(X)$ is a sufficient statistic for $\boldsymbol{\eta}$ for any given $\boldsymbol{\mu}$. Then, conditional distribution $f(X|T_\eta; \boldsymbol{\mu})$ does not depend on $\boldsymbol{\eta}$. The likelihood function corresponding to this conditional distribution is called the *conditional likelihood*. Note that estimates (e.g., MLE for μ) based on conditional likelihood may be different than for those based on full likelihood. This eliminates the nuisance parameter problem, if it can be done without too high a price.

For example, suppose we want to test the consistency of two Poisson distributed numbers. Such a question might arise concerning the existence of a signal in the presence of background. Our sampling distribution is

$$P(m,n) = \frac{\mu^m \nu^n}{m!n!} e^{-(\mu+\nu)}. \tag{2.63}$$

The null hypothesis is $H_0: \mu = \nu$, to be tested against alternative $H_1: \mu \neq \nu$. We are thus interested in the difference between the two means; the sum is effectively a nuisance parameter. A sufficient statistic for the sum is $N = m + n$. That is, we are interested in

$$\begin{aligned}P(n|m+n=N) &= \frac{P(N|n)P(n)}{P(N)} \\ &= \frac{\mu^{N-n}e^{-\mu}}{(N-n)!} \frac{\nu^n e^{-\nu}}{n!} \bigg/ \frac{(\mu+\nu)^N e^{-(\mu+\nu)}}{N!} \\ &= \binom{N}{n}\left(\frac{\nu}{\mu+\nu}\right)^n \left(\frac{\mu}{\mu+\nu}\right)^{N-n}.\end{aligned} \tag{2.64}$$

This probability now permits us to construct a uniformly most powerful test of our hypothesis (Lehmann and Romano, 2005). Note that it is simply a binomial distribution, for given N. The uniformly most powerful property holds independently of N, although the probabilities cannot be computed without N.

The null hypothesis corresponds to $\mu = \nu$, that is

$$P(n|m+n=N) = \binom{N}{n}\left(\frac{1}{2}\right)^N. \tag{2.65}$$

For example, with $N = 916$ and $n = 424$, the p-value is 0.027, assuming a two-tailed probability is desired. This may be compared with an estimate of 0.025 in the normal approximation. Note that for our binomial calculation we have included the endpoints (424 and 492). If we try to mimic more closely the normal estimate by subtracting one-half the probability at the endpoints, we obtain 0.025, essentially the normal number. We have framed this in terms of a hypothesis test, but confidence intervals on the difference $\nu - \mu$ may likewise be obtained. The estimation of the ratio of Poisson means is a frequently encountered problem that can be addressed similarly (Reid, 2003).

2.3
Fits for Small Statistics

Often we are faced with extracting parametric information from data with only a few samplings, that is, in the case of "small statistics". At large statistics, the central limit theorem can usually be counted on to provide an excellent normal approximation unless one is concerned with the far tails of the distribution. However, this can no longer be presumed with small statistics, and greater care in computing probabilities is required.

For example, in fits to a histogram the use of least-squares fitting becomes problematic. It can still be done, but biases may be introduced, and the interpretation of the sum-of-squares in terms of a χ^2 statistic becomes invalid. If the statistics are not too small, then neighboring bins can be combined in order to achieve a minimum level of counts in each bin. In this case, the usual least-squares procedure and χ^2 approximation can be applied. Rules-of-thumb for the minimum counts in each bin typically suggest around seven to ten. More precisely, the maximum likelihood method may be applied, using the Poisson distribution for the contents of each bin. If only the shape of the distribution is of interest, then the multinomial distribution over the bin contents may be used. This is the distinction between what are often referred to among physicists as "extended binned maximum likelihood" and "(nonextended) binned maximum likelihood" fits.

The maximum likelihood procedure may also be applied to the data without binning. This has the advantage of using the actual values of the samplings, which could be important if overly-coarse binning is used for binned fits. With small statistics, the computing burden is usually not oppressive. Because we are now in the small statistics regime, the nice asymptotic properties of maximum likelihood estimation may not hold. Studies, such as by Monte Carlo, are necessary to assess probabilities, for example to derive confidence intervals.

2.3.1
Sample Study of Coverage at Small Statistics

For an example of such a study, consider the common problem of estimating a signal from Poisson sampling with nuisance parameters for background and efficiency.

To make the example more explicit, suppose we are trying to measure a branching fraction for some decay process where we count decays (events) of interest for some period of time. We observe n events. However, we must subtract an estimated background contribution of $\hat{b} \pm \sigma_b$ events. Furthermore, the detection efficiency and integrated luminosity are estimated to give a scaling factor $\hat{f} \pm \sigma_f$.

For this illustration, instead of a true branching fraction, we will let B be the number of expected signal events produced. We wish to determine a (frequency) confidence interval for B.

- Assume n is sampled from a Poisson distribution with mean $\mu = E(n) = fB + b$.
- Assume background estimate $\hat{b}$ is sampled from a normal distribution $N(b, \sigma_b)$, with σ_b known.
- Assume scale estimate $\hat{f}$ is sampled from a normal distribution $N(f, \sigma_f)$ with σ_f known.

The likelihood function is

$$L(B, b, f; n, \hat{b}, \hat{f}) = \frac{\mu^n e^{-\mu}}{n!} \frac{1}{2\pi\sigma_b\sigma_f}$$

$$\times \exp\left[-\frac{1}{2}\left(\frac{\hat{b}-b}{\sigma_b}\right)^2 - \frac{1}{2}\left(\frac{\hat{f}-f}{\sigma_f}\right)^2\right]. \quad (2.66)$$

We are interested in parameter B. In particular, we wish to summarize the data relevant to B, for example, in the form of a confidence interval, without dependence on the uninteresting quantities b and f. We have seen that obtaining a confidence region in all three parameters (B, b, f) is straightforward. Unfortunately quoting a confidence interval for just one of the parameters is a hard problem in general.

This is a commonly encountered problem, with many variations. There are a variety of approaches that have been used over the years, often without much justification (also often with Bayesian ingredients). In a situation such as this, it is generally desirable to at least provide n, $\hat{b} \pm \sigma_b$, and $\hat{f} \pm \sigma_f$. This provides the consumer with sufficient information to average or interpret the data as they see fit. However, it lacks the compactness of a confidence interval, so let us explore one possible approach, using the profile likelihood. It is important to check the coverage properties of the proposed methodology to see whether it is acceptable.

To carry out the approach of the profile likelihood, we vary B, and for each trial B we maximize the likelihood with respect to the nuisance parameters b and f. We compute the value of $-2\log L$ and take the difference with the value computed at the maximum likelihood over all three parameters. We compare the difference in $-2\log L$ with the 68% probability point of a χ^2 for one degree of freedom (that is, with one). We summarize the algorithm as follows:

1. Write down the likelihood function in all parameters.
2. Find the global maximum.
3. Search for the value of the B parameter where $-\log L$ increases from the minimum by a specified amount (e.g., $\Delta \equiv -\Delta \log L = 1/2$), re-optimizing with respect to f and b.

With the method defined, we ask: does it work? To answer this, we investigate the frequency behavior of this algorithm. For large statistics (normal distribution), we know that for $\Delta = 1/2$ this method produces a 68% confidence interval on B. We need to check how far can we trust it into the small statistics regime.

Figure 2.4 shows how the coverage depends on the value of Δ. The coverage dependence for a normal distribution is shown for comparison. It may be seen that the Poisson sampling generally follows the trend of the normal curve, but that there can be fairly large deviations in coverage at low statistics. At larger statistics, either in signal or background, the normal approximation improves. The discontinuities arise because of the discreteness of the Poisson distribution. The small wiggles are artifacts of the limited statistics in the simulation.

Figure 2.5 shows the dependence of the coverage on the scale factor f and its uncertainty, for an expected signal of 1 event and an expected background of 2

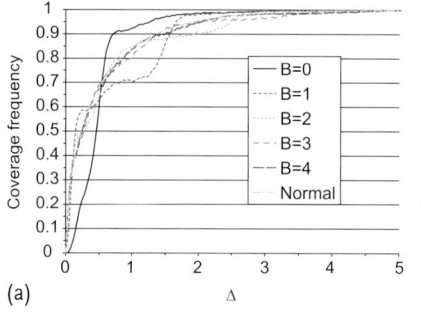
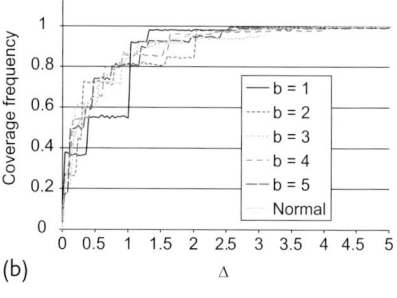

Figure 2.4 Dependence of coverage on $\Delta \log L$. The smoothest curves, labeled "Normal", show the coverage for a normal distribution. The curves with successively larger deviations from the normal curve are for progressively smaller statistics. (a) Curves for different values of B for $f = 1.0, \sigma_f = 0.1, b = 0.5, \sigma_b = 0.1$. (b) Curves for different values of b, for $B = 0, f = 1, \sigma_f = 0, \sigma_b = 0$.

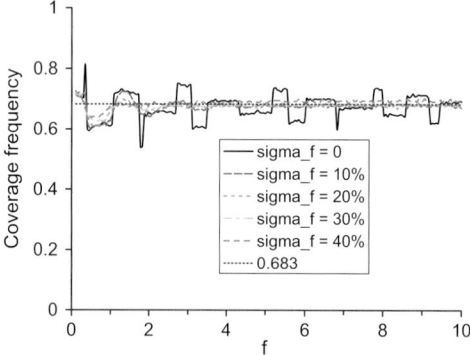

Figure 2.5 Dependence on f and σ_f for $B = 1, b = 2, \sigma_b = 0, \Delta = 1/2$. The horizontal line is at 68% coverage. Larger uncertainties in f produce successively smoother curves.

events. We can see how additional uncertainty in the scale factor helps the coverage improve; the same is true for uncertainty in the background (not shown).

Finally, Figure 2.6 shows what the intervals themselves look like, for a set of 200 experiments, as a function of the MLE for B. The MLE for B can be negative because of the background subtraction. The cluster of points around a value of $\hat{B} = -3$ for the MLE is what happens when zero events occur.

We have illustrated with this example how one can check the coverage for a proposed methodology. With today's computers, such investigations can often be performed without difficulty.

In the case of this example, we learn that the likelihood method considered works pretty well even for rather low expected counts, for 68% confidence intervals. The choice of $\Delta = 1/2$ is applicable to the normal distribution, hence we see the central limit theorem at work. To the extent there is uncertainty in b and f, the coverage tends to improve. It is important to recognize that being good enough for 68%

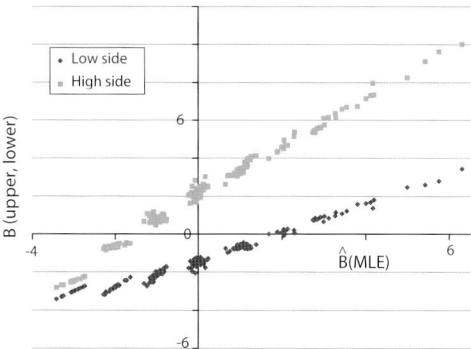

Figure 2.6 What the intervals for B look like as a function of the maximum likelihood estimator for B, for a sample of 200 experiments, where $\Delta = 1/2$, $B = 0$, $f = 1.0$, $\sigma_f = 0.1$, $b = 3.0$, $\sigma_b = 0.1$.

confidence interval doesn't mean good enough for a significance test (the subject of the next chapter), where we are usually concerned with the tails of the distribution.

2.3.2
When the pdf Goes Negative

A further difficulty in maximum likelihood estimation is that the technology for finding the maximum may fail to give a meaningful result, as we illustrate next. Consider a maximum likelihood fit for a set of events to some distribution, depending on parameters of interest. For example, suppose the sampling distribution consists of a flat background and a Gaussian signal of known mean and variance (see Figure 2.7):

$$f(x;\theta) = \frac{\theta}{2} + \frac{1-\theta}{A\sqrt{2\pi}\sigma} e^{-\frac{x^2}{2\sigma^2}}, \quad x \in (-1,1) ; \quad (2.67)$$

resulting in the likelihood function for a sample of size N:

$$L(\theta;\mathbf{x}) = \prod_{n=1}^{N} f(x_n;\theta) . \quad (2.68)$$

The constant A is simply to renormalize the Gaussian term on $(0, 1)$.

The maximum with respect to θ may be outside of the region where the pdf is defined. The function $f(x;\theta)$ may become negative in some regions of x. If there are no events in these regions, the likelihood is still "well-behaved". However, the resulting fit, as a description of the data, will typically look poor even in the region of positive pdf, as in Figure 2.8.

The practical resolution to this problem is to constrain the fit to remain within bounds such that the pdf is everywhere legitimate, see Figure 2.8. Note that the parameters may still be "unphysical", but this is generally not a concern in the

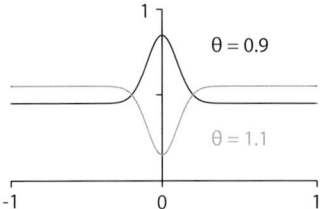

Figure 2.7 Possible pdfs given by (2.67).

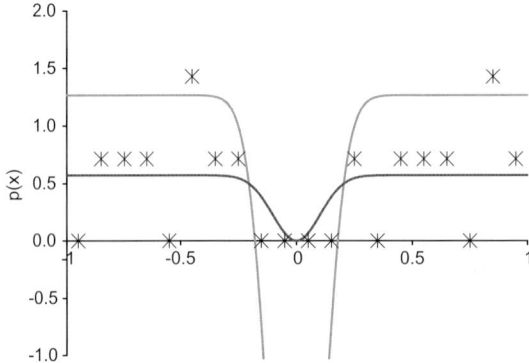

Figure 2.8 An example of an analysis based on (2.67). The points represent histogrammed data. The light curve occurs when the pdf is allowed to become negative in a maximum likelihood fit. The dark curve is the result when the fit is constrained so that the pdf is everywhere nonnegative.

context of frequency statistics. We have expressed this example as a problem in point estimation, but the same methodology applies in interval evaluation as well.

2.4
Results Near the Boundary of a Physical Region

An often troubling situation occurs when a fit produces results near the boundary of a physical region, for example a rate for some interesting process near zero. This can occur both for small and large statistics cases. Much of the reason this situation is troubling is psychological. Many estimation algorithms, including maximum likelihood, will return estimates outside the physical region if the data has fluctuated sufficiently. Whether this is okay or not depends on what you are trying to accomplish.

A related concern is that there is a long history in physics of deciding how to describe results based on what the result is. For example, if a signal is deemed to be significant, a two-sided interval is quoted. If it is deemed to be not significant, an upper limit is quoted. This can lead to bias. First "observations" tend to be biased high, reflecting the greater probability to make the claim if a positive fluctuation has occurred. This bias enters into subsequent averaging, because it is difficult to include limits in the average, and such information is then ignored. To avoid this, it is important to decide on a uniform procedure before the result is known. In particular, analysts should quote sufficient information to permit averaging with other results.

A further related issue is multiple goals in giving a result. At high statistics, results with a frequency interpretation are often quoted. But as statistics become smaller, there is a common tendency to switch over to a Bayesian procedure, usually with a uniform prior distribution (in the physical region) in whatever is being estimated. There are two reasons for this: (*i*) The Bayesian procedure is easier than figuring out how to get accurate coverage, and (*ii*) the Bayesian procedure naturally observes physical boundaries.

The switch-over reflects an unspoken mindset among most physicists – whatever the descriptive intent, they want the result to also serve an interpretive role. That is, the result "should" serve the inferential purpose of making a statement about the underlying truth. For high statistics, the two purposes are often satisfied by the same numbers, and the distinction is obscured. However, at low statistics the distinction is sharpened, and the analyst is faced with a dilemma concerning which path to take.

We recommend keeping the distinction clearly in mind. The result of a measurement should always be described in a meaningful way. Even if it is known that a downward (or upward) fluctuation has occurred, this should be reported. The point of a descriptive methodology is to clearly state the result of the measurement, along with providing a feeling for the "resolution" of the apparatus. It should not be obscured by an attempt at interpretation. The interpretation may also be desirable, but that should occur separately, once the result has been described.

Thus, in practice, we recommend that an analysis be designed to quote a two-sided confidence interval, independent of "physical" constraints. For example, an analysis that results in the observation of 890 events, with an expected background of 900 events, would quote a 68% confidence interval for a signal strength of -10 ± 30 events. Note that this accurately describes the result of the measurement in a way that can be readily combined with other results. It is unimportant here that the interval includes negative signal strengths. We are trying to describe the experimental result, not what we think the true value is. A Bayesian analysis to obtain a posterior degree-of-belief distribution may of course be carried out in addition. In the case of very low statistics, it may be difficult to quote a simple interval that provides sufficient description (e.g., permitting combination with other results). In this case, more details should be reported to provide a sufficiently complete description.

2.5
Likelihood Ratio Test for Presence of Signal

We discussed in Section 2.2 the use of likelihood ratios (or differences in log likelihood) for interval estimation. The same methodology may be used as well in testing hypotheses, such as the significance of a possible signal. We will discuss hypothesis tests at greater length under the rubric of goodness-of-fit in Chapter 3, but it is timely to introduce the likelihood ratio test here.

Consider first the simple test for hypothesis H_0 ("null") against hypothesis H_1 ("alternative"):

$$H_0: \boldsymbol{\theta} = \boldsymbol{\theta}_0 \tag{2.69}$$

$$H_1: \boldsymbol{\theta} = \boldsymbol{\theta}_1, \tag{2.70}$$

where $\boldsymbol{\theta}$ is a vector of parameters, completely specified by either $\boldsymbol{\theta}_0$ or $\boldsymbol{\theta}_1$. This complete specification of the hypotheses is what makes the test simple. The simple hypothesis problem is completely soluble with a most powerful test in classical statistics. The solution involves an ordering principle for the likelihood ratio statistic.

There are many possible critical regions R which give the same significance level. We wish to pick the "best critical region", by finding that region for which the power is greatest. Fortunately, this is straightforward for a simple test. We wish to maximize

$$1 - \beta = \int_R f(\mathbf{X}; \boldsymbol{\theta}_1) d\mathbf{X} \tag{2.71}$$

$$= \int_R \frac{f(\mathbf{X}; \boldsymbol{\theta}_1)}{f(\mathbf{X}; \boldsymbol{\theta}_0)} f(\mathbf{X}; \boldsymbol{\theta}_0) d\mathbf{X}, \tag{2.72}$$

subject to the constraint

$$\alpha = \int_R f(\mathbf{X}; \boldsymbol{\theta}_0) d\mathbf{X}. \tag{2.73}$$

Notice that

$$\frac{1-\beta}{\alpha} = E\left[\frac{f(\mathbf{X}; \boldsymbol{\theta}_1)}{f(\mathbf{X}; \boldsymbol{\theta}_0)}\right]_{(R;H_0)}, \tag{2.74}$$

where the subscript $(R; H_0)$ denotes an expectation restricted to the critical region, under hypothesis H_0. Thus, we build the critical region by including those values of $\mathbf{X}$ for which the ratio $\frac{f(\mathbf{X};\boldsymbol{\theta}_1)}{f(\mathbf{X};\boldsymbol{\theta}_0)}$ is largest. The region R contains all values $\mathbf{X}$ for which

$$\frac{f(\mathbf{X}; \boldsymbol{\theta}_1)}{f(\mathbf{X}; \boldsymbol{\theta}_0)} \geq \Lambda_\alpha, \tag{2.75}$$

where Λ_α is determined by the α constraint.

We may re-express this in the context of likelihood functions. Let

$$\lambda(\mathbf{X}) = \frac{L(\boldsymbol{\theta}_1; \mathbf{X})}{L(\boldsymbol{\theta}_0; \mathbf{X})} \tag{2.76}$$

be the *likelihood ratio* for the two hypotheses, for a sampled value **X**. Note that λ is itself a random variable. If $\lambda \geq \Lambda_\alpha$, the sample **X** is in the critical region. The likelihood ratio test is UMP for the test between simple hypotheses, meaning that for a given significance level α, it has the greatest power against any alternative $\boldsymbol{\theta}_1$.

We can turn this around, given a sampling **X** (and hence λ), and ask what is the significance level, or *p*-value for H_0, according to the value **X**. That is, what is the probability of getting the observed value, or more extreme (where "extreme" is defined in the sense of being in the direction toward favoring H_1), if H_0 is true? This is the probability with which we "rule out" H_0.

Let us try an example, motivated by the occasional need to test between two signs for an effect. Suppose

$$f(x; \theta) = \frac{1}{\sqrt{2\pi}} e^{-\frac{1}{2}(x-\theta)^2} . \tag{2.77}$$

We wish to test

$$H_0: \theta = \theta_0 = -1 , \tag{2.78}$$

against the alternative:

$$H_1: \theta = \theta_1 = +1 . \tag{2.79}$$

We sample a value x^* and form the likelihood ratio (we will also take the logarithm for convenience)

$$\log \lambda^* = -\frac{1}{2} \left[(x^* - \theta_1)^2 - (x^* - \theta_0)^2 \right]$$
$$= 2x^* . \tag{2.80}$$

This defines the critical region for the purpose of computing a *p*-value: $\log \Lambda_\alpha = 2x^*$. The critical region is thus given by $\log \lambda \geq 2x^*$. That is, we are trying to determine the probability for λ to exceed the observed value. Since $\log \lambda = 2x$, we want the probability that $x > x^*$:

$$p = \int_R f(x; \theta_0) dx = \int_{x^*}^\infty \frac{1}{\sqrt{2\pi}} e^{-\frac{1}{2}(x-\theta_0)^2} dx . \tag{2.81}$$

The greater the value of x^*, the smaller is α, and thus the more likely we are to rule out H_0. Our result is intuitive in this situation since everything is nicely monotonic.

Consider a specific example, with a sampled value of $x^* = 3$ from the distribution in (2.77). The hypotheses being compared are $\theta = \theta_0 = -1$ and $\theta = \theta_1 = 1$. We know that P will be given by the probability that $x > x^*$:

$$p = \int_{x^*}^{\infty} \frac{1}{\sqrt{2\pi}} e^{-\frac{1}{2}(x-\theta_0)^2} dx \tag{2.82}$$

$$= 3.2 \times 10^{-5} \tag{2.83}$$

Likewise, the power of the test, if x^* defines the critical region, is

$$1 - \beta = \int_{x^*}^{\infty} \frac{1}{\sqrt{2\pi}\sigma} e^{-\frac{1}{2}(\frac{x-\theta_1}{\sigma})^2} dx \tag{2.84}$$

$$= 0.023 \, . \tag{2.85}$$

The rather low power is just telling us that θ_0 and θ_1 are pretty close together on the scale of the standard deviation and the value of α. That is, for $\alpha = 3.2 \times 10^{-5}$, even if H_1 is correct, we will accept H_0 with 97.7% probability. In a sense, we were "lucky" with our actual sampling, to get such a small value for p (x^* is unlikely even if H_1 is true).

In general, we may not be able to analytically evaluate the sampling distribution for λ, under the hypothesis H_0. In this case, we resort to Monte Carlo simulation or other means to evaluate p. Care must be taken to simulate enough experiments to learn how the tails behave, since that is often where the action lies.

Now let us concern ourselves with the case of testing for the presence of a signal. For this test, we don't care how big the signal is, just that it be bigger than zero. Hence, the test we want is

$$H_0: \theta = 0 \tag{2.86}$$

against

$$H_1: \theta > 0 \, , \tag{2.87}$$

where θ is a parameter describing the strength of a possible signal. The situation may be more complicated now, because we often have nuisance parameters to deal with, and there may even be additional nuisance parameters under H_1 describing the signal (such as its location).

The likelihood ratio statistic for this problem is

$$\lambda(x) = \frac{L(\hat{\theta}, \hat{\boldsymbol{\eta}}_1; x)}{L(0, \hat{\boldsymbol{\eta}}_0; x)} \, , \tag{2.88}$$

where the $\hat{}$ notation refers to maximum likelihood estimates under H_1 in the numerator and under H_0 in the denominator. The nuisance parameters are $\boldsymbol{\eta}_1$ under

H_1 and $\boldsymbol{\eta}_0$ under H_0. In general, the evaluation of the probability distribution of $\lambda(\boldsymbol{x})$ is difficult. However, the problem is basically equivalent to the problem of determining confidence intervals, and the techniques for confidence intervals with nuisance parameters may be applied here as well.

A practical difficulty with significance tests compared with confidence intervals is the level of precision required. In the case of a confidence interval, we usually are content with a 68% confidence level, so we are insensitive to how well we know the low-probability tails of the distribution. However, for a test of significance, we find ourselves quite often making an evaluation of a probability far into a tail, and must work hard to be convinced that the estimate is sound. There is no magic way out of this; when people invent compute-nonintensive methods they can typically be traced to implicit assumptions such as normality.

The presence of a "signal" is not always encapsulated in the value of a signal parameter. It could instead refer to a region of a multidimensional parameter space, compared with some non-signal point. However, the basic approach with the likelihood ratio is the same. We will have more to say about the likelihood ratio statistic in the next chapter, in particular its asymptotic distribution.

2.6
✂ sPlots

It is often difficult to display the results of a fit in a multivariate problem in a visually informative way. In particular it is difficult or impossible to present the full multidimensional space at once. Thus we resort to "projections" in few dimensions, such as one or two.

The use of the density estimation technique known as *sPlots* (Pivk and Le Diberder, 2005) has proven useful in some physics applications. This is a multivariate technique that uses the distribution on a subset of variables to predict the distribution in another subset. It is based on a (parametric) model in the predictor variables, with different categories or classes (e.g., "signal" and "background"). It provides both a means to visualize agreement with the model for each category and a way to do "background subtraction".

Assume there are a total of $K + R$ parameters in the overall fit to the data: *(i)* The expected number of events (observations), $n_k, k = 1, \ldots, K$ in each class, and *(ii)* distribution parameters, $\{\theta_r, r = 1, \ldots, R\}$. We will denote these sets as $\boldsymbol{n}$ and $\boldsymbol{\theta}$, respectively. We use a total of N events to estimate these parameters via a maximum likelihood fit to the sample $\mathbf{x} = \boldsymbol{x}_1, \ldots, \boldsymbol{x}_N$, where $\boldsymbol{x}_e$ is a vector of measurements for event number e. We use e for the event subscript to keep its meaning clear here, since we are using n otherwise. The events are assumed to be i.i.d.

The likelihood function is

$$L(\boldsymbol{n}, \boldsymbol{\theta}; \mathbf{x}) = \prod_{e=1}^{N} \sum_{k=1}^{K} \frac{n_k}{N} f_k(\boldsymbol{x}_e; \boldsymbol{\theta}), \tag{2.89}$$

where f_k is the normalized probability density function for category k, and the constraint $\sum_{k=1}^{K} n_k = N$. Alternatively, including the overall Poisson factor (and no longer constraining $\sum_{k=1}^{K} n_k = N$):

$$L(\boldsymbol{n}, \boldsymbol{\theta}; \boldsymbol{x}) = \frac{e^{-\sum_{k=1}^{K} n_k}}{N!} \prod_{e=1}^{N} \sum_{k=1}^{K} n_k f_k(\boldsymbol{x}_e; \boldsymbol{\theta}) \,. \tag{2.90}$$

The goal is to find event weights $w_k(\boldsymbol{x}'_e)$, depending only on $\boldsymbol{x}'_e \subseteq \boldsymbol{x}_e$ (and implicitly on the unknown parameters), such that the asymptotic distribution in $\boldsymbol{Y}_e \notin \boldsymbol{X}'_e$ of the weighted events is the sampling distribution in $\boldsymbol{Y}_e$, for any chosen class k. The set relation on $\boldsymbol{x}_e$, and so on, refers to elements of the set of vector components. The possibility that $\boldsymbol{x}'_e = \boldsymbol{x}_e$ is included because $\boldsymbol{y}_e$ could refer to quantities that are not contained in $\boldsymbol{x}_e$. It is necessary in the derivation of the method described here to assume that $\boldsymbol{Y}_e$ and $\boldsymbol{X}'_e$ are statistically independent within each class. The $w_k(\boldsymbol{x}'_e)$ are the weights we obtain in an actual experiment of size N. They are random variables. We will introduce notation with an omega, $\omega_k(\boldsymbol{x}'_e)$, to stand for the asymptotic weights. The empirical frequency distribution for $\boldsymbol{y}$, in class k, is estimated using the weights according to

$$\hat{g}_k(\boldsymbol{y}) = \sum_{e=1}^{N} w_k(\boldsymbol{x}'_e) \delta(\boldsymbol{y} - \boldsymbol{y}_e) \,. \tag{2.91}$$

Optimizing the variance of $\hat{g}$ over all $\boldsymbol{y}$ (see exercises in Section 2.7), the asymptotic weights are (switching to matrix notation) (Pivk and Le Diberder, 2005):

$$\boldsymbol{\omega}(\boldsymbol{x}') = \frac{V \boldsymbol{f}}{\sum_{k=1}^{K} n_k f_k(\boldsymbol{x}'; \boldsymbol{\theta})} \,, \tag{2.92}$$

where V is a $K \times K$ matrix, given by

$$V^{-1} = \int d\boldsymbol{z} \, \frac{\boldsymbol{f}(\boldsymbol{z}; \boldsymbol{\theta}) \boldsymbol{f}^T(\boldsymbol{z}; \boldsymbol{\theta})}{\sum_{k=1}^{K} n_k f_k(\boldsymbol{z}; \boldsymbol{\theta})} \tag{2.93}$$

$$= E \left\{ \sum_{e=1}^{N} \frac{\boldsymbol{f}(\boldsymbol{x}'_e; \boldsymbol{\theta}) \boldsymbol{f}^T(\boldsymbol{x}'_e; \boldsymbol{\theta})}{\left[\sum_{k=1}^{K} n_k f_k(\boldsymbol{x}'_e; \boldsymbol{\theta})\right]^2} \right\} \,. \tag{2.94}$$

We estimate the inverse of V with the observed sample:

$$\widehat{V^{-1}} \equiv \sum_{e=1}^{N} \frac{\boldsymbol{f}(\boldsymbol{x}'_e; \boldsymbol{\theta}) \boldsymbol{f}^T(\boldsymbol{x}'_e; \boldsymbol{\theta})}{\left[\sum_{k=1}^{K} n_k f_k(\boldsymbol{x}'_e; \boldsymbol{\theta})\right]^2} \,. \tag{2.95}$$

Finally, we estimate weights $\omega_k(\boldsymbol{x}')$ with estimators $w_k(\boldsymbol{x}'_e)$, according to

$$w(\boldsymbol{x}'_e) \equiv \frac{\hat{V} \boldsymbol{f}(\boldsymbol{x}'_e; \boldsymbol{\theta})}{\sum_{k=1}^{K} n_k f_k(\boldsymbol{x}'_e; \boldsymbol{\theta})} \,. \tag{2.96}$$

On average, these weights reproduce ω. It should be remarked that the weights may be negative as well as positive.

An issue with (2.96) is that the parameters n and θ appear. In order to use this technique, we must provide estimates for these parameters. A natural choice is to use the maximum likelihood estimators. Where important, the variation of the sPlot due to the uncertainty in these parameters can be investigated.

A more subtle issue is whether one should use the parameter estimators based on the full fit to all variables or only from a fit to x'. To the extent of our assumption that y and x' are uncorrelated for any given category, the estimators for the relevant θ values don't depend on y (except in higher order) and may plausibly be taken from the full fit.

On the other hand, y and x' are in general correlated when the categories are taken together, and the parameters n may be different whether y is included in the fit or not. Pivk and Le Diberder (2005) recommend that n be estimated with a fit excluding y. The motivation is to keep the display "pure", in the sense that no y information is used in its construction, potentially obscuring problems. However, it might be that in some circumstances it is of more importance to have the best estimates for n, and violation of this recommendation could be entertained, with caution.

It is remarked in Pivk and Le Diberder (2005) that the matrix V may be estimated by the covariance matrix of the fit. Using (2.90), but omitting the dependence on y we find that

$$(\widehat{V^{-1}})_{mn} = \frac{\partial^2(-\log L)}{\partial n_m \partial n_n}(\hat{n}, \hat{\theta}; x') . \tag{2.97}$$

In this case, the y variable is excluded in the fit used to determine the covariance matrix.

Once the weights are determined, the sPlot is constructed by adding each event e with $y = y_e$ to the y-histogram (or scatter plot, etc., if y is multivariate), with weight $w_j(x'_e)$. The resulting histogram is then an estimator for the true distribution in y for class j. When using sPlots, it may be desirable to display all of the categories, not only the "signal", since each category provides distinct information on how well the data is understood.

Typically the sPlot error in a bin is estimated simply according to the sum of the squares of the weights. This sometimes leads to visually misleading impressions, due to fluctuations on small statistics. If the plot is being made for a distribution for which there is a prediction, then that distribution can be used to estimate the expected uncertainties, and these can be plotted. If the plot is being made for a distribution for which there is no prediction, it is more difficult, but a (smoothed) estimate from the empirical distribution may be used to estimate the expected errors (see Chapter 5).

The sPlot method assumes that all classes present in the data are included. A failure of this assumption, that is, the presence of additional classes, may show up as a deviation between the result of the fit (in terms of the assumed pdfs) and the sPlot. Thus, the sPlot can be a useful diagnostic tool. For example, consider the ΔE

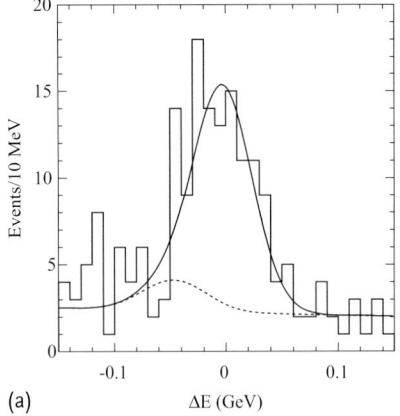 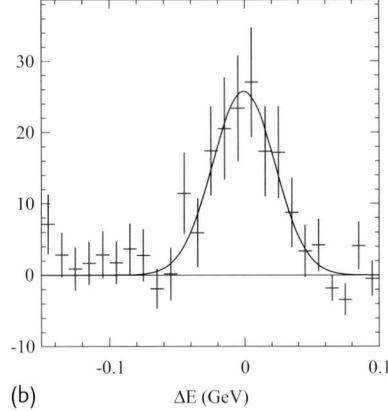

Figure 2.9 Illustration of the sPlot technique. (a) A non-sPlot, which uses a subset of the data, selected on signal likelihood, in an attempt to display signal behavior [reprinted with permission from Aubert *et al.* (2002), copyright 2002, American Physical Society]. (b) An sPlot for the signal category. The curve is the expected signal behavior. Note the excess of events at low values of ΔE. This turned out to be an unexpected portion of the signal distribution, which was found using the sPlot [reprinted with permission from Pivk (2006), copyright 2006, M. Pivk].

signal sPlot for a $B \to \pi^+\pi^-$ analysis, as shown in Figure 2.9b. It is important to understand that this is the estimated signal distribution in ΔE, based on the fit to the other variables (excluding ΔE). The curve shows the predicted signal distribution in ΔE. The systematic deviation at low ΔE is of interest! It indicates that events that are deemed to look like signal according to the x' variables contain a contribution that deviates from the expected signal shape in ΔE. Such deviations should be pursued further. In this case, the interpretation is in terms of radiative contributions which are not in the fit model.

It should also be remarked that the sPlot has other potential uses beyond the present focus as a presentation tool. For example, it may be employed to reconstruct Dalitz-plot distributions for signal, and hence to correct the signal yield for a selection efficiency varying across the Dalitz plot, providing the signal branching fraction without assumptions on the resonance structure of the signal. Figure 2.10 illustrates the use of the sPlot to obtain a background-subtracted mass spectrum.

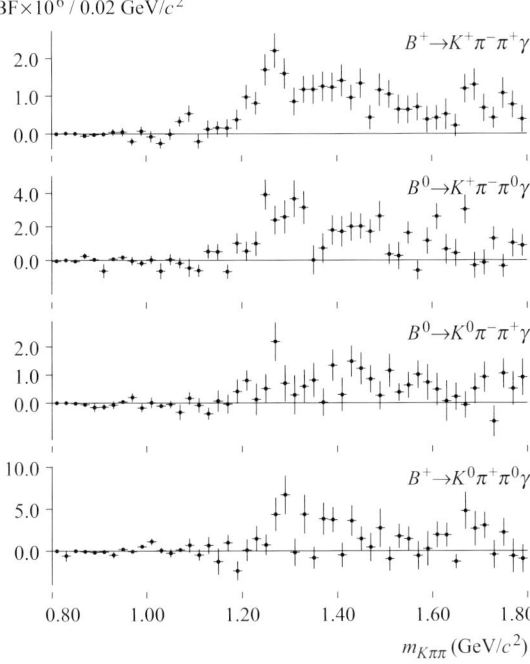

Figure 2.10 The sPlot technique used for background subtraction in mass spectra (reprinted with permission from Aubert *et al.* (2007), copyright 2007, American Physical Society). Plotted are the $K\pi\pi$ mass spectra in $B \to \gamma K\pi\pi$ after background subtraction using sPlots.

2.7
Exercises

1. Derive (2.4) and (2.6). Complete the proof of Theorem 2.2.
2. It is well known that the maximum likelihood estimator for the variance parameter of a normal distribution with unknown mean and variance is biased. Show that, if the nuisance parameter for the mean is marginalized, the maximum likelihood variance estimate is unbiased.
3. Consider a sample from the multivariate normal:

$$f(x; \theta) = \frac{1}{(2\pi)^{d/2}} \frac{1}{\sqrt{\det M}} \exp\left[-\frac{1}{2}(x - \theta)^T M^{-1}(x - \theta)\right], \quad (2.98)$$

where x and the corresponding location parameter vector are of dimension D. The covariance matrix M is known. Show that marginalizing the likelihood yields intervals with frequency interpretation for any desired location parameter of such a multivariate normal. Note that this is trivial except for the possible complication of correlations. The result should be intuitively clear, however.

4. Consider a sample of size N from a density with location and scale parameters:

$$f(x; \mu, \sigma) = \frac{1}{\sigma} f\left(\frac{x - \mu}{\sigma}\right). \quad (2.99)$$

Show that s/σ is a pivotal quantity, where

$$s = \frac{1}{N}\sum_{i=1}^{N}(x_i - m)^2, \quad m = \frac{1}{N}\sum_{i=1}^{N} x_i. \tag{2.100}$$

5. Physicists are sometimes taken with the notion that two-sided confidence intervals should be quoted for "significant" results, and one-sided (upper limits) intervals should be quoted when results are not significant. This comes from an inherently Bayesian mindset, and is fine in that context. However, let us examine this in the frequentist context. It will be sufficient to consider sampling from a univariate Gaussian distribution, $N(\mu, \sigma^2)$. We suppose, on physical grounds, that it is known that $\mu \geq 0$, but μ is otherwise unknown. Assume further that our experimental resolution, σ, is known; we may set it to one here for convenience.

a) Pretend you have done an experiment, producing observation $\hat{\mu}$, that corresponds to a sample from the above distribution, and wish to give a confidence interval for μ. You compute the p-value for the observation under the null hypothesis $\mu = 0$ (this should be a one-sided p-value towards positive μ). You decide to quote a 68% confidence upper limit if the p-value is more than 10% probable, and a two-sided 68% confidence interval if $p < 0.1$. Plot the coverage of your intervals as a function of μ.

b) In part (a) you probably found that the coverage was not always 68%, and depends on the true mean. The obvious fix is to decide on how you want to quote your results before you do the measurement. In the case of normal sampling, simply deciding to always quote the two-sided interval (including possibly negative values) works great – it has the quoted frequency interpretation, tells you the result of your measurement, how well it is measured (your resolution), and makes it easy to combine independent results.

However, a popular method was developed that preserves proper frequencies while making a decision on whether to give a "limit" or a two-sided interval (Feldman and Cousins, 1998). We will call this method "FC" after the authors last initials. The method is that of (2.38)–(2.40), with a twist: if there is a physical boundary to the value of parameter μ, then only those values of $\hat{\mu}$ that are in the physical region are considered in the construction. Thus, in the present example, if $\hat{\mu} < 0$ maximizes the likelihood, we set $\hat{\mu} = 0$.

Work out the FC formalism in detail for Gaussian sampling with physical restriction $\mu > 0$. Make a graph showing the 68% FC intervals as a function of $\hat{\mu}$. Also graph the conventional two-sided intervals described above (that is, without worrying over physical boundaries), and the Bayesian interval obtained with a uniform prior in the physically allowed region. Discuss the results.

6. Derivation of sPlots: Consider making an sPlot histogram. Derive the asymptotic relation in (2.92), showing that it provides the desired consistency and minimum variance properties. Hint: use the properties of the Poisson distri-

bution when constructing your variance. There is some similarity with demonstrating the Gauss–Markov theorem for best linear estimation, where Lagrange multipliers come in handy to incorporate the unbiased condition.

References

Aubert, B. et al. (2002) Measurements of branching fractions and CP-violating asymmetries in $B^0 \to \pi^+\pi^-, K^+\pi^-, K^+K^-$ decays. *Phys. Rev. Lett.*, **89**, 281802.

Aubert, B. et al. (2007) Measurement of branching fractions and mass spectra in $B \to K\pi\pi\gamma$. *Phys. Rev. Lett.*, **98**, 211804–211810.

Cox, D. and Mayo, D. (2011) A statistical scientist meets a philosopher of science. *Ration. Mark. Morals*, **2**, 103–114.

Feldman, G.J. and Cousins, R.D. (1998) A unified approach to the classical statistical analysis of small signals. *Phys. Rev. D*, **57**, 3873–3889.

James, F. and Roos, M. (1975) Minuit: A system for function minimization and analysis of the parameter errors and correlations. *Comput. Phys. Commun.*, **10**, 343–367, doi:10.1016/0010-4655(75)90039-9.

Lehmann, E.L. and Romano, J.P. (2005) *Testing Statistical Hypotheses*, Springer, 3rd edn.

Mayo, D.G. (2010) An error in the argument from conditionality and sufficiency to the likelihood principle, in *Error and inference: Recent exchanges on experimental reasoning, reliability and the objectivity and rationality of science* (eds D.G. Mayo and A. Spanos), Cambridge University Press, pp. 305–314.

Neyman, J. (1937) Outline of a theory of statistical estimation based on the classical theory of probability. *Philos. Trans. R. Soc. A*, **236**, 333–380.

Pivk, M. (2006) sPlot: a quick introduction. *arXiv:physics/0602023v1*, pp. 1–5.

Pivk, M. and Le Diberder, F.R. (2005) sPlot: a statistical tool to unfold data distributions. *Nucl. Instrum. Methods A*, **555**, 356–369.

Reid, N. (2003) Likelihood inference in the presence of nuisance parameters. *Proc. Conf. Stat. Probl. Particle. Phys. Astrophys. Cosmol.*, pp. 265–271.

Shao, J. (2003) *Mathematical Statistics*, Springer, 2nd edn.

3
Goodness of Fit

In physics, the goodness of fit (GOF) problem usually refers to the issue of whether a dataset is consistent with sampling from a model for the distribution. If the dataset is denoted by $\mathbf{X} = (x_1, \ldots, x_N)$, sampled i.i.d. from some cdf $F_X(x)$ and the model is denoted M, this is a hypothesis test of the form:

$$H_0: F = M \tag{3.1}$$

$$H_1: F \neq M. \tag{3.2}$$

This type of test is called a *one-sample test*. It is "one-sample" because we are comparing a dataset with a given (theoretical) distribution, not because $\mathbf{X}$ is a single number (it usually isn't).

However, the *two-sample test* is also commonly encountered in physics. In this case, we compare two datasets, $\mathbf{X}$ and $\mathbf{Y}$, sampled i.i.d. from cdfs F_X and G_Y, to see whether they are consistent with being drawn from the same distribution:

$$H_0: F_X = G_Y \tag{3.3}$$

$$H_1: F_X \neq G_Y. \tag{3.4}$$

Of course, this question can be extended to comparing more than two datasets. An important restricted example occurs when we ask whether a set of measurements are consistent with being drawn from distributions with the same mean. If we have a one-sample test, it can usually be translated into use for a two-sample test via replacing the model M with the second empirical distribution.

Goodness of fit tests may also be categorized as tests on *binned data* (e.g., histograms) or *unbinned data*. Statisticians call histograms *tables*. A simple histogram is a table with one row. A scatter plot may be binned into a table with multiple rows. It may be noted that categorizing a problem as binned or unbinned is somewhat artificial – a test designed to be used on unbinned data may usually be adapted to binned data (though the reverse may not be possible). For example, if an unbinned test is expressed in a form involving cdfs, the corresponding binned version may be obtained by working with the empirical binned cdfs from the histogram(s) under test.

Statistical Analysis Techniques in Particle Physics, First Edition. Ilya Narsky and Frank C. Porter.
©2014 WILEY-VCH Verlag GmbH & Co. KGaA. Published 2014 by WILEY-VCH Verlag GmbH & Co. KGaA.

A large number of goodness of fit tests exist; a few of the more common ones are described below. There is no "one size fits all" test. The choice of a good test depends on the details of the situation. "Goodness" of a test is measured in terms of its power for a specified significance level. To review, the specified significance level, denoted by α, of a test is the probability with which the null hypothesis is rejected, if the null hypothesis is in fact correct, or the probability of a *Type I error*. The power of a test is the probability that it will reject the null hypothesis if the alternative hypothesis is true (hence, power is one minus the probability of a *Type II error*). Clearly the power depends on the alternative distribution.

A digression is called for at this point. The practice of statistics is intimately tied up with philosophical issues. These extend beyond the famous Bayes versus frequency debate. The subject of hypothesis tests is also fertile ground, dating at least to the debate of Fisher and Neyman. Because physicists make use of both sides, it is appropriate to provide a glimpse into the distinctions (Hubbard and Bayarri, 2003).

We have reviewed the notion of a hypothesis test as that of a null hypothesis against an alternative hypothesis. Physicists make use of this paradigm, as well as another, the notion of a *p-value*. The *p*-value is defined, given an experimental result, as the probability that the null hypothesis will produce a result as extreme, or more, as the observed result. While the ideas of the significance level and power of a test are found in the context of comparing a hypothesis with an alternative, the *p*-value only refers to the null hypothesis – there is no explicit alternative. Advocates of this approach call the *p*-value the significance level, and reject the use of the term as describing the Type I error rate. One view into the difference is that the *p*-value is a random variable, while the critical region defined by the acceptable Type I error rate is not. The alternative hypotheses paradigm specifies a decision rule for accepting or rejecting the null against an alternative; the *p*-value states the result of the measurement in the context of a hypothesis without taking the step of a decision.

Physicists often think of the goodness of fit problem in terms of the *p*-value. The fit is deemed to be bad if the *p*-value is very small. However, it can be argued that there must be an implicit alternative hypothesis. For if there were no possible alternative to the null, then the null is the only allowed hypothesis, independent of the obtained *p*-value. An alternative must also be assumed in order to compute the power of a test, providing a comparison among possible tests.

A general cautionary remark concerning the distribution of test statistics is in order. Typically (but not always), only the asymptotic distribution (under H_0) of the test statistic for common tests such as those here is known. This does not mean that the test cannot be used when the asymptotic condition is not met. With sufficient computing capability, the distribution of the test statistic may be determined via simulations. However, this must be done with some care, as the null hypothesis is often not completely known. For example, one might wish to test the null hypothesis that two histograms are sampled from the same distribution. Unfortunately, the distribution itself may not be known. Instead, it must be estimated somehow from the available data. If suitable care is not taken, the estimate may not be robust against fluctuations, and badly erroneous results obtained. When this may be the case, suitable studies (e.g., with different estimates of H_0) should be undertaken to determine this sensitivity.

3.1
Binned Goodness of Fit Tests

Among physicists, the most commonly used goodness of fit test is the "chi-square" test. Motivation for this test may be found in the least-squares fitting process. We have a set of measurements, here sampled from a multivariate normal, $x_1, \ldots, x_D$, and a model to predict the corresponding means, $\mu_1, \ldots, \mu_D$. The model may depend on zero or more unknown parameters, $\theta_1, \ldots, \theta_R$. With the assumption of normality, the sampling distribution according to this model is

$$f_X(x) = \frac{1}{\sqrt{(2\pi)^D |M|}} \exp\left\{-\frac{1}{2}[x - \boldsymbol{\mu}(\boldsymbol{\theta})]^\mathsf{T} M^{-1}[x - \boldsymbol{\mu}(\boldsymbol{\theta})]\right\} \tag{3.5}$$

$$= \frac{1}{\sqrt{(2\pi)^D |M|}} \exp\left(-\frac{1}{2}\chi^2\right), \tag{3.6}$$

where M is the covariance matrix of the distribution. The least-squares fitting procedure is to find those values of $\boldsymbol{\theta}$ such that the χ^2 as defined by this equation (also known as the square of the Mahalanobis distance) is minimized. As long as no additional conditions are applied, the value of the minimum χ^2, $\chi^2_{\min}$, is drawn from a chi-square distribution with $D - R$ degrees of freedom (DOF). Comparing the observed $\chi^2_{\min}$ with the expected chi-square distribution provides the chi-square goodness of fit test.

As developed above, the least squares fit is equivalent to the maximum likelihood procedure – minimizing the χ^2 is identical to maximizing the likelihood. However, the least-squares procedure, just as the maximum likelihood procedure, can be applied even when the sampling distribution is not normal. In this case, the two procedures are generally not equivalent. Further, the $\chi^2_{\min}$ statistic will not in general be $\chi^2(D-R)$ distributed.

We consider the case of fitting a model to a histogram. This is the "binned goodness of fit test", because we have binned our data into a histogram. In this case, N is the number of histogram bins and x_n is the content of bin n. If the bin contents are simply counts from a Poisson process, then the bins are independent and the covariance matrix is diagonal with elements $\boldsymbol{\mu}$ along the diagonal. In the limit of large $\boldsymbol{\mu}$, the bin contents are approximately normal, and the least squares statistic, χ^2, is approximately $\chi^2(N-R)$ distributed. The statistic is known as the *Pearson chi-square* (Pearson, 1900), given by

$$\chi^2_\mathrm{P} = \sum_{n=1}^{N} \frac{(x_n - \mu_n)^2}{\mu_n}. \tag{3.7}$$

For specified $\boldsymbol{\mu}$, this is asymptotically $\chi^2(N)$ distributed (or $\chi^2(N-1)$ if the normalization is fixed to the observed total counts). In practice, we usually have parameters to estimate and replace $\boldsymbol{\mu}$ with $\hat{\mu}_i = \mu_i(\hat{\boldsymbol{\theta}})$ where $\hat{\boldsymbol{\theta}}$ is determined by minimizing χ_P^2. This reduces the degrees of freedom in the chi-square distribution, by one for each estimated parameter, subject to regularity conditions discussed below.

The Pearson chi-square test should not be used if any of the histogram bins have small statistics, since the $\chi^2(N-R)$ distribution will not apply. Various rules-of-thumb have been proposed for how many counts are needed in each bin for a good enough approximation, usually around 5–10. It is admissible to combine bins to meet the minimum requirement, though at the loss of sensitivity to possible structure in the data. Another approach is to use bins that have equal probability content under the null hypothesis, ensuring a uniform weighting of the intervals.

Especially at low statistics, there is a tendency for this algorithm to overestimate μ_n in the process of fitting parameters by minimizing χ_P^2. This may be avoided by instead estimating the parameters via maximum likelihood. An alternative approach defines a statistic (referred to as the Neyman modified chi-square):

$$\chi_N^2 = \sum_{n=1}^{N} \frac{(x_n - \mu_n)^2}{x_n} . \tag{3.8}$$

For sufficiently large statistics, this also works, but for low statistics it suffers a similar disease: fluctuations toward small values of x_i will be more highly weighted, now tending to bias towards small values of μ_n.

It is appropriate to mention here some other areas of common confusion involving the chi-square statistic. First, some care must be taken in counting "degrees of freedom". There are conditions for the validity of the $\chi^2(N-r)$ distribution of the test statistic. This often arises when using the chi-square to evaluate the statistical significance of a possible signal in a histogram.

The following situation often arises, with variations. We do two fits to the same dataset (say a histogram with N bins): Fit A has R_A parameters, with χ_A^2. Fit B has a subset R_B of the parameters in fit A, with χ_B^2, where the $R_A - R_B$ other parameters (call them $\boldsymbol{\theta}$) are fixed at zero. What is the distribution of $\Delta\chi^2 = \chi_B^2 - \chi_A^2$?

In the asymptotic limit (that is, as long as the normal sampling distribution is a valid approximation),

$$\Delta\chi^2 \equiv \chi_B^2 - \chi_A^2 \tag{3.9}$$

is the same as a likelihood ratio ($-2\log\lambda$) statistic for the test

$$H_0: \boldsymbol{\theta} = 0 \quad \text{against} \quad H_1: \text{some } \boldsymbol{\theta} \neq 0. \tag{3.10}$$

In this case, the $\Delta\chi^2$ is distributed according to a $\chi^2(R_A - R_B)$ distribution under the following conditions (Davies, 1977a,b; Demortier, 2008):

1. Parameter estimates in computing λ are consistent (converge to the correct values) under H_0.

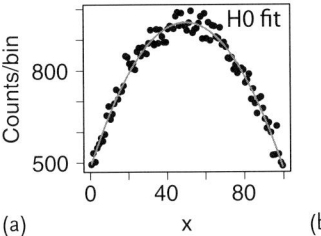 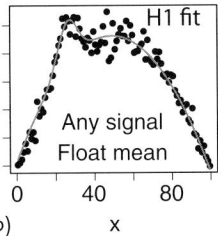

Figure 3.1 Examples of maximum likelihood fits: (a) Distribution generated under the background-only null hypothesis. Fit to the null hypothesis. (b) Distribution generated under the alternative hypothesis. Fit to the alternative hypothesis with unconstrained signal yield and location.

2. Parameter values in the null hypothesis are not boundary points of $H_0 \cup H_1$ (the *maintained hypothesis*). For example, if there is a single parameter θ, with H_0: $\theta = 0$ and H_1: $\theta > 0$, then the maintained hypothesis is $\theta \geq 0$ and the parameter value in the null hypothesis is a boundary point.
3. There are no nuisance parameters under the alternative hypothesis other than those present in H_0.

Unfortunately, commonly encountered situations may violate these requirements.

For example, consider fitting a spectrum to decide whether a bump is significant or not. In this case, the parameter of greatest interest is the signal strength. We compare fits with and without a signal component to estimate significance of the signal. Under the null hypothesis, the signal is zero. Under the alternative hypothesis, the signal is nonzero. If the signal fit has, for example, a parameter for location, this constitutes an additional nuisance parameter under the alternative hypothesis; that is, a nuisance parameter that is not defined under H_0. If the fit for signal constrains the signal yield to be nonnegative, this violates the interior point requirement. In such cases, the number of degrees of freedom is not a single integer. Let us illustrate this by example.

Figure 3.1 shows two sample results of maximum likelihood fits to mass spectra generated with and without a signal component.[1] In order to compute the distribution of $\Delta\chi^2$, we consider the spectrum generated under H_0, the background-only hypothesis. The difference in χ^2 between the H_0 and H_1 fits is calculated for this spectrum. The χ^2 is calculated as the value of $-2\log L$ at the maximum likelihood, but it is readily checked that this is in agreement, for this high statistics example, with an evaluation of the sum of the squared deviations between the fit and the histogram bins.

The distribution of the $\Delta\chi^2$ statistic is estimated by simulating each "experiment" many times, under the null hypothesis. The results are shown in Figure 3.2. When the location parameter is fixed in the fit, and the signal yield is allowed to be positive or negative, the distribution follows a χ^2 for 1 DOF, reflecting the difference of one parameter between the fits, with all of the above conditions satis-

---

1) The R function `optim` is used for the fits here.

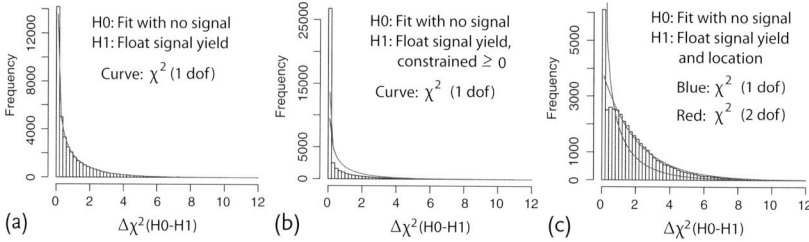

Figure 3.2 Distributions of $\Delta\chi^2$ for different alternative hypotheses fits. (a) Alternative fit for a fixed location, and signal yield unconstrained. The curve is the χ^2 distribution for 1 DOF. (b) Alternative fit for a fixed location, but with signal yield constrained to be positive. The curve is the χ^2 distribution for 1 DOF. (c) Alternative fit for a variable location, and signal yield unconstrained. The steeply falling curve is the χ^2 distribution for 1 DOF, and the other curve is for 2 DOF.

fied. However, when the fit constrains the yield to be nonnegative, the distribution becomes more sharply peaked towards zero than a χ^2 for 1 DOF. When both the signal yield and location are unconstrained, the distribution is somewhere between the curves for 1 and 2 DOF, and does not follow the 2 DOF that is often assumed. This is because the location parameter is an additional nuisance parameter under the alternative hypothesis.

A similar issue arises when the unknown parameters of a distribution are determined according to a maximum likelihood fit to the unbinned observations. If the data is then binned and a χ^2 statistic computed, this statistic will not in general be $\chi^2(N-R-1)$ (assuming the normalization is taken from the data) distributed. This is because the unbinned fit produces more efficient estimators than a binned fit in general. Chernoff and Lehmann (1954) show that, under regularity conditions, the asymptotic distribution lies between $\chi^2(N-1)$ and $\chi^2(N-R-1)$. If the $\chi^2(N-R-1)$ distribution is assumed, the result will be to reject the null hypothesis too often. Depending on the distribution, the error made may be minor or substantial, so this should be considered when taking this approach.

Another area of common confusion involves the use of the χ^2 statistic, and the χ^2 test, in the estimation of confidence intervals. We remark that a hypothesis test may be used to obtain a confidence interval. For any test T of a hypothesis H_0 versus alternative hypothesis H_1, we define statistic (*decision rule*) $T(\mathbf{X})$ with values 0 or 1. Acceptance of H_0 corresponds to $T(\mathbf{X}) = 0$, and rejection to $T(\mathbf{X}) = 1$. The set $A = \{\mathbf{X} : T(\mathbf{X}) = 0\}$ is called the *acceptance region*. The significance level of the test is the probability of rejecting H_0 when H_0 is true:

$$\alpha = P[T(\mathbf{X}) = 1 | H_0]. \tag{3.11}$$

Let $T_{\boldsymbol{\theta}_0}$ be a test for $H_0: \boldsymbol{\theta} = \boldsymbol{\theta}_0$ with significance level α and acceptance region $A(\boldsymbol{\theta}_0)$. Let, for each $\mathbf{X}$,

$$C(\mathbf{X}) = \{\boldsymbol{\theta} : \mathbf{X} \in A(\boldsymbol{\theta})\}. \tag{3.12}$$

Now, if $\boldsymbol{\theta} = \boldsymbol{\theta}_0$,

$$P[\mathbf{X} \notin A(\boldsymbol{\theta}_0)] = P(T_{\boldsymbol{\theta}_0} = 1) = \alpha . \tag{3.13}$$

That is, again for $\boldsymbol{\theta} = \boldsymbol{\theta}_0$,

$$1 - \alpha = P[\mathbf{X} \in A(\boldsymbol{\theta}_0)] = P[\boldsymbol{\theta}_0 \in C(\mathbf{X})] . \tag{3.14}$$

This holds for all $\boldsymbol{\theta}_0$, hence, for any $\boldsymbol{\theta}_0 = \boldsymbol{\theta}$,

$$P[\boldsymbol{\theta} \in C(\mathbf{X})] = 1 - \alpha . \tag{3.15}$$

That is, $C(\mathbf{X})$ is a confidence region for $\boldsymbol{\theta}$, at the $1 - \alpha$ confidence level.

For example, suppose we have a sample of size $N = 10$ from a $\mathsf{N}(\theta, 1)$ distribution. We wish to determine a 68% confidence interval for unknown parameter θ. We will compare the results of two different methods.

In the first method, we will use a pivotal quantity. Let $\Delta \chi^2(\theta)$ be the difference between the χ^2 estimated at θ and the minimum value, at $\hat{\theta} = \frac{1}{10} \sum_{n=1}^{10} x_n$:

$$\Delta \chi^2 = 10(\hat{\theta} - \theta)^2 . \tag{3.16}$$

Note that $\hat{\theta}$ is normally distributed with mean θ and variance $1/10$, and that $\hat{\theta} - \theta$ is pivotal, hence so is $\Delta \chi^2$. Finding the points where $\Delta \chi^2 = 1$ corresponds to our familiar method for finding the 68% confidence interval,

$$(\hat{\theta} - 1/\sqrt{10}, \; \hat{\theta} + 1/\sqrt{10}) .$$

This method is illustrated in the left plot in Figure 3.3. We note that this construction can also be expressed in terms of the likelihood ratio statistic, that is in terms of a likelihood ratio test.

For the second method, we will invert a test acceptance region based on the χ^2 statistic (equivalently here, a test based on the value of the likelihood). Consider the chi-square goodness of fit test for

$$H_0 : \theta = \theta_0 , \tag{3.17}$$

$$H_1 : \theta \neq \theta_0 . \tag{3.18}$$

At the 68% significance level, we accept H_0 if

$$\chi^2(\theta_0) < \chi^2_{\text{crit}} , \tag{3.19}$$

where χ^2_{crit} is defined by

$$P\left(\chi^2_{10} < \chi^2_{\text{crit}}\right) = 68\% \tag{3.20}$$

where χ^2_{10} is a chi-square RV with 10 DOF. Note that 10 DOF is used here, since θ_0 is specified.

3 Goodness of Fit

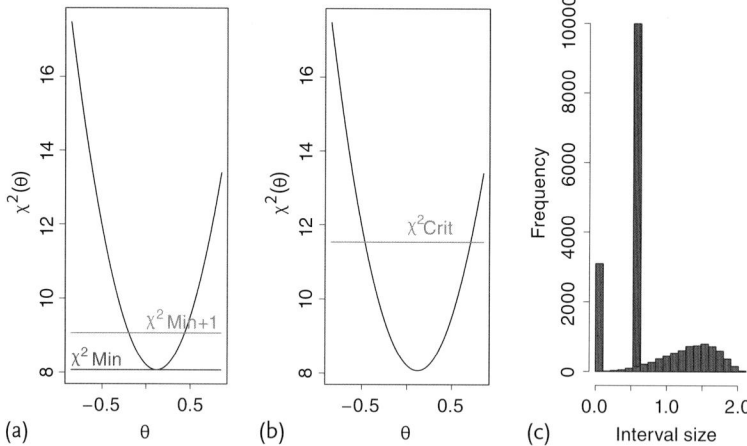

Figure 3.3 Comparison of two confidence intervals for θ for samples of size $n = 10$ from an $N(\theta = 0, 1)$ distribution. (a) The pivotal quantity method. (b) The method of inverting a test acceptance region. (c) Comparison of the distribution of the lengths of the intervals for the two methods. The darker histogram (with one nonzero bin) is for the pivotal quantity method. The other histogram is the method of inverting the χ^2 test acceptance region.

If $\chi^2_{\text{crit}} > \chi^2_{\text{min}}$, we have the confidence interval

$$\hat{\theta} \pm \sqrt{(\chi^2_{\text{crit}} - \chi^2_{\text{min}})/10},$$

and if $\chi^2_{\text{crit}} < \chi^2_{\text{min}}$, we have a null confidence interval. This method is illustrated in Figure 3.3b.

The distributions of the lengths of the confidence intervals obtained with these two methods are shown in the right plot of Figure 3.3. The intervals obtained using the pivotal quantity are always of the same length, reflecting the independence of the resolution of the measurement on the parameter value. The intervals obtained by inverting the χ^2 test acceptance region are of varying length, including often length zero. Both methods provide valid confidence intervals. However, the intervals from the first method are clearly preferable here. The χ^2 goodness of fit test acceptance intervals do a much poorer job of describing the precision of the measurement in this example. This has been observed as a significant issue in practice (Mueller and Madejski, 2009).

3.2
Statistics Converging to Chi-Square

There are three important test statistics that follow a χ^2 distribution for large samples, under certain assumptions. These are the likelihood ratio test, the Wald test, and the score test. In spite of their asymptotic commonality, they are quite different in general, yet each has an intuitive appeal.

In Section 2.5 we introduced the likelihood ratio test for the presence of a signal. A likelihood ratio statistic may be constructed more generally, including as a goodness of fit statistic. For a simple test, that is, a test in which H_0 and H_1 are completely specified, we showed in Section 2.5 that the likelihood ratio

$$\lambda = \frac{L(H_1; \mathbf{X})}{L(H_0; \mathbf{X})}, \qquad (3.21)$$

is uniformly most powerful. For composite tests, we maximize the likelihoods under the null and alternative hypotheses before taking the ratio. Taking the logarithm and multiplying by two, we have the statistic

$$2 \log \lambda = 2 \log \max_{H_1} L(H_1; \mathbf{X}) - 2 \log \max_{H_0} L(H_0; \mathbf{X}). \qquad (3.22)$$

The evaluation of $2 \log \lambda$ is generally in the context of parametric descriptions for H_0 and H_1, and we will assume parametric descriptions in the remainder of this discussion. In particular, we assume that H_0 and H_1 are described as regions in an r-dimensional parameter space, $\boldsymbol{\theta} = (\theta_1, \ldots, \theta_R)$. That is, the hypothesis being tested is

$$H_0: \boldsymbol{\theta} \in \Theta_0 \qquad (3.23)$$

$$H_1: \boldsymbol{\theta} \in \Theta_1. \qquad (3.24)$$

We assume that Θ_0 is not on the boundary of the space defined by $\Theta = \Theta_0 \cup \Theta_1$ (the *maintained hypothesis*, or just "the parameter space"). We have already seen in Section 3.1 how violating this assumption can affect the results.

As is typically the case, we will assume further that H_0 is a more restrictive statement than H_1 on the possible parameter values. Thus, we write

$$H_0: \boldsymbol{\theta} = \mathbf{q}(\boldsymbol{\vartheta}), \qquad (3.25)$$

where $\boldsymbol{\vartheta}$ is a parameter vector of dimension $V < R$, and $\mathbf{q}$ provides the mapping (assumed continuously differentiable) onto the R-dimensional $\boldsymbol{\theta}$. We may rewrite H_0 as the $R - V$ equations giving the kernel of the mapping, that is as $\mathbf{Q}(\boldsymbol{\theta}) = 0$. For example, if $H_0: \boldsymbol{\theta} = \boldsymbol{\theta}_0$, then $\mathbf{Q}(\boldsymbol{\theta}) = \boldsymbol{\theta} - \boldsymbol{\theta}_0$. In the following, we will represent the maximum likelihood estimator under H_0 as $\hat{\boldsymbol{\vartheta}}$, and the maximum likelihood estimator under H_1 as $\hat{\boldsymbol{\theta}}$.

The *Wald statistic* is

$$W = [\mathbf{Q}(\hat{\boldsymbol{\theta}})]^\top \left\{ \left[\frac{\partial \mathbf{Q}}{\partial \boldsymbol{\theta}} \right]_{\boldsymbol{\theta} = \hat{\boldsymbol{\theta}}}^\top I(\hat{\boldsymbol{\theta}})^{-1} \left[\frac{\partial \mathbf{Q}}{\partial \boldsymbol{\theta}} \right]_{\boldsymbol{\theta} = \hat{\boldsymbol{\theta}}} \right\}^{-1} \mathbf{Q}(\hat{\boldsymbol{\theta}}), \qquad (3.26)$$

where $I(\hat{\boldsymbol{\theta}})$ is the Fisher information matrix, estimated at $\boldsymbol{\theta} = \hat{\boldsymbol{\theta}}$:

$$I_{ij}(\boldsymbol{\theta} = \hat{\boldsymbol{\theta}}) = E \left[\frac{\partial^2 \log L(\boldsymbol{\theta}; \mathbf{X})}{\partial \theta_i \partial \theta_j} \right]_{\boldsymbol{\theta} = \hat{\boldsymbol{\theta}}}. \qquad (3.27)$$

For a normal distribution $I(\boldsymbol{\theta})$ is the inverse of the covariance matrix, and is independent of $\boldsymbol{\theta}$ if the parameters are functions of location only.

For example, suppose the null hypothesis is $H_0 \colon \boldsymbol{\theta} = \boldsymbol{\theta}_0$. Then $V = 0$ and $Q(\boldsymbol{\theta}) = \boldsymbol{\theta} - \boldsymbol{\theta}_0$, and we have the familiar-looking statistic

$$W = (\hat{\boldsymbol{\theta}} - \boldsymbol{\theta}_0)^\mathrm{T} I(\hat{\boldsymbol{\theta}})(\hat{\boldsymbol{\theta}} - \boldsymbol{\theta}_0). \tag{3.28}$$

The likelihood ratio compares two likelihood values; the Wald statistic compares two values of $\boldsymbol{\theta}$. Notice that the Wald statistic does not require evaluation of $\hat{\vartheta}$.

Alternatively, a *score statistic* based on the score may be computed, measuring the gradient of the likelihood under the null hypothesis. In this case it is not necessary to obtain the maximum likelihood estimates for $\boldsymbol{\theta}$. Instead, we compute

$$U = S(\boldsymbol{\theta}_0)^\mathrm{T} I^{-1}(\boldsymbol{\theta}_0) S(\boldsymbol{\theta}_0), \tag{3.29}$$

where S is the score function evaluated at $\boldsymbol{\theta}_0$. Comparing this with the Wald statistic, we see that the evaluation is made at the null hypothesis rather than at the peak of the likelihood, and the deviation measure is replaced by a slope.

Which of these approaches is best? Asymptotically, they are equivalent. More generally, there is no universal answer. One ingredient in deciding may be the different computational requirements. However, it may be noted that U often provides a more powerful smooth (Section 3.3.4) GOF test (Rayner and Best, 1989).

In considering the multinomial distribution (e.g., corresponding to describing a histogram) Cressie and Read (1984) introduced a class of test statistics known as the *power divergence family*:

$$T_\lambda = \frac{2}{\lambda(\lambda+1)} \sum_{d=1}^{D} X_i \left[\left(\frac{X_d}{\mu_d} \right)^\lambda - 1 \right], \tag{3.30}$$

for given total counts, where D is the dimension (number of bins) of multinomial random variable X and $\mu_d = E(X_d)$, and λ is a real number determining the element of the family. Both the Pearson χ^2 (with $\lambda = 1$) and the likelihood ratio test (with $\lambda \to 0$) are seen to belong to this family. Cressie and Read (1984) showed that, given some regularity conditions, any element of the power divergence family is asymptotically $\chi^2(D - r - 1)$ distributed, where r is the number of unknown (nuisance) parameters under the model hypothesis. Thus, we have available an infinite family of tests with known asymptotic behavior. These could be examined to determine which is best in a particular situation, but Cressie and Read (1984) offers some recommendations:

1. If the alternative is unknown, $\lambda \in [0, 1.5]$ is suggested, with $\lambda = 2/3$ a good compromise;
2. If the alternative is dipped, consider $\lambda = 0$;
3. If the alternative is peaked, consider $\lambda = 1$.

3.3
Univariate Unbinned Goodness of Fit Tests

If we have an i.i.d. sample of size N, $x_1, x_2, \ldots, x_N$, we have a univariate unbinned dataset, where we assume a continuous sampling distribution. This dataset may be used to test hypotheses concerning the nature of the sampling distribution. We discuss several popular (and not-so-popular in physics, but worthy of consideration) goodness of fit tests in the following sections.

3.3.1
Kolmogorov–Smirnov

Next to the chi-square and likelihood ratio tests, the Kolmogorov–Smirnov (KS) test is perhaps most familiar among physicists. This is the most straightforward test of the difference between two cumulative distributions. The test statistic, T, is formed simply as the maximum difference between the two cumulative distributions being compared.

We may set this in the general context of notions of distance between cumulative distributions. Given any pair of cdfs F and G on a sample space, it may be possible to define a distance or metric, $\rho(F, G)$, that returns a nonnegative number satisfying all the normal properties of a distance on a metric space. In particular we may define the distance

$$\rho(F, G) \equiv \sup_x |F(x) - G(x)|. \tag{3.31}$$

When $G = F_N$ is the empirical cdf of our dataset, and F is the H_0 cdf, ρ provides the *Kolmogorov–Smirnov* goodness of fit statistic. For a sample of size N and null hypothesis F, we denote the Kolmogorov statistic by $K_N(F)$.

The distribution of $K_N(F)$ is independent of F (exercise), for continuous F. Thus, the distribution of the Kolmogorov–Smirnov statistic has the convenient property that it is known and depends only on the sample size N (Figure 3.4). For small N, the distribution may be computed according to the exact formula (Shao, 2003):

$$P[K_N(F) \le t] = \begin{cases} 0 & t \le \frac{1}{2N} \\ N! \prod_{i=1}^{N} \left[\int_{\max\left(0, \frac{N-i+1}{N}-t\right)}^{\min\left(u_{N-i+2}, \frac{N-i}{N}+t\right)} du_1 \cdots du_N \right] & \frac{1}{2N} < t < 1 \\ 1 & t \ge 1, \end{cases} \tag{3.32}$$

with $u_{N+1} \equiv 1$. For large values of N it is more convenient to use the asymptotic form:

$$\lim_{N \to \infty} P[\sqrt{N} K_N(F) \le t] = 1 - 2 \sum_{j=1}^{\infty} (-1)^{j-1} e^{-2Nj^2 t^2}, \quad t > 0. \tag{3.33}$$

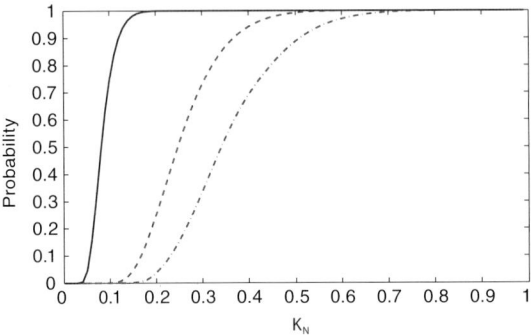

Figure 3.4 Cumulative distribution for the Kolmogorov–Smirnov statistic. From left to right, $N = 100, 10, 5$.

The Kolmogorov–Smirnov test is by construction a two-sided test. However, corresponding one-sided tests are obtained by dropping the absolute value signs and using either the minimum or the maximum, providing a signed measure of distance (Shao, 2003).

An obvious variation on the Kolmogorov–Smirnov approach is to replace the supremum distance function with another common measure of distance, the average squared deviation:

$$C_N^2(F) = \int_{-\infty}^{\infty} [F_N(y) - F(y)]^2 \, dF(y). \tag{3.34}$$

This is known as the *Cramér–von Mises test*. The distribution of $C_N^2(F)$ likewise does not depend on F.

3.3.2
Anderson–Darling

There are, of course, many test statistics one could invent based on the difference between cumulative distributions. The Kolmogorov–Smirnov test just described is an especially simple choice. It has the property that it tends to emphasize the region of most rapid change in the cdf (that is, the region of the peak of the pdf), as that is where the maximum difference under the null hypothesis is likely to occur.

The *Anderson–Darling test* is designed to give more weight to the tails of the distribution. It is thus particularly powerful in tests involving deviations in the tails. For example, sampling distributions are often approximately normal in the central region, but the tails may be significantly non-Gaussian. The Anderson–Darling test is known to be especially powerful in detecting such cases. This could be important in detecting signs of new physics if the signature is a few events in the tail, for example of a p_T distribution, where the KS is not especially sensitive.

The Anderson–Darling statistic is defined according to

$$A_N^2(\mathbf{x}) = N \int_{-\infty}^{\infty} \frac{[F_N(y) - F(y)]^2}{F(y)[1 - F(y)]} dF(y), \qquad (3.35)$$

where F_N is the empirical cdf of our dataset $\mathbf{x} = x_1, \ldots, x_N$ and F is the cdf under H_0. Note that the Anderson–Darling statistic modifies the Cramér–von Mises statistic with the addition of the denominator terms. These terms provide preferential weight to the regions near $F = 0$ and $F = 1$, that is, the tails of the distribution.

3.3.3
Watson

Another variation on the Cramér–von Mises approach is the *Watson test* (Watson, 1961):

$$U_N^2 = N \int_{-\infty}^{\infty} \left\{ F_N(x) - F(x) - \int_{-\infty}^{\infty} [F_N(y) - F(y)] dF(y) \right\}^2 dF(x). \qquad (3.36)$$

Here, the difference between empirical and theoretical distributions is "corrected" by subtracting the mean difference. Thus, this test ignores a simple shift and concentrates on higher order differences.

3.3.4
Neyman Smooth

Unsatisfied with the limitations of the χ^2 test (in particular, its inability to detect runs, for example, successive histogram bins improbably "running" higher than the model, as deviations from H_0), Neyman (1937) devised the *Neyman smooth test*[2]. The basic idea is to transform the data to uniformly-distributed quantities under the null hypothesis, and then use Legendre polynomials to frame the alternative hypothesis as a "smooth" pdf, with degree of smoothness determined by the order of the polynomials used. The observed moments with respect to the Legendre polynomials may then be compared with the values of zero expected under the null hypothesis. One feature of this approach is that it provides a framework to investigate in more detail the reason for a bad fit to the model.

If the density under the null hypothesis is $g(x|H_0)$, we make the transformation $x \to u$ according to

$$u = \int_{-\infty}^{x} g(x'|H_0) dx'. \qquad (3.37)$$

Under H_0, u is uniformly distributed on $(0, 1)$, with pdf $f(u) = 1$.

---

[2] Ironically, it can be shown that Pearson's chi-square is a type of smooth test (Rayner and Best, 1989).

The goodness of fit test of the null hypothesis is thus one of testing uniformity of the distribution for u. This is framed as a test against the alternative hypothesis:

$$H_1: f(u|\boldsymbol{\theta}) = \exp\left[-C(\boldsymbol{\theta}) + \sum_{k=1}^{K} \theta_k P_k(2u-1)\right], \quad u \in (0,1), \tag{3.38}$$

where $\boldsymbol{\theta} = (\theta_1, \ldots, \theta_k)$, $C(\boldsymbol{\theta})$ provides normalization, and P_k is the kth Legendre polynomial. The parameters $\boldsymbol{\theta}$ are expansion coefficients in a truncated Legendre series. P_0 is not explicitly included as it is redundant with C. The highest degree polynomial included, K, is called the "order" of the test. Notice that the test may be rephrased as

$$H_0: \boldsymbol{\theta} = 0, \tag{3.39}$$

$$H_1: \boldsymbol{\theta} \neq 0. \tag{3.40}$$

To complete the construction of a test statistic, a set of "optimal" values for $\boldsymbol{\theta}$ may be chosen, typically by maximizing the likelihood under H_1, yielding parameter estimates $\hat{\boldsymbol{\theta}}(\mathbf{x})$. Then either a likelihood ratio statistic or a Wald test statistic may be constructed. For example, the likelihood ratio statistic is

$$\lambda = \frac{L(\boldsymbol{\theta} = 0|\mathbf{x})}{L(\boldsymbol{\theta} = \hat{\boldsymbol{\theta}}|\mathbf{x})}. \tag{3.41}$$

In our discussion so far, the null hypothesis has been treated as a simple hypothesis. However, the method may be applied to composite null hypotheses as well. According to the above approach, the test may be expressed in terms of a statistic with χ^2 asymptotic distribution.

3.4
Multivariate Tests

We have described our tests so far in the context of univariate statistics, that is, the dataset $x_1, \ldots, x_N$ consists of N i.i.d. random samples from a one-dimensional distribution $F(X)$. If the $\mathbf{x}_n$ are themselves multidimensional, say with dimension D, further discussion is warranted. To some extent, we may modify the tests we have discussed in simple ways to address the multivariate problem. For example, we could construct a D-dimensional generalization of the histogram, and then apply the χ^2 test using the contents of our D-dimensional generalized bins. As long as the sample is large enough, we will have an approximate χ^2 distribution.

The problem with this simple view is the "as long as ..." If the number of dimensions is large, the contents of many bins will be sparse. If the appropriate segmentation in each dimension is 100 intervals, then we have a total of 100^D bins. It doesn't take many dimensions to get sparse bin populations, even with large datasets. The problem may be mitigated with an adaptive binning procedure, just as in the univariate case, Section 3.1. However, the power of the test suffers if there is important information in the distribution within the bin.

A variety of unbinned methods for dealing with the multivariate GOF problem are reviewed here. Williams (2010) studies several methods using the example of a high energy physics Dalitz plot analysis.

3.4.1
Energy Tests

An approach to the multivariate goodness of fit problem with a physical appeal is an *energy test* (Aslan and Zech, 2003, 2005), including its variants. The test is based on the quantity

$$\phi = \frac{1}{2} \iint g(x)g(x') R(\|x - x'\|) dx dx', \qquad (3.42)$$

where $R(y)$ is a "potential energy" function. For example, if $R = 1/|x - x'|$ we see that ϕ looks like the Coulomb energy of a charge distribution given by g. In our application, let g be the difference between a pdf being tested and the pdf under the null hypothesis. In this case, we have $\int g(x) dx = 0$. Smaller values of ϕ correspond to better agreement with the null. Exact agreement with the null hypothesis corresponds to $g = 0$.

This idea may be implemented with a Monte Carlo approach, in order to deal with the difficulty of multidimensional integrals. Thus, we compare our observations $x_1, \ldots, x_N$ with simulated data under H_0, $y_1, \ldots, y_M$. The energy statistic may then be expressed in the form (Aslan and Zech, 2003)

$$\phi_{NM} = \frac{1}{N^2} \sum_{j>i} R(\|x_i - x_j\|) - \frac{1}{NM} \sum_{i,j} R(\|x_i - y_j\|). \qquad (3.43)$$

Here, $N(N-1)$ is replaced by N^2 with better small sample properties, M is assumed to be large, and the term independent of x is omitted. The distribution of ϕ_{NM} under H_0 depends on R as well as H_0. It is not readily calculated, but may be estimated via simulations.

Various choices for function R have been proposed, with different choices suitable for different distributions. For example, a logarithmic form

$$R_{\log}(r) = \begin{cases} -\log r & r > a, \\ -\log a & r < a \end{cases} \qquad (3.44)$$

may be more appropriate for slowly varying distributions. The precise value of a is not important, its purpose is to prevent behavior at $r \to 0$ from overemphasizing fluctuations near $r = 0$ and may be chosen to be a typical distance between simulation points in the high density region. A Gaussian form may be more optimal for rapidly varying distributions. For example, Williams (2010) chooses

$$R(\|x_i - y_j\|) = \exp\left[-\frac{\|x_i - y_j\|^2}{2\sigma(x_i)\sigma(x_j)}\right], \qquad (3.45)$$

with $\sigma(x)$ varying as $1/f_0(x)$ where f_0 is the null hypothesis density, so that areas of high density are relatively highly weighted. With this weight function, Williams obtains powerful results for the Dalitz plot analysis in which rapid variations of the distribution occur.

3.4.2
Transformations to a Uniform Distribution

An intermediary step that is useful in some approaches to multivariate goodness of fit testing is to first make a transformation of the distribution under the null hypothesis. The idea is to transform the distribution to a uniform distribution on the unit D-cube. Then one can work on tests of uniformity in multidimensions. We assume continuous random variables in this discussion. The transformation is constructed as follows Rosenblatt (1952). Suppress the sampling index, and let the index on random variable X indicate the element in the multivariate space. The cdf for X is $F(x)$, with pdf $f(x)$. We begin by noticing that we can write the pdf in the form

$$f(x_1,\ldots,x_D) = f_1(x_1) f_2(x_2|x_1) \cdots f_D(x_D|x_1,\ldots,x_{D-1}). \tag{3.46}$$

The terms in this product may be computed as

$$f_k(x_k|x_1,\ldots,x_{k-1}) = \frac{\int_{-\infty}^{\infty} \cdots \int_{-\infty}^{\infty} f(x) dx_{k+1} \cdots dx_d}{\int_{-\infty}^{\infty} \cdots \int_{-\infty}^{\infty} f(x) dx_k \cdots dx_d}. \tag{3.47}$$

Let

$$F_k(x_k|x_1,\ldots,x_{k-1}) = P(X_k \le x_k | X_1 = x_1,\ldots,X_{k-1} = x_{k-1})$$
$$= \int_{-\infty}^{x_k} f_k(u|x_1,\ldots,x_{k-1}) du. \tag{3.48}$$

Finally, define the desired transformation according to

$$y_1 = F_1(x_1)$$
$$y_2 = F_2(x_2|x_1)$$
$$\vdots$$
$$y_d = F_d(x_d|x_1,\ldots,x_{d-1}). \tag{3.49}$$

It may be observed that the sampling distribution for Y is thus

$$P(Y_1 \le y_1 \cap \ldots \cap Y_D \le y_D) = \prod_{i=1}^{D} y_i, \quad 0 \le y_i \le 1. \tag{3.50}$$

The transformation is not unique; each possible ordering of the indices $1,\ldots,D$ produces a potentially different transformation. There may thus be $D!$ different transformations, and the goodness of fit statistics based on these transformations may not be equivalent. We recommend that when a problem is factorizable, as is often approximately the case in our physics problems, that it first be put into (approximately) factorized form (perhaps with a rotation of coordinates). Then the transformation to uniform coordinates is natural. When such factorization is not feasible, another approach is suggested. First, find subsets of the variables that factorize from each other. Within each subset, the variables are nonfactorizable. But with luck, the subsets are small enough that an exhaustive look through all permutations within each subset is possible.

With this transformation, we may design tests for uniformity on the D-cube. Whatever statistic is defined, its distribution under the null hypothesis of uniformity is completely determined by D and N. For example, many tests can be imagined using the distances between sampled data in this D-cube. Chapter 19 discusses specific implementations, including SLEUTH and nearest neighbors.

3.4.3
Local Density Tests

The multivariate GOF problem can be thought of as the problem of testing whether the spatial distribution of observed points in our observation space is consistent with the hypothesized model. Thus, we may form an estimate of the local density of observations around any given point and compare this estimate with the prediction of the model. The idea is similar to the notion of nearest neighbors, except that we compare densities of observations, not distributions of distances. It may be applied to either transformed or nontransformed variables. We proceed to set up the formalism:

Let $I(\text{statement})$ be an *indicator function*, that is

$$I(\text{statement}) = \begin{cases} 1 & \text{if statement is true}, \\ 0 & \text{if statement is false}. \end{cases} \quad (3.51)$$

Let $|x_i - x_j|$ be the distance between observations x_i and x_j. Any metric could be tried, but the Euclidean distance is probably a good choice. Then

$$N_i \equiv \sum_{j \neq i}^{N} I(\|x_i - x_j\| < r) \quad (3.52)$$

counts the number of observations in our dataset within a distance r of observation i. The quantity N_i/V_r, where V_r is the volume of the (hyper-) sphere of radius r, is thus a measure of the local density of observations in the vicinity of observation i. If the underlying sampling distribution is uniform, then we have expectation value $E(N_i) = (N-1) V_r / V$, where V is the total volume of our sampling space (assumed finite for this discussion), and we assume for simplicity that our sampling space is continuous.

If the data are sampled from a uniform distribution, the observations will tend to be "maximally spread out" in some statistical sense, compared with a distribution with peaks.[3] That is, the values of N_i/V_r will tend to cluster around the average density of points (excluding one), $(N-1)/V$. On the other hand, if the sampling distribution is not uniform, there must be clustering around other values than $(N-1)/V$, that is, regions of higher density. This results in N_i values that are higher than for the uniform case, on average. Hence, a candidate statistic is the sum of the N_is over the dataset, for a given r, or, in practice (Williams, 2010; Ripley, 1977):

$$K(r) = \frac{V}{N^2} \sum_{i=1}^{N} \frac{N_i}{a(i,j)}. \tag{3.53}$$

Here, $a(i, j)$ is a correction factor for the fact that some portion of a sphere of radius $|x_i - x_j|$ may lie outside the sampling space, which may be taken as the fraction of the sphere lying within V.

Statistic K may be computed for different values of r, giving sensitivity at different scales, and a plot of K against r provides a useful visual tool. This method provides a simple way to test for uniformity. It can be generalized for other distributions by "dividing out" the the hypothetical local density of observations. Thus, if $f_0(x)$ is the pdf for the null hypothesis, we insert a denominator of $V^2 f_0(x_i) f_0(x_j)$ into the sum in (3.53). Williams (2010) applies this to the example of the Dalitz plot. The method is especially powerful when there are large local deviations from the model. It is not so useful for small datasets, where the local density estimates have large variance. Dixon (2012) provides review of the intimately related subject of nearest neighbor methods.

3.4.4
Kernel-based Tests

If we can use our data to form an estimator $\hat{f}(x)$ of the true distribution, $f(x)$, then we may compare this estimate with the null hypothesis $f_0(x)$ in order to obtain a GOF statistic:

$$G = \int [\hat{f}(x) - f_0(x)]^2 dx. \tag{3.54}$$

To apply this idea, we need to form the estimate $\hat{f}(x)$. In Chapter 5 we discuss several techniques for estimating a density. Here, we consider applying kernel density estimation (Section 5.3).

With a normalized kernel function of form $K(x - x_n)$, we have the density estimate:

$$\hat{f}(x) = \frac{1}{N} \sum_{n=1}^{N} K(x - x_n). \tag{3.55}$$

---

[3] We remark that the other extreme, where the data may be sampled from a regular grid of values, will be more spread out than for a uniform distribution, and may also be of interest.

The kernel K could, for example, be a multivariate Gaussian. Thus, we estimate the distribution by adding Gaussians of width specified by parameter w, centered on each observation. The larger the w, the smoother our estimate.

To apply the test, simply plug the kernel estimate into (3.54) to find G. The distribution of G under the null hypothesis may in some cases be approximated well enough by asymptotic methods. In general, however, it must be simulated or estimated using a resampling technique (Chapter 4). A p-value may then be estimated. In applying this method to a Dalitz plot example, Williams (2010) finds disappointing performance from this method compared with other methods studied.

3.4.5
Mixed Sample Tests

We sometimes wish to compare two datasets to see whether they are consistent with being drawn from the same population. A common case is comparing an experimentally observed dataset with a Monte Carlo simulation. One approach to the multivariate problem combines the nearest neighbor idea with *pooling* the data, that is, combining the two datasets (Shilling, 1986).

Let $x_1, \ldots, x_{N_x}$ and $y_1, \ldots, y_{N_y}$ be the two datasets we wish to compare, and let $z_1, \ldots, z_N$ be the pooled data where $N = N_x + N_y$, $z_i = x_i$ for $1 \leq i \leq N_x$ and $z_{N_x+i} = y_i$ for $1 \leq i \leq N_y$. Let n_k denote the index in the pooled data of kth nearest neighbor to z_n. Define the indicator function:

$$I(n, n_k) = \begin{cases} 1 & z_n \text{ and } z_{n_k} \text{ both from } \{x\} \text{ or both from } \{y\} \\ 0 & \text{otherwise} \end{cases} \quad (3.56)$$

Form the statistic:

$$T_{k,N} = \frac{1}{kN} \sum_{n=1}^{N} \sum_{r=1}^{k} I(n, n_r). \quad (3.57)$$

That is, $T_{k,N}$ is the average fraction of nearest neighbors in the pooled sample, up through the kth, that are in the same original unpooled dataset.

The idea behind this statistic is that, if the null hypothesis is correct, there will be no preference for which dataset contains the nearest neighbor to a given point. On the other hand, if the null hypothesis is incorrect, then the nearest neighbors will tend to come from the same dataset at the given point. Hence, large values of $T_{k,N}$ indicate a poor match between the two datasets. Shilling (1986) shows that this statistic (under H_0) is asymptotically (large N_x, N_y, k, and D, although in practice D may not need to be very large) distribution-free and normal with mean

$$\left(\frac{N_x}{N}\right)^2 + \left(\frac{N_y}{N}\right)^2 \quad (3.58)$$

and variance (Williams, 2010)

$$\text{Var}(T_{k,N}) \sim \frac{1}{kN}\left(\frac{N_x N_y}{N^2} + 4\frac{N_x^2 N_y^2}{N^4}\right). \quad (3.59)$$

Williams (2010) applies this method for the Dalitz plot example and finds a power generally comparable to the χ^2 test, but without having to know anything about the pdf.

3.4.6
Using a Classifier

Another approach to multivariate goodness of fit tests is based on the fact that many of the classifier techniques described in following chapters of this book are well suited to multivariate problems. The input to a trained classifier is a data sample where each sample is a vector of random variables. The output of the classifier, for each input vector, is a single number, a *scoring function*, that may be used in deciding the category (e.g., "signal" or "background") of the input. Thus, we have a transformation from a vector to a number, or from a multivariate sampling to a univariate sampling. The idea is to use this transformation, plus one of the many available univariate goodness of fit tests, to construct a multivariate goodness of fit test.

In more detail, consider the construction of a two-sample test (Friedman, 2003). Let the sample one data be denoted $\{x_n, n = 1, \ldots, N\}$, and the sample two data be denoted $\{y_m, m = 1, \ldots, M\}$. Let $\{u_i : u_i = x_i, i \leq N, u_i = y_{i-N}, i > N\}$ be the union of the two samples. A binary classifier is trained on this dataset, according to response variable $r_i = -1$ for the first sample, and $r_i = +1$ for the second sample. Once the training is complete, we have a set of scores, $s_i, i = 1, \ldots, N+M$ according to the classifier scoring function. The first N scores correspond to the first sample, while the remainder correspond to the second sample. This provides us with two univariate sets of random samplings from the two samples under test. We may then compare these two sets with a univariate two-sample test (e.g., a Kolmogorov–Smirnov test). The obvious appeal of this algorithm is that it seeks out possible differences between the samples.

The algorithm just described has the drawback that the distribution of the test statistic may not follow the distribution of the chosen univariate test statistic. This is because we have used the same dataset for training as for conducting the test. Never try this with a powerful classifier, such as a decision tree ensemble, which may be flexible enough to achieve zero training error. This problem may be avoided by training on a subset, and using the remaining data for the test. If this is done, then the known, or computable, distribution of the test statistic under the null hypothesis applies. Throwing away some of the data to perform the training reduces the potential power. Cross-validation, introduced in Section 4.5, may be used to mitigate this.

It is, however, also possible to train on the whole dataset and compute the distribution of the test statistic under H_0 (Friedman, 2003). Let $T = T(\{s_i, i = 1, \ldots, N\}, \{s_i, i = N+1, \ldots, N+M\})$ be the chosen univariate test statistic defined on the scores. This is a random variable, which takes on some particular value t for our given sampling. Suppose $p(i)$ is a random permutation of the index values $\{1, \ldots, N+M\}$. Apply this permutation to the u values, $u_i \rightarrow u_{p(i)}$, but retain the classification of the first N elements of the permuted u values with response value −1, and the last M elements with response value +1. Train on this dataset, compute the values of the scores and thence the test statistic T. Repeat many times. The ensemble of samplings of T thus obtained provides an estimate of the distribution of T under H_0 that can be used to accept/reject H_0, or to compute a p-value.

This general technique can also be applied for a one-sample (goodness of fit) test, where the null hypothesis is a specified distribution. Simply produce a simulated dataset according to the H_0 distribution and apply the above two-sample method. The simulated dataset may be made large ($M \gg N$) in order to reduce statistical uncertainty, and hence increase power. If this unbalanced dataset is a concern (Section 9.5), an alternative is to generate many smaller simulated datasets and use these to estimate the distribution of the test statistic.

3.5 Exercises

1. Obtain the result in (3.16).
2. Show that the likelihood ratio test is uniformly most powerful for a simple hypothesis test.
3. You have a histogram giving an observed mass distribution. You are interested in the possibility that there is a resonance at $x = 0$ on a flat background (to get to $x = 0$, suppose you have shifted the distribution by the mass of the resonance you are looking for, assumed known). Including the possibility of interference, model the pdf for the mass distribution according to

$$f(x; a, \theta) = B|1 + ae^{i\theta}/(x + i)|^2, \qquad (3.60)$$

where B is a normalization constant, and a and θ are resonance parameters. Note that you are assuming that both the mass and width of the resonance are known, and that the experimental mass resolution is negligible.
You want to test the hypothesis:

$$H_0: a = 0 \qquad (3.61)$$

$$H_1: a \neq 0. \qquad (3.62)$$

To do this, you perform two fits to the histogram, one with $a = 0$, and one with a and θ allowed to float. You compute the change in log likelihood, Δ, between the two fits. Assume that you have large bin contents in your histogram.

a) Considering the χ^2 distribution, in order to compute a p-value for your test, how many degrees of freedom should you use for statistic 2Δ?
b) Now a troublemaker comes along, points you to the discussion in Section 3.1, and asks "are you sure?" Confused, you decide to check. Generate a large number of Monte Carlo datasets under the null hypothesis, and repeat your

analysis on each of them. Compute 2Δ for each experiment and make a histogram of 2Δ. Overlay a curve showing the distribution you expected according to part (a). If you got it right, congratulations! If not, see if you can find something that works. For concreteness, assume $x \in (-10, 10)$, and an experiment has 50 000 events on the average.

4. Consider the context of a least squares fit with R parameters $\boldsymbol{\theta} = \{\theta_1, \ldots, \theta_R\}$ and M measurements $\boldsymbol{x} = \{x_1, \ldots, x_M\}$, with sum of squares given by

$$\chi^2 = [\boldsymbol{x} - \boldsymbol{g}(\boldsymbol{\theta})]^T \boldsymbol{\Sigma}^{-1} [\boldsymbol{x} - \boldsymbol{g}(\boldsymbol{\theta})], \tag{3.63}$$

where $E[\boldsymbol{X}] = \boldsymbol{g}(\boldsymbol{\theta})$, and $\boldsymbol{\Sigma}$ is the covariance matrix for $\boldsymbol{X}$. Assume $\boldsymbol{\theta}$ is unknown and estimate it with least squares estimator $\hat{\boldsymbol{\theta}}$. Suppose that to a good approximation the sampling distribution is multivariate normal, that is, $f(\boldsymbol{x}; \boldsymbol{\theta}) \sim \exp(-\chi^2/2)$. Here we are interested in obtaining confidence regions for one or more parameters. Referring, for example, to our discussion in Section 3.1, we have two approaches that work: (i) Find where χ^2 increases by an appropriate amount from its minimum value; (ii) Use the method of inverting a test acceptance region, which corresponds to finding where the χ^2 is equal to a given "critical" value.

a) Consider the first approach. Determine the required change in χ^2 for a confidence region in the R-dimensional parameter space corresponding to a desired confidence level $1 - \alpha$. Also determine the required change in χ^2 for a confidence interval for a selected parameter corresponding to a desired confidence level $1 - \alpha$.
b) Do the same for the second approach. Here, instead of considering changes in χ^2, you must find critical χ^2 values that give regions with the specified (α) probability content.
c) Consider $\alpha = 0.05$ for 95% confidence regions. For both parts above, perform simulations to check whether you obtain the desired frequency coverage. If you don't, you should rethink your answers above!

5. Demonstrate the assertion that $\lambda = 1$ corresponds to the Pearson χ^2 statistic and $\lambda \to 0$ gives the likelihood ratio statistic in the power divergence family defined in (3.30)

6. Consider the signed Kolmogorov–Smirnov statistics for continuous CDF F:

$$K_N^+(F) = \sup_x [F_N(x) - F(x)] \tag{3.64}$$

$$K_N^-(F) = \sup_x [F(x) - F_N(x)]. \tag{3.65}$$

Show that the distribution of $K_N^{\pm}(F)$ is independent of F, and thence, the distribution of $K_N(F)$ is independent of F.

7. Investigate whether the Anderson–Darling statistic satisfies the properties of a distance.

8. There are many tests that have been developed, with different performance on different problems. It is thus important to think a little about your problem before picking a test. Performance is characterized by power at a given level of significance. Compare the power of the Anderson–Darling test and the Kolmogorov–Smirnov test on the null hypothesis that a dataset is sampled from a normal distribution with mean zero and variance one. The alternative hypothesis is that it is not sampled from such a normal distribution. For a significance level of 0.01, compute the power of these two tests for a dataset of size $N = 100$ sampled from a Cauchy distribution with location parameter zero and full width at half maximum (FWHM) equal to the FWHM of the normal distribution in the null hypothesis.

References

Aslan, B. and Zech, G. (2003) A new class of binning-free, multivariate goodness of fit tests: the energy tests. *arXiv:hep-ex/0203010v5*.

Aslan, B. and Zech, G. (2005) Statistical energy as a tool for binning-free, multivariate goodness of fit tests, two-sample comparison and unfolding. *Nucl. Instrum. Methods Phys. Res. A*, **537**, 626–636.

Chernoff, H. and Lehmann, E.L. (1954) The use of maximum likelihood estimates in χ^2 tests for goodness of fit. *Ann. Math. Stat.*, **25**, 579–586.

Cressie, N. and Read, T. (1984) Multinomial goodness of fit tests. *J. R. Stat. Soc. B (Methodological)*, **46**, 440–464.

Davies, R.B. (1977a) Hypothesis testing when a nuisance parameter is present only under the alternative. *Biometrika*, 74 (1), 33–43.

Davies, R.B. (1977b) Hypothesis testing when a nuisance parameter is present only under the alternative. *Biometrika*, 64 (2), 247–254.

Demortier, L. (2008) P values and nuisance parameters, in *Proceedings of the PHYSTAT LHC workshop* (eds H. Prosper, L. Lyons, and A. De Roeck), CERN-2008-001, 23–33.

Dixon, P.M. (2012) Nearest neighbor methods. http://www.public.iastate.edu/~pdixon/stat406/NearestNeighbor.pdf (accessed 24 July 2013).

Friedman, J. (2003) On multivariate goodness of fit and two-sample testing. *PHYSTAT 2003, SLAC*.

Hubbard, R. and Bayarri, M.J. (2003) Confusion over measures of evidence (p's) versus errors (α's) in classical statistical testing. *Am. Stat.*, **57**, 171–178, doi:10.1198/0003130031856.

Mueller, M. and Madejski, G. (2009) Parameter estimation and confidence regions in the method of light-curve simulations for the analysis of power density spectra. *Astrophys. J.*, **700**, 243.

Neyman, J. (1937) 'Smooth' test for goodness of fit. *Skandinavisk Aktuarietidskrift*, **20**, 150–199.

Pearson, K. (1900) On the criterion that a given system of deviations from the probable in the case of a correlated system of variables is such that it can be reasonably supposed to have arisen from random sampling. *Philos. Mag. 5*, **50**, 157–175.

Rayner, J. and Best, D. (1989) *Smooth Tests of Goodness of Fit*, Oxford University Press.

Ripley, B.D. (1977) Modelling spatial patterns. *J. R. Stat. Soci. B (Methodological)*, **39**, 172–212.

Rosenblatt, M. (1952) Remarks on a multivariate transformation. *Ann. Math. Stat.*, **23**, 470–472.

Shao, J. (2003) *Math. Stat.*, Springer, 2nd edn.

Shilling, M.F. (1986) Multivariate two-sample tests based on nearest neighbors. *J. Am. Stat. Assoc.*, **81**, 799–806.

Watson, G.S. (1961) goodness of fit tests on a circle. *Biometrika*, **48**, 109.

Williams, M. (2010) How good are your fits? Unbinned multivariate goodness of fit tests in high energy physics. *arXiv:1006.3019v2 [hep-ex]*.

4
Resampling Techniques

With the advent of modern computing resources, new (or at least formerly little used) statistical methods of analysis have become feasible. This is really the motivation for this book. Among these methods is the approach to estimation afforded by *resampling*. This class of technique avoids the need to derive formulas for the properties (such as variance) of a statistic (such as a point estimator). Indeed, it permits us to avoid imposing potentially invalid assumptions (a model) about the distribution. Instead, we use the data in hand to produce a statistical ensemble with desired properties. Often, Monte Carlo modeling is used to answer questions about the distribution of a statistic. Resampling offers an alternative approach that avoids assumptions in the Monte Carlo model. Various resampling methods have become popular, and may be used to address different problems. We will see these methods heavily used when we discuss the classification problem in later chapters of this book. Here, we introduce several resampling techniques, illustrated with nonclassification applications.

You will notice that the various resampling methods overlap considerably in application. However, as a general guide, the permutation methods are generally used for hypothesis testing; the bootstrap and jackknife are used to estimate bias, variance, and confidence intervals; while cross-validation is used to estimate the accuracy of a predictive model. We give some flavor of the considerations concerning the pros and cons of the different approaches. It may be noted that the theoretical basis for these methods is in asymptotic properties such as consistency and convergence. A thorough discussion is well outside our scope, and indeed this remains an active area of research. While the basic ideas are rather simple and elegant, care should be exercised in the execution. References for further study are noted.

4.1
Permutation Sampling

A conceptually simple situation is the following: Suppose we have two datasets, A and B, of sample sizes N_A and N_B. We wish to test whether the two populations are consistent with arising from the same underlying distribution by comparing some statistic computed on each dataset. Call this statistic S_A or S_B; for example it could be the sample mean, or the median, or the variance, and so on. Under the null hypothesis, that the sampling distribution for A and B is the same, we must have $E(S_A) = E(S_B)$. Thus the difference $\Delta S = S_B - S_A$ is a measure of the difference between the distributions.

In order to apply a test based on the observed ΔS, we need to know its distribution under the null hypothesis. The method of permutation resampling permits us to estimate this distribution without any assumptions (e.g., normality) of the sampling distribution. Under the null hypothesis, all samples in A and B are equivalent, and therefore, any permutation of the labeling A and B has equal probability. Thus, the distribution of ΔS is determined by considering all possible permutations of the A and B labels, grouping the $N_A + N_B$ samplings into sets of size N_A and N_B and computing ΔS for each permutation. The exhaustive set of permutations provides precisely all of the information in the data about the distribution of ΔS, or any other statistic.

The set of values of ΔS from all permutations provides the estimated distribution of ΔS. The actual ΔS may be compared with this distribution to obtain a *p*-value for the null hypothesis (or to accept/reject H_0 at a specified significance level). With additional effort, the test may be inverted to derive confidence intervals. Since the underlying assumption for the null hypothesis is that the two samples are identically distributed, the test may have limited utility if the datasets differ even in uninteresting ways. Because there are no model assumptions, the *permutation test* is called an *exact* test. This is probably the greatest incentive for using the permutation test.

Permutations within a given A and B repartition are unnecessary. Even so, the computing required to evaluate all remaining $\binom{N_A+N_B}{N_A}$ permutations grows rapidly with sample size and may be prohibitive. For larger samples, we may randomly sample permutations without being exhaustive (providing a *Monte Carlo permutation test*). This is called a *conditional Monte Carlo* test (Pesarin and Salmaso, 2010) because it is a Monte Carlo simulation conditioned on the empirical data set. We have already encountered this idea in Section 3.4.6.

As an example, we apply a Monte Carlo permutation test to the distribution in Figure 4.1. A sample of size $N_A = 1000$ is compared with a sample of size $N_B = 20$, and we ask whether the mean of sample A is greater than the mean of sample B. If the sampling were from a normal distribution, the statistic

$$t = (\bar{x}_A - \bar{x}_B) \bigg/ \sqrt{\frac{(N_A-1)s_A^2 + (N_b-1)s_B^2}{N_A + N_B - 2} \frac{N_A + N_B}{N_A N_B}} \qquad (4.1)$$

is distributed as the Student *t*-distribution with $N_A + N_B - 2$ degrees of freedom and may be used to test the desired hypothesis. Here, $\bar{x}$ refers to the sample mean, and *s* refers to the sample variance (computed with $N-1$). However, for nonnormal distributions the *t*-distribution may not apply. In the case of the distribution of Figure 4.1, a test at nominal 1% significance rejects the (true) null hypothesis with only 0.57% probability. The permutation test rejects the null hypothesis at closer

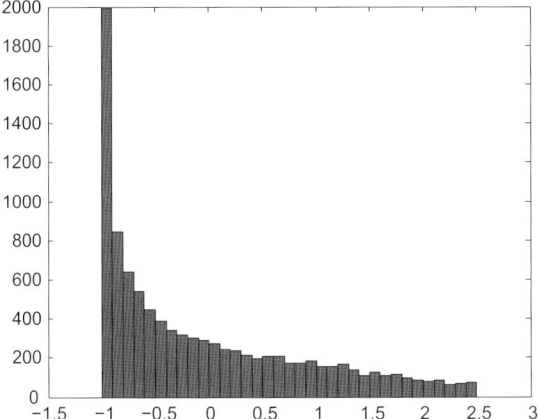

Figure 4.1 Distribution used in demonstration of the permutation resampling test.

to the desired 1% probability (1.07% in a simulation using 10 000 permutations). The point here is that large errors may result from erroneous model assumptions, whereas the permutation test avoids making such assumptions.

The permutation test is a type of nonparametric test, as it makes no assumptions about the form of the sampling pdf. It is thus more robust than parametric tests which depend on the validity of the sampling model. However, it should be remembered that such an advantage may come at the price of power. If a reliable model is available, more powerful tests can generally be constructed. In the permutation test it is also important to exercise care that there really is exchange symmetry under the null hypothesis.

4.2
Bootstrap

A popular resampling technique is the *bootstrap* (Efron, 1979), motivated as a means to estimate the variance of an estimator for a population parameter. It can be especially useful when the sampling distribution is unknown, but is not limited to this situation. The basic bootstrap algorithm is as follows:

Suppose that we have the problem of estimating parameter θ (which may be a vector) with an i.i.d. sample of size N, $X_1, \ldots, X_N$. Each of the X_i represents a vector of random variables. For example, X_i could describe an event in a particle physics dataset. Denote the estimator for θ by $\hat{\theta}(X)$. Perhaps this is a maximum likelihood estimator, but that is not required. Now form a set of B *bootstrap replicas* or samples by randomly sampling sets of size N from X, with replacement. For example, this can be achieved with the R statement:

```
for (i in 1:B) xr[i] = sample(x,replace=TRUE)
```

For replication i, form replicated estimator $\hat{\theta}(i)$, where the argument is now the replication index. This procedure may be called the *Monte Carlo bootstrap*, because of the Monte Carlo approach to choosing replications.

These replications can be used to estimate the variance of the estimator. Simply take the sample mean and variance of the bootstrap estimators:

$$\bar{\theta} = \frac{1}{B} \sum_{i=1}^{B} \hat{\theta}(i), \tag{4.2}$$

$$s_\theta^2 = \frac{1}{B-1} \sum_{i=1}^{B} (\hat{\theta}(i) - \bar{\theta})^2. \tag{4.3}$$

Then s_θ is the estimated standard deviation of the estimator $\hat{\theta}$. If we have multiple parameters, the covariances may also be estimated with the bootstrap:

$$\text{Cov}(\hat{\theta}_m, \hat{\theta}_n) = \frac{1}{B-1} \sum_{i=1}^{B} \left[\hat{\theta}_m(i) - \bar{\theta}_m(i)\right]\left[\hat{\theta}_n(i) - \bar{\theta}_n(i)\right]. \tag{4.4}$$

What does the bootstrap accomplish? It provides a way to approximately sample from the actual parent distribution for X. Instead of repeatedly sampling from the actual distribution, which is likely to be impractical, we sample from the empirical distribution.

As an example, consider the estimation of the median parameter of a Cauchy distribution (noting that the mean is undefined: this is a case where an estimate of location such as the median, or alternatively a trimmed mean, is required). Of course, we use the sample median as our estimator. We may then use the bootstrap to estimate the variance of the sample median. Figure 4.2 shows the bootstrap estimate for the cdf of the median estimator, for two different samplings of size 1001 from the Cauchy distribution. This may be compared with the actual cdf, given by the overlaid curve. The shapes of the distributions are similar, illustrating the applicability of the bootstrap for estimating variance. The translations of the bootstrap cdfs are expected from fluctuations in the median estimation, but do not affect the estimation of variance. However, the variance estimates will also fluctuate around the true variance. The convergence of the bootstrap estimator for variance may be examined by plotting the estimator against the number of bootstrap samples, as in Figure 4.2b; again for our Cauchy median example.

The bootstrap samples are based on the empirical distribution, hence will reflect any fluctuations that may be present. Figure 4.3 shows the distribution of the bootstrap estimator for the standard deviation of the median, again for our Cauchy example. This distribution may be compared with the actual value of the standard deviation, 0.050.

In this example, the sampling distribution is known, and the variance of the median can be precisely calculated. This should of course be done when possible. However, when the sampling distribution is not known the bootstrap provides a straightforward method for estimating the variance.

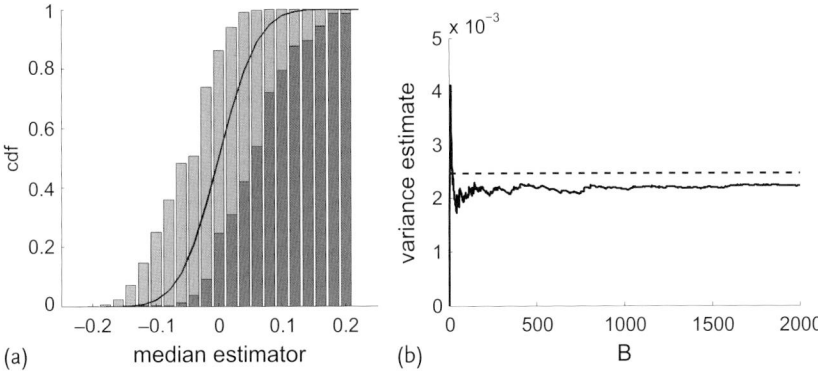

Figure 4.2 (a) Bootstrap estimator for the cdf of the median of a Cauchy distribution for two samples. The curve shows the true cdf. (b) Convergence of the bootstrap estimator for the variance of the median of a Cauchy distribution, as a function of number of bootstrap samples. The dashed line shows the actual variance.

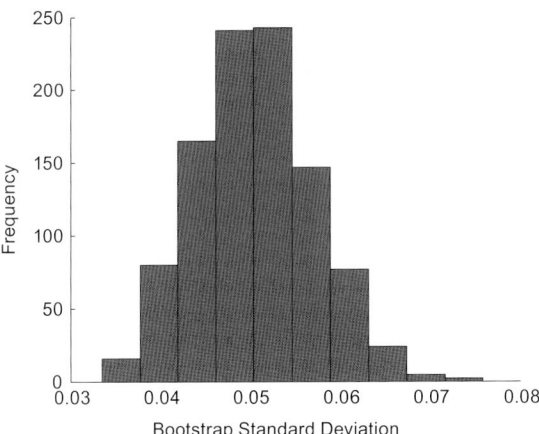

Figure 4.3 Distribution of the bootstrap standard deviation estimator.

The bootstrap may also be used to estimate the bias of a parameter estimator (Efron and Tibshirani, 1994). Suppose $\theta = \theta(F)$ is a parameter (e.g., the variance) of distribution F. Estimator $\hat{\theta} = \hat{\theta}(\mathbf{x})$ is a statistic computed from a dataset of i.i.d. scalars $x_1, \ldots, x_N$. Denote the bias of $\hat{\theta}$ by

$$b_F(\theta) = E_F[\hat{\theta}(\mathbf{x})] - \theta(F), \tag{4.5}$$

where the subscript F denotes expectation value with respect to distribution F. If we don't know what F is, even up to unknown θ, then we cannot evaluate the bias. However, our data provides an approximation to F, the empirical distribution $\hat{F}$, and we can use this to obtain a bootstrap estimate of bias:

$$b_{\hat{F}} = E_{\hat{F}}[\hat{\theta}(\mathbf{x}^*)] - \theta(\hat{F}). \tag{4.6}$$

Here, **x*** is a (bootstrap) sample drawn from empirical distribution $\hat{F}$. Notice that since the empirical distribution $\hat{F}$ is used, the bias estimate depends only on that, and not on θ itself. For example, if we estimate the variance (θ) of F with $\hat{\theta} = \frac{1}{N}\sum_{n=1}^{N}(x_n - \bar{x})^2$, where $\bar{x}$ is the sample mean, we find (exercise) a bootstrap bias estimate of $-\frac{1}{N}\hat{\theta}$, which approximates the known bias of $-\theta/N$ for this estimator.

In general, we cannot evaluate the expectation in (4.6) analytically, and must resort to Monte Carlo bootstrap sampling. Thus, we obtain a sequence of bootstrap estimators $\hat{\theta}^*(1),\ldots,\hat{\theta}^*(B)$, where B is the number of bootstrap replications. We approximate the desired expectation with the average of these, $\hat{\theta}^*(\cdot) \equiv \frac{1}{B}\sum_{i=1}^{B}\hat{\theta}^*(i)$, obtaining

$$\hat{b}_{\hat{F};B} = \hat{\theta}^*(\cdot) - \theta(\hat{F}). \tag{4.7}$$

The distinction between permutation and bootstrap tests is somewhat subtle. They both involve sampling from the empirical dataset, in order to learn something about the distribution of a statistic. However, they are not identical. The bootstrap has more flexibility in application, as there is no assumption about invariance required for a null hypothesis. That is, the bootstrap assumes that all samplings are independent, and distributed, under H_0, with the same value of whatever parameter is being tested. The permutation test makes the stronger assumption that every sampling is identically distributed under H_0. On the other hand, unlike the permutation, the bootstrap is not an exact test; the reader is referred to Pesarin and Salmaso (2010) for discussion of this fine point.

The largest use of the bootstrap in particle physics has so far been in classification, as will be discussed in later chapters. However, it is beginning to be used in parametric error analysis as well (e.g., Lees *et al.*, 2013), and it may be expected that this usage will grow.

4.2.1
Bootstrap Confidence Intervals

The bootstrap may be used to estimate confidence intervals. This is particularly useful when the sampling distribution is not known, or approximately known. In this case many of our familiar methods don't apply. Suppose that we wish to find a confidence interval for a population parameter, θ. We have a statistic $\hat{\theta} = \hat{\theta}(\mathbf{X})$ to estimate θ. Bootstrap sampling from $\mathbf{X}$ corresponds to obtaining an i.i.d. sample $\mathbf{X}^*$ of size N from the empirical distribution $\hat{F}$, our surrogate for the actual distribution. Given a set of B such bootstrap replicas, we compute the corresponding estimators $\hat{\theta}^*(1),\ldots,\hat{\theta}^*(B)$. Let $P_B(u)$ be the empirical distribution corresponding to this set of $\hat{\theta}^*$s. Unless we apply smoothing, this is a monotonic step function. To obtain, say, an estimated upper confidence bound at the $1 - \alpha$ confidence level, we solve for $\hat{u}_\alpha$ in

$$\hat{u}_\alpha = P_B^{-1}(\alpha). \tag{4.8}$$

Because of the discontinuities, this may be only approximately soluble, so let us be more explicit. Begin by ordering all of the $\hat{\theta}^*$s. Then count up until reaching a

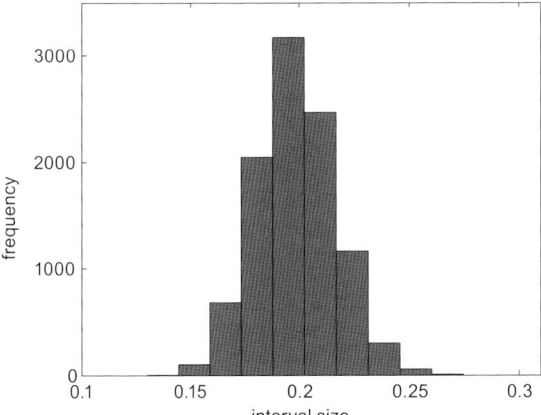

Figure 4.4 Size of confidence interval estimated with the bootstrap.

fraction of at least $1 - \alpha$ of them. The smallest such $\hat{\theta}^*$ value is $\hat{u}_\alpha$. That is, the *bootstrap percentile* method corresponds to finding the appropriate percentile of P_B. This method can readily be adapted to estimating two-sided confidence intervals.

For an example, we estimate the population mean θ and estimate a 68% confidence interval. Our example uses a dataset of size $N = 100$ from a N(0, 1) distribution. The estimator is the sample mean, $\hat{\theta} = \frac{1}{N}\sum_{i=1}^{N} x_i$. If we knew that we were sampling from a Gaussian, we would typically quote the interval $(\hat{\theta} - 0.1, \hat{\theta} + 0.1)$. However, supposing that we don't know anything about the sampling distribution, we use the bootstrap. In MATLAB, we may use:

```
ci = bootci(nbootstrap,{@mean,x},'alpha',alpha,'type','per');.
```

Here, @mean is a function that computes the sample mean, alpha ≈ 0.32, and per specifies that we use the "percentile" method described above.

We perform the calculation with nbootstrap = 300 bootstrap samples (sufficient for the illustration, but a bit small in practice). The method is found to cover the true mean with a probability of $\sim$ 67.8%, in good agreement with the desired 68.3%. A histogram of the size of the confidence interval, for 10 000 experiments is shown in Figure 4.4. This may be compared with the fixed size of 0.2 in the usual approach when the distribution is known to be Gaussian. We find that the bootstrap may be used to estimate confidence intervals with accurate coverage. The cost of not knowing the sampling distribution shows up as variation of the interval size.

The method can also be applied to multivariate sampling spaces and multidimensional parameter spaces. On the other hand, some requirements need to be met for this method to be strictly valid. In particular, we must have it that (Shao, 2003)

$$P[\phi(\hat{\theta}) - \phi(\theta) \geq u] = \Psi(u) \tag{4.9}$$

holds for any sampling distribution F. Here, Ψ is a continuous, strictly increasing cdf symmetric about 0 ($\Psi(u) = 1 - \Psi(-u)$), and $\phi(\theta)$ is an increasing transformation on θ. This requirement restricts the type of statistic for which the method can be applied to to those for which such a transformation exists. It must hold for any F because we don't know which F we might be sampling from. Improved methods have been developed that generalize on this requirement, see Section 4.4 for a good choice, as well as references (Efron, 1987; DiCiccio and Efron, 1996).

4.2.2
Smoothed Bootstrap

The bootstrap corresponds to a particular estimate for the sampling distribution, given by the empirical distribution. This estimate is discrete, while the underlying distribution may be continuous. If it is desired to sample from a continuous estimate, various techniques for smoothing may be applied, referred to as the *smoothed bootstrap*. We discuss some methods of smoothing in density estimation in the next chapter.

4.2.3
Parametric Bootstrap

The traditional bootstrap is a nonparametric method in that it does not make any model assumptions about the sampling distribution, which is estimated using the empirical distribution. However, if a reliable model is available, involving some unknown parameters, then the model may be used to eliminate some of the uncertainty inherent in the empirical approach. Thus, our parametric bootstrap sample is achieved by randomly sampling from the model distribution with some values for the parameters. Of course, if the parameters are unknown, we use some estimate for their values. Our bootstrap samples are now samples of size N from this model distribution, and we use these to estimate the distribution of any statistic of interest. This brings us into the realm of well-known uses of Monte Carlo simulation in physics analysis. Calling it the parametric bootstrap is just a new name.

4.3
Jackknife

An algorithm known as the *jackknife* may be used to estimate both variance and bias of a parameter estimator. To get the idea for how it works, consider the simple situation in which we use our sample $x_1, \ldots, x_N$ to estimate the mean of the parent population. Naturally, we will use the sample mean for our estimator:

$$\hat{\theta}(\mathbf{x}) = \frac{1}{N} \sum_{n=1}^{N} x_n . \tag{4.10}$$

What is the variance, $\sigma_{\hat\theta}^2$, of our estimator? We may estimate this using $\frac{1}{N}$ times the sample variance:

$$s^2 = \frac{1}{N(N-1)} \sum_{n=1}^{N} \left(x_n - \hat\theta\right)^2 . \tag{4.11}$$

Now let us try something. Let $\mathbf{x}_{-i} = \{x_1, \ldots, x_{i-1}, x_{i+1}, \ldots, x_N\}$ be the dataset obtained by removing x_i from our original sample. The set $\mathbf{x}_{-i}$ is called a *jackknife sample*. Let

$$\hat\theta_{-i} = \frac{1}{N-1} \sum_{n \neq i} x_n , \tag{4.12}$$

be the sample mean for set $\mathbf{x}_{-i}$. We find

$$s^2 = \frac{1}{N(N-1)} \sum_{i=1}^{N} \left(N\hat\theta - \sum_{n \neq i} x_n - \hat\theta\right)^2$$

$$= \frac{N-1}{N} \sum_{i=1}^{N} \left(\hat\theta - \hat\theta_{-i}\right)^2 . \tag{4.13}$$

Now define $\hat\theta'$ as the sample mean over all the $\hat\theta_{-i}$s:

$$\hat\theta' = \frac{1}{N} \sum_{i=1}^{N} \hat\theta_{-i} . \tag{4.14}$$

Use this to rewrite the estimated variance in the form

$$s^2 = \frac{N-1}{N} \sum_{i=1}^{N} \left(\hat\theta - \hat\theta' + \hat\theta' - \hat\theta_{-i}\right)^2$$

$$= \frac{N-1}{N} \sum_{i=1}^{N} \left(\hat\theta' - \hat\theta_{-i}\right)^2 + (N-1)(\hat\theta' - \hat\theta)^2 \tag{4.15}$$

The first term provides the variance with respect to the (jackknife) sample mean, that is an estimate of the variance of $\hat\theta$ about its mean with the $(N-1)/N$ scale factor. The second term compares $\hat\theta'$ with $\hat\theta$, and is related to the bias. We can make this concrete as follows:

We assume that our estimator is consistent in the sense that $E\hat\theta \xrightarrow{N \to \infty} \theta$. If this limit is approached at leading order as $1/N$, then $b_{N-1} \propto \frac{N}{N-1} b_N$, for large enough N, where b_N denotes the bias for a sample of size N:

$$b_N(\theta) = E\hat\theta - \theta . \tag{4.16}$$

Hence, we can use our jackknife samples to estimate the bias of $\hat\theta$, noting that

$$E(\hat\theta' - \hat\theta) = b_{N-1} - b_N = \frac{1}{N-1} b_N . \tag{4.17}$$

We summarize the jackknife method for estimating variance and bias: The jackknife estimate for the variance of estimator $\hat{\theta}$ with respect to its expectation value is

$$s'^2 = \frac{N-1}{N} \sum_{i=1}^{N} \left(\hat{\theta}' - \hat{\theta}_{-i}\right)^2 . \tag{4.18}$$

The jackknife estimate for the bias of estimator $\hat{\theta}$ with respect to θ is

$$\hat{b}_N = (N-1)(\hat{\theta}' - \hat{\theta}) . \tag{4.19}$$

A bias correction may be applied to estimator $\hat{\theta}$ to obtain the improved (bias-corrected jackknife estimate) estimate for θ:

$$\hat{\theta}^* = N\hat{\theta} - (N-1)\hat{\theta}' . \tag{4.20}$$

The algorithm readily generalizes beyond our construction. Instead of the sample mean, we may substitute any statistic of the form

$$\hat{\theta}(\mathbf{x}) = a + \sum_{n=1}^{N} b(x_n) . \tag{4.21}$$

This is called a *linear statistic* – it is here simply a linear function of a transformation from our original set of i.i.d. RVs to another set of i.i.d. RVs. Our discussion above in this section goes through without difficulty for this case. For nonlinear statistics, we may still apply the method, but with due caution.

For a simple example with known properties, consider the estimation of a population variance, σ^2. Along the lines of (4.10), we consider the estimate

$$\hat{\sigma}^2(X) = \frac{1}{N} \sum_{n=1}^{N} (x_n - \bar{x})^2 , \tag{4.22}$$

where $\bar{x}$ is the sample mean. We then construct the jackknife samples, and using (4.18) and (4.19), calculate the jackknife estimates of variance and bias.

We carry out this exercise on a sample of size $N = 100$ drawn from a uniform distribution on $(0, 1)$. The variance estimates from the jackknife samples are shown in Figure 4.5. The estimated variance for this sample from (4.22) is 0.0821. The jackknife estimates the bias of this estimator as $-0.000\,829$, which may be compared with the known bias of $-1/12N = -0.000\,833$. The jackknife estimate for the variance of the estimator, (4.18), gives 0.0077^2. Hence our bias-corrected estimate of the variance of the sampling distribution is 0.0829 ± 0.0077, which may be compared with the known variance $1/12 = 0.0833$. In this case the bias correction is small compared with the statistical uncertainty. In fact, this statistic is not linear, but our specific example is "close" to linear, so it is not unexpected that the method works.

The jackknife and the newer bootstrap can both be used to estimate variance and bias. It is thus natural to ask which is better. As "better" is vague, the answer is "it depends". We therefore illustrate some considerations.

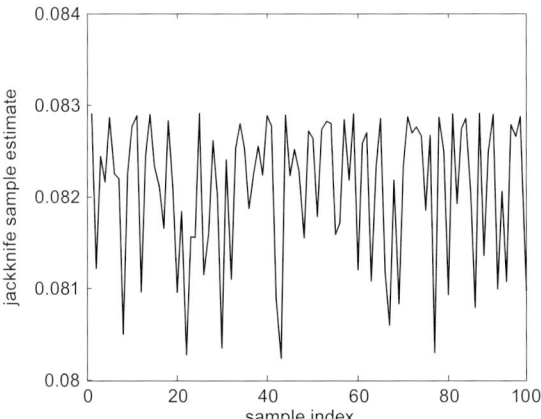

Figure 4.5 Estimation of variance using the jackknife.

A place where the jackknife does better is in the estimation of bias, at least for linear statistics, such as the mean of a distribution. We know that the sample mean is an unbiased estimator for this parameter, and that is just what the jackknife tells us (see exercises in Section 4.7). However, estimating this bias with a set of B randomly drawn bootstrap samples according to (4.7) will in general produce a nonzero estimate of bias due to the random fluctuations in the samples. An exhaustive search over all possible bootstrap samples will result in zero bias, but the computation time required is prohibitive even for fairly modest sample sizes.

This raises the issue of computing requirements. The jackknife requires examining a set of N samples of size $N - 1$, while the bootstrap requires examining some rather large number B of bootstrap datasets each of size N. Unless N is itself large, the computation required in the jackknife is much more manageable. However, the jackknife does not make use of all of the available information in the case of nonlinear statistics – it actually represents a linear approximation to the (exhaustive) bootstrap and may be comparatively inefficient (Efron and Tibshirani, 1994).

A specific situation where the jackknife runs into trouble involves "nonsmooth" statistics. Smoothness captures a notion of continuity on a dataset – small changes in the data are reflected as small changes in the statistic. An example of a nonsmooth statistic is the median. To illustrate this, suppose we have a dataset of size 3 with $x, a, b = 1, 2, 3$. The median is $a = 2$. Now imagine increasing x. Nothing happens to the median until x passes two. Then the median is equal to x until x becomes greater than 3. While the median is a continuous function of x, its derivative is discontinuous, that is, not smoothly dependent on x.

This kind of nonsmoothness can show up as trouble in jackknife estimation. In a dataset of any size, the leave-one-out jackknife samples will have at most three different values for the medians. If we attempt to estimate the variance of the median estimator using the jackknife, we obtain unreliable (i.e., inconsistent) results. The bootstrap does much better. An example is shown in Figure 4.6. A sample of size 100 is drawn from a uniform distribution between 0 and 100. For 1000 such ex-

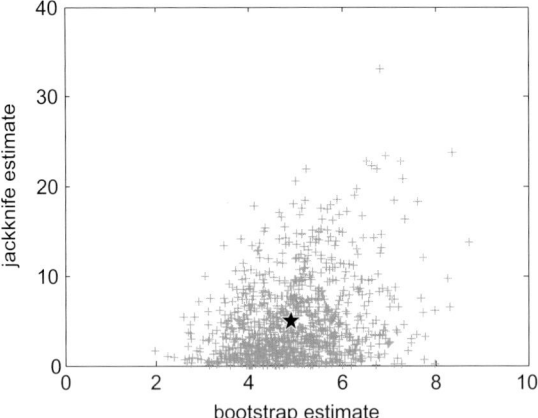

Figure 4.6 Estimation of standard deviation for the median of a 100 sample size from a $U(0, 100)$ sample. Vertical axis: Jackknife estimate; Horizontal axis: Bootstrap estimate. The star indicates the true value.

periments, the jackknife estimate of the standard deviation of the median is shown on the vertical axis, against the bootstrap estimate (from 2000 bootstrap replicas) on the horizontal axis. The star gives the location of the true value of the standard deviation. While the bootstrap estimates cluster around the true value, most of the jackknife estimates are below the true value, with a long tail up to quite high estimates. The high estimates occur when the two samplings on either side of the median are far apart.

This problem can be mitigated, at the cost of additional computing power (and perhaps bias and variance in more typical situations), with the *delete-d* jackknife. In this algorithm one leaves out d samples at a time, instead of one. The value of d should be between $\sqrt{N}$ and N for consistency (Efron and Tibshirani, 1994), and the variance is estimated according to

$$s'^2 = \frac{N-d}{d\binom{N}{d}} \sum_{i=1}^{\binom{N}{d}} \left(\hat{\theta}' - \hat{\theta}_{-d_i} \right)^2 , \qquad (4.23)$$

where the d_i subscript indicates a jackknife sample leaving out d samples, and $\hat{\theta}'$ is the average of the jackknife values. We will encounter the delete-d jackknife again when we discuss bias in density estimation in the next chapter.

The delete-d jackknife generalizes the leave-one-out jackknife. With this notion, another generalization is possible, known as the *generalized jackknife*. In the delete-d jackknife, the bias-corrected estimator for parameter θ is obtained with the appropriate modification of (4.20):

$$\hat{\theta}^* = \frac{\left[N\hat{\theta}_0 - (N-d)\bar{\theta}_d \right]}{d} , \qquad (4.24)$$

where we introduce the notation $\hat{\theta}_d$ to indicate an estimator based on a subsample of size $N-d$, and $\bar{\theta}_d$ to indicate an average over values of $\hat{\theta}_d$ from some number of subsamples of size $N-d$. Ideally, there are $\binom{N}{d}$ such subsamples to average over, but computing limitations may require using a subset of these. In particular, it is common to form j disjoint subsets with $j(N-d) = N$, and average over those disjoint subsets.

When we obtained (4.20), or the delete-d generalization (4.24), we made an assumption about the leading order rate where the bias decreases as the inverse sample size. The generalized jackknife provides a way to use the data to estimate the rate of convergence, including higher orders. For example, the expression (Schucany et al., 1971; Sharot, 1976)

$$\hat{\theta}^{(k)} = \frac{\begin{vmatrix} \hat{\theta}_0 & \hat{\theta}_1 & \cdots & \hat{\theta}_k \\ 1/N & 1/(N-1) & \cdots & 1/(N-k) \\ \vdots & \vdots & \vdots & \vdots \\ 1/N^k & 1/(N-1)^k & \cdots & 1/(N-k)^k \end{vmatrix}}{\begin{vmatrix} 1 & 1 & \cdots & 1 \\ 1/N & 1/(N-1) & \cdots & 1/(N-k) \\ \vdots & \vdots & \vdots & \vdots \\ 1/N^k & 1/(N-1)^k & \cdots & 1/(N-k)^k \end{vmatrix}}$$

$$= \frac{N^k}{k!} \sum_{j=0}^{k} (-)^j \binom{k}{j} \left(1 - \frac{j}{N}\right)^k \hat{\theta}_j \qquad (4.25)$$

produces an unbiased estimator from any estimator with bias of the form

$$E\hat{\theta} - \theta = \sum_{i=1}^{k} \frac{a_i}{N^i} b_i(\theta). \qquad (4.26)$$

A constraint on our treatment of the bootstrap and jackknife is the i.i.d. assumption. Resampling techniques exist that may be used for situations where this assumption does not hold. We discuss the nonidentical case of heteroscedasticity in Section 4.6. Here, we give an example of a technique addressing lack of independence. Suppose we have a sequence of random variables (perhaps a "time series"), $X_1, \ldots, X_N$. We assume the sequence is *stationary*, meaning every subsequence of the form $X_i, X_{i+1}, \ldots, X_{i+k-1}$ has the same joint distribution as any other such subsequence (when marginalized over all other variables). The samplings need not be independent, for example the correlation may be nonzero for any pair of RVs. It just has to be the same correlation for any pair separated by the same difference in indices.

Now suppose we have a parameter of interest estimated using a sampled sequence, with value $\hat{\theta}_N = \hat{\theta}_N(x_1, \ldots, x_N)$, where $\hat{\theta}$ is a linear statistic. We would like to estimate the variance of this estimator. A technique called *subsampling* (Politis et al., 1999) may be employed in this case. In this technique, one samples without replacement, to obtain datasets of size less than N, as in the jackknife. In the present case, however, we subsample blocks of consecutive elements of the sequence. Here is an algorithm. Assume that the variance is proportional to the inverse of the sample size:

$$\text{Var}(\hat{\theta}_N) = \frac{a}{N}. \qquad (4.27)$$

Pick a subsample size, k, and form the $N - k + 1$ possible subsequences:

$$x_i, x_{i+1}, \ldots, x_{i+k-1}, \quad i = 1, \ldots, N - k + 1. \qquad (4.28)$$

On each subsequence, evaluate the estimator to obtain $\hat{\theta}_{ki}$, $i = 1, \ldots, N - k + 1$. Use these estimators to estimate the variance of $\hat{\theta}_k$ (Politis et al., 1999):

$$s_k^2 = \frac{k}{N - k + 1} \sum_{i=1}^{N-k+1} \left(\hat{\theta}_{ki} - \hat{\theta}_N\right)^2. \qquad (4.29)$$

Then the variance of interest is then given by

$$\text{Var}(\hat{\theta}_N) = \frac{k}{N} s_k^2. \qquad (4.30)$$

4.4
BC$_a$ Confidence Intervals

The estimation of confidence intervals by the bootstrap percentile method has appropriate asymptotic behavior, but improvements in coverage and stability are possible for practical situations. We describe here an improved algorithm, called the *bias-corrected and accelerated* method, or BC$_a$. Further improvements are possible, but the BC$_a$ will satisfy many typical needs. An analytic and computationally less demanding approximation to BC$_a$, known as ABC (for "approximate bootstrap confidence" intervals) is also implemented in popular frameworks.

Consider the estimation of parameter θ with estimator $\hat{\theta} = \hat{\theta}(\mathbf{X})$. We draw B bootstrap replications of our dataset and form the bootstrap estimators $\hat{\theta}^*(1), \ldots, \hat{\theta}^*(B)$. We wish to construct a $1 - \alpha = 1 - 2\epsilon$ confidence interval $(\underline{\theta}, \overline{\theta})$ for θ, with the property that (presuming a continuous sampling distribution here):

$$P(\underline{\theta} > \theta) = P(\overline{\theta} < \theta) = \epsilon. \qquad (4.31)$$

We have already seen how to obtain estimates $(\hat{\underline{\theta}}, \hat{\overline{\theta}})$ for $(\underline{\theta}, \overline{\theta})$ according to the bootstrap percentile method in Section 4.2.1. There we used percentiles of the bootstrap cdf for $\hat{\theta}^*$, $P_B(u)$. The BC$_a$ method also uses percentiles of $P_B(u)$, but with different percentiles.

Let $N(z)$ denote the standard normal cdf (i.e., Gaussian with mean zero and standard deviation one). The bootstrap percentile method could be expressed tautologically as finding the percentile points with percentiles given by

$$\epsilon = N(z_\epsilon) \qquad (4.32)$$

$$1 - \epsilon = N(z_{1-\epsilon}), \quad (4.33)$$

where z_ϵ is the ϵ cumulative probability point of the normal distribution. The BC_a algorithm generalizes this by asking for the percentile points corresponding to ϵ_1 and $1 - \epsilon_2$, where

$$\epsilon_1 = N\left[\hat{z}_0 + \frac{(\hat{z}_0 + z_\epsilon)}{1 - \hat{a}(\hat{z}_0 + z_\epsilon)}\right] \quad (4.34)$$

$$1 - \epsilon_2 = N\left[\hat{z}_0 + \frac{(\hat{z}_0 + z_{1-\epsilon})}{1 - \hat{a}(\hat{z}_0 + z_{1-\epsilon})}\right]. \quad (4.35)$$

Parameter z_0, called the *bias correction*, is used to correct for a possible median bias in estimator $\hat{\theta}$. A bootstrap estimator for z_0, $\hat{z}_0$, is used, with

$$\hat{z}_0 = N^{-1}\left[P_B(\hat{\theta})\right]. \quad (4.36)$$

Note that $P_B(\hat{\theta})$ is the bootstrap probability that $\hat{\theta}^* < \hat{\theta}$. Parameter a, called the *acceleration* is used to correct for possible dependence of the standard deviation (of a function of $\hat{\theta}$) on the value of $\hat{\theta}$. It may be estimated with the help of the jackknife, using the same notation as in Section 4.3:

$$\hat{a} = \frac{1}{6} \frac{\sum_{i=1}^{N}\left(\hat{\theta}' - \hat{\theta}_{-i}\right)^3}{\left[\sum_{i=1}^{N}\left(\hat{\theta}' - \hat{\theta}_{-i}\right)^2\right]^{3/2}}. \quad (4.37)$$

The reader is referred to Efron and Tibshirani (1994) for further justification and development. The BC_a algorithm is implemented as the default in MATLAB's bootci and as type="bca" in the R function boot.ci.

For our earlier example of confidence intervals for the mean of a normal distribution (but without assuming normality), both the percentile and BC_a bootstrap methods perform similarly. We thus try a comparison in a more challenging situation: the estimation of confidence intervals for the variance of a distribution (again a normal, but estimated nonparametrically without assuming anything about the distribution). We generate a sample of size 20 and observe the intervals as a function of the number of bootstrap replicas as shown in Figure 4.7. The convergence properties are similar for both methods, but the actual intervals are quite different.

To compare performance in terms of coverage, we perform 10 000 experiments, each sampling $N = 20$ values from a $N(0, 1)$ distribution. The percentile and BC_a methods are applied to each experiment, and the coverage statistics obtained. These are summarized in Table 4.1. We find that the overall coverage of the estimated confidence intervals from our two methods is similar, although rather smaller than the desired 68% probability. However, the BC_a intervals produce much better coverage symmetry between the low and high tails.

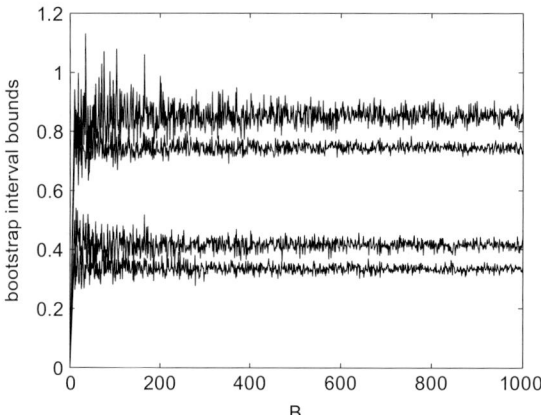

Figure 4.7 Dependence of percentile and BC_a bootstrap intervals (68% CL) on the number of bootstrap samples. The top pair of curves are the estimated upper limit of the interval, and the bottom pair the lower limit. In each pair, the upper curve is from the BC_a method, and the lower is from the percentile method. All bootstrap replicas are independent, so the local scatter of the points indicates the statistical variation from the resampling process. The sample is size 20 from $N(0, 1)$. The sample variance is 0.568 for this sample.

Table 4.1 Comparing coverage of estimated 68% confidence intervals from the bootstrap percentile and BC_a methods.

	Percentile	BC_a
Target tail probability	0.1587	
Low tail probability	0.0826	0.2093
High tail probability	0.3035	0.1889
Target coverage	0.6827	
Coverage	0.6139	0.6018

4.5
Cross-Validation

In later chapters we develop a variety of classification algorithms. We can view the typical situation as one where we have a set of understood data that we wish to learn from, and thence make predictions about data with some unknown characteristic. It is important to know how good our prediction is. Another situation occurs when we wish to make a prediction in the familiar regression setting. Again, we wish to know how good our prediction is. We may attempt to provide a measure for this with the *expected prediction error* (EPE), defined below. The EPE requires knowing the sampling distribution, which is in general not available. Thus, we must find a means to estimate EPE. A technique for doing this is the resampling method known as cross-validation.

To define the expected prediction error, consider the regression problem (it could also be a classification problem, as we shall discuss in Section 9.4.1) where we wish to predict a value for random variable Y, depending on the value for random variable X. In general, X is a vector of random variables, but we will treat it as one-dimensional for the moment. The joint probability distribution for X and Y is $F(x, y)$, with pdf $f(x, y)$. We wish to find the "best" prediction for Y, given any $X = x$. That is, we look for a function $r(x)$ providing our prediction for Y. What do we mean by "best"? Well, that is up to us to decide; there are many possibilities. However, the most common choice is an estimator that minimizes the expected squared deviation, and this is how we define the expected (squared) prediction error:

$$\text{EPE}(r) = E\{[Y - r(X)]^2\} = \int [y - r(x)]^2 f(x, y) dx\, dy. \tag{4.38}$$

Note that the expectation is over both X and Y; it is the expected error (squared) over the joint distribution. The function r that minimizes EPE is the regression function

$$r(x) = E(Y)_{X=x} = \int y f(x, y) dy, \tag{4.39}$$

that is, the conditional expectation for Y, given $X = x$. Fortunately, this is a nicely intuitive result. We remark again that x and y may be multivariate.

How might we estimate the EPE? For a simple case, consider a bivariate dataset $\mathcal{D} \equiv \{(X_n, Y_n), n = 1, \ldots, N\}$. Suppose we are interested in finding the best straight line fit. In this case, our regression function is $\ell(x) = ax + b$. We estimate parameters a and b by finding $\hat{a}$ and $\hat{b}$ that minimize (assuming equal weights for simplicity)

$$\sum_{n=1}^{N} \left(Y_n - \hat{a} X_n - \hat{b}\right)^2. \tag{4.40}$$

The value of a new sampling is predicted given X_{N+1}:

$$\hat{Y}_{N+1} = \hat{a} X_{N+1} + \hat{b}. \tag{4.41}$$

We wish to estimate the EPE for Y_{N+1}.

A simple approach is to divide our $\{(X_i, Y_i), i = 1, \ldots, N\}$ dataset into two pieces, perhaps two halves. Then one piece (the training set) could be used to determine the regression function, and the other piece (the testing set) could be used to estimate the EPE. However, this seems a bit wasteful, since we are only using half of the available data to obtain our regression function, and we could do a better job with all of the data. The next thing that occurs to us is to reverse the roles of the two pieces and somehow average the results, and this is a pretty good idea. But let us take this to an extreme, known as *leave-one-out cross-validation*.

The algorithm for leave-one-out cross-validation is as follows:

1. Form N subsets of the dataset $\mathcal{D}$, each one leaving out a different datum, say (X_k, Y_k). We will use the subscript $-k$ to denote quantities obtained omitting datum (X_k, Y_k). Likewise, we let $\mathcal{D}_{-k}$ be the dataset leaving out (X_k, Y_k).
2. Do the regression on dataset $\mathcal{D}_{-k}$, obtaining regression function r_{-k}.
3. Using this regression predict the value for the missing point:

$$\hat{Y}_k = r_{-k}(X_k). \tag{4.42}$$

4. Repeat this process for $k = 1, \ldots, N$. Estimate the EPE according to

$$\frac{1}{N}\sum_{k=1}^{N}(\hat{Y}_k - Y_k)^2. \tag{4.43}$$

Let us try an example application. Suppose we wish to investigate polynomial regression models for a dataset (perhaps an angular distribution or a background distribution). For example, we have a dataset and wish to consider whether to use the straight line relation

$$Y = aX + b, \tag{4.44}$$

or the quadratic relation

$$Y = aX^2 + bX + c. \tag{4.45}$$

We know that the fitted errors for the quadratic model will always be smaller than for the linear model. A common approach is to compute the sum of squared fitted errors, and apply some criterion on the difference in this quantity between the two models, often resorting to an approximation with a χ^2 distribution. That is, we use the Snedecor F distribution to compare the χ^2 values from fits for the two models. However, we may not wish to rely on the accuracy of this approximation. In this case, cross-validation may be applied.

The predictive error is not necessarily smaller with the additional adjustable parameters. We may thus use our estimated prediction errors as a means to decide between models. Suppose, for example, that our data is actually sampled from the linear model, as in the filled circles in Figure 4.8a. We do cross-validation estimates of the prediction error for both the linear and quadratic fit models, and take the difference (linear EPE minus quadratic EPE). The distribution of this difference, for 100 such "experiments", is shown in Figure 4.8b. Choosing the linear model when the difference is larger than zero gets it right in 84 out of 100 cases. The MATLAB function `crossval` is used to perform the estimates, with calls of the form:

```
crossval('mse',x,y,'Predfun',@linereg,'leaveout',1);
```

Alternatively, suppose that our data is sampled from a quadratic model (with a rather small quadratic term), as in the plus symbols in Figure 4.8a. We do cross-

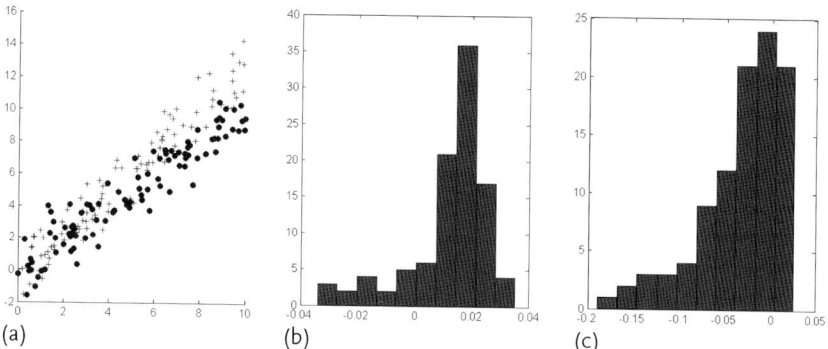

Figure 4.8 Leave-one-out cross-validation example. (a) Data samples, each of size 100, generated according to a linear model $Y = X$ (filled circles) or a quadratic model $Y = X + 0.03X^2$ (plus symbols). (b) Distribution of linear model minus quadratic model EPE for data generated according to a linear model. (c) Distribution of linear model minus quadratic model EPE for data generated according to a quadratic model.

validation estimates of the prediction error for both the linear and quadratic fit models, and again take the difference. The distribution of this difference, for 100 such experiments, is shown in Figure 4.8c. Choosing the quadratic model when the difference is less than zero gets it right in 79 out of 100 cases.

Leave-one-out cross-validation is particularly suitable if the size of our dataset is not large. However, as N becomes large, the required computer resources may be prohibitive. We may back off from the extreme represented by leave-one-out cross-validation and obtain $\mathcal{K}$-*fold cross-validation*. In this case, we divide the dataset into $\mathcal{K}$ disjointed subsets, of essentially equal size $m \approx N/\mathcal{K}$. Leave-one-out cross-validation corresponds to $\mathcal{K} = N$.

In $\mathcal{K}$-fold cross-validation, the model is trained on the dataset consisting of everything except the "held-out" sample of size m. Then the prediction error is obtained by applying the model to the held-out validation sample. This procedure is applied in turn for each of the $\mathcal{K}$ held-out samples, and the results for the squared prediction errors averaged.

We have a choice for $\mathcal{K}$; how can we decide? To get some insight, we note that there are three relevant issues: computer time, bias, and variance in our estimated prediction error. The choice of $\mathcal{K}$ may require compromises among these. In particular, for large datasets, we may be driven to small $\mathcal{K}$ by limited computing resources.

To understand the bias consideration, we introduce the *Learning Curve*, illustrated in Figure 4.9. This curve shows how the expected prediction error decreases as the training set size is increased.[1] We may use it to understand how different choices of folding can lead to different biases in estimating the EPE. Except for very small dataset sizes, if we use N-fold cross-validation, we use essentially the whole dataset

---

1) It is conventional in the classification problem to show one minus the error rate as the learning curve.

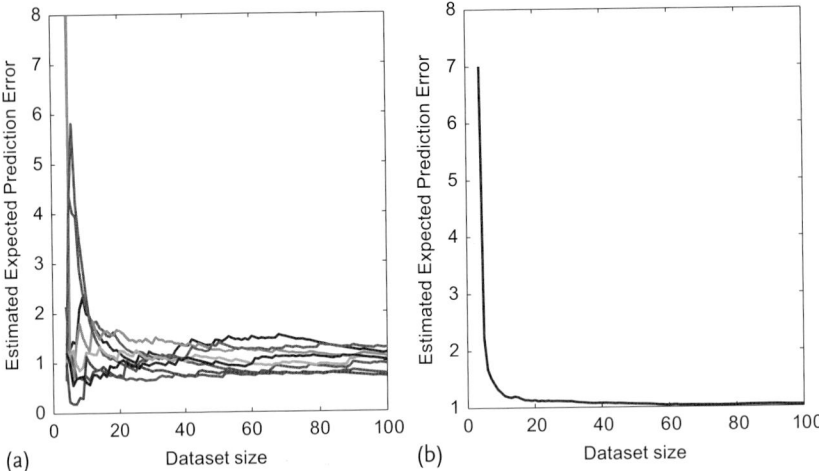

Figure 4.9 Computing the learning curve for cross-validation. (a) The dependence of estimated predicted error for leave-one-out cross-validation on sample size for several data samples. The data is generated according to the linear model as in Figure 4.8. (b) The learning curve, estimated by averaging together the results from 100 data samples. That is, 100 curves of the form illustrated in the left plot are averaged together.

for each evaluation of the EPE, and our estimator is close to unbiased (though as we see in Figure 4.9a, it may have a large variance). However, consider what happens if we go to smaller $\mathcal{K}$ values. To make the point, consider a dataset of size 20, and $\mathcal{K} = 2$. In this case, our estimated EPE will be based on a training sample of size 10. Looking at the learning curve, we see that yields an overestimate of the EPE for $N = 20$.

On the other hand, the larger $\mathcal{K}$ is, the more computation is required, so the lowest acceptable $\mathcal{K}$ is preferred. A much more subtle issue is the variance of the estimator (Breiman and Spector, 1992). As a rule of thumb, it is proposed that $\mathcal{K}$ values between 5 and 10 are typically approximately optimal. Performance on our regression example above degrades somewhat from leave-one-out to $\mathcal{K} = 5$ or 10.

We will see in the next chapter another application of cross-validation, to the problem of optimizing smoothing in density estimation.

4.6
✄ Resampling Weighted Observations

It is not uncommon to deal with datasets in which the samplings are accompanied by a nonnegative "weight", w. For example, we may have samples corresponding to different luminosities that we wish to incorporate into an analysis. Hence, we describe our dataset as a set of pairs, $(x_n, w_n; n = 1, \ldots, N)$, where w_n is the weight associated with the sampling x_n (x_n could also be a vector). The question arises: How can we apply resampling methods to such a dataset?

If we consider the bootstrap, a simple answer is apparent. Instead of assigning an equal resampling probability to each datum, we assign a probability proportional to the weight. That is, the datum x_n is resampled with a probability $P(x_n) = w_n / \sum_j w_j$. Let us assume that the weights are scaled such that $\sum_n w_n = N$. Then our bootstrap resampling corresponds to regarding our empirical distribution to be

$$\hat{f} = \sum_{n=1}^{N} w_n \delta(x - x_n). \tag{4.46}$$

It is important to realize that any statistic $S(\mathbf{x}^*)$ computed from the bootstrap replica is computed as if the weights were all identical (equal to $1/N$). This is because the weights are already included in forming the replica.

However, it is not so obvious what to do in the case of the jackknife algorithm, since we are looping through removing one of the datums and its weight at a time, and the interpretation as sampling from the empirical distribution is less clear. For example, how do we estimate the acceleration parameter a in the BC_a bootstrap confidence interval for weighted data?

To address this question, we may begin by noticing that the appearance of a quantity such as $\hat{\theta}' - \hat{\theta}_{-i}$, as in (4.37) is a measure of the "influence" that leaving out one datum will have. Statisticians formalize this concept and define an *influence function* corresponding to a population parameter $s = s(F)$ and distribution F:

$$U(x, F) \equiv \lim_{\epsilon \to 0} \frac{s[(1-\epsilon)F + \epsilon \delta(x)] - s(F)}{\epsilon}. \tag{4.47}$$

The reader is referred to standard statistics textbooks (e.g., Shao, 2003) for a more rigorous treatment; we take the intuition that $U(x, F)$ tells us something about how sensitive our statistic is to the addition of a small point source of probability at x. For this reason, it appears in discussions of robust statistics.

The influence function is useful in solving our "influence" estimation problem. We noted in Section 4.3 that the jackknife has difficulty for nonlinear statistics, so let us assume linearity here, and resign ourselves to thinking harder if confronted with a nonlinear problem. Thus, consider a population parameter of the form

$$s(F) = \int g(x) \, dF(x). \tag{4.48}$$

Then the influence function is simply

$$U(x, F) = g(x) - s(F). \tag{4.49}$$

Since $E_F U(x, F) = 0$, the central moments of U are

$$\mu_m = \int [U(x, F)]^m \, dF(x). \tag{4.50}$$

Suppose we have an unweighted dataset; let us approximate the influence function by using the empirical distribution. The empirical equivalent of (4.48) is

$$s(\hat{F}) = \int g(x) d\hat{F}(x) = \frac{1}{N} \sum_{n=1}^{N} g(x_n) . \qquad (4.51)$$

Hence,

$$U(x_k, \hat{F}) = g(x_k) - s(\hat{F})$$
$$= g(x_k) - \frac{1}{N} \sum_{n \neq k} g(x_n) . \qquad (4.52)$$

With, from (4.12) and (4.14),

$$\hat{\theta}_{-k} = \frac{1}{N-1} \sum_{n \neq k} g(x_n) \qquad (4.53)$$

$$\hat{\theta}' = \frac{1}{N} \sum_{n} \hat{\theta}_{-k} , \qquad (4.54)$$

and some manipulation, we find

$$U(x_k, \hat{F}) = (N-1)\left(\hat{\theta}' - \hat{\theta}_{-k}\right) . \qquad (4.55)$$

The a acceleration parameter is given by $\frac{1}{6} E[(\hat{\theta} - \theta)^3]/E[(\hat{\theta} - \theta)^2]^{3/2}$. In (4.37) we estimate the moments with the jackknife. We learn in (4.55) that we can equivalently use the empirical influence function:

$$\hat{a} = \frac{1}{6} \frac{\sum_n U(x_n, \hat{F})^3}{\left[\sum_n U(x_n, \hat{F})^2\right]^{3/2}} . \qquad (4.56)$$

Let us see what we need to do to modify this for weighted data. Our empirical equivalent of (4.48) is now

$$s(\hat{F}) = \int g(x) d\hat{F}(x) = \sum_{n=1}^{N} w_n g(x_n) . \qquad (4.57)$$

Thus,

$$U(x_k, \hat{F}) = g(x_k) - s(\hat{F})$$
$$= (1 - w_k) g(x_k) - \sum_{n \neq k} w_n g(x_n) \qquad (4.58)$$

$$= (N-1) \left(\sum_{n=1}^{N} w_n \hat{\theta}_{-n} - \hat{\theta}_{-k} \right) , \qquad (4.59)$$

keeping the same definition for $\hat{\theta}_{-n}$ as in (4.53). Thus, parameter a may be estimated with

$$\hat{a} = \frac{1}{6\sqrt{N}} \frac{\sum_n w_n U(x_n, \hat{F})^3}{\left[\sum_n w_n U(x_n, \hat{F})^2\right]^{3/2}}. \tag{4.60}$$

Another scenario where observations may be fruitfully weighted occurs when the data is *heteroscedastic*, that is, sampled with different variances. The familiar "inverse-square" weighting in forming averages is a well-known case demonstrating the utility of this. In this case, datum x_i may be given a weight $w_i = 1/\sigma_i^2$, where σ_i^2 is the variance of the sampling distribution for x_i (taken as a scalar for this discussion). A number of resampling methods have been developed to address this situation.

For example, consider the problem of performing a least squares linear regression. The model is

$$X = A\boldsymbol{\theta} + \epsilon, \tag{4.61}$$

where $\boldsymbol{\theta} = \theta_1, \ldots, \theta_R$ is a vector of parameters to be estimated, $X = X_1, \ldots, X_D$ is a vector of random variables, A is a known $D \times R$ matrix, and $\epsilon = \epsilon_1, \ldots, \epsilon_D$ is a vector of independent errors, assumed distributed with mean zero and $\text{Var}(\epsilon_i) = \sigma_i^2$. The least-squares estimator, $\hat{\boldsymbol{\theta}}$, for $\boldsymbol{\theta}$ is obtained as the solution to

$$\min_{\boldsymbol{\theta}} \|X - A\boldsymbol{\theta}\|^2. \tag{4.62}$$

We will assume for simplicity here that a unique solution exists, in which case it is

$$\hat{\boldsymbol{\theta}} = HA^T X, \tag{4.63}$$

where $H \equiv (A^T A)^{-1}$. It is readily checked that this estimate is unbiased, $E(\hat{\boldsymbol{\theta}}) = \boldsymbol{\theta}$.

If we knew the σ_is, we could compute the variance of this estimator, obtaining

$$\text{Var}(\hat{\boldsymbol{\theta}}) = H \sum_{i=1}^{D} \sigma_i^2 A_i^T A_i H, \tag{4.64}$$

where A_i is a vector (of dimension R) with components given by the ith row of A. Notice that this is just the special case of the general least-squares result with a general covariance matrix: $\text{Var}(X) = M$:

$$\text{Var}(\hat{\boldsymbol{\theta}}) = HA^T MAH. \tag{4.65}$$

However, we may have no idea what the σ_is are. The jackknife can be used to obtain a consistent estimate for the variance of $\hat{\boldsymbol{\theta}}$ (Wu, 1986; Shao, 2003), even though the data are not i.i.d. In this case, one can use the jackknife to compute an analytic expression, for example, an expression for the estimated variance, $\hat{V}$, is given in Shao (2003):

$$\hat{V} = \sum_{d=1}^{D}(1 - h_d)\left(\hat{\boldsymbol{\theta}}_{-d} - \hat{\boldsymbol{\theta}}\right)^2 = H\sum_{d=1}^{D}\frac{r_d^2 A_d A_d^T}{1 - h_d}H, \qquad (4.66)$$

where $r_d \equiv X_d - A_d^T \hat{\boldsymbol{\theta}}$ is the fit residual, $h_d = A_d^T H A_d$, and $\hat{\boldsymbol{\theta}}_{-d}$ is the LSE for $\boldsymbol{\theta}$ from the jackknife dataset omitting X_d (and the corresponding row of A).

4.7
Exercises

1. Consider θ the variance of a distribution F. Show that the bootstrap estimate for the bias of the estimator $\hat{\theta} = \frac{1}{N}\sum_{n=1}^{N}(x_n - \bar{x})^2$, where $\bar{x}$ is the sample mean, is

$$b_{\hat{F}} = -\frac{1}{N^2}\sum_{n=1}^{N}(x_n - \bar{x})^2. \qquad (4.67)$$

2. Show that the jackknife estimate of the bias of the sample mean, as an estimator of the mean, is zero.
3. We discussed the use of the jackknife in the estimate of the BC_a acceleration parameter a, and how it can be modified to incorporate weights. If we really had been thinking about a weighted jackknife, we might have defined:

$$\hat{\theta} = \sum_{n=1}^{N} w_n g(x_n) \qquad (4.68)$$

$$\hat{\theta}_{-i} = \frac{\sum_{n \neq i} w_n g(x_n)}{\sum_{n \neq i} w_n}. \qquad (4.69)$$

Show that, in this case, the influence function for the empirical distribution is

$$U(x_k, \hat{F}) = \frac{1 - w_k}{w_k}\left[\frac{1}{N}\sum_{n=1}^{N}\frac{1 - w_n}{1 - 1/N}\hat{\theta}_{-n} - \hat{\theta}_{-k}\right]. \qquad (4.70)$$

4. Derive the formula for the variance of a LSE in (4.65) and thence (4.64).

References

Breiman, L. and Spector, P. (1992) Submodel selection and evaluation in regression. The X-random case. Int. Stat. Rev., 60 (3), 291–319.

References

DiCiccio, T.J. and Efron, B. (1996) Bootstrap confidence intervals. *Stat. Sci.*, *11* (3), 189–228.

Efron, B. (1979) Bootstrap methods: Another look at the jackknife. *Ann. Stat.*, *7* (1), 1–26.

Efron, B. (1987) Better bootstrap confidence intervals. *J. Am. Stat. Assoc.*, *82* (397), 171–185.

Efron, B. and Tibshirani, R.J. (1994) *An introduction to the bootstrap*, Chapman&Hall/CRC.

Lees et al. (2013) Measurement of an excess of $\bar{B} \to D^{(*)} \tau^- \bar{\nu}_\tau$ decays and implications for charged Higgs bosons. Submitted to *Phys. Rev. D*.

Pesarin, F. and Salmaso, L. (2010) *Permutation Tests for Complex Data: Theory, Applications, and Software*, John Wiley & Sons, Ltd.

Politis, D.N., Romano, J.P., and Wolf, M. (1999) *Subsampling*, Springer.

Schucany, W.R., Gray, H.L., and Owen, D.B. (1971) On bias reduction and estimation. *J. Am. Stat. Assoc.*, *66* (335), 524–533.

Shao, J. (2003) *Mathematical Statistics*, Springer, 2nd edn.

Sharot, T. (1976) The generalized jackknife: Fintie samples and subsample sizes. *J. Am. Stat. Assoc.*, *71* (354), 451–454.

Wu, C.F.J. (1986) Jackknife, bootstrap, and other resampling methods in regression analysis. *Ann. Stat.*, *14* (4), 1261–1295.

5
Density Estimation

Density estimation deals with the problem of estimating probability density functions based on some data sampled from the pdf. It may use assumed forms of the distribution, parameterized in some way (parametric statistics), or it may avoid making assumptions about the form of the pdf (nonparametric statistics). We have discussed parametric statistics, now we are concerned more with the nonparametric case. In some ways these aren't such distinct concepts, as we shall see.

Nonparametric estimates may be useful when the form of the distribution (up to a small number of parameters) is not known or readily calculable. They may be useful for comparison with models, either parametric or not. For example, the ubiquitous histogram is a form of nonparametric density estimator (if normalized to unit area). Nonparametric density estimators may be easier or better than parametric modeling for efficiency corrections or background subtraction. They may be useful in "unfolding" experimental effects to learn about some distribution of more fundamental interest. These estimates may also be useful for visualization (again, the example of the histogram is notable, as is the scatter plot). Finally, such estimates can provide a means to compare two sampled datasets.

The techniques of density estimation may be useful as tools in the context of parametric statistics. For example, suppose we wish to fit a parametric model to some data. It might happen that the model is not analytically calculable. Instead, we simulate the expected distribution for any given set of parameters. For each set of parameters, we need a new simulation. However, simulations are necessarily performed with finite statistics, and the resulting fluctuations in the prediction may lead to instabilities in the fit. Density estimation tools may be helpful here as "smoothers", to smooth out the fluctuations in the predicted distribution.

We frame our discussion in terms of a set of observations (dataset) from some "experiment". This dataset consists of the values x_n; $n = 1, 2, \ldots, N$. Our dataset consists of repeated samplings from a (presumed unknown) probability distribution. The samplings are here assumed to be i.i.d., although we will note generalizations here and there. Order is not important; if we are discussing a time series, we could introduce ordered pairs $\{(x_n, t_n), n = 1, \ldots, N\}$, and call it two-dimensional, but this case may not be i.i.d., due to correlations in time. In general, our quanti-

5 Density Estimation

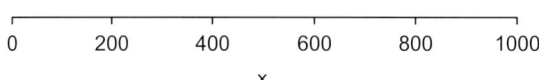

Figure 5.1 Example of an empirical probability density function. The vertical axis is such that the points are at infinity, with the "area" under each infinitely narrow point equal to $1/N$. In this example, $N = 100$. The actual sampling distribution for this example is a $\Delta(1232)$ Breit–Wigner (Cauchy; with pion and nucleon rest masses subtracted) on a second-order polynomial background. The probability to be background is 50%.

ties can be multidimensional; but we will use one-dimensional notation for now, changing to multivariate notation when we need to.

When we discussed point estimation, we introduced the hat notation for an estimator. Here we will be concerned with estimators for the density function itself, hence $\hat{f}(x)$ is a random variable giving our estimate for density $f(x)$.

5.1
Empirical Density Estimate

The most straightforward density estimate is the *empirical probability density function*, or *epdf*: Place a delta function at each data point. Explicitly, this estimator is:

$$\hat{f}(x) = \frac{1}{N} \sum_{n=1}^{N} \delta(x - x_n).$$

Figure 5.1 illustrates an example. Note that x could be multidimensional here; the familiar scatter plot presents a representation of the epdf in two dimensions. We have already made practical use of the epdf as the sampling density for the bootstrap procedure.

5.2
Histograms

Probably the most familiar density estimator is based on the histogram:

$$h(x) = \sum_{n=1}^{N} I(x - \tilde{x}_n; w), \tag{5.1}$$

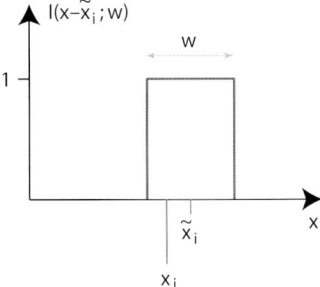

Figure 5.2 An indicator function, used in the construction of a histogram.

where $\tilde{x}_n$ is the center of the bin in which observation x_n lies, w is the bin width, and

$$I(x; w) = \begin{cases} 1 & x \in [-w/2, w/2] \\ 0 & \text{otherwise} . \end{cases} \quad (5.2)$$

This function is called the *indicator function*. Figure 5.2 illustrates the idea in the histogram context.

This is written for uniform bin widths, but may be generalized to differing widths with appropriate relative normalization factors. Figure 5.3 provides an example of a histogram, for the same sample as in Figure 5.1.

Given a histogram, the estimator for the probability density function is

$$\hat{f}(x) = \frac{1}{Nw} h(x) . \quad (5.3)$$

There are some drawbacks to the histogram: (*i*) It is discontinuous even if the pdf is continuous; (*ii*) It is dependent on the bin size and bin origin; (*iii*) The information from the location of a datum within a bin is lost.

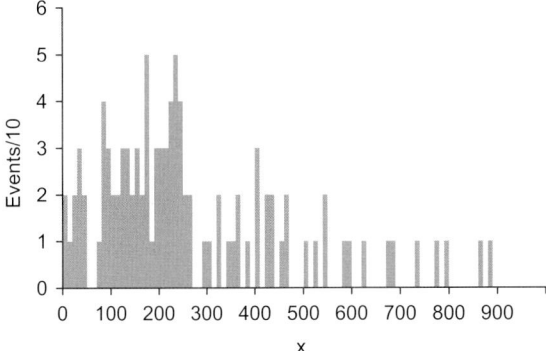

Figure 5.3 Example of a histogram for the data in Figure 5.1, with bin width $w = 10$.

5.3
Kernel Estimation

We may generalize the idea of the histogram by replacing the indicator function I with something else:

$$\hat{f}(x) = \frac{1}{N} \sum_{n=1}^{N} k(x - x_n; w), \tag{5.4}$$

where $k(x, w)$ is the *kernel function*, normalized to unity:

$$\int_{-\infty}^{\infty} k(x; w) dx = 1. \tag{5.5}$$

This provides a means to avoid some of the drawbacks of histograms, for example, we can obtain a continuous density estimate. We are usually interested in kernels of the form

$$k(x - x_n; w) = \frac{1}{w} K\left(\frac{x - x_n}{w}\right). \tag{5.6}$$

The kernel estimator for the pdf is then

$$\hat{f}(x) = \frac{1}{nw} \sum_{n=1}^{N} K\left(\frac{x - x_n}{w}\right), \tag{5.7}$$

The role of parameter w as a "smoothing" parameter is apparent. The delta functions of the empirical distribution are spread over regions of order w.

Often, the particular form of the kernel used doesn't matter very much. This is illustrated with a comparison of several kernels (with commensurate smoothing parameters) in Figure 5.4. The Gaussian is probably the most popular, and is smooth. Optimization criteria and error estimation that may be applied to kernel methods are discussed in Sections 5.6 and 5.7.

5.3.1
Multivariate Kernel Estimation

Kernel estimation may easily be applied to multivariate situations. For example, in $D = 2$ dimensions:

$$\hat{f}(x, y) = \frac{1}{N w_x w_y} \sum_{n=1}^{N} K\left(\frac{x - x_n}{w_x}\right) K\left(\frac{y - y_n}{w_y}\right). \tag{5.8}$$

This is a "product kernel" form, with the same kernel in each dimension, except for possibly different smoothing parameters. It does not have correlations. The kernels we have introduced are classified more explicitly as "fixed kernels": the smoothing parameters are independent of x and y.

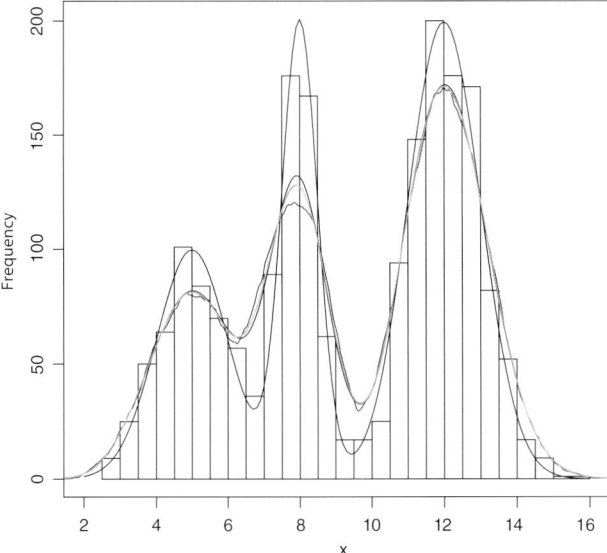

Figure 5.4 Comparison of density estimates using different kernels. The curve with the largest excursions is the sampling distribution, the next largest is the estimate with Gaussian kernel, followed by indistinguishable triangular and cosine kernel estimates, and finally a rectangular kernel estimate.

5.4
Ideogram

A simple variant on the kernel idea is to permit the kernel to depend on additional knowledge in the data. Physicists call this an *ideogram*. Most common is the *Gaussian ideogram*, in which each data point is entered as a Gaussian of area one and standard deviation appropriate to that datum. This addresses a way that the i.i.d. assumption might be broken.

The Particle Data Group uses ideograms as a means to convey information about possibly inconsistent measurements. Figure 5.5 shows an example of this.

5.5
Parametric vs. Nonparametric Density Estimation

The distinction between parametric and nonparametric is somewhat murky. A histogram is nonparametric, in the sense that no assumption about the form of the sampling distribution is made. Often an implicit assumption is made that the distribution is "smooth" on a scale smaller than the bin size. For example, we might know something about the resolution of our apparatus and adjust the bin size to be commensurate. But the estimator of the parent distribution made with a histogram is parametric – the parameters are populations (or frequencies) in each bin. The es-

5 Density Estimation

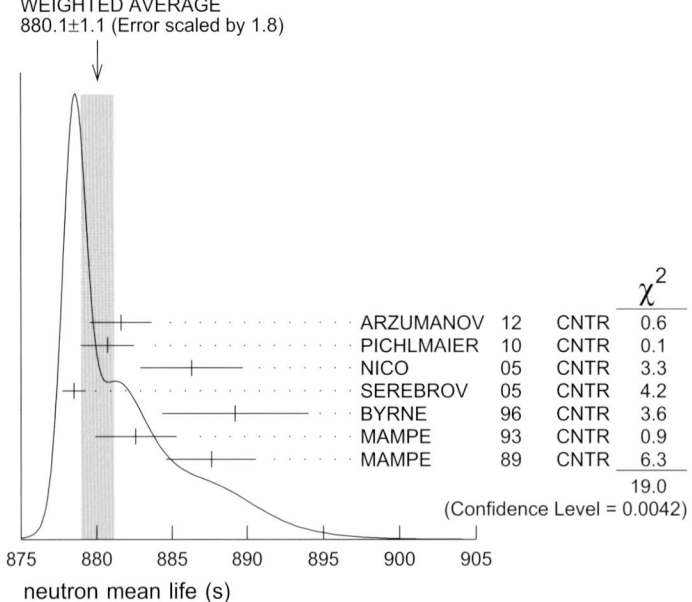

Figure 5.5 Example of a Gaussian ideogram, showing measurements of the mean lifetime of the neutron, from the particle data listings of the Particle Data Group's Review of Particle Properties (reprinted with permission from Beringer et al. (2012), copyright 2012, American Physical Society).

timators for those parameters are the observed histogram populations. There are even more parameters than in a typical parametric fit!

The essence of the difference may be captured in notions of "local" and "nonlocal". If a datum at x_n influences the density estimator at some other point x, this is nonlocal. A nonparametric estimator can be defined as one in which the influence of a point at x_n on the estimate at any x at a distance $d(x_n, x) > \epsilon$ vanishes, asymptotically. For example, for a kernel estimator, the bigger the smoothing parameter w, the more nonlocal the estimator,

$$\hat{f}(x) = \frac{1}{Nw} \sum_{n=1}^{N} K\left(\frac{x - x_n}{w}\right). \tag{5.9}$$

As we will discuss, the "optimal" choice of smoothing parameter decreases with increasing sample size N.

5.6 Optimization

We would like to make an optimal density estimate from our data. But we need to know what this means. We need a criterion for "optimal". In practice, the choice of criterion may be subjective; it depends on what you want to achieve.

As a plausible starting point, we may compare the density estimator $\hat{f}(x)$ with the true density, at point x:

$$\Delta(x) = \hat{f}(x) - f(x). \tag{5.10}$$

For a good estimator, we aim for small $\Delta(x)$, called the *error* in the estimator at point x.

We have seen that a common choice in point estimation is to minimize the sum of the squares of the deviations, as in a least-squares fit. We may take this idea over here, and form the *Mean Squared Error* (MSE):

$$\text{MSE}[\hat{f}(x)] \equiv E\{[\hat{f}(x) - f(x)]^2\} = \text{Var}[\hat{f}(x)] + \text{Bias}^2[\hat{f}(x)], \tag{5.11}$$

where

$$\text{Var}[\hat{f}(x)] \equiv E[(\hat{f}(x) - E[\hat{f}(x)])^2] \tag{5.12}$$

$$\text{Bias}[\hat{f}(x)] \equiv E[\hat{f}(x)] - f(x). \tag{5.13}$$

Since this isn't quite our familiar parameter estimation, let us take a little time to make sure it is understood. Suppose $\hat{f}(x)$ is an estimator for the pdf $f(x)$, based on data $\{x_n; n = 1, \ldots, N\}$, i.i.d. from $f(x)$. Then

$$E[\hat{f}(x)] = \int \cdots \int \hat{f}(x; \{x_i\}) \prod_{n=1}^{N} [f(x_n) dx_n]. \tag{5.14}$$

As an example, we will derive (5.11):

$$\text{MSE}[\hat{f}(x)] = E[(\hat{f}(x) - f(x))^2]$$

$$= \int \cdots \int \left[\hat{f}(x; \{x_i\}) - f(x)\right]^2 \prod_{n=1}^{N} [f(x_n) dx_n]$$

$$= \int \cdots \int \left[\hat{f}(x; \{x_i\}) - E(\hat{f}(x)) + E(\hat{f}(x)) - f(x)\right]^2$$
$$\times \prod_{n=1}^{N} [f(x_n) dx_n]$$

$$= \int \cdots \int \left\{\left[\hat{f}(x; \{x_i\}) - E(\hat{f}(x))\right]^2 + [E(\hat{f}(x)) - f(x)]^2 \right.$$
$$\left. - 2\left[\hat{f}(x; \{x_i\}) - E(\hat{f}(x))\right][E(\hat{f}(x)) - f(x)]\right\} \prod_{n=1}^{N} [f(x_n) dx_n]$$

$$= \text{Var}[\hat{f}(x)] + \text{Bias}^2[\hat{f}(x)] + 0. \tag{5.15}$$

Often with parametric statistics we assume unbiased estimators, hence the "Bias" term is zero. However, this is not a good assumption here, as we now demonstrate. This is a fundamental difficulty with smoothing.

Theorem 5.1. *(Scott, 1992; Rosenblatt, 1956) A uniform minimum variance unbiased estimator for $f(x)$ does not exist.*

To be unbiased, we require

$$E[\hat{f}(x)] = f(x), \quad \forall x .$$

To have uniform minimum variance, we require

$$\text{Var}[\hat{f}(x)|f(x)] \leq \text{Var}[\hat{g}(x)|f(x)], \quad \forall x ,$$

for all $f(x)$, where $\hat{g}(x)$ is any other estimator of $f(x)$.

To illustrate this theorem, suppose we have a kernel estimator

$$\hat{f}(x) = \frac{1}{N} \sum_{n=1}^{N} k(x - x_n; w) ,$$

Its expectation is

$$E[\hat{f}(x)] = \frac{1}{N} \sum_{n=1}^{N} \int k(x - x_n; w) f(x_n) dx_n$$

$$= \int k(x - y) f(y) dy . \tag{5.16}$$

Unless $k(x - y) = \delta(x - y)$, $\hat{f}(x)$ will be biased for some $f(x)$. But $\delta(x - y)$ has infinite variance.

Thus, the nice properties we strive for in parameter estimation (and sometimes achieve) are beyond reach. The intuition behind this limitation may be understood as the effect that smoothing lowers peaks and fills in valleys. We see this effect at work in Figure 5.6, where both a histogram and a Gaussian kernel estimator smooth out some of the structure in the original sampling distribution. Figure 5.7 shows how the effect may be mitigated, though not eliminated, with choice of binning or smoothing parameter.

The MSE for a density is a measure of uncertainty at a point. It is useful to somehow summarize the uncertainty over all points in a single quantity. We wish to establish a notion for the "distance" from function $\hat{f}(x)$ to function $f(x)$. This is a familiar subject, we are just dealing with normed vector spaces. The choice of norm is a bit arbitrary; obvious extremes are

$$\|\hat{f}(x) - f(x)\|_{L_\infty} \equiv \sup_x |\hat{f}(x) - f(x)|$$

$$\|\hat{f}(x) - f(x)\|_{L_1} \equiv \int |\hat{f}(x) - f(x)| dx . \tag{5.17}$$

As is commonly done, we will use the L_2 norm, more precisely here the *Integrated Squared Error* (ISE):

$$\text{ISE} \equiv \int [\hat{f}(x) - f(x)]^2 dx . \tag{5.18}$$

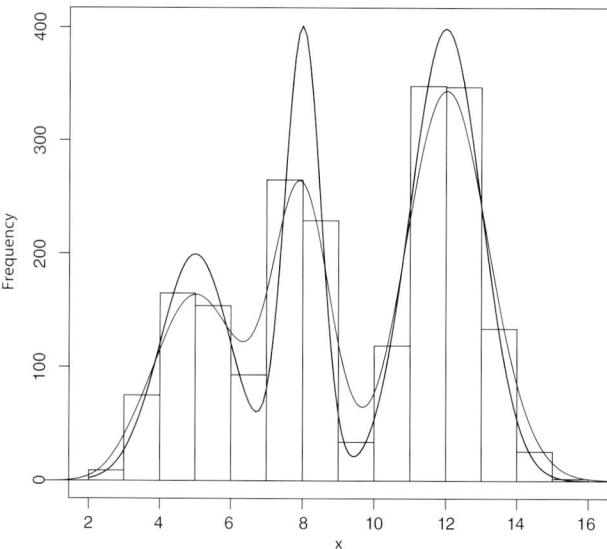

Figure 5.6 Effect of smoothing. The curve with the largest excursions is the sampling pdf. The histogram is made with data sampled from this pdf, and the remaining curve shows a Gaussian kernel estimator for the pdf.

In fact, the ISE is still a difficult object, as it depends on the true density, the estimator, and the sampled data. We may remove this latter dependence by evaluating the *Mean Integrated Squared Error* (MISE), or equivalently, the "integrated mean square error" (IMSE):

$$\text{MISE} \equiv E[\text{ISE}] = E\left[\int [\hat{f}(x) - f(x)]^2 dx\right] \qquad (5.19)$$

$$= \int E[(\hat{f}(x) - f(x))^2] dx = \int \text{MSE}[\hat{f}(x)] dx \equiv \text{IMSE} . \qquad (5.20)$$

A desirable property of an estimator is that the error decreases as the number of samples increases. This is a familiar notion from parametric statistics.

Definition 5.2. *A density estimator $\hat{f}(x)$ is consistent iff:*

$$MSE[\hat{f}(x)] \equiv E[\hat{f}(x) - f(x)]^2 \to 0 \qquad (5.21)$$

as $N \to \infty$.

5.6.1 Choosing Histogram Binning

Considerations such as minimizing the MSE may be used to choose an "optimal" bin width for a histogram. Noting that the bin contents of a histogram are binomially distributed, we could show (exercise) that, for the histogram density estimator

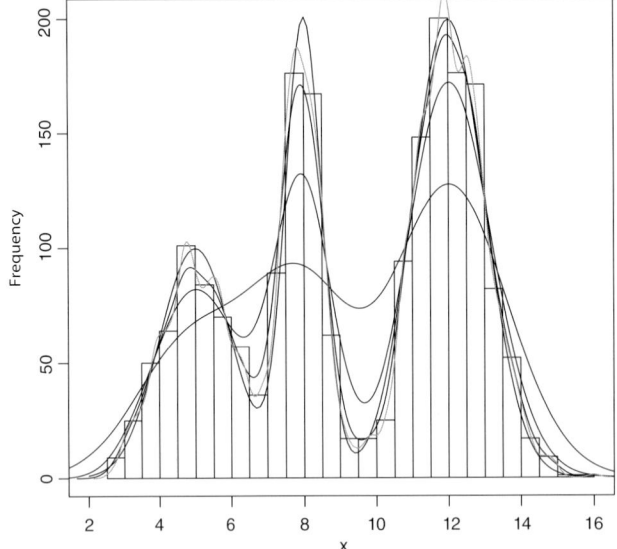

Figure 5.7 Plot showing effect of histogram binning and choice of smoothing parameter, for the same pdf as in Figure 5.6. The smooth curve with the largest excursions is the sampling pdf. The other curves, in order of progressive smoothness are Gaussian kernel estimates with smoothing parameters $w/4$, $w/2$, w, $2w$.

$\hat{f}(x) = h(x)/(Nw)$:

$$\text{Var}[\hat{f}(x)] \leq \frac{f(x_j^*)}{Nw}$$

$$|\text{Bias}[\hat{f}(x)]| \leq \gamma_j w, \tag{5.22}$$

where:

- $x \in \text{bin } j$,
- $x_j^* \in \text{bin } j$ is defined (and exists by the mean value theorem) by

$$\int_{\text{bin } j} f(x)dx = w f(x_j^*); \tag{5.23}$$

- γ_j is a positive constant such that

$$|f(x) - f(x_j^*)| < \gamma_j |x - x_j^*|, \quad \forall x \in \text{bin } j;$$

- equality is approached as the probability to be in bin j decreases (e.g., by decreasing bin size).

Thus, we have a bound on the MSE for a histogram:

$$\text{MSE}[\hat{f}(x)] = E[\hat{f}(x) - f(x)]^2 \leq \frac{f(x_j^*)}{Nw} + \gamma_j^2 w^2. \tag{5.24}$$

Theorem 5.3. *The MSE of the histogram estimator $\hat{f}(x) = h(x)/(Nw)$ is consistent if the bin width $w \to 0$ as $N \to \infty$ such that $Nw \to \infty$.*

Note that the $w \to 0$ requirement ensures that the bias will approach zero, according to our earlier discussion. The $Nw \to \infty$ requirement ensures that the variance asymptotically vanishes.

Theorem 5.4. *The MSE(x) bound above is minimized when*

$$w = w^*(x) = \left[\frac{f(x_j^*)}{2\gamma_j^2 N}\right]^{1/3} . \tag{5.25}$$

This theorem suggests that the optimal bin size decreases as $1/N^{1/3}$. The $1/N$ dependence of the variance is our familiar result for Poisson statistics. The optimal bin size depends on the value of the density in the bin. This suggests an "adaptive binning" approach with variable bin sizes. However, we caution that, according to Scott (1992), "...in practice there are no reliable algorithms for constructing adaptive histogram meshes."

Alternatively, for a Gaussian kernel the MISE error is minimized (asymptotically, for normally distributed data) when (Taylor, 1989)

$$w^* = \left(\frac{4}{3}\right)^{1/5} \sigma N^{-1/5} . \tag{5.26}$$

An early and popular choice is *Sturges' rule*, which says that the number of bins should be

$$k = 1 + \log_2 N , \tag{5.27}$$

where N is the sample size. This is the rule that was used in making the histogram in Figure 5.6. It is the default choice when making a histogram in R.

However, the argument behind this rule has been criticized (Hyndman, 1995). Indeed we see in our example that we probably would have "by hand" selected more bins; our histogram is "over-smoothed". There are other rules for optimizing the number of bins. For example, *Scott's rule* (Scott, 1992) for the bin width is:

$$w = 3.5 s N^{-1/3} , \tag{5.28}$$

where s is the sample standard deviation. The standard rules are typically based on some assumptions about the distribution and in practice often leave the visual impression that the binning could usefully be finer: see Figure 5.8. If your data are unimodal, then the rules may reasonably apply. However, if the data are more complicated then a more adaptive approach is required to obtain optimal results. If additional information is available, such as the experimental resolution, this can help to inform the bin width choice.

100 | 5 *Density Estimation*

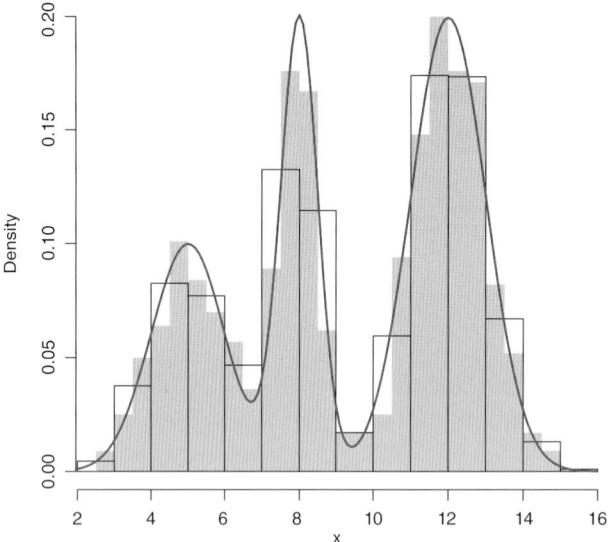

Figure 5.8 Illustration of histogram binning. The curve is the sampling pdf. The "standard rules" (Sturges, Scott, Freedman–Diaconis (Scott, 1992)) correspond roughly to the coarser binning above. The shaded histogram seems like a better choice, illustrating that blind application of the rules to complicated data may not yield the desired result.

5.7
Estimating Errors

We revisit now the bootstrap, which provides a means to evaluate how much to trust our density estimate. The bootstrap algorithm in this context is as follows:

1. Form density estimate $\hat{f}$ from data $\{x_n; n = 1, \ldots, N\}$.
2. Resample (uniformly) N values from $\{x_n; n = 1, \ldots, N\}$, with replacement, obtaining $\{x_n^*; n = 1, \ldots, N\}$ (bootstrap data). Note that in resampling with replacement, our bootstrap dataset may contain the same x_n multiple times.
3. Form density estimate $\hat{f}^*$ from data $\{x_n^*; n = 1, \ldots, N\}$.
4. Repeat steps 2&3 many (B) times to obtain a family of bootstrap density estimates $\{\hat{f}_b^*; b = 1, \ldots, B\}$
5. The distribution of $\hat{f}_b^*$ about $\hat{f}$ mimics the distribution of $\hat{f}$ about f.

Consider, for a kernel density estimator, the expectation of the bootstrap dataset:

$$E\left[\hat{f}^*(x)\right] = E\left[k(x - x_n^*; w)\right] = \hat{f}(x), \tag{5.29}$$

where the demonstration is left as an exercise. Thus, the bootstrap distribution about $\hat{f}$ does not reproduce the bias which may be present in $\hat{f}$ about f. However, it does properly reproduce the variance of $\hat{f}$, hence the bootstrap is a useful tool for estimating the variance of our density estimator. An illustration of the distribution of bootstrap samples is shown in Figure 5.9a.

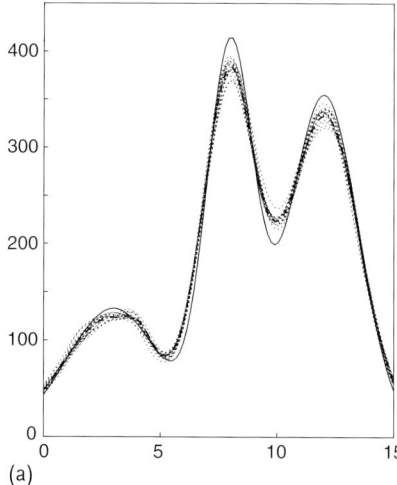 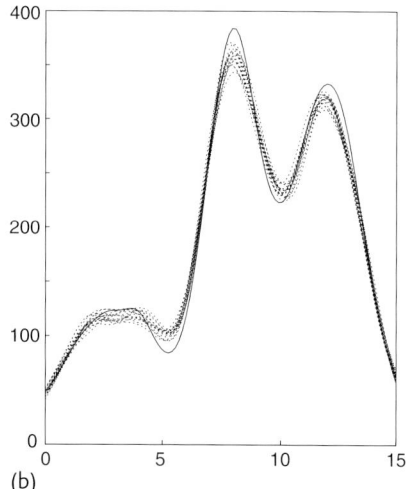

Figure 5.9 Use of the bootstrap to determine the variance and bias of a Gaussian kernel density estimator. The sample is size $N = 1000$. (a) The solid curve shows the sampling distribution, the heavy dashed curve shows the kernel estimator, and the lighter dotted curves show 15 bootstrap replica kernel estimators. (b) The solid curve is the kernel density estimator from the dashed curve in (a) and the lighter dotted curves show 15 smoothed bootstrap replica kernel estimators.

It is possible to use a variation of the bootstrap, called the *smoothed bootstrap*, to obtain an estimate for the bias. In this case, the replicas are sampled from the (kernel) estimate for the density, instead of from the empirical density. Denote the kernel estimate by $\hat{f}_w(x)$, where w indicates the dependence on the smoothing parameter. Suppose we draw a smoothed bootstrap replica x^* from this distribution. We can make a kernel density estimate from this replica, $\hat{f}_w^*(x)$. Now the bias of $\hat{f}_w^*(x)$ compared with $\hat{f}_w(x)$ will mimic the bias of $\hat{f}_w(x)$ compared with $f(x)$. Thus, using the smoothed bootstrap we may estimate the full MSE.

For example, we show the smoothed bootstrap in Figure 5.9b. It can be seen that the bias of the smoothed bootstrap replicas about the kernel density estimator indeed mimics the bias of the bootstrap replicas in Figure 5.9a about the true distribution. We may thus use the smoothed bootstrap to apply a bias correction to the kernel density estimator. We could further construct confidence intervals (e.g., using the percentile method in Chapter 4) including the effect of bias. As noted in Taylor (1989), the smoothed bootstrap introduces some extra variance, which may be corrected for.

Besides the bootstrap, the jackknife and cross-validation methods may also be applied to error estimation (and hence smoothing optimization) in kernel estimation. For example, we may use the jackknife to estimate bias. The idea behind this method is that bias depends on sample size. If we can assume that the bias vanishes asymptotically, we may use the data to estimate the dependence of the bias on sample size.

5 Density Estimation

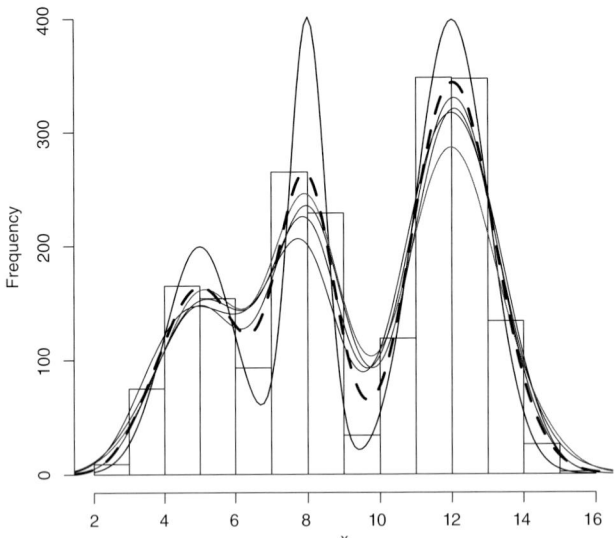

Figure 5.10 Jackknife estimation of bias. The curve with the largest excursions shows the sampling pdf. The heavy dashed curve shows a Gaussian kernel estimate to the entire dataset of size $N = 2000$. The remaining four curves show kernel estimates on disjoint jackknife subsamples of size 500. The histogram is made with the full sample.

We discussed the jackknife in Chapter 4. For example, we may use a delete-d version here, with $k = N/d$. A simple algorithm we may use is as follows:

1. Divide the data into k random disjoint subsamples.
2. Evaluate the estimator for each subsample.
3. Compare the average of the estimates on the subsamples with the estimator based on the full dataset.

Figure 5.10 illustrates this jackknife technique. In this case, we see the shift from the full data kernel estimate to the jackknifed kernel estimates, indicating a sample-size-dependent bias. As in Chapter 4, the jackknife may be used to reduce bias (see Scott, 1992).

5.8
The Curse of Dimensionality

While in principle problems with multiple dimensions in the sample space are not essentially different from a single dimension, there are very important practical differences. This is referred to as the *curse of dimensionality*. The most obvious difficulty is in displaying and visualizing as the number of dimensions increases. Typically one displays projections in lower dimensions, but this loses the correlations that might be present in the undisplayed dimensions.

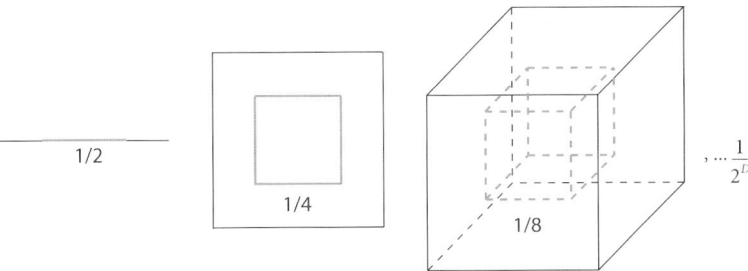

Figure 5.11 Demonstration of the "curse of dimensionality".

Another difficulty is that "all" the volume (of a bounded region) goes to the boundary (exponentially!) as the dimensions increase. This is illustrated in Figure 5.11. A unit cube in D dimensions is illustrated as D increases. A central cube with edges $\frac{1}{2}$ unit in length is shown. The fraction of the volume contained in this central cube decreases as 2^{-D}. As a consequence, data becomes "sparse".

To adequately represent a high-dimensional distribution of independent variables, the size of data must grow exponentially with dimensionality. One way to visualize this proposition is to plot the fraction of points contained within a unit sphere centered inside a uniformly populated unit cube, as shown in Figure 5.12. We use 10 000 points for this simulation. In one dimension, the cube and sphere are represented by the same interval and this fraction is 1. In 10 dimensions, no points lie inside the sphere. In high dimensions, the cube resembles a spherical "hedgehog" with 2^D spikes, one spike per cube corner, each of length $\sqrt{D}/2$ measured from the unit sphere. Other interesting facts about the high-dimensional geometry of simple bodies can be found in Koppen (2002).

Another way of understanding the curse of dimensionality is to plot the standard deviation of the distance between two randomly selected points normalized to the distance mean. If the ratio of the standard deviation over the mean decreases with dimensionality, the ratio of the maximal to minimal distance approaches one. A rigorous proof of this statement can be found in Beyer et al. (1999). We show the deviation–mean ratio in Figure 5.12 using the Euclidean distance for points uniformly distributed in a unit cube.

5.9
Adaptive Kernel Estimation

We saw in our discussion of histograms that it is probably more optimal to use variable bin widths. This applies to other kernels as well. Indeed, the use of a fixed smoothing parameter, deduced from all of the data introduces a nonlocal, hence parametric, aspect into the estimation. It is more consistent to look for smoothing which depends on data locally. This is *adaptive kernel estimation*.

We argue that the more data there is in a region, the better that region can be estimated. Thus, in regions of high density, we should use narrower smoothing. In

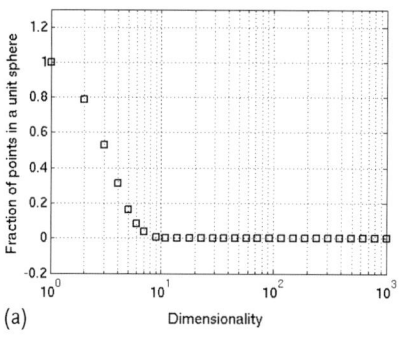 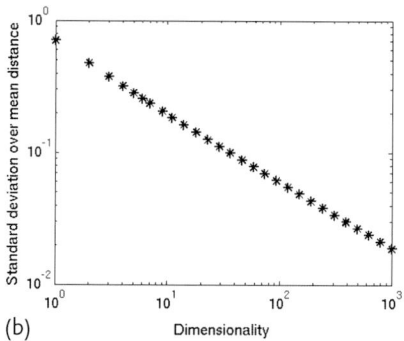

Figure 5.12 Effect of increasing dimensionality on points uniformly distributed in a hypercube bound between -1 and $+1$ in each dimension. (a) Fraction of points contained within a unit hypersphere centered at the origin. (b) Ratio of the standard deviation and the observed mean of the pairwise distance for the Euclidean distance. When we add dimensions, the standard deviation remains flat, and the mean scales as $\sqrt{D}$.

Poisson statistics (e.g., histogram binning), the relative uncertainty scales as

$$\frac{\sqrt{N}}{N} \propto \frac{1}{\sqrt{p(x)}}. \tag{5.30}$$

Thus, in the region containing x_n, the smoothing parameter should be

$$w(x_n) = \frac{w^*}{\sqrt{p(x_n)}}. \tag{5.31}$$

There are two issues with implementing this:

- What is w^*?
- We don't know $p(x)$.

For $p(x)$, we may try substituting our fixed kernel estimator, call it $\hat{p}_0(x)$. For w^*, we use dimensional analysis:

$$D[w(x_n)] = D[x]; \quad D[p(x)] = D\left[\frac{1}{x}\right] \Rightarrow D[w^*] = D[\sqrt{x}] = D[\sqrt{\sigma}].$$

Then, for example, using the "MISE-optimized" choice earlier, we iterate on our fixed kernel estimator to obtain

$$\hat{p}_1(x) = \frac{1}{N} \sum_{n=1}^{N} \frac{1}{w_n} K\left(\frac{x - x_n}{w_n}\right), \tag{5.32}$$

where

$$w_n = w(x_n) = \left(\frac{4}{3}\right)^{1/5} \sqrt{\frac{\rho \sigma}{\hat{p}_0(x_n)}} N^{-1/5}. \tag{5.33}$$

ρ is a factor which may be further optimized, or typically set to one.

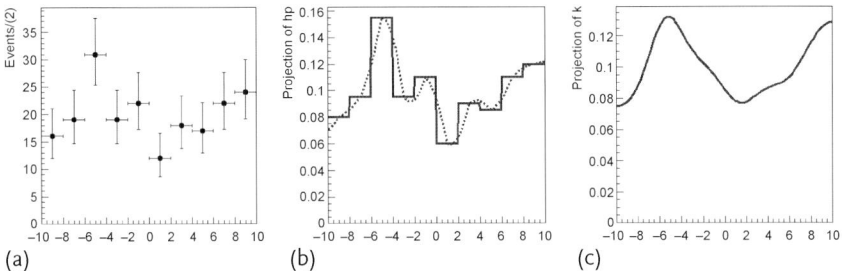

Figure 5.13 Example of the KEYS adaptive kernel estimation. (a) Input data; (b) histogram-based estimate with second order interpolation; (c) KEYS adaptive kernel estimate. (Reprinted with permission from Verkerke and Kirkby (2006), copyright 2006, W. Verkerke and D. Kirkby).

The iteration on the fixed kernel estimator nearly removes the dependence on our initial choice of w. The boundaries pose some complication in carrying this out.

There are packages for adaptive kernel estimation, for example, the KEYS ("Kernel Estimating Your Shapes") package (Cranmer, 2001). Figure 5.13 illustrates the use of this package.

5.10
Naive Bayes Classification

We anticipate the classification chapters of this book with the introduction of a simple yet often very satisfactory classification algorithm, the *naive Bayes* classifier. The problem is to correctly classify an observation x into one of K classes, $C_k, k = 1, \ldots, K$. We must learn how to do this using a dataset of N observations $\mathbf{X} = x_1, \ldots, x_N$ with known classes. We assume that observation x is sampled from the same distribution as the "training set" $\mathbf{X}$. For example, the training set may be produced by a simulation where the true classes are known.

The idea of the naive Bayes algorithm is to use Bayes' theorem to form an estimate for the probability that the x belongs to class k. With such probability estimates for each class, the class with the highest probability is chosen. To apply Bayes' theorem, we need two ingredients: the marginal (prior) probability to be in class k, $P(C_k)$; and the probability to sample observation x given class k, $P(x|C_k)$. That is,

$$P(C_k|x) \propto P(x|C_k)P(C_k). \tag{5.34}$$

The normalization isn't needed as it is the same for all classes.

The training set is used to estimate $P(x|C_k)$, by evaluating the local density for class C_k in the neighborhood of x. In general, this may be very difficult to do precisely if there are many dimensions. So a great simplification is made in the naive

Table 5.1 Confusion matrices corresponding to Figure 5.14.

		Predicted class			
		(a)		(b)	
		Plus	Dot	Plus	Dot
Actual class	Plus	215	185	143	257
	Dot	40	1960	23	977

Bayes algorithm: Assume that the probability $P(x|C_k)$ factorizes, for each class like

$$P(x|C_k) = \prod_{d=1}^{D} P(x_d|C_k) . \tag{5.35}$$

That is, we assume that the variables in x are statistically independent when conditioned on class. This makes the density estimation problem much easier, since we have only to estimate D one-dimensional densities (for each class) instead of a D-dimensional density. Typically, this is implemented with D one-dimensional kernel density estimators.

In spite of the simplicity, the naive Bayes algorithm often performs rather well, even in some cases when the assumption of class-conditioned independence is not correct. Examples of naive Bayes classification are illustrated in Figure 5.14. Figure 5.14a shows separation for a class with a bimodal distribution from a class with a unimodal distribution, but for which the assumption of independence is strictly correct. Figure 5.14b shows an example of the trained class boundary for a dataset in which the assumption of independence within a class is not correct. The algorithm is clearly having trouble in this case. Other values for the smoothing parameter do not improve the performance in this example. The separation boundary of a more general classifier, a neural net, on this dataset may be seen for comparison in Figure 12.3.

Quantitative measures of performance of a classifier are developed in coming chapters. However, a simple picture is given by the *confusion matrices*, given in Table 5.1. These are computed by applying the classifiers as trained by the data in the two sides of Figure 5.14 to new datasets generated with the same distributions. We could also compute the matrices directly on the data used for training, but the results would tend to be biased towards overstating the performance.

5.11
Multivariate Kernel Estimation

Besides the curse of dimensionality, the multidimensional case introduces the complication of covariance. When using a product kernel, the local estimator has

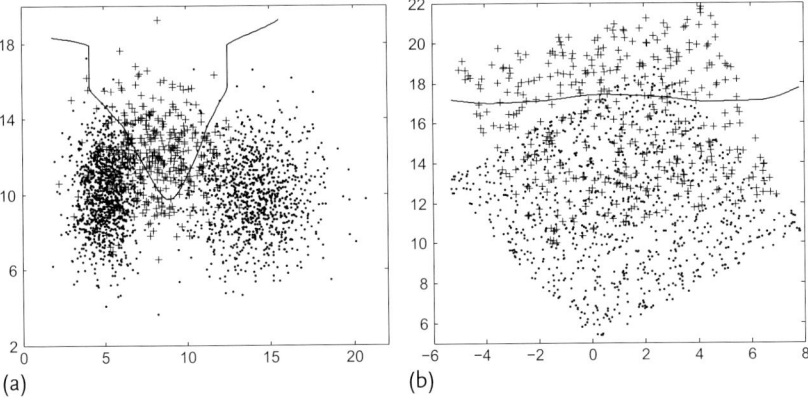

Figure 5.14 Naive Bayes classifier examples in two feature dimensions and two classes. The two classes are indicated with dots and pluses. The solid line shows the prediction boundary between the two classes as determined by the naive Bayes algorithm using the dataset in each figure. Training is performed with the MATLAB NaiveBayes.fit method, using default Gaussian kernel smoothing for the density estimation.

diagonal covariance matrix. In principle, we could apply a local linear transformation of the data to a coordinate system with diagonal covariance matrices. This amounts to a nonlinear transformation of the data in a global sense, and may not be straightforward. However, we can at least work in the system for which the overall covariance matrix of the data is diagonal.

If $\{y_n\}_{n=1}^N$ is the suitably diagonalized data, the product fixed kernel estimator in D dimensions is

$$\hat{p}_0(y) = \frac{1}{N} \sum_{n=1}^{N} \left[\prod_{d=1}^{D} \frac{1}{w_d} K\left(\frac{y^{(d)} - y_n^{(d)}}{w_d} \right) \right], \tag{5.36}$$

where $y^{(d)}$ denotes the dth component of the vector y. The asymptotic, normal MISE-optimized smoothing parameters are now

$$w_d = \left(\frac{4}{D+2} \right)^{1/(D+4)} \sigma_d N^{-1/(D+4)}. \tag{5.37}$$

The corresponding adaptive kernel estimator follows the discussion as for the univariate case. An issue in the scaling for the adaptive bandwidth arises when the multivariate data is approximately sampled from a lower dimensionality than the dimension D.

Figure 5.15 shows an example in which the sampling distribution has diagonal covariance matrix (locally and globally).

Applying kernel estimation to this distribution yields the results in Figure 5.16, shown for two different smoothing parameters.

For comparison, Figure 5.17 shows an example in which the sampling distribution has nondiagonal covariance matrix. Applying the same kernel estimation to this distribution gives the results shown in Figure 5.18. It may be observed (this is the same data as in Figure 5.15, just rotated to give a nondiagonal covariance matrix) that this is more difficult to handle.

5.12
Estimation Using Orthogonal Series

We may take an alternative approach, and imagine expanding the pdf in a series of orthonormal functions (not to be confused with the notion of orthogonality in the statistical sense of a diagonal covariance matrix):

$$f(x) = \sum_{k=0}^{\infty} a_k \psi_k(x), \qquad (5.38)$$

where

$$a_k = \int \psi_k(x) f(x) \rho(x) dx = E[\psi_k(x) \rho(x)], \qquad (5.39)$$

$$\int \psi_k(x) \psi_\ell(x) \rho(x) dx = \delta_{k\ell}, \qquad (5.40)$$

and $\rho(x)$ is a positive "weight function".

Since the expansion coefficients are expectation values of functions, it is natural to substitute sample averages as estimators for them. This corresponds to using

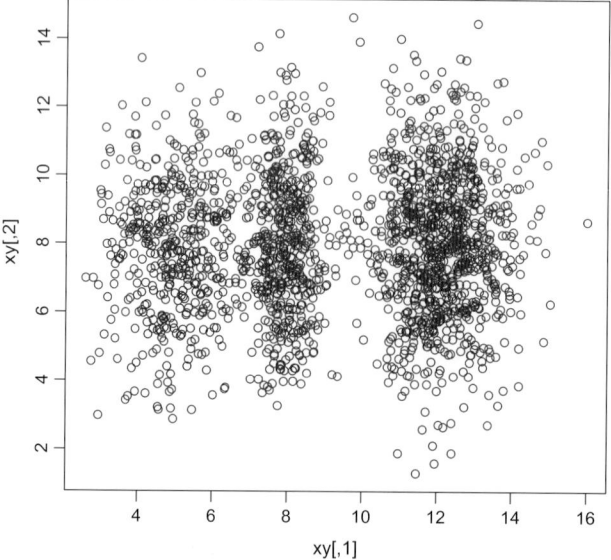

Figure 5.15 A two-dimensional distribution with diagonal covariance matrix.

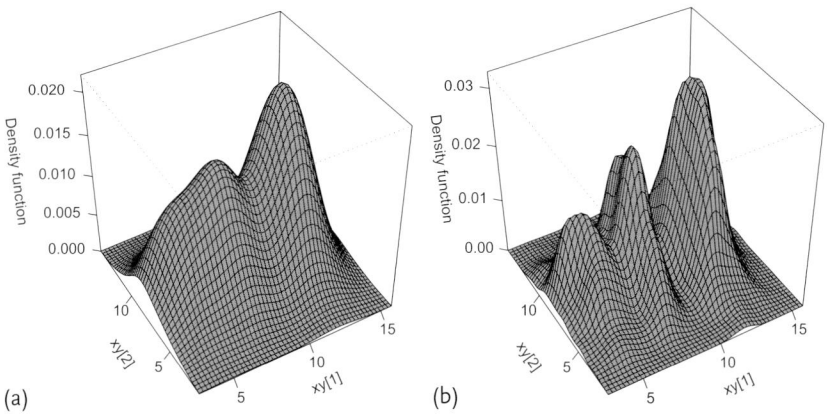

Figure 5.16 Kernel estimation applied to the two-dimensional data in Figure 5.15. (a) Default smoothing parameter. (b) Using one-half of the default smoothing parameter.

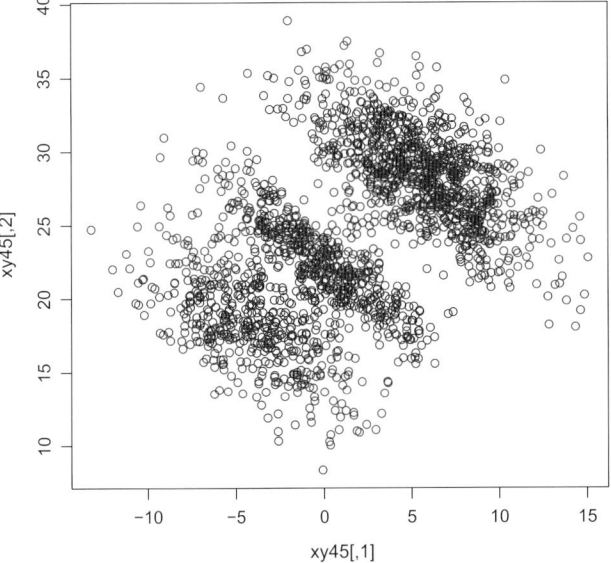

Figure 5.17 A two-dimensional distribution with nondiagonal covariance matrix.

the empirical probability distribution:

$$\hat{a}_k = \int \psi_k(x) \hat{f}(x) \rho(x) dx$$

$$= \int \psi_k(x) \frac{1}{N} \sum_{n=1}^{N} \delta(x - x_n) \rho(x) dx$$

$$= \frac{1}{N} \sum_{n=1}^{N} \psi_k(x_n) \rho(x_n) , \qquad (5.41)$$

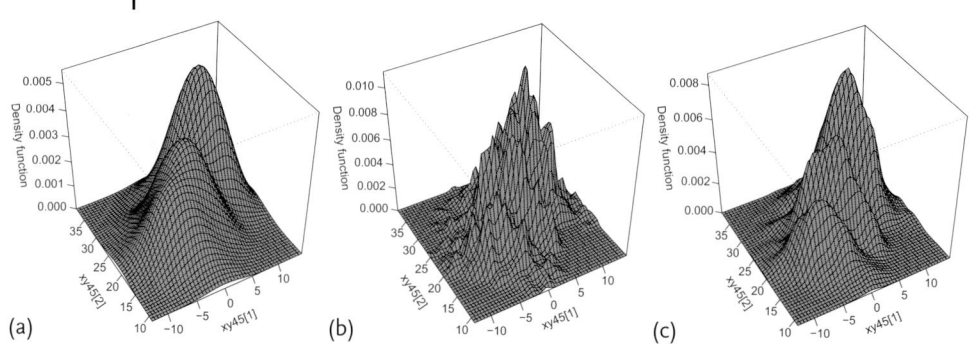

Figure 5.18 Kernel estimation applied to the two-dimensional data in Figure 5.17. (a) Default smoothing parameter. (b) Using one-half of the default smoothing parameter. (c) Intermediate smoothing.

and thus,

$$\hat{f}(x) = \sum_{k=1}^{m} \hat{a}_k \psi_k(x), \tag{5.42}$$

where the number of terms m is chosen by some optimization criterion.

Note the analogy between choosing m and choosing smoothing parameter w in kernel estimators, and between choosing K and choosing $\{\psi_k\}$. This method is often used in angular distributions, where the orthogonal functions are Legendre polynomials or spherical harmonics.

We may try an example in a two-dimensional sampling space. A dataset of size $N = 1000$ is generated according to density:

$$f(\cos\theta, \phi) = \frac{1}{4\pi}\left(1 + \frac{1}{2}\cos\theta + \frac{1}{2}\sin\theta\cos\phi\right). \tag{5.43}$$

A series in real linear combinations of $Y_{\ell m}$ spherical harmonics is used to fit this data, according to the prescription of (5.41). Since we know the true distribution, we may compute the error in our density estimate. This is shown in Figure 5.19. Figure 5.19a shows the fit with exactly the same angular functions in the series as used to generate the data. In Figure 5.19b, we have added two additional linearly independent terms, $\sin\theta\sin\phi$ and $\cos^2\theta$ (in the form of suitable additional orthonormal $Y_{\ell m}$s). The errors made become more serious when we add the terms. While we are fitting the empirical distribution more accurately with the additional freedom, we are actually doing a poorer job estimating the actual distribution. The extreme of this is the limit of keeping infinite terms in our expansion, in which we recover the empirical distribution.

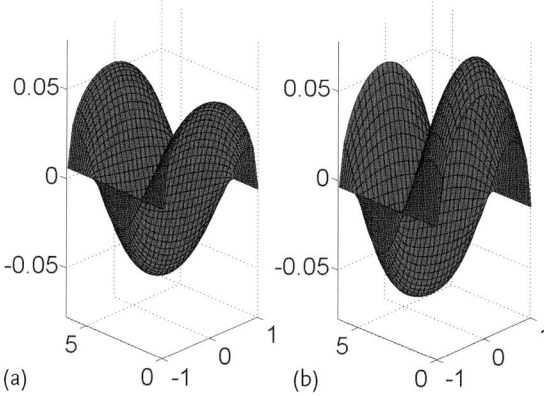

Figure 5.19 The error in the density estimate as a function of angular variables $\cos\theta, \phi$. (a) Error for a series that has the correct angular functions. (b) Error for a series that has extra terms.

5.13
Using Monte Carlo Models

We often build up a data model using Monte Carlo computations of different processes, which are added together to get the complete model. This may involve weighting of observations, if relatively more simulated data is generated for some processes than for others. The overall simulated empirical density distribution is then (x_n has components $x_n^{(d)}, d = 1, \ldots, D$):

$$\hat{f}(x) = \sum_{n=1}^{N} \rho_n \delta(x - x_n), \tag{5.44}$$

where $\sum \rho_n = 1$ (or N to correspond with a sample of some total size).

The weights must be included in computing the sample covariance matrix:

$$V_{k\ell} = \sum_{n=1}^{N} \rho_n \frac{(x_n^{(k)} - \hat{\mu}_k)(x_n^{(\ell)} - \hat{\mu}_\ell)}{\sum_j \rho_j}, \tag{5.45}$$

where $\hat{\mu}_d = \sum_n \rho_n x_n^{(d)} / \sum_j \rho_j$ is the sample mean in dimension d.

The simulated data is discrete. We may use kernel smoothing to obtain a continuous model. Assuming we have transformed to a diagonal system using the sample covariance matrix, the product kernel density based on our simulation is then

$$\hat{f}_0(x) = \frac{1}{\sum_j \rho_j} \sum_{n=1}^{N} \rho_n \prod_{d=1}^{D} \frac{1}{w_d} K\left(\frac{x^{(d)} - x_n^{(d)}}{w_d}\right). \tag{5.46}$$

This could be iterated to obtain an adaptive kernel estimator as discussed earlier.

5.14
Unfolding

We may not be satisfied with merely estimating the density from which our sample X was drawn. The interesting physics may be obscured (smeared) by transformation with uninteresting functions, for example efficiency dependencies, resolution, classification errors, or radiative corrections. This is very similar to the problem of reconstructing an image from a blurred representation (MATLAB provides several functions deconv... in this context). We refer to the problem as one of *unfolding* the interesting distribution from the sampled (i.e., smeared) distribution (e.g., Cowan, 1998). The term *deconvolution* is also used. Strictly speaking, the term "unfolding" is more general, not restricted to smearing in the form of a convolution ((5.55) below). A theoretical treatment of this subject appears in Meister (2009). Puetter et al. (2005) provides a review of methods of image reconstruction in astrophysics. As we shall see, there are pitfalls to unfolding, so if all you really want is a comparison of data with theory, it is better to smear the theory than to unfold the data.

We suppose that there is a "fundamental" sampling distribution, $f(x)$, and a transformation K that maps this distribution to the "experimental" distribution, $e(x)$:

$$e(x) = \int K(x, y) f(y) dy .\tag{5.47}$$

We sample from $e(x)$, but want to learn about $f(x)$. If $f(x)$ were known up to a small number of parameters, we would address this problem in the context of parametric statistics. Here we consider the context of nonparametric density estimation. We assume that the transformation kernel $K(x, y)$ is known. It is often called the *point spread function* because it gives the density produced at x by a point density "source" at y. It must be cautioned that this may not be exactly true. The kernel may itself be estimated from some auxiliary data. In that case, one must consider the uncertainties ("systematic errors") so introduced.

When we sample from e, we obtain an estimator $\hat{e}$ for density e. For example, $\hat{e}$ could be the empirical density estimator. In principle, the estimation of f from this is easy:

$$\hat{f}(y) = \int K^{-1}(y, x) \hat{e}(x) dx ,\tag{5.48}$$

where

$$\int K^{-1}(x, y) K(y, x') dy = \delta(x - x') .\tag{5.49}$$

If $\hat{e}(x) = \frac{1}{N} \sum_{n=1}^{N} \delta(x - x_n)$ is the empirical distribution, then we obtain

$$\hat{f}(y) = \frac{1}{N} \sum_{n=1}^{N} K^{-1}(y, x_n) .\tag{5.50}$$

If we don't know how to invert K, we may try an iterative (e.g., Neumann series) solution. For example, consider the problem of unfolding radiative corrections in a cross section measurement. The observed cross section, $\sigma_E(s)$, as a function of center-of-mass energy-squared s, is related to the "interesting" cross section σ according to

$$\sigma_E(s) = \sigma(s) + \delta\sigma(s), \tag{5.51}$$

where

$$\delta\sigma(s) = \int K(s, s')\sigma(s')ds'. \tag{5.52}$$

We form an iterative estimate for $\sigma(s)$ according to

$$\hat{\sigma}_0(s) = \sigma_E(s) \tag{5.53}$$

$$\hat{\sigma}_i(s) = \sigma_E(s) - \int K(s, s')\hat{\sigma}_{i-1}(s')ds', \quad i = 1, 2, \ldots \tag{5.54}$$

An important case is when measurement results in the addition of a stochastic error to a statistical sampling from the fundamental distribution. In this case, we are concerned with the addition of two independent random variables. If the fundamental distribution is f and the resolution density is g, then the distribution of the sum of the two contributions is just the convolution:

$$e(x) = \int g(x - y)f(y)dy. \tag{5.55}$$

That is, our kernel is a function of the difference between its arguments. This suggests working in Fourier transform space, since then the convolution simplifies to a product:

$$e_{FT}(t) = g_{FT}(t)f_{FT}(t), \tag{5.56}$$

where, for example,

$$f_{FT}(t) = \int e^{it \cdot x} f(x)dx \tag{5.57}$$

is the Fourier transform of $f(x)$.

To implement such an approach, we could use our data to form an empirical estimate $\hat{e}_{FT}$ of e_{FT}:

$$\hat{e}_{FT}(t) = \int e^{it \cdot x} \hat{e}(x)dx = \frac{1}{N}\sum_{n=1}^{N} e^{it \cdot x_n}. \tag{5.58}$$

With g_{FT} known, we simply apply (5.56) to obtain estimator $\hat{f}_{FT}$. Then we take the inverse Fourier transform to get $\hat{f}$. A difficulty emerges in this last step – we may need to do something (*regularization*) to make sure the inverse transform exists and is not erratic (i.e., sensitive to small changes in the data). We will see a related issue in our next example. This approach has long been used in fields such as image reconstruction.

We may equally well apply unfolding to a histogram with Poisson-distributed bin contents $x = x_1, \ldots, x_b$. Suppose the fundamental distribution is

$$f(x) = \prod_{i=1}^{b} \frac{\mu_i^{x_i} e^{-\mu_i}}{x_i!}, \quad x_i = 0, 1, \ldots \tag{5.59}$$

and the experimental distribution is

$$e(x) = \prod_{i=1}^{b} \frac{\nu_i^{x_i} e^{-\nu_i}}{x_i!}, \quad x_i = 0, 1, \ldots, \tag{5.60}$$

where the ν_i are related to μ by some transformation. Given a sampling x, the empirical (maximum likelihood) distribution is

$$\hat{e}(y) = \prod_{i=1}^{b} \frac{x_i^{y_i} e^{-x_i}}{y_i!}, \quad y_i = 0, 1, \ldots \tag{5.61}$$

Now suppose that the transformation is of the form

$$\nu = A\mu + B. \tag{5.62}$$

Since A is a matrix, this includes the possibility that events get assigned to the wrong bin, as well as possible inefficiencies. The B term allows for possible "background" contributions. Our estimate for ν is x. Thus, assuming A is invertible, we estimate the unfolded distribution according to

$$\hat{\mu} = A^{-1}(\hat{\nu} - B). \tag{5.63}$$

In practice it is better to use Gaussian elimination or other methods (for example, the Moore–Penrose pseudo-inverse) when solving numerically. This provides an unbiased estimator for μ, in fact the maximum likelihood estimator, which is minimum variance since our distribution is in the exponential family.

Let us try an example, illustrated in Figure 5.20. We consider in this example a fundamental distribution with three Gaussian peaks, and a total expected sample size of 5000. Figure 5.20a shows the mean counts expected in each bin for the fundamental distribution. The fundamental distribution is modified to the experimental distribution with expected counts shown in 5.20b. The matrix A was taken to be

$$A_{ij} = 0.3/2^{|i-j|}, \tag{5.64}$$

mimicking a situation with an overall inefficiency plus migration among bins, where the migration probability decreases as bins are further apart. The "background" vector B is taken to be a vector with the number five in each position.

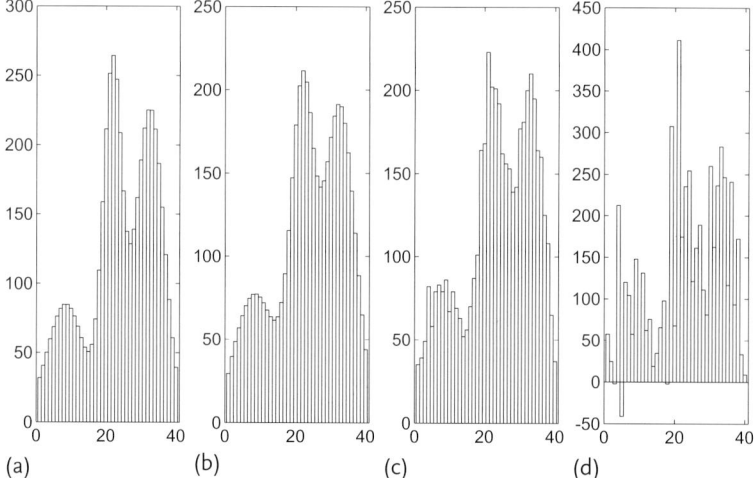

Figure 5.20 Illustration of unfolding a histogram. (a) The fundamental expected bin contents. (b) The expected bin contents after passing through the detector and analysis. (c) A sampling from (b), according to Poisson statistics. (d) The unfolded distribution using the sampling in (c).

Figure 5.20c shows a sampling from the distribution of Figure 5.20b, according to Poisson statistics. Finally, Figure 5.20d shows the result of applying our unfolding procedure to the data in Figure 5.20c.

Figure 5.20c is our (maximum likelihood) estimate for the expectation values in Figure 5.20b, and the unfolded Figure 5.20d is our estimate for the expectation values in Figure 5.20a (also maximum likelihood). While Figure 5.20c looks like a familiar sort of estimate for Figure 5.20b, Figure 5.20d is comparatively disappointing as an estimate for Figure 5.20a. Figure 5.20d replicates the basic features of Figure 5.20a, but the bin-to-bin fluctuations will be disconcerting to most readers. Rest assured that these are indeed minimum variance unbiased estimators for the expectations in Figure 5.20a. There is nothing wrong with our math.

In spite of its correctness, many people will conclude that the unfolding is unsatisfactory because of the large fluctuations. What is the problem? We are paying a large price in variance for an unbiased result. The effect of the transformation to the experimental distribution is to smooth the data, as may be seen comparing Figure 5.20b with Figure 5.20a. This tends to introduce bias and reduce variance. We take a data sample from the smoothed distribution, and apply the "unsmoothing" transformation. The smallish fluctuations in the dataset are amplified in the process. We eliminate the bias, but increase the variance.

It is often desirable to obtain a result with less erratic behavior, essentially taking into account our expectation that the true sampling distribution has some level of smoothness. We may accomplish this by accepting some bias. We apply some smoothing or interpolation in our unfolding process. This is referred to as *regularization* in this context.

Another remark will be helpful. In our histogram example, we might have imagined that our experiment corresponds to a sampling x from the fundamental distribution f, which is transformed to the data we actually observe with $y = Ax + B$. Of course, if this is correct, we can simply compute $x = A^{-1}(y - B)$ and exactly recover the sampling x from the fundamental distribution and there is no additional variance introduced. However, this is not correct. Our sampling is from the transformed distribution e, with whatever fluctuations that allows. To make this obvious, imagine a transformation that takes an expected signal of one million events down to an expectation of just ten events. We must deal with the statistics of ten instead of a million. Effectively, the transformation has introduced noise into the measurement.

5.14.1
Unfolding: Regularization

There are many possible approaches to regularization in unfolding. All represent ways to trade bias for variance to achieve some optimal balance, perhaps as measured by MISE. The smoothing techniques such as kernels or orthogonal series that we have discussed already in this chapter are suitable candidates for regularization.

For example, let us consider the use of kernel density estimation in the context of a convolution kernel and the Fourier transform approach. Instead of the empirical density estimate, we use kernel estimator

$$\hat{e}'(x) = \frac{1}{Nw} \sum_{n=1}^{N} K\left(\frac{x - x_n}{w}\right). \tag{5.65}$$

The Fourier transform is

$$\hat{e}'_{FT}(t) = \frac{1}{Nw} \sum_{n=1}^{N} \int e^{itx} K\left(\frac{x - x_n}{w}\right)$$

$$= \hat{e}_{FT}(wt) K_{FT}(wt), \tag{5.66}$$

where e_{FT} is the empirical Fourier transform estimator of (5.58). Thus, our smoothed estimator for $f(x)$ becomes

$$\hat{f}'(x) = \frac{1}{2\pi N} \sum_{n=1}^{N} \int e^{-it(x-x_n)} \frac{K_{FT}(wt)}{g_{FT}(t)} dt. \tag{5.67}$$

This is known as the *standard deconvolution kernel density estimator* (Meister, 2009). The kernel K may be chosen to ensure that this transform exists, at least if $g_{FT}(t) \neq 0$. For example, we may achieve this with a kernel whose Fourier transform is bounded with compact support. A promising choice is $K(x) = \sin x/\pi x$, with transform $K_{FT}(t) = 1$ if $t \in [-1, 1]$ and zero otherwise.

An alternative and popular approach is to add a penalty function to the optimization problem. In the context of a least-squares minimization, where we find $\boldsymbol{\mu} = \hat{\boldsymbol{\mu}}$ that minimizes $(A\boldsymbol{\mu} + B - x)^2$, we may add a term and instead minimize:

$$(A\boldsymbol{\mu} + B - x)^2 + \lambda^2 (L\boldsymbol{\mu})^2, \tag{5.68}$$

where L is some linear operator that measures lack of smoothness in $\boldsymbol{\mu}$ and λ is a regularization or smoothing parameter. Larger values of λ result in smoother density estimates. This technique is usually referred to as *Tikhonov regularization*. As derivatives measure (lack of) smoothness, L may suitably be chosen as a differential operator.

For example, we return to our histogram example. Since this involves discrete binning, our derivative operator becomes a discrete approximation. The simplest choices, for first- and second-derivatives, are [D_1 is $(N-1)\times N$ and D_2 is $(N-2)\times N$]:

$$D_1 = \begin{pmatrix} 1 & -1 & 0 & 0 & \cdots & 0 \\ 0 & 1 & -1 & 0 & \cdots & 0 \\ \vdots & & & & & \vdots \\ 0 & \cdots & 0 & 0 & 1 & -1 \end{pmatrix},$$

$$D_2 = \begin{pmatrix} -1 & 2 & -1 & 0 & \cdots & 0 \\ 0 & -1 & 2 & -1 & \cdots & 0 \\ \vdots & & & & & \vdots \\ 0 & \cdots & 0 & -1 & 2 & -1 \end{pmatrix} \tag{5.69}$$

Figure 5.21 shows the results of applying Tikhonov regularization to our sample histogram, both for first and second derivative operators, and for two values of the regularization parameter λ. We see that this technique substantially reduces the variance from the unregulated estimator, while adding bias. At least in this case, the choice of regulator is not very important, but there is a clear, and plausible, dependence on the parameter λ. The parameter λ may be chosen, for example, to minimize the MISE, or an estimate thereof. Although it did not happen in this case, Tikhonov regularization permits estimators with negative values (if we had picked smaller values for λ, we would have seen this). This isn't bad or wrong, but may present a difficulty if it is intended to use the estimator as a density in further sampling, for example in Monte Carlo simulations.

The unfolding problem may also be addressed from a Bayesian perspective. We may have some prior belief $\pi(f)$ concerning f. As prior belief relates f at different points, the prior contributes a smoothing effect in forming the posterior distribution, given by $P(f|x) \propto P(x|f)\pi(f)$. Let us try this out for the case of a histogram. We wish to estimate the expected bin contents $\boldsymbol{\mu}$. We pick as the best estimates those for which the posterior distribution is maximal:

$$P(\hat{\boldsymbol{\mu}}|x) = \max_{\boldsymbol{\mu}} P(x|\boldsymbol{\mu})\pi(\boldsymbol{\mu}). \tag{5.70}$$

Notice that, without the prior function π, this just gives us the maximum likelihood estimator that we obtained earlier in (5.63). The prior provides regulation.

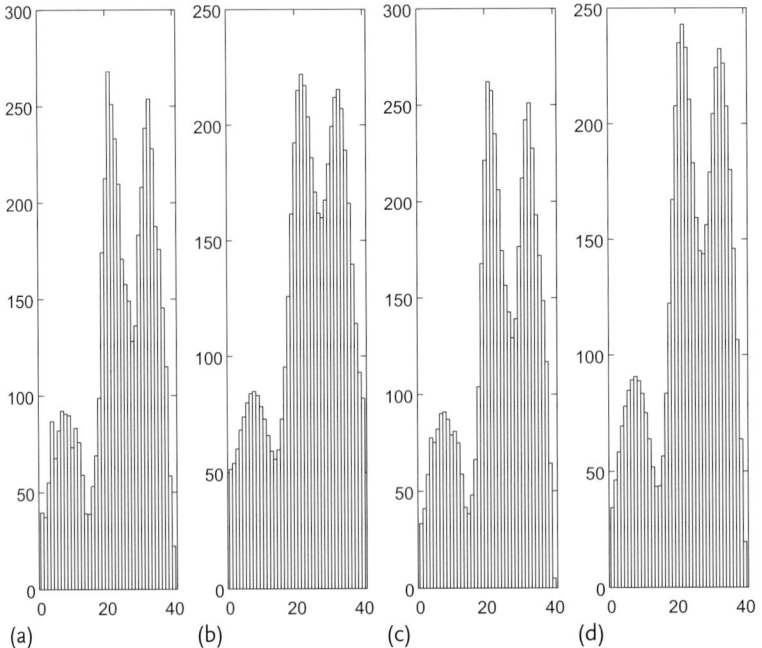

Figure 5.21 Illustration of unfolding a histogram with Tikhonov regularization: (a) and (b) use first derivative smoothing; (c) and (d) use the second derivative; (a) and (c) use smoothing parameter $\lambda = 0.2$; and (b) and (d) use $\lambda = 1$.

If we have some principled choice for π, we can insert it and use it. But what if we wish to start with a notion of "complete ignorance"? One view of complete ignorance is that a sampling is just as likely to land in one bin as any other. That is, the probability to land in bin i is

$$p_i = \frac{\mu_i}{\mu_T} = p_j = \frac{\mu_j}{\mu_T} = p. \tag{5.71}$$

Thus, our prior distribution for $\boldsymbol{\mu}$ is

$$\pi(\boldsymbol{\mu}) = \frac{\mu_T!}{\prod_{i=1}^{b} \mu_i!} \frac{1}{b^{\mu_T}}, \tag{5.72}$$

where $\mu_T = \sum_{i=1}^{b} \mu_i$.

It is instructive to introduce here the notion of *entropy*, defined (in the notation of the case at hand) by

$$H = -\sum_{i=1}^{b} p_i \log p_i. \tag{5.73}$$

With the constraint $\sum_{i=1}^{b} p_i = 1$, this is maximized when $p_i = p, i = 1, \ldots, b$. We recover our notion of complete ignorance as the notion of the distribution with maximum entropy.

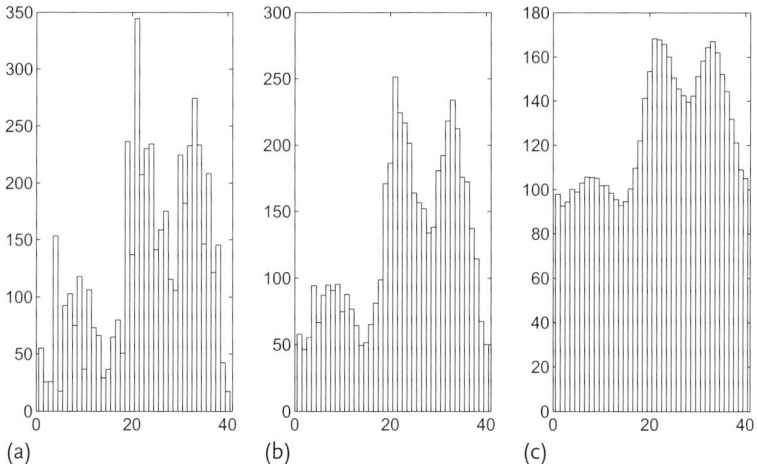

Figure 5.22 Illustration of unfolding a histogram with entropy regularization: (a) $\lambda = 0.01$; (b) $\lambda = 0.1$; (c) $\lambda = 1$;.

We apply this technique to our three-peak spectrum, with the result in Figure 5.22c. Specifically, we have maximized (dropping terms independent of $\boldsymbol{\mu}$):

$$\log L(\boldsymbol{\mu}; \boldsymbol{x}) + \log \pi(\boldsymbol{\mu}) = \sum_{i=1}^{b}(x_i \log \nu_i - \nu_i) + \log \Gamma(\mu_T + 1)$$

$$- \sum_{i=1}^{b} \log \Gamma(\mu_i + 1) - \mu_T \log b . \quad (5.74)$$

We see that the smoothing works, but is rather smoother than we probably wanted. Unless one wants to insist that this is the "correct prior", things are getting pulled too far towards the ignorance prior (or towards maximum entropy). We can easily salvage the situation if we are willing to give up on a strict Bayesian interpretation. We accomplish this by multiplying the prior (or the entropy, if we use that for our regulating function) by a regularization parameter, λ. That is, we find $\boldsymbol{\mu}$ maximizing:

$$\log L(\boldsymbol{\mu}; \boldsymbol{x}) + \lambda \log \pi(\boldsymbol{\mu}) . \quad (5.75)$$

The results for $\lambda = 0.01$ and $\lambda = 0.1$ are shown in Figure 5.22a and b, respectively. As before, the value of λ may be optimized to minimize the MISE, in the context of some reasonable approximate model.

A way to check and possibly tune the unfolding procedure is to use the estimated unfolded distribution as the "actual" distribution in a simulated experiment. Then the unfolding procedure can be used on the simulated data and compared with the distribution used in the simulation.

It is possible to imagine refinements, such as adaptive regularization. For example, in image reconstruction there may be true sharp edges that shouldn't be smoothed the same as softer regions. However, unfolding has its pitfalls, and caution should be exercised lest one creates a significant-looking peak where none really exists. It should also be kept in mind that the result of unfolding includes correlations. In an unfolded histogram the bin contents are not independent random variables.

5.15
Exercises

1. Demonstrate the claim in (5.22).
2. To get a quantitative impression for the performance of histogram binning choices, make a plot of the MISE for a sample of size $N = 1000$ as a function of histogram bin width w. Do this for sampling from a normal distribution and for sampling from a Cauchy distribution on a finite interval. Indicate on your graph the expected value of w according to Scott's rule.
3. Demonstrate (5.29).
4. Demonstrate the result in (5.66).

References

Beringer, J. et al. (2012) Review of particle physics (rpp). *Phys. Rev. D*, **86**, 010001, doi:10.1103/PhysRevD.86.010001.

Beyer, K., Goldstein, J., Ramakrishnan, R., and Shaft, U. (1999) *When Is "Nearest Neighbor" Meaningful?*, Lecture Notes in Computer Science, vol. 1540, Springer, pp. 217–235.

Cowan, G. (1998) *Statistical Data Analysis*, Oxford University Press.

Cranmer, K.S. (2001) Kernel estimation in high-energy physics. *Comput. Phys. Commun.*, **136**, 198–207, doi:10.1016/S0010-4655(00)00243-5.

Hyndman, R. (1995) The problem with Sturges rule for constructing histograms. http://www-personal.buseco.monash.edu.au/~hyndman/papers/sturges.pdf (accessed 25 July 2013).

Koppen, M. (2002) The curse of dimensionality. *Most*, **1** (34), e723–4.

Meister, A. (2009) *Deconvolution Problems in Nonparametric Statistics*, Springer.

Puetter, R.C., Gosnell, T.R., and Yahill, A. (2005) Digital image reconstruction: Deblurring and denoising. *Annu. Rev. Astron. Astrophys.*, *43*, 139–194.

Rosenblatt, M. (1956) Remarks on some nonparametric estimates of a density function. *Ann. Math. Stat.*, *27* (3), 832–837.

Scott, D.W. (1992) *Multivariate Density Estimation*, John Wiley & Sons, Inc.

Taylor, C.C. (1989) Bootstrap choice of the smoothing parameter in kernel density estimation. *Biometrika*, *76*, 705–712.

Verkerke, W. and Kirkby, D. (2006) Roofit users manual v2.07. http://roofit.sourceforge.net/docs/RooFit_Users_Manual_2.07-29.pdf (accessed 25 July 2013).

6
Basic Concepts and Definitions of Machine Learning

Wikipedia defines *machine learning* as "a scientific discipline concerned with the design and development of algorithms that allow computers to evolve behaviors based on empirical data, such as from sensor data or databases." Although not incorrect, this definition would not tell you what this book is about.

We view "machine learning" as a list of subjects. Some of these subjects are well known among physicists. Others are known to some extent, perhaps in one specific context. Others are not known at all. This book aims at explaining the seemingly well-known subjects and introducing less known material. We draw a line in the sand between less known algorithms with great potential for modern data analysis in physics and less known algorithms which may not be appreciated by the physics community anytime soon. Algorithms of the latter kind are not included in this book.

The two most popular machine learning applications in physics are classification and density estimation. These are the main subjects of this book. The material in this section briefly describes the broader domain of machine learning.

6.1
Supervised, Unsupervised, and Semi-Supervised

Suppose we have a set of N multivariate observations $\mathbf{X} = \{x_n\}_{n=1}^{N}$. The goal of *unsupervised learning* is to discover interesting structures in the data. Density estimation, reviewed in the previous chapters, represents the strongest form of unsupervised learning. It aims at modeling the full distribution in the multivariate space. Clustering (not included in this book) and bump hunting are examples of weaker unsupervised analysis. The goal of clustering is to identify groups of similar observations in $\mathbf{X}$, and the goal of bump hunting is to find a group of observations in $\mathbf{X}$ most inconsistent with the expected distribution. Clustering algorithms such as, for instance, k-means have been occasionally used in analysis of astrophysics and particle physics data. Bump hunting, including nonparametric techniques, has gained popularity in new signal searches.

Suppose we also know categories, or classes, for the observations in $\mathbf{X}$. For supervised learning, we assume that the data are drawn from an unknown distribution

$P(x, y)$. We use the known class labels, $y = \{y_n\}_{n=1}^N$, and the known **X** to build a predictive model $y = f(x)$. A good model should predict well not just for the data it is built on (training data), but for any data drawn from the distribution $P(x, y)$. To measure disagreement between the true class and model prediction, we choose a loss function, $\ell(y, f(x))$. The goal of *supervised learning* is then to minimize the expected loss $E_{X,Y}\ell(Y, f(X))$. Since the true distribution is not known, this expectation cannot be computed. In practice, we estimate the expected loss by computing the empirical loss on data not used for training such as, for instance, an independently obtained test set. If $f(x)$ is a consistent estimator of y, the empirical loss approaches the expected loss asymptotically for large N.

Semi-supervised learning (SSL) is between supervised and unsupervised. The true class labels are known only for a fraction of observations in **X**. In real-world applications, this fraction can vary from almost zero to almost one. Ideally, an SSL algorithm should provide a smooth transition from one edge case to the other.

It may be tempting to approach SSL from either the unsupervised or supervised perspective. In the first approach, we learn without supervision and then use the known labels to assign classes to the discovered structures. In the second approach, we learn using only observations with known class labels and then label the rest according to the obtained boundaries between the classes. Either approach can be successful in certain conditions; one of such conditions would be the fraction of known labels. Clearly, there are better ways of accommodating available information than reducing the problem to one of the two edge cases.

Consider, for example, data grouped into several multivariate normal distributions. In the fully supervised case, this problem is most efficiently solved by discriminant analysis described in Section 11.1. In the fully unsupervised case, this problem is solved by finding an appropriate Gaussian mixture of distributions. This clustering algorithm, described in many textbooks, works as follows. Assume a fixed number of components in the mixture, express the likelihood as a function of their means and covariance matrices, and optimize the likelihood by an expectation-maximization (EM) algorithm. The same approach works in the semi-supervised setting as long as the optimization algorithm keeps observations from the same class in the same Gaussian component. Labeled observations are used to estimate the number of classes and their distributions, and unlabeled observations are used to improve these distribution estimates.

In the example above, we include unlabeled observations in the training set. Our goal is, again, to build a model minimizing the expected loss, but the problem is different from that of supervised learning because the training set contains observations without known y. Alternatively, we could assume one of the allowed labels for every unlabeled observation in **X** and minimize the empirical loss for the new fully labeled training set. Then we could assume a different configuration of labels for the unlabeled observations in **X** and minimize the empirical loss for this training set. And so on. For K classes and U unlabeled unique observations in **X**, we would end up with K^U models. Then we would select the one with the minimal empirical loss. This approach is called *transductive learning*.

This transductive learning algorithm and the SSL algorithm with EM can be viewed as related. Indeed, minimizing the empirical loss is similar to maximizing the likelihood, and the EM algorithm optimizes class memberships for unlabeled observations too. Beware of an important philosophical divide between the two methodologies. In the traditional inductive approach, we minimize the expected loss over the entire domain of $P(x, y)$. In transductive learning, we optimize the predictive power for the *observed* unlabeled data. We thus do not care about classifying data that *could be* observed. Similar lines of reasoning can be seen in the context of frequentist and Bayesian approaches to parameter estimation. A frequentist constructs a confidence interval for the parameter by imagining an infinite set of independent identically distributed measurements. A Bayesian constructs a credible interval using the actual measurement only.

If transductive learning were to be applied to physics data today, it would likely raise strong objections from some physicists, especially those who religiously practice blind analysis. According to the philosophy of blind analysis, one should not attempt to measure the real signal before polishing the analysis routine on simulated data. We refrain from taking sides on the issue and merely note that there was a time when some modern analysis techniques applied to physics data were not kept in high regard either.

Unsupervised and supervised learning are described in many textbooks. Chapelle *et al.* (2006) give a decent overview of the subject but offer enough technical material unrelated to physics analysis. To the curious reader, we recommend the preface, the article on the taxonomy for SSL methods by Seeger, and contributions on transductive support vector machines by Joachims and Vapnik.

6.2
Tall and Wide Data

Modern statistics textbooks often explain material in matrix notation. It is customary to collect input data in a matrix **X**. Here, we have two obvious choices: store observations in rows and variables in columns, or do the opposite. This is where statistics textbooks differ. We choose the first option. Our **X** is of size $N \times D$ for N observations and D variables. "Observations" are often referred to as "events" by physicists and "cases" or "examples" by statisticians. "Variables" are often called "predictors" by statisticians and "features" or "attributes" by researchers in machine learning.

Physics data typically contain many more observations than variables. For $N \gg D$, the matrix **X** is tall. This is not necessarily the case in other disciplines. In DNA microarray analysis, experimental samples are typically formed of a few dozen subjects, either healthy or diagnosed with some disease. Input variables are thousands of microarray probes exposing various gene segments. In this case, D (thousands) is much greater than N (a few dozen), and the matrix **X** is wide.

In this book, we hardly describe analysis techniques for wide data because this is an unlikely scenario for physics analysis. Some techniques can be extended to wide data in a straightforward manner. Others require nontrivial modifications. Decision tree, for instance, can be grown on wide data using the same algorithm as for tall data. In contrast, application of linear discriminant analysis requires inversion of the pooled-in covariance matrix. This matrix is of size $D \times D$ and at most of rank N; therefore, it must be singular for wide data. Some solutions to this problem are discussed in Section 11.1.5.

6.3
Batch and Online Learning

A particle physics detector processes data continuously in controlled runs. Every event is processed by detector components and either discarded for the lack of interesting information or stored for future analysis. This choice, discard or store, is made by hardware and software triggers. This decision must be made in a short period of time. Modern particle detectors deal with unprecedented rates of data collection, and these rates keep growing from one generation of detectors to the next. For example, the Compact Muon Solenoid must record a proton–antiproton collision every 25 nanoseconds during normal operation. If a software trigger is used, it must operate under strict memory constraints. Algorithms used for triggering must be simple, memory-efficient, and fast.

The stored data are then analyzed in big chunks. For example, the BaBar Collaboration searched for the rare leptonic decay $B^+ \to K^+ \nu \bar{\nu}$ using 351 million $B\bar{B}$ pairs collected in Runs 1–5. Then the BaBar Collaboration searched again for this decay after adding 108 million $B\bar{B}$ pairs from Run 6. In either case, the analysis was applied to the collected data as a whole. The search for the $B^+ \to K^+ \nu \bar{\nu}$ decay uses a multistage approach; at every step, a large fraction of background is removed to retain a few signal candidates, with the complexity of the selection algorithm increasing from one step to the next. Although the selection criteria for separating signal candidates from background are similar for the two analyses, they differ, even for the data from Runs 1–5 shared between the two samples. Some variance in selection criteria is typical when the analysis is carried out by different people, as is the case here. Besides, the later analysis is based on improved simulation of the detector and better particle reconstruction. The two analyses were completed one year apart, and each had been worked on for two or more years.

These two examples represent the two extreme regimes of data analysis in particle physics: *online* processing and *batch* learning. We describe the online regime as "processing" because machine learning does not occur. For example, the mentioned trigger algorithms do not evolve by themselves. Physicists can adjust the trigger settings to optimize collection or rejection of events of a certain type, but every time this adjustment is done by a human. In the batch regime, physicists often trust sophisticated algorithms such as neural nets or boosted decision trees, that is, apply machine learning.

The described distinction between learning and processing is subjective. Indeed, a trigger can be programmed to adjust data collection rates automatically in response to the changing environment. These adjustments could be simple "if-then" statements conditioned on various parameters of the detector; yet these statements can be viewed as elements of a self-learning algorithm. On the other hand, the trained neural net is typically scrutinized by the analyst and perhaps re-trained several times until a desired effect is achieved. Nevertheless, online and batch learning represent two different paradigms. The key notion behind online learning is a fast incremental upgrade. The algorithm needs to update the learned information using the small bit of data collected most recently, just in time before the next bit of data is received.

Some algorithms can be used for batch and online learning with equal ease. Neural net with backpropagation described in Chapter 12 is one of such algorithms. When a new observation passes through layers of a net, the net updates its link weights and node bias values. You can stop and resume sending observations to the neural net at any time.

In contrast, decision tree described in Chapter 14 is a batch algorithm by design. To find the optimal decision split on a variable, the tree needs to consider all available data at once. Updating the split after new data are received is possible but not straightforward. Decision tree extensions for online learning (Hulten *et al.*, 2001) came to light years after the pioneering book by Breiman *et al.* (1984) on decision trees was published.

Software suites such as MATLAB, R, StatPatternRecognition and TMVA generally provide only algorithms for batch learning. Off-the-shelf software for online learning is hard to find because it is run as part of dedicated online processing systems.

6.4
Parallel Learning

Algorithms reviewed in this book were developed for a single processor. The most common modern implementation of a machine learning algorithm loads data into computer memory and then processes data entries one by one. Modern computers with multiple cores are capable, thanks to the demise of frequency scaling, of executing computationally demanding parallel tasks. You can get even more power by utilizing multiple-node clusters. Parallelism can help, to a degree, by speeding up execution of simple sub-tasks. For example, imagine that your algorithm at a certain step needs to compute the mean of every input variable. This part can be easily parallelized. Perhaps, it already is. If the utility for computing means can run in a parallel mode and if your data are in a structure suitable for parallel computation, the execution can be carried out in parallel without any effort on your part. The speed-up from this low-level parallelization can be minor. To get the full benefits of parallel learning for complex algorithms, you need to find their parallelized implementations compatible with your computing environment.

We define *parallel learning* as executing a single learning task, that is, learning a model on input data, in parallel. Computing clusters let you fire off many data processing jobs at once, with every job executing a single-processor algorithm. This is not "parallel learning."

We discuss neither hardware platforms nor software tools for parallel computing. We briefly discuss conceptual approaches to parallelizing machine learning algorithms, mostly for classification.

Usually, parallel learning consists of four steps:

1. Partition the data and send each partition to a separate processing unit.
2. Run a learning algorithm on every partition.
3. Optionally share the learned information across the partitions.
4. Combine the learned information from all units into a final model.

Steps 1 through 3 can be repeated many times. Every step can be implemented in various ways. Not surprisingly, there are numerous realizations of this strategy.

For classification, training data has N observations, D variables, and K classes. Each of these components can be used for partitioning. These partitions can overlap or not. They can be drawn at random or using some pre-learned knowledge of the data structure. The partitions can contain unique observations only or they can be drawn by sampling with replacement. One simple strategy, used for cross-validation, is to partition N observations randomly into disjoint subsets of equal size. Random subspace selection partitions D variables, often into subsets with dimensionality much lower than that of the full data. Reducing multiple classes to a set of binary problems, in particular error correcting output code (ECOC) reviewed in Chapter 16, provides schemes for class-based partitioning. Division can be based on pre-learned information obtained by clustering data, across N or across D, or both. Every subset of the data then gets a fixed fraction of every cluster and of every class.

Steps 2 and 3 are specific to the parallelized algorithm, and it is hard to describe them in general terms. It is easy to parallelize algorithms capable of learning on subsets without exchanging information across the learners. An example of such an algorithm is bagging, or bootstrap aggregation. Every unit learns on a bootstrap replica of the full data obtained by sampling N out of N with replacement. If the data are big, you can sample rN out of N for $0 < r < 1$. Step 4 for bagging reduces to simple averaging over the learners independently obtained in step 2. In contrast, boosting is inherently sequential as it needs to adjust weights of the training observations based on the prediction accuracy for the learners grown so far. Parallelization schemes for boosting inevitably include exchanging information across the partitions in step 3. Boosting and bagging are described in Chapter 15.

Step 4, usually described as *meta learning*, combines learners from the individual partitions into a single model. This combining can be as simple as taking the majority vote across all learners. Alternatively, one could take a weighted majority vote in which the weights are determined from the learner accuracies. Another approach is *stacked generalization* which involves training a global classifier on the individual learner predictions. Yet another alternative is *arbiter trees*. An arbiter tree is grown bottom up. At the first level, arrange all individual learners in pairs. For every pair of learners, merge the datasets used for their training and compute the predictions for both learners on the merged set. Train an arbiter on these predictions. Now arrange all arbiters trained at the first level in pairs and apply the same procedure. Continue recursively until the tree is collapsed to its root.

In Section 15.2, we discuss bagging and random subspace, two popular algorithms for ensemble learning. Parallel learning and these algorithms have a lot in common. Otherwise we refrain from discussing parallelization through the rest of the book. For a good review of similar approaches, we recommend Rokach (2010), especially sections on ensemble diversity, mixture of experts, and meta learning.

6.5
Classification and Regression

Supervised learning comes in two flavors, classification and regression. Classification aims at separating observations of different kinds such as signal and background. The fitted function in this case takes nominal values. Fitting a function with continuous values is addressed by regression. Fitting a scalar function of a scalar argument by least-squares is a well-known regression tool. In case of a vector argument, classification appears to be used more often in modern physics analysis.

In contrast, statisticians and practitioners in other fields use regression a lot. Linear multiple regression (univariate response) and linear multivariate regression (multivariate response) are common analysis tools. Nonlinear regression has been a popular subject in recent machine learning research.

Many real-world problems can be cast as either classification or regression, and sometimes it is hard to know in advance what formulation is most efficient. Machine learning algorithms often come in two flavors, one for regression and one for classification, differing in minor implementation details. Sometimes there is no distinction between the two formulations – they describe the same learning algorithm.

Consider, for example, a binary decision tree. At every node, the tree finds the optimal decision, splitting this node in two children. Suppose we grow a decision tree to separate signal from background. This is a classification problem with two classes. Every split is found by minimizing the sum of the Gini indices from the two child nodes weighted by the node probabilities. The Gini index is defined as $2pq$, where p is the fraction of observations of one class and q is the fraction of observations of the other class. The Gini index for a node with s signal and b background observations is $2sb/(s+b)^2$.

Let us cast this problem as regression. Take the signal response to be $+1$ and the background response to be -1. A regression tree finds splits by minimizing the mean squared error (MSE), $\sum_{n=1}^{N}(y_n - \bar{y})^2/N$. The mean response per node with s signal and b background observations is $(s-b)/(s+b)$. The MSE for this node is then

$$\frac{s\left(1 - \frac{s-b}{s+b}\right)^2 + b\left(-1 - \frac{s-b}{s+b}\right)^2}{s+b}.$$

After some algebra, this expression simplifies to $2sb/(s+b)^2$. The optimal split minimizes the sum of the MSE values weighted by the node probabilities. We obtain the same optimization criterion as the one for classification tree.

We have established that a classification tree optimizing the Gini index is equivalent to a regression tree optimizing MSE for the $\{-1, +1\}$ response. This equivalence of classification and regression holds for relatively few problems. If we replaced the Gini index with cross-entropy, $p \log p + q \log q$, or if we replaced MSE with mean absolute deviation (MAD), $\sum_{n=1}^{N} |y_n - \bar{y}|/N$, the two algorithms would be no longer equivalent.

We focus on classification through the rest of the book. We describe a few regression techniques as well, because these techniques can be used for classification. If you are looking for in-depth coverage of regression, two classic books, Seber and Lee (2003) and McCullagh and Nelder (1983), are a good place to start.

References

Breiman, L., Friedman, J., Stone, C., and Olshen, R. (1984) *Classification and Regression Trees*, Chapman & Hall.

Chapelle, O., Scholkopf, B., and Zien, A. (eds) (2006) *Semi-Supervised Learning*, The MIT Press.

Hulten, G., Spencer, L., and Domingos, P. (2001) Mining time-changing data streams, in *Proceedings of the seventh ACM SIGKDD international conference on Knowledge discovery and data mining*, ACM New York, NY.

McCullagh, P. and Nelder, J. (1983) *Generalized Linear Models*, Chapman & Hall.

Rokach, L. (2010) *Pattern Classification Using Ensemble Methods, Machine Perception and Artificial Intelligence*, vol. 75, World Scientific.

Seber, G. and Lee, A. (2003) *Linear Regression Analysis*, Wiley Series in Probability and Statistics, Wiley-Interscience, 2nd edn.

7
Data Preprocessing

Before learning can begin, data need to be cleaned up and organized in a suitable way. Three commonly encountered issues are categorical variables, data with missing values, and outliers. These issues may be addressed by preprocessing the data. Each categorical variable can be represented by a vector numeric variable. Missing values can be deleted or replaced by hypothesized common values. Outlier observations can be deleted or downweighted to reduce their influence.

Sometimes such preprocessing steps do not suffice. In that case, treatment of categorical variables, missing values, or outliers becomes an essential part of data modeling.

In this chapter, we discuss both techniques applicable at the preprocessing stage and techniques integrated with predictive modeling.

7.1
Categorical Variables

Numeric or continuous variables, most common in physics analysis, are defined on an interval scale. This scale allows comparison operations such as $x_1 < x_2$ and computation of meaningful intervals: x_2 is twice as far from x_0 as x_1 is if $|x_2 - x_0|/|x_1 - x_0| = 2$. In contrast, distance between values of a categorical variable is undefined.

There are two types of categorical variables: *ordinal* and *nominal*. The ordinal scale allows greater, less, and equal comparison. The nominal scale only allows a test of equality. An example of a nominal variable is classification of events in an electron–positron collider: Bhabha, muon-pair, $B\bar{B}$, and so on. An example of an ordinal variable is classification of charged tracks into sorted categories. For instance, the particle identification analysis described in Bevan *et al.* (2013) aims at separating particles of 4 types: e, K, π, and p. For each particle type, six identification categories are considered: super loose, very loose, loose, tight, very tight and super tight. Note that the distance between "tight" and "very tight" cannot be compared to distance between "very tight" and "super tight" in a meaningful way.

We distinguish between two types of variables: predictors and response. For numerical analysis, categorical predictors are often converted to numeric vectors or

Statistical Analysis Techniques in Particle Physics, First Edition. Ilya Narsky and Frank C. Porter.
©2014 WILEY-VCH Verlag GmbH & Co. KGaA. Published 2014 by WILEY-VCH Verlag GmbH & Co. KGaA.

matrices. Take a nominal predictor with K states (levels). We can represent level k as a vector with 1 in kth place and 0 elsewhere. MATLAB function dummyvar does exactly this:

```
>> dummyvar([1 2 3 1])
ans =
     1 0 0
     0 1 0
     0 0 1
     1 0 0
```

Each row of the returned matrix corresponds to one element of the input vector. This technique is often referred to as making *dummy variables*. The last column (3rd dummy predictor) can be dropped because its value is uniquely defined by the first two columns. Thus, a nominal predictor with K levels is reduced to $K-1$ dummy predictors. Above, we use $K=3$.

Substituting the obtained matrix for the original vector, we replace one predictor variable with two. This may or may not pose a problem for the numerical analysis. We may need to take extra steps if

- We use a distance-based method such as, for instance, nearest neighbor rules. The Hamming loss described in Section 16.1 can be used to measure distance between observations if all predictors are binary (a binary predictor takes values 0 and 1). Distance estimation for a mix of categorical and numeric (continuous) predictors is not trivial. See more in Section 13.7.
- We need to treat the dummy predictors as separate entities. Suppose we use one of the techniques discussed in Chapter 18 to rank these predictors by importance. We learn that predictor 1 is important but predictor 2 is not. We want to drop predictor 2 from the model, but we cannot discard it without dropping predictor 1.

If the relative distance between categorical levels is unimportant, levels of an ordinal predictor can be represented by integers. This representation is sufficient for decision trees. Every binary split divides the ordinal levels in two groups preserving the level order. For most classifiers discussed in this book, the same trick does not work because distance between the levels is used for modeling.

If the simple-minded representation fails, we can "dummify" ordinal predictors as well: Represent level k as a vector with 1 in positions from 1 to $k-1$ and 0 in positions from k to the end. If we treat the same vector [1 2 3 1] as ordinal, this conversion gives (with one matrix row for each element in the original vector):

```
     0 0 0
     1 0 0
     1 1 0
     0 0 0
```

The last column is filled with zeros only and therefore not needed. Just like a nominal predictor, an ordinal predictor with K levels is wrapped in $K-1$ dummy variables.

What about categorical response? Classification is the main subject of the remainder of this book, and we refrain from describing methods applicable to regression here. Classifiers such as decision trees, neural nets, support vector machines and others are designed to work with nominal class labels. Ordinal classification is less popular. With few exceptions, we discuss it in this chapter only.

Frank and Hall (2001) use the same conversion as above to reduce ordinal classification with K classes to $K-1$ binary classification problems. For example, for 3 ordinal classes train a classifier to separate label 1 from labels 2 and 3 and then train a classifier to separate labels 1 and 2 from label 3. This approach fits into the unified framework for reducing multiclass to binary, to be discussed in Chapter 16. There is a caveat, however. While training a multiclass learner is straightforward, the prediction part is less clear. We can compute confidence values for assigning a new observation into K classes, one confidence value per class. Then we can assign this observation into the class with the highest confidence value. So far so good. Now suppose we need the class with the second largest confidence value. In this case, we may find ourselves in a paradoxical situation: the first and second most likely classes can be far apart. Our model predicts that the track is a "super tight" kaon, and if not, then a "loose" kaon. This does not make sense. If the kaon is most likely "super tight", the second best guess ought to be "very tight". In other words, the prediction should be monotone.

Alternatively, we could represent ordinal classes by scalar numbers and fit a regression model. The simplest choice is an integer ranging from 1 to K. These numbers are arbitrary since we cannot measure the distance between the categories; yet our arbitrary choice influences the regression model. The fitted response now satisfies the monotonicity requirement because the confidence of the predicted label is inversely proportional to the distance between this label and the predicted response.

Da Costa *et al.* (2008) propose a neural network modified for ordinal classification. Their network guarantees monotonicity by construction. They also propose a generic regression method. The key assumption in both proposals is a *unimodal* distribution of the predicted response over class labels. "Unimodal" means "one maximum". Suppose the predicted response $F(x)$ for K classes is a binomial random variable, $F \sim \text{Bin}(K-1, p(x))$, for $K-1$ trials and the binomial parameter $p(x)$ varies across the predictor space. The predicted class at x is then $\hat{k}(x) = f(x) + 1$. We refer the reader to da Costa *et al.* (2008) for details on the classification neural network and briefly mention the regression approach here. Da Costa *et al.* (2008) show that, if cast as a regression problem, the binomial parameter $p(x)$ can be estimated by regressing $(k - 0.5)/K$; $k = 1, \ldots, K$; on x. Thus, replacing the arbitrary assignment of K classes to integers by the arbitrary assumption of the binomial distribution of the predicted response over class labels, da Costa *et al.* (2008) produce an almost identical regression model.

In Section 11.2.2, we mention the *proportional odds model* developed for logistic regression. This model fits the logarithms of $K-1$ probability ratios $\log[P(Y \leq k|x)/P(Y > k|x)]$ to a linear function of x and does not guarantee unimodal response. Here, Y is a random variable for the ordinal class label.

A nonexhaustive list of other approaches to ordinal classification includes modifications of specific learners such as support vector machines (Cardoso et al., 2005; Chu and Keerthi, 2007), neural networks (Cheng et al., 2008), and decision trees (Xia et al., 2006); as well as methods based on nonstandard loss functions (Dembczynski et al., 2008) and general frameworks for reduction to binary (Li and Lin, 2006) or multiclass (Xia et al., 2007) problems. We are not familiar with software tools implementing these methods. The approach by Frank and Hall (2001) and the regression technique described above can be carried out using tools described in Chapter 20.

7.2
Missing Values

Suppose you record data with D input variables. Every time you obtain a new observation, you record measured values for D quantities. Occasionally, for reasons beyond your control, some of these D variables are not measured and you get an incomplete record. Analysis of data with such incomplete records is a popular exercise in social and medical studies, where cases are rare and collection of data is expensive and messy. In physics, tools for analysis of missing data are not in high demand, partly because experimental data are more abundant and less ambiguous than elsewhere. Nevertheless, researchers working on data with a high fraction of missing values may find this section useful.

The classic textbook (Little and Rubin, 2002) focuses on treatment of missing data in parametric models. If you wish to fit a parametric likelihood function to data with missing values, you will find this book tremendously useful. For nonparametric density estimation and classification, you would have to seek wisdom elsewhere.

In this section, we briefly review the fundamentals and describe modern research on treatment of missing values in classification.

Categorization of missing data was introduced in Rubin (1976). Data can be *missing completely at random* (MCAR), *missing at random* (MAR), or *missing not at random* (MNAR). The names are somewhat misleading. Let us clarify them with examples.

Suppose your particle detector has a drift chamber and a CsI electromagnetic calorimeter. The detector is immersed in a magnetic field. Track curvatures measured in the drift chamber are used to identify charged particles, and electromagnetic showers measured in the calorimeter are used to detect photons and electrons. The detector has been recording data for several years, and you want to search for a signature of an interesting decay through all records. During some intervals of the detector operation, the calorimeter was turned off and the data were recorded using the drift chamber only. If the drift chamber operated under similar conditions with the calorimeter turned off or on, you can reasonably believe that the calorimeter information is MCAR. Randomness is complete as far as the incidence of missing data does not correlate with the observed data.

Consider another example. Both drift chamber and calorimeter record data, but some tracks observed in the drift chamber do not leave trace in the calorimeter. This can occur, for instance, if a charged track exits the drift chamber into space not covered by the calorimeter or hits the support structure before entering a CsI crystal. The calorimeter data are missing, and we can predict if the calorimeter information is missing or not, based on the track measurements from the drift chamber. This is a case of MAR: the pattern of "missingness" can be predicted based on the observed data.

Suppose now that one of the CsI crystals is attached to a malfunctioning piece of electronics. If the energy deposited by an electromagnetic shower in this crystal is below some threshold, this piece of electronics returns a measurement consistent with noise. On the other hand, if the energy deposit is large enough, you get a solid reading. In principle, this is how all CsI crystals operate, but in this case the threshold is unreasonably high and a good fraction of deposits is not detected by this crystal. The pattern of missingness depends on the missing information. This is the hardest case, MNAR.

In the real world, you may observe a mix of two or more patterns of missingness at once. There are almost no formal test procedures for identifying the type of the missing data mechanism. To test the MCAR hypothesis, you can take every variable with missing values and split the rest of the data in two groups: one in which this variable is missing and another one in which this variable has been measured. You can then compare the distributions of all variables but this one between the two groups. Most usually, analysts compare marginal distributions for the individual variables. This strategy faces the usual problem of multiple testing: when you run many tests at once, you have to adjust the test significance level. We discuss the problem of multiple testing in Chapters 10 and 18. Little (1988) proposes a global test of MCAR based on the assumption of multivariate normality. Of course, if the global test rejects MCAR, you get no indication of what variable is responsible for violating the MCAR assumptions. Moreover, rejecting MCAR does not help you learn if the data are MAR or MNAR. To distinguish one pattern from another, you have to rely on logic and intimate knowledge of your data.

In practice, the important distinction is often between MCAR or MAR on one side and MNAR on the other. If the data are MCAR or MAR and if the mechanism by which data are missing does not interact with the mechanism by which the observed data are generated, the analysis can be carried out using just the observed data. In this case, the mechanism of missingness is said to be *ignorable*. If these assumptions do not hold, you need to model the mechanism by which data are missing. If you do not find these statements obvious, we offer some math to back them up below. Otherwise take our word for it and skip the next paragraph.

Formally, we write down the probability of observing data x with missing values indicated by i as $P(x, i; \theta, \phi)$. Here, X is a random vector with D variables representing full data, and I is a random indicator vector filled with zeros and ones. If the dth element of i is 0, the value of x_d is missing (cannot be measured), and if i_d is 1, the value of x_d is available. Parameters $\boldsymbol{\theta}$ are what we hope to estimate by analyzing the data, and $\boldsymbol{\phi}$ is the mechanism by which the missing data are generated. Assuming that parameters $\boldsymbol{\theta}$ are distinct from parameters $\boldsymbol{\phi}$, we factorize this probability as

$$P(x, i; \boldsymbol{\theta}, \boldsymbol{\phi}) = P(x; \boldsymbol{\theta}) P(i|x; \boldsymbol{\phi}). \tag{7.1}$$

The full data X is a combination of observed and missing data, $X = \{X_{\text{obs}}, X_{\text{mis}}\}$, and the probability of observing x_{obs} is therefore

$$P(x_{\text{obs}}, i; \boldsymbol{\theta}, \boldsymbol{\phi}) = \int P(x_{\text{obs}}, x_{\text{mis}}; \boldsymbol{\theta}) P(i|x_{\text{obs}}, x_{\text{mis}}; \boldsymbol{\phi}) dx_{\text{mis}}. \tag{7.2}$$

If the data are MCAR, the mechanism for missing data is independent of the observed and missing data, $P(i|x_{\text{obs}}, x_{\text{mis}}; \boldsymbol{\phi}) = P(i; \boldsymbol{\phi})$, and the equation above simplifies to

$$P(x_{\text{obs}}, i; \boldsymbol{\theta}, \boldsymbol{\phi}) = P(x_{\text{obs}}; \boldsymbol{\theta}) P(i; \boldsymbol{\phi}). \tag{7.3}$$

If the data are MAR, the mechanism for missing data is completely explained by the observed data, $P(i|x_{\text{obs}}, x_{\text{mis}}; \boldsymbol{\phi}) = P(i|x_{\text{obs}}; \boldsymbol{\phi})$, leading to a slightly more general factorization

$$P(x_{\text{obs}}, i; \boldsymbol{\theta}, \boldsymbol{\phi}) = P(x_{\text{obs}}; \boldsymbol{\theta}) P(i|x_{\text{obs}}; \boldsymbol{\phi}). \tag{7.4}$$

In either case, the overall probability depends on parameters $\boldsymbol{\theta}$ only through $P(x_{\text{obs}}; \boldsymbol{\theta})$. Therefore the observed data carries sufficient information about the parameters of interest.

For classification, the usual goal is to estimate the amount of every class (signal and background) in the data. In this formalism, the magnitudes of the class contributions are included in parameters $\boldsymbol{\theta}$. If the pattern of missingness depends on the class, $\boldsymbol{\theta}$ and $\boldsymbol{\phi}$ cannot be separated and the factorizations above do not apply.

From now on we focus on data with an ignorable mechanism of missingness. Problems with nonignorable missingness are harder to describe in general terms. We do not mean to discourage you from analyzing missing data with nonignorable mechanisms. In fact, sometimes such data can be easier to analyze. For example, if the frequency of missing measurements is much larger for one class than for the other, this missingness can be used as a powerful separation variable.

7.2.1
Likelihood Optimization

If the joint probability $P(x_{\text{obs}}, i; \boldsymbol{\theta}, \boldsymbol{\phi})$ is readily factorized in the form (7.4), and if you can parametrize the observed part $P(x_{\text{obs}}; \boldsymbol{\theta})$, the analysis should be straightforward. For independent observations, the likelihood is a product of probabilities

$P(x_{\text{obs}}; \boldsymbol{\theta})$ over individual observations, and you can find $\boldsymbol{\theta}$ by using one of the standard likelihood optimization techniques.

Unfortunately, even if the missing mechanism is ignorable, often it is hard to factorize the joint probability $P(x_{\text{obs}}, i; \boldsymbol{\theta}, \boldsymbol{\varphi})$ in the form (7.4). Consider, for example, a multivariate normal distribution. The logarithm of its density is essentially $(x - \boldsymbol{\mu})^\top \boldsymbol{\Sigma}^{-1}(x - \boldsymbol{\mu})$. If the covariance matrix is not diagonal, you end up with cross-terms $x_i x_j$ for $i \neq j$. If x_i is observed and x_j is missing, the density is not immediately factorizable. You can integrate the missing data out as in (7.2). For a multivariate normal distribution, this integration is straightforward because you can obtain an analytical expression in a closed form. In general, you may not be so lucky and would have to carry out a numerical integration for every set of parameters $\boldsymbol{\theta}$.

A popular approach to such problems is the *expectation-maximization* (EM) algorithm. Let us apply EM to estimate $\boldsymbol{\mu}$ and $\boldsymbol{\Sigma}$ of the multivariate normal distribution. First, compute estimates of $\boldsymbol{\mu}$ and $\boldsymbol{\Sigma}$ using complete observations only. Then compute the expected value of the log-likelihood $E \log L$ by integrating over the missing observations (E step). Suppose the data are collected in an $N \times D$ matrix $\mathbf{X}$ for N observations (including missing data) and D variables. For a multivariate normal distribution, the E step amounts to computing the expectation $E(X_{ni} X_{nj})$ for every observation n with missing values in either variable i or j, or both. This expectation is a function of available (measured) values in $\mathbf{X}$ and parameters $\boldsymbol{\mu}$ and $\boldsymbol{\Sigma}$. The E step effectively integrates the probability of missing data conditioned on the observed data $P(x_{\text{mis}}|x_{\text{obs}}; \boldsymbol{\theta})$. Now maximize $E \log L$ over $\boldsymbol{\mu}$ and $\boldsymbol{\Sigma}$ to obtain their new estimates (M step). Continue these two-step iterations until convergence.

The EM algorithm and its extensions are reviewed in Little and Rubin (2002). These techniques apply to models based on likelihood functions. In such models, the likelihood function may not be easily parametrized, but it is possible to estimate the probability of observing data for a set of model parameters. Such estimation usually can be reliably carried out only in low dimensions and without too much data missing.

Below we review a number of heuristic techniques applicable to high-dimensional data with many missing values. Their main advantage is speed. Our review is focused on algorithms for classification.

7.2.2
Deletion

The simplest approach to dealing with missing data, known as *casewise deletion* in the statistics literature, is to discard observations with incomplete measurements. For example, if your analysis relies on calorimeter information, you can throw away data collected when the calorimeter is turned off. In general, discarding incomplete data leads to loss of precision and bias in parameter estimates if the data are not MCAR. You should use this approach only when the fraction of missing data is small or when the missing data are MCAR and influential (it is not possible to perform an accurate analysis without measuring what is missing).

A better alternative is *lazy evaluation* (Friedman et al., 1996), also known as the *reduced model* approach. Analysis of classification problems typically consists of two steps: training and prediction. First, train a classifier on data with known class labels. Then apply the trained classifier to new unlabeled data. The training data are often obtained by simulation, and the unlabeled data are observed in an experiment. Lazy evaluation defers training until the unlabeled data are observed. When you observe the unlabeled data, you can find what variables are missing and construct a classifier using the training data with these variables excluded. A new classifier needs to be trained for every combination of missing values. Consider data with 5 variables. If the 5th variable is missing in the observed data, train a classifier on the first 4 variables. If the 4th variable is missing, train a classifier on variables 1–3 and 5. For D variables, you would have to train at most $2^D - 1$ classifiers. Contrary to its name, "lazy" evaluation is computationally expensive. Use it when the number of combinations with missing values is low or when constructing a classifier is cheap.

Empirical studies in Saar-Tsechansky and Provost (2007) and Polikar et al. (2010) show that such reduced models can be superior to imputation (reviewed in Section 7.2.4) in terms of classification accuracy if the data are MCAR. Ensemble techniques such as averaging classifiers built on subsets of variables (reviewed in Section 15.2.3) can improve the accuracy of such models even further. Saar-Tsechansky and Provost (2007) consider relatively small ensembles in which classifiers exclude only variables with missing values in the new unlabeled data. Polikar et al. (2010) grow large ensembles on randomly selected subsets of input variables. The number of variables per subset ranges from 15 to 50% of the full number of dimensions. Their Learn^{++}.MF algorithm keeps record of how often a particular variable is selected by classifiers in the ensemble and reduces its sampling probability every time this variable is used by an individual classifier. Prediction for a new unlabeled observation is computed by aggregating *usable* classifiers, that is, classifiers not trained on variables missing for this observation. Polikar et al. (2010) give analytical formulas for computing the expected number of usable classifiers for known fraction of missing data and ensemble size. They demonstrate the predictive superiority of Learn^{++}.MF over mean imputation on tall datasets with a few hundred to a few thousand observations, a few to a few hundred variables, two to twelve classes, and up to 30% of data missing. In their analysis, all variables have equal probabilities of being amiss, and most real-world datasets do not conform to this assumption.

Both studies, Saar-Tsechansky and Provost (2007) and Polikar et al. (2010), address only handling of missing values in unlabeled (test) data. If you have missing values in labeled (training) data, you can use casewise deletion for every low-dimensional learner in the ensemble. In that case, you should optimize the subset dimensionality based on a trade-off between the number of observations and variables used for every classifier. Both numbers would need to be reasonably large to generate individual classifiers with acceptable accuracy.

Reduced models are most efficient when you have substantial correlations between variables. Imagine data with two identical variables. Missing a value for one of them does not lead to loss of information, and a classifier built on the reduced space is as good as the same classifier built on both variables. Now imagine data with two independent variables, both important for classification. If one of them is missing, classification error must increase. Using one variable instead of two is still better than deleting this observation entirely, but the advantage over imputation techniques is not as impressive as in case of strong correlation. Physicists sometimes go to a great length to choose a few most powerful and weakly correlated variables out of many. In this regime, reduced models are likely not going to help much. They are best suited for working with dozens or hundreds of variables. In that case, an ensemble of reduced models can choose the best variables for you.

7.2.3
Augmentation

Data with missing values can be modified by introducing new information to account for missingness. We refer to such methods as *augmentation*.

For every variable with missing values, we can introduce an indicator variable to record when the value is missing. We can then replace every missing value in the original variable with some unphysical value outside the allowed range. If the original variable were continuous, as most variables in physics analysis are, the new variable would have a discrete peak at the unphysical value. The success of this strategy would depend on the used learning model. Decision trees can easily deal with semi-continuous variables with oddly shaped distributions. Methods based on nearest neighbors could handle such odd data to a certain degree, but some care would be needed. Parametric models such as discriminant analysis based on the assumption of multivariate normality would be poor candidates for this job.

We refrain from discussing other augmentation techniques such as complete-case reweighting because we have not seen these techniques used in supervised learning. The most popular approach to handling missing data in classification is imputation, reviewed in the next section.

7.2.4
Imputation

Imputation is defined as substituting assumed values for missing measurements. The simplest form of imputation is replacing every missing value with a constant, one constant per variable. Sometimes this constant is set to an unphysical value to emphasize its made-up nature. Sometimes this constant is set to a physical value representing the lack of an interesting measurement. For example, if you cannot measure the energy deposition in a CsI crystal, you can set its value to zero. Sometimes this constant is set to the mean or median of the observed sample for this variable. Imputation by a constant provides less information than augmentation by an indicator variable, and for that reason should have a lower predictive power for a sufficiently flexible learner. In general, imputation by a constant is a poor choice, although better than casewise deletion.

Hot deck imputation replaces the missing value by an average of the measured values from similar observations. For every observation with missing values, Batista and Monard (2003) find several nearest neighbors by minimizing distance in the subspace defined by the nonmissing variables. This imputation improves the classification accuracy of the C4.5 decision tree (Quinlan, 1993) on a variety of tall datasets under MCAR with up to 60% of data missing both in training and test sets.

Sharpe and Solly (1995) and Gupta and Lam (1996) apply neural networks to *imputation by regression*. Each variable in the training set is regressed on the remaining variables using complete cases (observations without missing values). If a variable has missing values in the unlabeled set, these are replaced by the predicted regression response. The two papers demonstrate good performance of this technique on small low-dimensional datasets with 10–20% of data missing. At the same time, Sharpe and Solly (1995) show that network-based reduced models discussed in Section 7.2.2 outperform imputation on their datasets. Note that classification of imputation methods is somewhat arbitrary: hot deck imputation by nearest neighbors could very well be included in the imputation by regression category because averaging over closest observations amounts to regression by nearest neighbors.

Decision tree is one of the earliest nonparametric machine learning models in which treatment of missing values was convincingly incorporated. Before you proceed, you might want to read an introduction to decision trees in Chapter 14. The issue of missing values arises for decision trees at three stages:

1. Computing split criterion values such as the decrease in the Gini index.
2. Partitioning the training set in the process of growing a tree.
3. Classifying a new observation with a fully grown tree.

Quinlan (1989) reviews several proposals for every stage of the analysis and investigates the predictive performance for a good fraction of their combinations. For this comparison, he uses data with a few hundred observations, up to 25 variables, and 5–70% of missing values per variable. He recommends the following algorithms for each stage, respectively:

1. When you search for the optimal split on a given variable at a tree node, consider only observations with known values of this variable. If the optimal split is chosen by maximizing the gain ΔI in a measure of the class separability such as the Gini index, reduce ΔI by the proportion of observations with missing values for this variable at this node.
2. When you partition the training data, assign a fraction of the observation missing the optimal split variable to each child node. This fraction should be proportional to the number of observations with known values of this variable in the respective child node.
3. When you classify new data, send an observation missing the optimal split variable down all branches. Take the tree prediction to be a weighted sum of predictions from all leaves reached by this observation.

These choices are included in the C4.5 implementation of decision trees (Quinlan, 1993). We refer to this procedure as *probabilistic splits*.

Another popular treatment of missing data by decision trees is based on surrogate splits (Breiman *et al.*, 1984). *Surrogate splits* are reviewed in detail in Section 14.7, but we briefly explain the idea here. When you grow a tree, compute ΔI for a given split using only observations with known values for this variable. Choose the best decision split out of all possible splits on all variables at this node. In addition, find best splits on variables other than the best split variable. These are surrogate splits. For every surrogate split, compute a measure of predictive association with the best split on the best variable; a formal definition of this predictive association can be found in Section 14.7. Sort the surrogate splits in descending order by the predictive association. To classify a new observation, use the best split on the best variable. If this observation is missing the best split variable, use the best surrogate split. If the best surrogate split variable is also missing, use the second best surrogate split, and so on. If all surrogate split variables are missing, stop at this node and return the majority class.

Expect that reduced models, probabilistic splits, and surrogate splits be more accurate than simple-minded imputation. It is less clear how these three methods compare among themselves. Surrogate splits perform best for data with high redundancy. For two identical variables, a tree with surrogate splits gives identical predictions for unlabeled data with and without missing values. In contrast, if your data are composed of a few thoroughly selected and largely uncorrelated variables, probabilistic splits may be more accurate. Evidence in Saar-Tsechansky and Provost (2007) and Ding and Simonoff (2010) suggests that reduced models never perform worse than probabilistic or surrogate splits. Perhaps, reduced models should be the default choice if you can accept the computational cost.

7.2.5
Other Methods

We have reviewed most popular methods for missing data used in machine learning. Of course, there are more. To the curious reader, we recommend Garcia-Laencina *et al.* (2010) and references therein.

7.3
Outliers

The term *outliers* in the statistics literature generally means unusual observations. Their "unusualness" calls for special treatment. In this book, we deal mostly with supervised learning. We therefore discuss handling outliers at two major stages of a supervised analysis – training and prediction.

For training, outliers are most influential observations. The outcome of a learning process exhibits some sensitivity to every observation in the training set. Examples of such an outcome could be the learned optimal configuration of a classifier or the parameter estimate in a parametric model. The most influential observations are those affecting the outcome most. Thus, "outlier" does not by any means imply "something to be discarded". In some analyses, outlier observations can give the main publishable effect. An example of such an analysis could be a likelihood fit to a mixture of signal and background distributions, in which the "golden" signal observation right under the modeled signal peak drives the signal significance just above the 5σ level.

Sometimes the only thing you can do with outliers is to throw them away. Suppose you estimate the mean of a univariate normal distribution using a small dataset. Compute the sample mean $\bar{x}$ and variance S^2 and treat them as estimates of the true mean μ and variance σ^2. Suppose one of the observed points in the data lies $10S$ away from $\bar{x}$. This point is highly unlikely under normality. If you include this point in your analysis, the estimated mean will be biased away from the true mean of the normal distribution. If you discard the strange point, you will get a better estimate. You still have to convince your colleagues that your model is good, even though it does not fully explain the observed data. Leaving this no small issue aside, removing the outlier facilitates your analysis.

In this example, S^2 has been computed including the outlier. Removing the outlier would reduce the variance estimate. Observations viewed as "normal" earlier may now be classified as outliers. This effect is called *masking*: the extreme outlier discovered in the first step masks mild outliers discovered later. We can continue trimming data and identifying outliers iteratively until we stop discovering new candidates. Alternatively, we can use leave-one-out cross-validation to identify all outliers in the first pass. These two solutions are widely used in practice. Either strategy can fail to detect a group of outliers.

Finding outliers in multivariate data is more challenging. The same two strategies can be used with some success, especially for parametric models. If you are willing to assume multivariate normality, outliers can be identified using the squared Mahalanobis distance $(x-\mu)^\mathsf{T} \Sigma^{-1}(x-\mu)$ for the distribution mean μ and covariance matrix Σ. Just like for univariate data, the estimates of μ and Σ are not reliable in the presence of outliers; this effect is even more pronounced in many dimensions. The *minimal volume ellipsoid* (MVE) (Rousseeuw, 1984) and *minimal covariance determinant* (MCD) (Rousseeuw and van Driessen, 1999) methods find outliers by iterative fitting. MVE draws an ellipsoid with at least $N/2 + 1$ observations over the data and estimates the mean and covariance matrix using the center and dimensions of the ellipsoid. MCD finds a subset of fixed size, typically 75% of the data, with the minimal determinant of the covariance matrix; this subset is then used to estimate μ and Σ.

For prediction, outliers are suspicious observations. The constructed model may not apply to them, although usually you can get a prediction of some sort. For example, discriminant analysis reviewed in Section 11.1 assumes multivariate normality and assigns every observation to the class with the largest posterior probability. If the observation is far away from the class means, it is unlikely to originate from any class. Yet it can be assigned to one of the classes since the multivariate normal densities can be computed everywhere. You can find such improbable observations and assign them to the abnormal class, one not found in the training data.

How can you identify abnormal observations? If you are willing to assume multivariate normality, you can use the squared Mahalanobis distance. For known $\boldsymbol{\mu}$ and known full-rank $\boldsymbol{\Sigma}$, the squared Mahalanobis distance is distributed as χ^2 with D degrees of freedom for data in D dimensions. You can use the upper tail of the χ^2_D distribution to measure the degree of abnormality for an observation. This upper tail can be interpreted as a p-value for testing the "no abnormality" hypothesis. Observations with low p-values can be flagged as abnormal and inspected closely.

Many papers have been published on analysis of outliers in the past decade. The curious reader is encouraged to work through a journal review (Hadi et al., 2009) and two tutorials (Chawla and Sun, 2006; Kriegel et al., 2010) presented at data mining conferences.

7.4
Exercises

1. Consider quark colors red, green, and blue. These colors were chosen by an arbitrary convention. Is quark color a nominal or ordinal variable? Now consider star colors. The color of a star is an indicator of its temperature: stars are classified into 7 categories by their color, or equivalently by their temperature. Is star color a nominal or ordinal variable?
2. Is unimodal response always a desirable property for ordinal classification? Think of an example in which unimodality must be violated.
3. Consider the regression approach by da Costa et al. (2008) for prediction of ordinal class labels described in Section 7.1. Let F be a random variable for the response of an ordinal classification model with support $0, 1, \ldots, K - 1$. Weigh pros and cons of the binomial model with $K - 1$ trials and a scalar binomial parameter $p(x)$, $F \sim \text{Bin}(K-1, p(x))$, versus multinomial model with a vector of multinomial probabilities $\boldsymbol{p}(x)$ with K elements, $F \sim \text{Mult}(\boldsymbol{p}(x))$. How would you enforce unimodality in the multinomial model?
4. Choose a classification model, a software toolkit implementing this classifier, and a dataset without missing values. Split the data in halves, for training and testing. Train the classifier using the training set and measure its classification error using the test set. Now discard approximately 10% of data in the training set at random. To discard 10% of data at random, form a matrix **X** of size $N \times D$ for N observations and D variables, flatten this matrix into a vector of length ND, label 10% of randomly chosen elements in this vector as "unknown" and reshape this vector into a matrix of size $N \times D$ again. (In MATLAB, you can label a value as "unknown" by replacing this value with NaN, where NaN stands for "not a number". For example, to discard approximately 10% of matrix **X** at random, execute X(rand(N,D)<0.1)=NaN.) Use deletion, augmentation and mean imputation to obtain three new training sets. Train the classifier on each

of them and estimate its error using the test data. Study the effectiveness of the three techniques for treatment of missing values in three scenarios:

a) The training data have missing values, but the test data do not.
b) The test data have missing values, but the training data do not.
c) Both training and test data have equal fractions of missing values.

Repeat this study, gradually increasing the fraction of missing data in each scenario from 10 to 90%. Plot the classification error against the fraction of missing data. Summarize your findings.

5. Repeat the study in the previous exercise using at least one of the following techniques: an ensemble of reduced models, hot-deck imputation by nearest neighbor rules, imputation by linear regression, an ensemble of C4.5 decision trees with probabilistic splits, and an ensemble of CART trees with surrogate splits. Compare the results with those obtained in the previous exercise. (In MATLAB, you can use the `fitensemble` function. The `Subspace` method provides reduced models. The `Bag` method and various boosting methods provide ensembles of decision trees. Set the `surrogate` parameter to `on` for the decision tree template.)

6. Consider estimating the mean of a normal random variable using an empirical sample drawn from a mixture of a normal distribution and an unknown distribution with tails heavier than normal. Assume that most observations are drawn from the normal distribution and a few are drawn from the heavy-tailed distribution. The usual estimator, the sample mean, can be rather inaccurate due to the outliers. A robust estimator of the mean is the sample median. Use this idea to design a robust estimator of variance and higher moments of the normal random variable.

References

Batista, G. and Monard, M. (2003) A study of k-nearest neighbour as an imputation method, CiteSeerX, doi=10.1.1.14.3558.

Bevan, A. et al. (ed.) *Physics of the B Factories*, Springer Verlag (to be published in 2013).

Breiman, L., Friedman, J., Stone, C., and Olshen, R. (1984) *Classification and Regression Trees*, Chapman and Hall, Chapter 5.3 Surrogate splits and their uses.

Cardoso, J., da Costa, J., and Cardoso, M. (2005) Modelling ordinal relations with SVMs: An application to objective aesthetic evaluation of breast cancer conservative treatment. *Neural Netw.*, **18**, 808–817.

Chawla, S. and Sun, P. (2006) Outlier detection: Principles, techniques and applications, Tutorial at 10th Pacific-Asia Conf. Knowl. Discov. Data Min, (PAKDD), Singapore.

Cheng, J., Wang, Z., and Pollastri, G. (2008) A neural network approach to ordinal regression. IEEE World Congr. Comput. Intell., Hong Kong, pp. 1279–1284.

Chu, W. and Keerthi, S. (2007) Support vector ordinal regression. *Neural Comput.*, **19** (3).

da Costa, J., Alonso, H., and Cardoso, J. (2008) The unimodal model for the classification of ordinal data. *Neural Netw.*, **21**, 78–91.

Dembczynski, K., Kotlowski, W., and Slowinski, R. (2008) Ordinal classification with decision rules, in *Mining Complex Data*, Lecture Notes in Computer Science, vol. 4944, pp. 169–181.

Ding, Y. and Simonoff, J. (2010) An investigation of missing data methods for classification trees applied to binary response data. *J. Mach. Learn. Res.*, **11**, 131–170.

Frank, E. and Hall, M. (2001) A simple approach to ordinal classification, in *Proc. 12th Eur. Conf. Mach. Learn.*

Friedman, J., Kohavi, R., and Yun, Y. (1996) Lazy decision trees, in *Proc. 13th Natl. Conf. Artif. Intell.*, AAAI Press.

Garcia-Laencina, P., Sancho-Gomez, J.L., and Figueiras-Vidal, A. (2010) Pattern classification with missing data: a review. *Neural Comput. Appl.*, **19** (2).

Gupta, A. and Lam, M. (1996) Estimating missing values using neural networks. *J. Oper. Res. Soc.*, **47** (2), 229–238.

Hadi, A., Imon, A.R., and Werner, M. (2009) Detection of outliers. *WIREs Comput. Stat.*, **1**, 57–70.

Kriegel, H.P., Kroger, P., and Zimek, A. (2010) Outlier detection techniques, Tutorial at SIAM Int. Conf. Data Min., Columbus OH.

Li, L. and Lin, H.T. (2006) Ordinal regression by extended binary classification, in *Advances in Neural Information Processing Systems 19* (eds B. Scholkopf, J.C. Platt, and T. Hofmann), MIT Press, Vancouver, Canada, pp. 865–872.

Little, R. (1988) A test of missing completely at random for multivariate data with missing values. *J. Am. Stat. Assoc.*, **83** (404), 1198–1202.

Little, R. and Rubin, D. (2002) *Statistical Analysis with Missing Data*, Wiley Series in Probability and Statistics, John Wiley & Sons, 2nd edn.

Polikar, R., DePasquale, J., Mohammed, H., Brown, G., and Kuncheva, L. (2010) Learn^{++}.MF: A random subspace approach for the missing feature problem. *Pattern Recognit.*, **43** (11), 3817–3832.

Quinlan, J. (1989) Unknown attribute values in induction, in *Proc. 6th Int. Workshop Mach. Learn.*, Morgan Kaufmann Publishers Inc., San Francisco, CA.

Quinlan, J. (1993) *C4.5: programs for machine learning*, Morgan Kaufmann Publishers, San Francisco, CA.

Rousseeuw, P. (1984) Least median of squares regression. *J. Am. Stat. Assoc.*, **79** (388), 871–880.

Rousseeuw, P. and van Driessen, K. (1999) A fast algorithm for the minimum covariance determinant estimator. *Technometrics*, **41** (3), 212–223.

Rubin, D. (1976) Inference and missing data. *Biometrika*, **63**, 581–592.

Saar-Tsechansky, M. and Provost, F. (2007) Handling missing values when applying classification models. *J. Mach. Learn. Res.*, **8**, 1625–1657.

Sharpe, P. and Solly, R. (1995) Dealing with missing values in neural network-based diagnostic systems. *Neural Comput. Appl.*, **3** (2), 73–77.

Xia, F., Zhang, W., and Wang, J. (2006) An effective tree-based algorithm for ordinal regression, CiteSeerX, doi:10.1.1.76.5420.

Xia, F., Zhou, L., Yang, Y., and Zhang, W. (2007) Ordinal regression as multiclass classification. *Int. J. Intell. Control Syst.*, **12** (3), 230–236.

8
Linear Transformations and Dimensionality Reduction

In this chapter, we review various linear transformations applied to input data. Standardization of variable distributions, accomplished by an appropriate univariate transformation, is a preprocessing step recommended for some learning algorithms discussed in this book. Principal and independent component analysis are multivariate techniques often used for data interpretation and dimensionality reduction. They can be used as the centerpiece of your analysis or as a preprocessing step.

8.1
Centering, Scaling, Reflection and Rotation

Centering stands for subtracting means. Take observed values $\{x_n\}_{n=1}^N$ for random variable X, compute the mean $\hat{\mu} = \sum_{n=1}^N x_n/N$, and subtract it from every value. The rationale is to obtain a new random variable with zero expectation. Since the estimated mean can differ from the true expectation of X, centered variables do not necessarily have zero expectation values. The Central Limit Theorem stipulates that the mean of N values drawn from X asymptotically converges to the normal distribution $N(\mu, \sigma^2/N)$ for the expected value $\mu = EX$ and true variance $\sigma^2 = E(X^2) - (EX)^2$. Equivalently, the standardized difference between the estimated and true means is asymptotically distributed as a normal random variable with zero mean and unit variance: $(\hat{\mu}-\mu)/(\sigma/\sqrt{N}) \sim N(0,1)$. If we know neither μ nor σ, a more useful asymptotic approximation is $(\hat{\mu}-\mu)/(S/\sqrt{N}) \sim t_{N-1}$. Here, $S^2 = \sum_{n=1}^N (x_n - \hat{\mu})^2/(N-1)$ is the observed variance of the sample and t_{N-1} is the Student t distribution with $N-1$ degrees of freedom.

Scaling stands for multiplying a variable by a positive constant (multiplying a variable by negative one is called *reflection*). By doing so, you can force all variables to have the same variance. Centering and scaling to unit variance is called *standardization*. Scaling serves two purposes. It can simplify numerical execution of various algorithms described in this book. It can also convert all measurement units to the same scale, a recommended step when these measurement units do not carry useful information. To apply scaling, compute the observed variance S^2 and

Statistical Analysis Techniques in Particle Physics, First Edition. Ilya Narsky and Frank C. Porter.
©2014 WILEY-VCH Verlag GmbH & Co. KGaA. Published 2014 by WILEY-VCH Verlag GmbH & Co. KGaA.

divide every value of the centered variable by its square root, the observed standard deviation. If you wish to put bounds on the true variance σ^2, use the fact that $(N-1)S^2/\sigma^2$ asymptotically converges to χ^2_{N-1}, the chi-square distribution with $N-1$ degrees of freedom.

Linear transformations other than centering, scaling and reflection are usually carried out to reveal hidden relations among variables. These revealed features can be used to gain better understanding of the data. We loosely use the term "rotation" for all linear transformations other than the univariate transformations described above. In our definition, rotation is not necessarily orthogonal. We discuss two such rotations, principal and independent component analysis, in the following sections.

When do you need to apply a linear transformation for preprocessing data? The answer depends on the next analysis step. For example, decision trees are not sensitive to centering, scaling or reflection. Nearest neighbors and other distance-based methods are sensitive to scaling, but not centering or reflection. Neural networks can handle noncentered and nonscaled variables to some extent, but show an improved training stability if the inputs are centered and scaled to the same range.

8.2
Rotation and Dimensionality Reduction

The two rotation techniques, principal component analysis (PCA) and independent component analysis (ICA), search for the most informative data projections. PCA finds a new set of orthogonal variables called "principal components". By construction, the first component defines the projection with largest variance, the second component defines the projection with second largest variance, and so on. ICA finds a new set of nonorthogonal variables (called "independent components") in which the variables are, similarly, ordered by their explanatory power.

Before describing these techniques in detail, we illustrate them using a simple example. Prepare two bivariate normal distributions. Start by placing a symmetric bivariate normal pdf at the origin. To obtain the first distribution, stretch this symmetric pdf along the vertical axis. Stretching does not introduce correlation, and the two variables in the first distribution are independent. To obtain the second distribution, stretch the original symmetric pdf along the horizontal axis and slightly rotate counterclockwise. Due to this rotation, the two variables in the second distribution correlate. There are twice as many points in the first normal distribution as in the second.

Components found by PCA and ICA are shown in Figure 8.1. PCA finds the projection with the largest variance shown by a solid line. For bivariate data, the second PCA component is fully defined by the first component. ICA finds two variables with maximal independence. The first independent component has a larger variance.

Components found by either technique are ordered by importance. Eliminating the least important components, we could find an efficient low-dimensional rep-

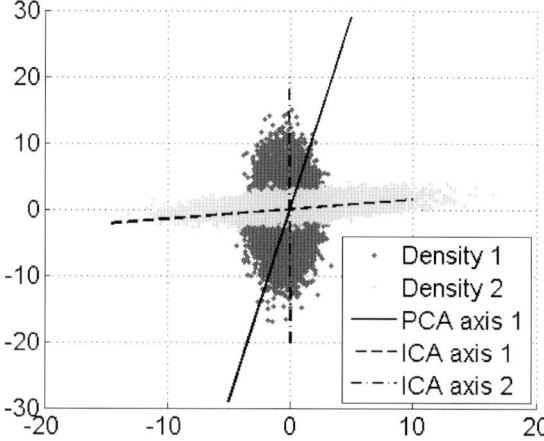

Figure 8.1 Principal and independent components for two bivariate normal densities. The second PCA axis is orthogonal to the one shown. The two ICA axes are aligned with the stretch directions.

resentation of the data. Whether such elimination is possible must be decided on the merits of a specific analysis. Techniques for choosing the optimal number of principal components are described in Section 8.3.4. Here, dimensionality reduction occurs in the transformed space. Reducing the number of original, nontransformed variables is discussed in Chapter 18.

The two rotations described in this chapter disregard class labels. They apply to the entire data, signal and background, and therefore are not used for supervised learning. Sometimes analysts apply principal component analysis and claim that it improves separation of signal and background by a consequent classification algorithm. Such an improvement can occur by accident but is not guaranteed; the effect of the rotation could be just the opposite. Class-conscious linear techniques are described in Chapter 11.

8.3
Principal Component Analysis (PCA)

Principal component analysis is one of the oldest statistical tools. It was proposed in Pearson (1901). Similar formalism was later developed in other fields. The terms "principal component analysis", "Hotelling transform" (Hotelling, 1933) and "Karhunen–Loeve transform" sometimes mean the same thing and sometimes mean slightly different things, depending on what is viewed as the "standard" PCA. We follow the approach by Hotelling using modern notation. PCA is described in many textbooks.

Nonlinear PCA, factor analysis (PCA with noise), and other extensions of PCA, although of potential interest to physics analysis, are not covered here.

In-depth reviews of PCA can be found in Jolliffe (2002) and Abdi and Williams (2010).

8.3.1
Theory

Suppose X is a random column vector with D elements. Without loss of generality, we can assume that the expectation of any element in X is 0: $EX = \mathbf{0}_{D\times 1}$. If this is not the case, we can center X by subtracting its expectation.

We seek a linear transformation $\mathbf{W}$ such that the transformed variables in vector Z,

$$Z = \mathbf{W}X, \tag{8.1}$$

are uncorrelated. Here, $\mathbf{W}$ is a $D \times D$ matrix, and Z has the same dimensionality as X. If the expectation of any element in X is 0, so is the expectation of any element in Z. Two scalar random variables with zero expected values are uncorrelated if their expected product is zero. In matrix notation, we require that the covariance matrix

$$\mathbf{\Sigma}_{ZZ} \equiv E(ZZ^\mathsf{T}) \tag{8.2}$$

be diagonal. Since Z is a column vector, ZZ^T is a $D \times D$ matrix. If we substitute $Z = \mathbf{W}X$ in the equation above, we obtain

$$\mathbf{\Sigma}_{ZZ} = \mathbf{W}\mathbf{\Sigma}_{XX}\mathbf{W}^\mathsf{T}, \tag{8.3}$$

or equivalently

$$\mathbf{\Sigma}_{XX} = \mathbf{V}\mathbf{\Sigma}_{ZZ}\mathbf{V}^\mathsf{T}, \tag{8.4}$$

where $\mathbf{V}$ is the inverse of $\mathbf{W}$. To avoid writing the sub-indices from now on, we define $\mathbf{\Sigma}_{XX} \equiv \mathbf{\Sigma}$ and $\mathbf{\Sigma}_{ZZ} \equiv \mathbf{\Lambda}$ and finally obtain

$$\mathbf{\Sigma} = \mathbf{V}\mathbf{\Lambda}\mathbf{V}^\mathsf{T}. \tag{8.5}$$

The right-hand side gives an eigenvalue decomposition (EVD) of the covariance matrix $\mathbf{\Sigma}$. Matrices $\mathbf{\Sigma}$ and $\mathbf{\Lambda}$ have identical eigenvalues and are said to be similar. Matrix $\mathbf{V}$ is orthogonal, $\mathbf{V}^\mathsf{T}\mathbf{V} = \mathbf{V}\mathbf{V}^\mathsf{T} = \mathbf{I}_{D\times D}$, where $\mathbf{I}$ is an identity matrix with 1 on the main diagonal and 0 elsewhere.

The covariance matrix $\mathbf{\Sigma}$ is symmetric and therefore can be always decomposed in this form. It is trivial to show that the covariance matrix is positive semidefinite, that is, $a^\mathsf{T}\mathbf{\Sigma}a \geq 0$ for any vector a. The eigenvalues of $\mathbf{\Sigma}$, or equivalently the diagonal elements of $\mathbf{\Lambda}$, are nonnegative. It is easy to see that the mth diagonal element of $\mathbf{\Lambda}$ gives the variance of X along the mth principal component. In the real world, EVD is subject to numerical errors. Sometimes your software application can find negative eigenvalues, especially if you work in high dimensions.

Base vectors in the Z space are called *principal components*. To get the mth principal component, set the mth element of z to 1 and the rest to 0. Directions of the principal components in the X space, known as *loadings* in the statistics literature, are given by the columns of matrix V. Projections of a column vector x onto the principal axes, known as *scores* in the statistics literature, can be found by taking D dot products, $x^T V$.

The eigenvalue decomposition is defined up to a permutation of diagonal elements in Λ with a corresponding permutation of columns in V. If we sort the diagonal elements in Λ in descending order, $\lambda_{11} \geq \lambda_{22} \geq \ldots \geq \lambda_{DD}$, we commit to a unique ordering of eigenvalues. The decomposition is still not unique as we can flip the sign of any column in V.

PCA can be applied to the covariance matrix Σ or to the correlation matrix $C = \Omega^{-1/2} \Sigma \Omega^{-1/2}$, where $\Omega = \text{diag}(\Sigma)$ is a matrix with the main diagonal set to that of Σ and all off-diagonal elements set to zero. The choice between *covariance PCA* and *correlation PCA* should be based on the nature of analyzed variables. If the variables are measured in different units and have substantially different standard deviations, correlation PCA should be preferred. If the variables are measured in a similar fashion and their standard deviations can be compared to each other in a meaningful way, covariance PCA would be appropriate.

If X is drawn from a multivariate normal distribution, its principal components are aligned with the orthogonal normal axes. If two normal random variables are uncorrelated, they are independent. On occasion you can hear people say that PCA requires multivariate normality and finds independent variables. This merely describes one particular, although important, case of PCA. In general, PCA is not restricted to normal distributions. Independence is a stronger requirement than zero correlation. For normal random variables, these two requirements happen to be equivalent. For nonnormal variables, a PCA transformation does not necessarily produce new independent variables. A technique that attempts to obtain independent variables by a linear transformation is called Independent Component Analysis and described in Section 8.4.

The derivation of PCA shown here is due to Hotelling. Originally, PCA was proposed by Pearson who approached from a different angle. Define a rotation $Z = WX$. Transform Z back to the original space $X = W^T Z$ and take $E(||X - W^T W X||^2)$ to be the reconstruction error induced by the transformation W. Set the reconstruction error to zero by choosing $W = V^T$ with V defined in (8.5).

8.3.2
Numerical Implementation

In practice, the covariance matrix Σ usually needs to be estimated from data. Let X be an $N \times D$ matrix with one row per observation and one column per variable. Think of X as a set of observed instances of the random column vector X, transposed and concatenated vertically. First, center X by subtracting the mean of every column from all elements in this column. Since most usually the true mean is not known, use the observed mean instead. Then estimate the covariance matrix by putting $\hat{\Sigma} = \mathbf{X}^T\mathbf{X}/(N-1)$. The $1/(N-1)$ factor is needed to get an unbiased estimate of the covariance matrix assuming a multivariate normal distribution. It is absorbed in the definition of Λ and has no effect on the principal components.

Numerically, PCA can be carried out in various ways. Modern software packages often use singular value decomposition (SVD). One advantage of SVD is not having to compute the covariance matrix $\mathbf{X}^T\mathbf{X}$. Instead we decompose

$$\frac{\mathbf{X}}{\sqrt{N-1}} = \mathbf{U}\mathbf{S}\mathbf{V}^T . \tag{8.6}$$

In the full SVD decomposition, $\mathbf{U}$ is of size $N \times N$, $\mathbf{S}$ is of size $N \times D$, and $\mathbf{V}$ is of size $D \times D$. Matrices $\mathbf{U}$ and $\mathbf{V}$ are orthogonal, and matrix $\mathbf{S}$ is diagonal. A nonquadratic diagonal matrix is defined by putting $s_{ij} = 0$ for any pair i and j except $i = j$.

In physics analysis, $\mathbf{X}$ is often very tall, $N \gg D$. Computing an $N \times N$ matrix $\mathbf{U}$ in this case can consume a lot of memory and time. Fortunately, this is not necessary; because $N - D$ bottom rows of $\mathbf{S}$ are filled with zeros, the last $N - D$ columns of $\mathbf{U}$ can be discarded. In the thin version of SVD, $\mathbf{U}$ is $N \times D$ and $\mathbf{S}$ is $D \times D$. Note that $\mathbf{U}$ is no longer orthogonal: $\mathbf{U}^T\mathbf{U} = \mathbf{I}_{D \times D}$ holds, but $\mathbf{U}\mathbf{U}^T$ is not an identity matrix.

Substituting (8.6) into (8.5), we obtain a simple relation between eigen and singular value decompositions,

$$\Lambda = \mathbf{S}^2 , \tag{8.7}$$

for a square matrix $\mathbf{S}$. The elements of $\mathbf{S}$ are standard deviations along principal components. The matrix $\mathbf{V}$ is the same in both decompositions.

When we square a diagonal matrix, we also square its condition number, defined as the ratio of the largest and smallest eigenvalues. Generally, matrices with large condition numbers pose problems for numerical analysis. If the condition number is too large, the matrix is said to be ill-conditioned. For this reason, PCA implementations often prefer SVD of $\mathbf{X}$ over EVD of $\mathbf{X}^T\mathbf{X}$.

PCA is included in many data analysis software suites. One example is function pca available from the Statistics Toolbox in MATLAB (or function `princomp` in older MATLAB releases).

Numerical issues in matrix operations are described in many books. We recommend Press et al. (2002) and Moler (2008).

8.3.3
Weighted Data

In physics analysis, observations are often weighted. Let w be a vector with weights for observations (rows) in matrix $\mathbf{X}$. Here, vector w has nothing to do with the transformation matrix $\mathbf{W}$; unfortunately, both entities are most usually denoted by the same letter. Suppose the weights have been normalized to sum to 1. The observed weighted mean for variable d; $d = 1, \ldots, D$; is then

$$\hat{\mu}_d = \sum_{n=1}^{N} w_n x_{nd} . \tag{8.8}$$

The observed weighted covariance for variables i and j is given by

$$\hat{\sigma}_{ij} = \sum_{n=1}^{N} w_n (x_{ni} - \hat{\mu}_i)(x_{nj} - \hat{\mu}_j). \tag{8.9}$$

This estimate of the covariance is biased. For unweighted data drawn from a multivariate normal distribution, the unbiased estimate of the covariance matrix is given by $\mathbf{X}^T\mathbf{X}/(N-1)$. But if we set all weights in (8.9) to $1/N$, we would get $\mathbf{X}^T\mathbf{X}/N$. A small corrective factor fixes this discrepancy:

$$\hat{\sigma}_{ij} = \frac{\sum_{n=1}^{N} w_n (x_{ni} - \hat{\mu}_i)(x_{nj} - \hat{\mu}_j)}{1 - \sum_{n=1}^{N} w_n^2}. \tag{8.10}$$

Multiplying all elements of the covariance matrix by the same factor does not change the principal components.

An equivalent way of estimating the covariance matrix would be to use $\mathbf{WX}$, where $\mathbf{W}$ is a diagonal $N \times N$ matrix. The nth element on its main diagonal is set to

$$\sqrt{\frac{w_n}{1 - \sum_{n=1}^{N} w_n^2}},$$

and off-diagonal elements are set to zero. PCA of weighted data can be then carried out just like the ordinary PCA, either by EVD of $\hat{\boldsymbol{\Sigma}} = \mathbf{X}^T\mathbf{W}^2\mathbf{X}$ or by SVD of $\mathbf{WX}$.

8.3.4
How Many Principal Components Are Enough?

For N observations and D variables, we can find, after centering, $M_{\max} = \max(N-1, D)$ principal components at most. The number of principal components is limited by the rank of matrix $\mathbf{X}$ and can be smaller than $M_{\max}$ if some variables (or observations) are linear combinations of other variables (or observations). Yet for sufficiently large N and D the number of principal components can be impractically high.

Often it is possible to keep just a few largest principal components and discard the rest. This strategy is justified if the condition number of the estimated covariance matrix $\mathbf{X}^T\mathbf{X}$ for covariance PCA or estimated correlation matrix for correlation PCA is large. Under multivariate normality, a formal procedure described in Bartlett (1950) can be used to test the equality of all eigenvalues for covariance PCA. In practice, a number of heuristic techniques, not requiring the normality assumption, can be deployed for deciding how many principal components deserve to be kept.

One simple approach is to select as many components as needed to keep the fraction of the total variance explained by the first M components,

$$\delta_M = \frac{\sum_{m=1}^{M} \lambda_m}{\sum_{m=1}^{M_{\max}} \lambda_m}, \tag{8.11}$$

above a specified threshold. Here, λ_m is the mth diagonal element of matrix Λ. Jolliffe (2002) recommends setting this threshold to a value between 0.7 and 0.9.

Another simple approach is to plot eigenvalues λ_m versus m and find where the slope of the plot goes from steep to flat, implying that the extra components add little information. This point is called an *elbow*, and this plot is called an *elbow* or *scree plot*. In a slightly modified version of this technique, plot the difference between the adjacent eigenvalues $\lambda_m - \lambda_{m+1}$ to detect the point where the difference approaches zero.

These two simple techniques are subjective and can produce inconclusive results. We apply these techniques to the ionosphere data available from the UCI repository (Frank and Asuncion, 2010) and show the results in Figure 8.2. There is no well-defined elbow on the scree plot. Based on the two plots, we could decide to retain at least seven components.

The two simple techniques described above could be applied to either covariance or correlation PCA. Peres-Neto et al. (2005) evaluate a number of less subjective approaches for correlation PCA aimed at discovering nontrivial principal components. A component is *nontrivial* if it has sizable contributions from two or more variables. The case of trivial components corresponds to an identity correlation matrix when all variables are normally distributed and independent. The presence of nontrivial components would be seen in a departure of some eigenvalues from one. The algorithms reviewed in Peres-Neto et al. (2005) search for nontrivial components with large eigenvalues.

Even if data were drawn from a perfectly spherical pdf, unequal eigenvalues would be observed due to random fluctuations. The largest eigenvalue would be always above one. To estimate the significance of the mth component, we could compare the observed mth eigenvalue with the distribution for the mth eigenvalue found in data with trivial components only. One of the more accurate algorithms for identifying nontrivial components in data with trivial and nontrivial components mixed combines this approach and a permutation test. This algorithm works as fol-

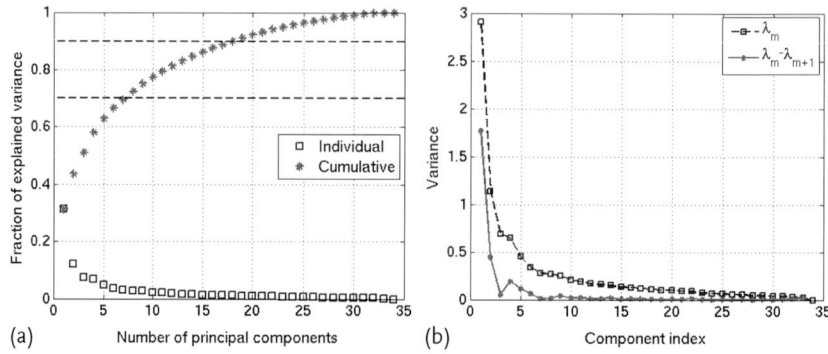

Figure 8.2 (a) Variance of an individual component normalized to the total variance versus component index (squares) and explained fraction of total variance versus the number of components (stars). (b) Two versions of the scree plot. Both plots are for the ionosphere data.

lows. Run PCA on the input data to obtain the eigenvalues $\lambda_1 \geq \lambda_2 \geq \cdots \geq \lambda_{M_{\max}}$. Generate R replicas of the input data for sufficiently large R. In each replica, shuffle values for each variable at random. Apply PCA to each replica r and record the eigenvalues, $\lambda_1^{(r)} \geq \lambda_2^{(r)} \geq \cdots \geq \lambda_{M_{\max}}^{(r)}$. Compute p-values for $M_{\max}$ null hypotheses "the mth component is trivial". To compute the p-value for the mth hypothesis, count the number of eigenvalues found in the shuffled data $\{\lambda_m^{(r)}\}_{r=1}^R$ above the observed eigenvalue λ_m. A low p-value indicates that λ_m is statistically large and likely produced by a nontrivial component. For large values of $M_{\max}$, we should account for effects of multiple testing, to be discussed in Sections 10.4 and 18.3.2.

We apply this algorithm to the ionosphere data using $R = 1000$ replicas. The p-values for the five largest components are exactly zero, and so is the p-value for the last (34th) component. The p-values for all other components are exactly one. We conclude that the five largest components are not trivial. The 34th component does not represent a significant effect. The second variable in the ionosphere data has zero variance, and the rank of the input matrix is therefore at most 33. The respective eigenvalue differs from zero by a value of the order of 10^{-31} due to floating-point error. The 34th eigenvalue in every shuffled replica is equally meaningless and should be ignored.

Note that trivial components can be essential for explaining the data. For instance, one variable with large variance and small correlation with the rest of the variables could be responsible for most observed variance. The described algorithm, as well as the other algorithms in Peres-Neto et al. (2005), by design would fail to detect its significance.

An alternative approach is to keep as many components as needed to satisfy bounds on the reconstruction error. These bounds could be set by the physics or by engineering tolerance constraints on the measured data. Here is one way to define the reconstruction error. Project data $\mathbf{X}$ onto the principal components, $\mathbf{Z} = \mathbf{XV}$. Then project the obtained scores $\mathbf{Z}$ back onto the original variables, $\hat{\mathbf{X}} = \mathbf{XVV}^\top$. The reconstructed matrix $\hat{\mathbf{X}}$ equals $\mathbf{X}$ because $\mathbf{V}$ is orthogonal. Let $\mathbf{V}_M$ be a matrix of PCA loadings for the first M components. To obtain this matrix, delete columns with indices $M + 1$ and higher in the full loading matrix $\mathbf{V}$. The reconstructed matrix $\hat{\mathbf{X}}_M = \mathbf{XV}_M \mathbf{V}_M^\top$ may not equal $\mathbf{X}$ because $\mathbf{V}_M$ may not be orthogonal. Let $\mathbf{R}^{(M)} = \hat{\mathbf{X}}_M - \mathbf{X}$ be a matrix of residuals. The Frobenius norm,

$$\|\mathbf{R}^{(M)}\|_F = \sqrt{\sum_{n=1}^N \sum_{d=1}^D \left(r_{nd}^{(M)}\right)^2}, \qquad (8.12)$$

divided by the square root of the total number of elements in $\mathbf{X}$ can be used as the average reconstruction error. The maximal reconstruction error is given by the element of $\mathbf{R}^{(M)}$ with the maximal magnitude.

If we used the same data $\mathbf{X}$ to estimate $\mathbf{V}$ and to compute the reconstruction error, the error estimate would be biased low. We will discuss this phenomenon again in Chapter 9 in the context of supervised learning. If a large amount of data is available, we can find the loadings $\mathbf{V}$ using one dataset and estimate $\mathbf{R}^{(M)}$ using another set. If there is not enough data, we can use cross-validation. Split **X** into $\mathcal{K}$ disjoint subsets with $N/\mathcal{K}$ observations (rows) per subset, on average. Take the first subset out of **X**. Run PCA on the remaining $\mathcal{K}-1$ subsets to estimate $\mathbf{V}^{(1)}$. Apply the found loadings $\mathbf{V}^{(1)}$ to the held-out subset to obtain $\hat{\mathbf{X}}_M^{(1)}$; $M = 1, \ldots, M_{\max}$. Repeat for the remaining $\mathcal{K}-1$ subsets. Form the reconstructed matrix $\hat{\mathbf{X}}_M$ by concatenating the reconstructed subset matrices $\hat{\mathbf{X}}_M^{(k)}$; $k = 1, \ldots, \mathcal{K}$. This concatenation is unambiguous because every observation (row) in **X** can be found in one subset matrix only. The residual matrix $\mathbf{R}^{(M)}$ is then defined in the usual way.

We compute the average and maximal reconstruction error for the ionosphere data by 10-fold cross-validation. The results are shown in Figure 8.3. The average error steadily decreases as the number of principal components grows. The maximal error shows no improvement over the value obtained using just the first component until the number of components exceeds 20.

A more sophisticated technique for optimizing the number of components, based on the reconstruction error, is described in Krzanowski (1987).

8.3.5
Example: Apply PCA and Choose the Optimal Number of Components

Contents
- Load data
- Center the data
- Perform PCA
- Plot the explained variance
- Make a scree plot
- Find nontrivial eigenvalues
- Partition the data in 10 folds for cross-validation
- Estimate reconstruction error by cross-validation.

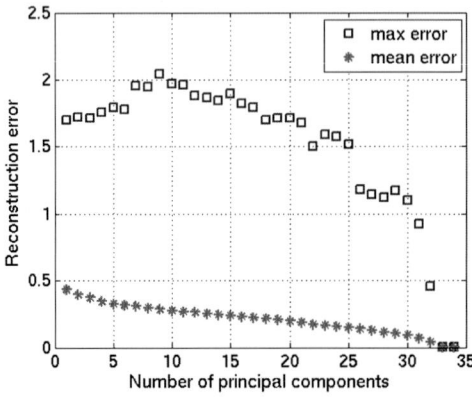

Figure 8.3 Average and maximal reconstruction error versus the number of principal components for the ionosphere data.

8.3 Principal Component Analysis (PCA)

Load data

We use the ionosphere data from the UCI repository http://archive.ics.uci.edu/ml/datasets/Ionosphere. These data are primarily used for binary classification. In this example, we ignore class labels and apply PCA. The data have 351 observations and 34 variables.

```
load ionosphere
[N,D] = size(X)

N =
   351
D =
   34
```

Center the Data

The data are readings from a phased array of 16 high-frequency antennas. All 34 variables are produced in a similar fashion, measured using the same units and have, roughly, the same variance. One exception is the second variable whose variance is exactly zero. We therefore choose to apply covariance PCA. To ensure correct computation of the reconstructed error in later steps, we center the entire data by subtracting the mean of every column from the respective column in the input matrix.

```
mu = mean(X);
X = bsxfun(@minus,X,mu);
```

Perform PCA

We execute the pca function available from the Statistics Toolbox. This function returns:

1. Loadings (matrix **V**, in our notation)
2. Scores (matrix **X** * **V**, in our notation)
3. Eigenvalues (the main diagonal of matrix Λ, in our notation)
4. Hotelling's T-squared statistic for each observation in **X** (not discussed in the book)
5. Percentage of variance explained by the respective component

We need only the eigenvalues and the fraction of explained variance. The fraction of explained variance could be easily obtained from the eigenvalues. We thus replace the first two output arguments with the tilde symbol and do not ask for the 4th and 5th arguments.

```
[~,~,lambda] = pca(X);
```

Plot the Explained Variance

We plot the variance for each component normalized to the overall variance versus the component index. We then plot the cumulative variance versus the number of components.

```
figure;
plot(lambda/sum(lambda),'bs','MarkerSize',8);
hold;
plot(cumsum(lambda)/sum(lambda),'r*','MarkerSize',8);
hold off;
grid on;
xlim([0 D+1]);
line([0 D+1],[0.7 0.7],'Color','k','LineStyle','-');
line([0 D+1],[0.9 0.9],'Color','k','LineStyle','-');
legend('Individual','Cumulative','Location','E');
ylabel('Fraction of explained variance');
```

Make a Scree Plot

This reproduces the figure in the book.

```
figure;
plot(lambda,'bs-','MarkerSize',6);
hold;
plot(-diff(lambda),'r*-','MarkerSize',6);
grid on;
xlabel('Component index');
ylabel('Variance');
legend('\lambda_m','\lambda_m-\lambda_{m+1}','Location','NE');
xlim([0 D+1]);
```

Find Nontrivial Eigenvalues

We generate 1000 replicas of the ionosphere data by shuffling every variable (column) in the input matrix at random. We store the eigenvalues for every run on the shuffled data in the lambdaShuffled matrix. Although the conclusions should not depend in a significant way on how exactly the data are shuffled, we set the random number generator seed for reproducibility by executing the rng function.

```
R = 1000;
lambdaShuffled = zeros(D,R);
Xperm = zeros(N,D);
rng(1);
for r=1:R
    for d=1:D
        Xperm(:,d) = X(randperm(N),d);
    end
    [~,~,lambdaShuffled(:,r)] = pca(Xperm);
end
```

We estimate the *p*-values for the eigenvalues observed in the ionosphere data by computing the fraction of replicas in which the respective eigenvalue exceeds the one observed in the same position. The first five *p*-values equal zero indicating the nontrivial nature of the first five principal components. The rest of the *p*-values are one. We show the first ten *p*-values only.

8.3 Principal Component Analysis (PCA)

```
pval = sum(bsxfun(@gt, lambdaShuffled, lambda), 2)/R;
pval(1:10)'

ans =
   0 0 0 0 0 1 1 1 1 1
```

Partition the Data in 10 Folds for Cross-Validation

The cv object returned by cvpartition stores the data partition. We use the training and test methods of this object in the next step to access the indices of observations in the respective folds.

```
K = 10;
cv = cvpartition(N,'kfold',K);
```

Estimate Reconstruction Error by Cross-Validation

Increasing the number of principal components from 1 to the number of variables, we compute the reconstruction error. For an assumed number of principal components, m, we copy the first m loadings from matrix V to matrix Vreduced. We then compute the associated scores, X*Vreduced, and transform them back to the original variables using X*Vreduced*Vreduced' to obtain an estimate, Xhat, of the input matrix X. We set the matrix of residuals, residual, to the difference of the reconstructed and original data Xhat-X. We then compute two kinds of reconstruction error:

1. meanErr, the mean of all residuals
2. maxErr, the maximal magnitude of a single residual value.

To compute meanErr, we divide the Frobenius norm of the residual matrix by the square root of N*D, the total number of elements in this matrix. To compute maxErr, we take the maximal magnitude of all residual values.

```
M = numel(lambda);
meanErr = zeros(1,M);
maxErr = zeros(1,M);
Xhat = zeros(N,M);
%
for m=1:M
    %
    for k=1:K
        itrain = training(cv,k);
        itest = test(cv,k);
        V = pca(X(itrain,:));
        Vreduced = V(:,1:m);
        Xhat(itest,:) = X(itest,:)*(Vreduced*Vreduced');
    end
    %
    residual = Xhat - X;
    meanErr(m) = norm(residual,'fro')/sqrt(N*D);
    maxErr(m) = max(abs(residual(:)));
end
```

Reproduce the figure in the book.

```
figure;
plot(maxErr,'bs','MarkerSize',8);
hold;
plot(meanErr,'r*','MarkerSize',8);
hold off;
legend('max error','mean error','Location','NE');
grid on;
xlabel('Number of principal components');
ylabel('Reconstruction error');
```

8.4
Independent Component Analysis (ICA)

Independent Component Analysis is much younger than PCA. The first publications on ICA date back to the early 1980s. Just like PCA, ICA seeks a linear transformation to reduce the dependence of input variables. Unlike PCA, ICA seeks a nonorthogonal transformation to independent (not just uncorrelated) variables.

Historically, ICA was developed in the context of blind source separation. Here is how Hyvarinen and Oja (2000) introduce ICA: "Imagine that you are in a room where two people are speaking simultaneously. You have two microphones, which you hold in different locations. The microphones give you two recorded time signals, which we could denote by $x_1(t)$ and $x_2(t)$, with x_1 and x_2 the amplitudes, and t the time index. Each of these recorded signals is a weighted sum of the speech signals emitted by the two speakers, which we denote by $s_1(t)$ and $s_2(t)$." At any fixed time t we can write $X = AS$, where A is the mixing matrix. We measure x and would like to find s. We seek transformation $S = WX$, where W is the unmixing matrix. In this paragraph, we use the conventional ICA notation and terminology. For consistency with the PCA section, from now on we will use z instead of s and V instead of A.

Below we describe linear noiseless ICA, the closest cousin of the classical PCA. Linear ICA with noise and nonlinear ICA are discussed elsewhere.

Stone (2004) and Hyvarinen and Oja (2000) give good introductions to the subject. For in-depth reading, we recommend Hyvarinen et al. (2001).

8.4.1
Theory

We seek a linear transformation of the form (8.1) to map dependent variables in X onto independent variables in Z. Random variables Z_d; $d = 1, \ldots, D$; are independent iff their joint pdf is factorizable:

$$f_Z(z) = \prod_{d=1}^{D} f_{Z_d}(z_d). \tag{8.13}$$

8.4 Independent Component Analysis (ICA)

Equivalently, the joint distribution of Z is a product of marginal distributions.

Posed in this form, the problem does not lend itself to an obvious solution. Estimating a joint probability in high-dimensional space is a nontrivial task. Even if we master this part, how do we test if the condition (8.13) is satisfied? We work with finite samples, and the equality would never hold exactly.

What we need is a measure of independence for a set of random variables. One popular measure is mutual information, also known as the Kullback–Leibler divergence. Define the *differential entropy* of a continuous random variable A as the negative expectation of its natural logarithm:

$$H(A) \equiv -E \log f_A(A) = -\int f_A(a) \log(f_A(a)) da . \tag{8.14}$$

Similarly, for the differential entropy of two random variables use

$$H(A,B) \equiv -E \log f_{A,B}(A,B) = -\int f_{A,B}(a,b) \log(f_{A,B}(a,b)) da db. \tag{8.15}$$

Above, f_A is a pdf for A and $f_{A,B}$ is a joint pdf for A and B. Mutual information between variables A and B is then

$$J(A,B) = H(A) + H(B) - H(A,B) . \tag{8.16}$$

Observe that if A and B are independent, then $H(A) = H(A|B)$, where $H(A|B)$ is the conditional entropy of A at fixed B. On the other hand, if A and B are not independent, then $H(A) > H(A|B)$. Observe also that $H(A,B) = H(A|B) + H(B)$. Putting these two equalities and one inequality together, we conclude that the mutual information $J(A,B)$ is always nonnegative and equal to zero only when A and B are independent. The same proof can be generalized to any number of random variables. Thus, we seek transformation of the form (8.1) minimizing the mutual information for the transformed variables

$$J(Z) = \sum_{d=1}^{D} H(Z_d) - H(Z) . \tag{8.17}$$

Beware of an important caveat. The differential entropy (8.14) is defined for a continuous variable A. Its discrete analogue,

$$H(A) = -\sum_{a \in \mathcal{A}} P(a) \log P(a) , \tag{8.18}$$

is called *Shannon entropy*. Above, $P(a)$ is a pmf defined on the set $\mathcal{A}$. For a discrete variable A defined on a real-valued set, the differential entropy (8.14) does *not* approach the Shannon entropy (8.18) as the integration step approaches zero. The two entropy definitions differ by an infinite constant. Here, we discuss the ICA formalism for continuous variables only.

A linear transformation from X to Z leads to a simple relation between their joint pdfs:

$$|W| f_Z(z) = f_X(x) , \tag{8.19}$$

where $|\mathbf{W}|$ is the determinant of the transformation matrix. Therefore

$$H(\mathbf{Z}) = H(\mathbf{X}) + \log |\mathbf{W}|. \tag{8.20}$$

For minimization of (8.17), we can treat $H(\mathbf{X})$ as a constant. Therefore we need to minimize

$$\sum_{d=1}^{D} H(Z_d) - \log |\mathbf{W}|. \tag{8.21}$$

Since we know neither $\mathbf{Z}$ nor $\mathbf{W}$, the optimal transformation $\mathbf{Z} = \mathbf{W}\mathbf{X}$ is defined up to arbitrary scale factors. Indeed, if we multiply Z_d and the dth row in $\mathbf{W}$ by the same constant, we get another perfectly valid solution. For uniqueness, let us require that every transformed variable Z_d has unit variance. In addition, independence implies that the covariance of any two variables Z_i and Z_j is zero. Setting the covariance matrix of $\mathbf{Z}$ to identity, $E(\mathbf{Z}\mathbf{Z}^\mathsf{T}) = \mathbf{I}_{D\times D}$, and substituting $\mathbf{W}\mathbf{X}$ for $\mathbf{Z}$ we obtain

$$|\mathbf{W}| \cdot |E(\mathbf{X}\mathbf{X}^\mathsf{T})| \cdot |\mathbf{W}^\mathsf{T}| = |\mathbf{I}_{D\times D}| = 1. \tag{8.22}$$

This means that $|\mathbf{W}|$ is constant and can be dropped from (8.21). The minimization problem becomes

$$\text{minimize } \sum_{d=1}^{D} H(Z_d) \quad \text{subject to} \quad E(\mathbf{Z}\mathbf{Z}^\mathsf{T}) = \mathbf{I}. \tag{8.23}$$

All we need now is to figure out how to estimate $H(Z_d)$ from observed data.

The brute force approach would be to obtain $\hat{f}_{Z_d}(z_d)$ by nonparametric density estimation and approximate the expected value by a sum. As usual, we collect values of $\mathbf{Z}$ into an $N \times D$ matrix $\mathbf{Z}$ with one observation per row and one variable per column. Then we put

$$\hat{H}(Z_d) = -\frac{1}{N} \sum_{n=1}^{N} \log \left(\hat{f}_{Z_d}(z_{nd}) \right). \tag{8.24}$$

In practice, this approach is not feasible. To solve our minimization problem, we would have to devise an iterative algorithm of some kind. This algorithm would need to vary $\mathbf{W}$ and compute D density estimates $\hat{f}_{Z_d}(z_d)$ for every value of $\mathbf{W}$. Accurate nonparametric density estimation such as, for instance, kernel approximation requires plenty of CPU time, and quick and dirty methods such as histograms lack accuracy. A more promising approach is to estimate entropy $H(Z_d)$ directly based on some broad assumptions for the shape of $f_{Z_d}(z_d)$.

As it turns out, a good approximation to the entropy of Z_d can be obtained from

$$H(\Phi) - H(Z_d) \approx c(E[G(\Phi)] - E[G(Z_d)])^2, \tag{8.25}$$

where c is a positive constant and $\Phi \sim N(0,1)$ is a normal random variable with zero mean and unit variance. The left-hand side of this equation is nonnegative because a normal variable with unit variance has the largest entropy among all random variables with unit variance. One of the best choices for function G is

$$G(z) = \frac{1}{\alpha} \log \cosh(\alpha z), \tag{8.26}$$

where $1 \leq \alpha \leq 2$ is an appropriately chosen constant. The values of c, $H(\Phi)$ and $E[G(\Phi)]$ can be computed analytically, and we estimate

$$E[G(Z_d)] \approx \frac{1}{N} \sum_{n=1}^{N} G(z_{nd}) \tag{8.27}$$

in the usual way. Equations (8.25) and (8.26) have appeared out of a hat; it is not obvious why they would provide a good approximation. We refer the reader to Hyvarinen (1998) and Hyvarinen (1999) for an answer to this puzzle.

Finally, our minimization problem is given by (8.23), in which $H(Z_d)$ is estimated using (8.25) and (8.26).

8.4.2
Numerical implementation

ICA does not require that the transformation (unmixing) matrix W be orthogonal. Consequently the mixing matrix $V = W^{-1}$ does not have to be orthogonal either. We would rather work with orthogonal matrices W and V because they are easier to estimate. To force the transformation matrix to be orthogonal, we apply PCA described in the previous section. Center the random variable X, take EVD $E(XX^T) = V\Lambda V^T$ and set $\tilde{X} = V\Lambda^{-1/2}V^T X$ (raising a diagonal matrix to power q means raising every element of this matrix to power q). It is easy to verify that $E(\tilde{X}\tilde{X}^T) = I$. Transforming X to $\tilde{X}$ is called *whitening*. Matrix $\tilde{W}$ in $Z = \tilde{W}\tilde{X}$ transforms one uncorrelated basis to another and therefore must be orthogonal. Instead of searching for the mixing matrix W directly, search for $\tilde{W}$. Then take $W = \tilde{W}V\Lambda^{-1/2}V^T$ to get the transformation from X to Z.

To minimize (8.23), we construct column vectors $\tilde{w}_d$ as follows. Search for $\tilde{w}_1$ minimizing the entropy of $\tilde{w}_1^T \tilde{X}$ subject to $\text{Var}(\tilde{w}_1^T \tilde{X}) = 1$, then search for $\tilde{w}_2$ orthogonal to $\tilde{w}_1$ and minimizing the entropy of $\tilde{w}_2^T \tilde{X}$ subject to $\text{Var}(\tilde{w}_2^T \tilde{X}) = 1$, then search for $\tilde{w}_3$ orthogonal to the plane defined by $\tilde{w}_1$ and $\tilde{w}_2$, and so on. After all vectors $\tilde{w}_d$; $d = 1, \ldots, D$; are found, transpose and concatenate them vertically to obtain the unmixing matrix $\tilde{W}$. Each component $\tilde{w}_d$ is found by an iterative procedure producing a sequence of estimates $\tilde{w}_d^{(t)}$; $t = 1, 2, \ldots$; converging to the true value of $\tilde{w}_d$. Gram–Schmidt orthogonalization, termed *deflation* in the ICA papers,

$$\tilde{w}_d^{(t+1)} = \tilde{w}_d^{(t)} - \sum_{i=1}^{d-1} \left(\tilde{w}_d^{(t)}\right)^T \tilde{w}_i \tilde{w}_i \tag{8.28}$$

can be used to ensure that vector $\tilde{w}_d$ at step $t+1$ be orthogonal to the orthogonal set $\{\tilde{w}_i\}_{i=1}^{d-1}$ obtained earlier. Further details of this iterative procedure can be found in the literature.

Our description has closely followed the ICA presentation in Hyvarinen and Oja (2000) and Hyvarinen (1999). Continuing in this vein, it makes perfect sense to introduce FastICA (2005), a popular MATLAB implementation of ICA by the same authors. You can find wrappers for R, C++ and Python at the same link.

8.4.3
Properties

In PCA we could order the components by their variance. For ICA we require that every component have unit variance, and the PCA-like ordering is lost. The implicit order of the ICA components is given by the entropy values $H(Z_d)$. The ICA algorithm can be viewed as minimization of one entropy at a time subject to the growing number of constraints due to orthogonalization as the variable index d increases. In that sense, ICA is similar to the projection pursuit introduced by Friedman in the 1970s and described in popular textbooks such as Hastie et al. (2008). It would be natural to expect that the entropy for a new ICA component added to the model be larger than the entropies of all previously found components.

How many independent components are enough? One could envision selecting the optimal number of components by a procedure similar to those described in Section 8.3.4. We are not aware of such formal prescriptions. To solve this problem in practice, you need in-depth understanding of the analyzed data. Dimensionality reduction for ICA can occur at the preprocessing stage when you perform PCA. If not, add independent components one by one and re-analyze. Stop when adding a new component does not improve the analysis objective.

A normal random variable has the largest entropy among all random variables with fixed variance. This is why minimizing the entropy can be thought of as maximizing non-Gaussianity. Some books and papers present ICA as a technique for making variables as nonnormal (non-Gaussian) as possible. This interpretation leads to alternative numerical implementations. For example, a popular measure of nonnormality is *kurtosis*. If random variable X has zero mean and unit variance, its kurtosis is $E(X^4)$. If X is normal, its *kurtosis excess*, $E(X^4) - 3$, is zero. The reverse is not true in theory (you can come up with examples of nonnormal variables with zero kurtosis excess) but usually holds in practice. Random variables with negative kurtosis excess are called sub-Gaussian and those with positive kurtosis excess are called super-Gaussian. ICA finds a transformation to maximize the magnitude of the kurtosis excess for each component. This setup is very similar to seeking a transformation maximizing the variance of each component – the problem solved by PCA. Given this simple formulation, it is possible to write a one line implementation of ICA in MATLAB. We leave to the advanced reader to figure this out.

8.5 Exercises

1. Let X be a random column vector with D elements, let W be a constant $D \times D$ matrix, and let $\mathbf{X}$ be an $N \times D$ matrix of i.i.d. observations sampled from X. Write down an estimator of $E[WX]$ using matrix notation.
2. Show that the square of the Frobenius norm $\|\mathbf{X}\|_F^2$ can be expressed as $\text{tr}(\mathbf{X}^T\mathbf{X})$ or $\text{tr}(\mathbf{XX}^T)$ for a matrix $\mathbf{X}$.
3. Prove that the estimator (8.10) is unbiased. Treat the weights w_n as constant factors.
4. Let A and B be continuous scalar random variables. Let $H(A)$ be the entropy of A and let $H(A|B)$ be the entropy of A at fixed B. Prove that $H(A) \geq H(A|B)$ and the equality is attained when A and B are independent.
5. Consider the Gram–Schmidt orthogonalization procedure (8.28). Assume that the estimate $\tilde{\boldsymbol{w}}_d^{(t)}$ at step t is not orthogonal to some vectors in the orthogonal set $\{\tilde{\boldsymbol{w}}_i\}_{i=1}^{d-1}$, that is, $(\tilde{\boldsymbol{w}}_d^{(t)})^T \tilde{\boldsymbol{w}}_i \neq 0$ for some i. Show that $(\tilde{\boldsymbol{w}}_d^{(t+1)})^T \tilde{\boldsymbol{w}}_i = 0$ for any i; $1 \leq i \leq d$.

References

Abdi, H. and Williams, L. (2010) Principal component analysis. *Comput. Stat.*, **2** (4), 433–459.

Bartlett, M. (1950) Tests of significance in factor analysis. *Br. J. Stat. Psychol.*, **3** (2), 77–85.

FastICA (2005) The FastICA package for MATLAB (version 2.5). http://www.cis.hut.fi/projects/ica/fastica/index.shtml (accessed 13 July 2013).

Frank, A. and Asuncion, A. (2010) UCI machine learning repository. http://archive.ics.uci.edu/ml (accessed 13 July 2013).

Hastie, T., Tibshirani, R., and Friedman, J. (2008) *The Elements of Statistical Learning*, Springer Series in Statistics, Springer, 2nd edn.

Hotelling, H. (1933) Analysis of a complex of statistical variables into principal components. *J. Educ. Psychol.*, **24**, 417–441.

Hyvarinen, A. (1998) New approximations of differential entropy for independent component analysis and projection pursuit, in *Advances in Neural Information Processing Systems*, vol. 10, MIT Press, pp. 273–279.

Hyvarinen, A. (1999) Fast and robust fixed-point algorithms for independent component analysis. *IEEE Trans. Neural Netw.*, **10** (3), 626–634.

Hyvarinen, A., Karhunen, J., and Oja, E. (2001) *Independent Component Analysis*, Wiley-Interscience.

Hyvarinen, A. and Oja, E. (2000) Independent component analysis: Algorithms and applications. *Neural Netw.*, **13**, 411–430.

Jolliffe, I. (2002) *Principal Component Analysis*, Springer Series in Statistics, Springer, 2nd edn.

Krzanowski, W. (1987) Cross-validation in principal component analysis. *Biometrics*, **43** (3), 575–584.

Moler, C. (2008) *Numerical Computing with MATLAB*, SIAM. http://www.mathworks.com/moler/ (accessed 13 July 2013).

Pearson, K. (1901) On lines and planes of closest fit to systems of points in space. *Philos. Mag.*, **2**, 559–572.

Peres-Neto, P., Jackson, D., and Somers, K. (2005) How many principal components? stopping rules for determining the number of non-trivial axes revisited. *Comput. Stat. Data Anal.*, **49**, 974–997.

Press, W., Teukolsky, S., Vetterling, W., and Flannery, B. (2002) *Numerical Recipes in C++: The Art of Scientific Computing*, Cambridge University Press, 2nd edn.

Stone, J. (2004) *Independent Component Analysis, A Tutorial Introduction*, MIT Press.

9
Introduction to Classification

Classification is about separating observations of different categories from each other. If observations are grouped in just two categories, or classes, this is a problem of *binary* classification. If the number of possible categories exceeds two, this is a *multiclass* problem. In binary classification, one category is often more interesting than the other. The first category is then described as "signal", and the second category is described as "background".

Formally, we have a scalar random variable Y and a vector random variable X. At any point X in the multivariate space, class label Y is distributed according to a mass function $P(y|x)$, the probability of observing y at x. The goal of *statistical classification* is to learn the distribution $P(y|x)$. This learning is accomplished by building (training) a predictive model on data with known class labels (*labeled data*). The constructed model can predict y for data without known class labels (*unlabeled data*). We assume that the class label is a nominal variable. For discussion of nominal variables, refer to Section 7.1.

In this chapter, we discuss common approaches and workflows for statistical learning.

9.1
Loss Functions: Hard Labels and Soft Scores

Suppose we train a predictive model on labeled data. How do we know if the predicted distribution $\hat{P}(y|x)$ is a good approximation to the true distribution $P(y|x)$? This is a tough problem. Some classifiers do not offer a straightforward way of computing $\hat{P}(y|x)$ from the learned model. Others use assumptions about the functional form of $\hat{P}(y|x)$ which may not hold for the analyzed data. Even if we knew $\hat{P}(y|x)$, we would need to summarize our confidence in the learned model using one number. It is not immediately obvious what this number would be.

In practice, the quality of the learned model at point x is measured using a *loss function*, $\ell(y, f(x))$. You can think of it as a distance between the true class label y and predicted response $f(x)$. Classification loss for the learned model $f(x)$ is the expected distance,

$$L(X, Y) \equiv E_{X,Y} \ell(Y, f(X)) = \sum_{y \in \mathcal{Y}} \int_{\mathcal{X}} \ell(y, f(x)) P(x, y) dx, \qquad (9.1)$$

over the entire domain of $\mathcal{Y}$ and $\mathcal{X}$ for the joint probability density function (pdf) $P(x, y)$. A small loss indicates a good predictive power of the learned model $f(x)$.

The expected loss (9.1) is usually estimated by averaging $\ell(y, f(x))$ over labeled data $\{(x_n, y_n)\}_{n=1}^{N}$ drawn from the joint pdf $P(x, y)$,

$$\hat{L} = \frac{1}{N} \sum_{n=1}^{N} \ell(y_n, f(x_n)). \qquad (9.2)$$

Recipes for estimation of $\hat{L}$ are discussed in Section 9.3. Here, we emphasize one important technical point. To draw data from the joint pdf $P(x, y)$, we must know what $P(x, y)$ is. But if we knew $P(x, y)$, the problem would be already solved. In the standard classification setting, the conditional distributions $P(x|y)$ can be modeled for all possible labels y. Assuming a distribution $P(y)$, we can model the joint pdf using $P(x, y) = P(x|y) P(y)$. Often, the analyst would simulate signal and background observations by Monte Carlo, based on the known forms of $P(x|y)$, and then mix the two sets in a certain proportion. This proportion effectively defines $P(y)$, called *prior probability*, and mixing the two sets in this proportion amounts to simulating $P(x, y)$. The empirical loss (9.2) thus depends on the assumed prior distribution $P(y)$. The sensitivity of the learned model $f(x)$ to the assumed prior distribution $P(y)$ can be studied empirically, by learning several models for several assumed prior distributions.

What exactly is the predicted response $f(x)$? At the very least, every classifier must predict the most probable label $\hat{y}$ at x. In addition to the hard classification label $\hat{y}$, many models compute soft classification scores. A soft score always expresses the level of classification confidence, but its nature varies from one model to another. For linear discriminant analysis (LDA) reviewed in Chapter 11, the score for class y is the posterior probability $\hat{P}(y|x)$ estimated under the assumption of multivariate normality. For binary classification by support vector machines (SVM) reviewed in Chapter 13, the score is the signed distance to the hyperplane separating the two classes: the score is $+1$ if x is a support vector for the positive class, -1 if x is a support vector for the negative class, and 0 if x lies on the hyperplane of optimal separation. We discuss the nature of soft scores for specific models in various places in this book.

For binary classification, the soft score is most usually scalar. For example, LDA predicts posterior probabilities for the two classes, but we need to know only one of them because they add up to one. For binary SVM, the sign of the distance to the separating hyperplane is defined by a convention: we call one class "positive", and the other one "negative". If we swap the names, we must flip the sign of the predicted score. For multiclass problems, the widely adopted convention is to compute one score per class. You can think of binary classifiers as computing one score per class too, but one of the two scores is redundant. For multiclass problems, it may not be possible to obtain the score for one class from the scores for the rest of the classes. For simplicity, we assume that every multiclass learner predicts a vector of soft scores with one scalar element per class. The hard label is set to the class with the largest score.

Response $f(x)$ predicted by a classification model can therefore be one of: a nominal variable, a numeric scalar, or a vector with K elements for K classes. The exact meaning of $f(x)$ depends on the nature of the problem and properties of the classification model. In this book, we use $\hat{y}$ for the predicted class label, $f(x)$ for scalar response, and $f(x)$ for vector response.

If a classifier returns only hard labels, there is only one good choice for the loss function. The simplest measure of the predictive power is *0-1 loss*, or *classification error*:

$$\ell(y, \hat{y}) = \begin{cases} 0 & \text{if } y = \hat{y} \\ 1 & \text{if } y \neq \hat{y} \end{cases} . \tag{9.3}$$

The expected loss $L(X, Y)$ is then minimized by classifying every x into the most probable class, $y^*(x) = \arg\max_{y \in y} P(y|x)$. This loss equals the probability of observing one of the less probable classes,

$$\epsilon^* = 1 - \int_X P(y^*|x) P(x) dx .$$

In the statistics literature, $P(y|x)$ is often called *posterior probability* and the minimal classification error is often called *Bayes error*.

The 0-1 loss is a crude measure of performance. If soft scores are available, we can use a more accurate measure of the predictive power. We compile a list of loss functions used for various binary classifiers in Table 9.1. With the exception of the 0-1 loss, these functions are used for specific classifiers and are intimately tied to their learning algorithms.

In physics analysis, a common goal is to optimize a figure of merit expressed as a function of the expected signal and background, s and b. An example of such

Table 9.1 Classification loss functions. Class label y is -1 for the negative class and $+1$ for the positive class. Operator $[\ ]_+$ evaluates to its argument if this argument is positive and 0 if it is not.

Name	Function	Range of f	Example of usage
0-1	$\ell(y, \hat{y}) = \begin{cases} 0 & \text{if } y = \hat{y} \\ 1 & \text{if } y \neq \hat{y} \end{cases}$	N/A	any classifier
quadratic	$\ell(y, f) = (y - f)^2$	$[-1, +1]$	neural net
exponential	$\ell(y, f) = \exp(-yf)$	$(-\infty, +\infty)$	AdaBoost.M1
binomial	$\ell(y, f) = \log[1 + \exp(-yf)]$	$(-\infty, +\infty)$	LogitBoost
hinge	$\ell(y, f) = [1 - yf]_+$	$(-\infty, +\infty)$	SVM

a figure of merit is precision, $s/\sqrt{s+b}$. Minimizing one of the theory-driven loss functions in Table 9.1 does not guarantee minimization of the figure of merit in question. Of course, if we learn the posterior distribution $P(y|x)$, we obtain, in a sense, the complete solution to the classification problem. There are two caveats, however. First, as already noted, not every classifier is capable of estimating $P(y|x)$. Second, even if the classifier can estimate $P(y|x)$, the analyst may not be interested in this information. The question is then: Can we design a more efficient optimization algorithm which, instead of learning the complete solution we don't need, would accurately and quickly learn the partial solution we are looking for? Crafting such an optimization algorithm is not trivial. In practice, one can construct a classifier by minimizing one of the loss functions from Table 9.1 and then find the best threshold on the soft score by optimizing the chosen figure of merit. This strategy works reasonably well. There have been attempts at minimizing physics-driven figures of merit directly, with modest improvements. Such techniques are rarely applied and difficult to summarize.

What we have described above is a statistical approach to supervised learning. In this approach, X and Y are random variables and learning aims at finding a statistical relationship between them. In statistical estimation, it is often useful to think in terms of bias and variance. We introduce these concepts now.

9.2
Bias, Variance, and Noise

Before introducing bias, variance, and noise for classification, let us take a step back and describe their definitions for regression of univariate response. For multiple regression, the observed y is a numeric scalar and the predicted $f(x)$ is a scalar function. The predictive power of a regression model is often measured by the mean squared error (MSE). Its expectation, $E_{X,Y}[(Y - f(X))^2]$, is what we aim to minimize.

So far, we have been treating X and Y as random variables and computing the expected loss by integrating over their joint probability $P(x, y)$, while keeping $f(x)$ fixed. If we had an infinitely large training set, we could learn $f(x)$ with infinite precision. In reality, we work with training data of limited size, and the learned model $f(x)$ varies from one training set to another. In addition, randomness can be introduced by the learning algorithm itself. For example, optimization algorithms often show sensitivity to the initial choice of parameters. Hence, we distinguish between a model $f(x)$ learned under specific conditions and a regression method producing models learned under various conditions. In the latter case, the learned model, F, is a random variable too; we emphasize this by using the upper case letter. The expected loss $E_{X,Y}\ell(Y, f(X))$ speaks to the quality of the specific model $f(x)$. The expected loss $E_{X,Y,F}\ell(Y, F(X))$ is a measure of the accuracy of the regression method. Averaging over F implies averaging over all random components of the learning process. At the same time, factors affecting the learning process in a systematic way should be, ideally, excluded. For example, the accuracy of the learned model often depends on the size of the training set, and this averaging should be therefore carried out over training sets of the same size.

We are interested in the *generalization error* of the regression method. To estimate this error, we learn F using data with known response values (*training data*) and then compute the loss using data not known to the learned model (*test data*). If we computed the loss using the data on which this model is trained, we would likely obtain an overly optimistic estimate. Because we learn F on one set of data and compare its predictions to the observed response Y on another set, we treat Y and F as independent random variables below.

Let $y^*(x)$ and $f^*(x)$ be the expectations of Y and F at x, respectively. Formally we put $y^*(x) = E_{Y|X}[Y|X=x]$ and $f^*(x) = E_{F|X}[F|X=x]$. Quadratic loss at x can be expressed as

$$[Y - F]^2 = [(y^* - f^*) + (f^* - F) + (Y - y^*)]^2 .$$

Now take the expectation of both sides at fixed X:

$$\begin{aligned} E_{Y,F|X}[(Y - F)^2] &= E_{Y|X}[(Y - y^*)^2] \\ &+ (y^* - f^*)^2 \\ &+ E_{F|X}[(f^* - F)^2] . \end{aligned} \quad (9.4)$$

The first term on the right side is *irreducible noise*, the second term is the square of the *regression bias*, and the third term is the *regression variance*. The irreducible noise is a property of the data, and we can do nothing to eliminate it. All we can hope to do is build a predictive model with minimal bias and variance.

In the decomposition above, all three components are positive and independent of each other. Ideally, we would like to obtain a similar decomposition for classification. This proves difficult because Y is nominal. Several decompositions have been proposed in Breiman (1998); Tibshirani (1996); Kohavi and Wolpert (1996); Domingos (2000); Friedman (1997); Valentini and Dietterich (2004) and James (2003), but none of them is perfect.

Let us take a look at the simplest case, 0-1 loss for binary classification. Let $y^*(x)$ be the most probable class at x and therefore the optimal classification at x. The minimal possible error (Bayes error), the equivalent of irreducible noise, at x is $1 - P(y^*|x)$. If $\hat{y}(x)$ is the predicted label, the error at x is $\epsilon(x) = 1 - P(\hat{y} = y|x)$. Let $P_F(y|x)$ be the probability of predicting into y, taken over all possible realizations of the learned model $F(x)$. To estimate $P_F(y|x)$, we generate many training sets, learn $f(x)$ on every set, and use a large independent test set to count how often the learned model predicts into y. Let $P(y|x)$ be the true probability of observing y at x. Let $\hat{y}^*$ be the class most often predicted at x, that is, the class maximizing $P_F(\hat{y}|x)$. Let $y^{**}(x)$ and $\hat{y}^{**}(x)$ be the labels complementary to $y^*(x)$ and $\hat{y}^*(x)$, respectively; that is, y^{**} is the less probable class at x and $\hat{y}^{**}$ is the class less often predicted at x. Then we have

$$P(\hat{y} = y|x) = P(Y = \hat{y}^* \cap \hat{y} = \hat{y}^*|x) + P(Y = \hat{y}^{**} \cap \hat{y} = \hat{y}^{**}|x) .$$

Using independence of Y and $\hat{y}$ in the test data, we reduce this probability to

$$P(\hat{y} = y|x) = P(\hat{y}^*|x)P_F(\hat{y}^*|x) + P(\hat{y}^{**}|x)P_F(\hat{y}^{**}|x).$$

Since the probabilities $P_F(y|x)$ must sum to one, we can put

$$1 - P(\hat{y} = y|x) = [1 - P(y^*|x)] + P(y^*|x)\left[P_F(\hat{y}^*|x) + P_F(\hat{y}^{**}|x)\right]$$
$$- \left[P(\hat{y}^*|x)P_F(\hat{y}^*|x) + P(\hat{y}^{**}|x)P_F(\hat{y}^{**}|x)\right].$$

Re-arranging the terms, we obtain the decomposition proposed in Breiman (1998),

$$\epsilon(x) = [1 - P(y^*|x)]$$
$$+ [P(y^*|x) - P(\hat{y}^*|x)]P_F(\hat{y}^*|x)$$
$$+ [P(y^*|x) - P(\hat{y}^{**}|x)]P_F(\hat{y}^{**}|x). \qquad (9.5)$$

The first term on the right side is *irreducible noise*, the second term is *bias* and the third term is *variance*. Bias and variance are nonnegative because $P(y^*|x) \geq P(y^{**}|x)$. A classification method is unbiased if it predicts into the most probable class most of the time. If the label is always assigned into the same class, the variance of prediction is zero.

Although it has appealing properties, this decomposition also has a few flaws. It cannot be generalized to any loss function other than 0-1 loss. At any fixed x, the entire reducible error is assigned either to bias or to variance, and assigning the entire error of a biased learner to bias seems naive. To generalize this definition to multiple classes, we would have to replace $P(\hat{y}^{**}|x)$ with $\sum_{y \neq \hat{y}^*} P(y|x)$ in the variance term, with a similar substitution for $P_F(\hat{y}^{**}|x)$. Since $P(y^*|x)$ is not guaranteed to be greater than the sum of probabilities for the rest of the classes, the variance term could be negative. The variance of an estimator is traditionally defined in statistics textbooks as a positive quantity.

James (2003) observes that *variance* (measure of randomness) is different from *variance effect* (contribution to the model error). He decomposes the classification loss into a sum of irreducible noise, bias effect, and variance effect:

$$E_{Y,F|x}\ell(Y, F) = E_{Y|x}\ell(Y, y^*)$$
$$+ E_{Y|x}[\ell(Y, f^*) - \ell(Y, y^*)]$$
$$+ E_{Y,F|x}[\ell(Y, F) - \ell(Y, f^*)]. \qquad (9.6)$$

For binary classification with 0-1 loss, he obtains

$$\epsilon(x) = [1 - P(y^*|x)]$$
$$+ [P(y^*|x) - P(\hat{y}^*|x)]$$
$$+ [2P(\hat{y}^*|x) - 1][1 - P_F(\hat{y}^*|x)]. \qquad (9.7)$$

The bias-effect term looks similar to the bias term in (9.5), except it is not scaled by $P_F(\hat{y}^*|x)$. The variance-effect term is negative if the method predicts more often into the less probable class at x. If the classifier is unbiased, the entire reducible error is due to the variance effect. On the other hand, if the classifier is biased, the reducible error gets contributions both from bias and variance, and under some circumstances the variance effect can *decrease* the error. This decomposition can be generalized to an arbitrary number of classes.

The decomposition in (9.6) could be applied to all loss functions in Table 9.1 with one caveat: it may not be obvious what $\ell(y, y^*)$ means for functions other than 0-1 loss. The loss $\ell(y, f)$ requires a numeric function f, but y^* is nominal. Conceptually, $E_{Y|X}\ell(Y, y^*)$ is the minimal generalization loss at x attained by the best classifier. In some cases, it would be possible to replace y^* with a (vector) score predicted by the ideal classifier for this class, for example, $f = +\infty$ for the exponential loss at $y^* = 1$. In practice, $E_{X,Y}\ell(Y, y^*(X))$ is the asymptotic limit of the learning curve as the size of the training set approaches infinity. In some problems, it would be possible to estimate this limit by simulation.

We demonstrate bias and variance of a decision tree using the same dataset as in Section 8.2. The data are composed of two classes, each drawn from a bivariate normal distribution. The size of the training set is fixed at 10 000 observations with 2/3 originating from class 1. One of the most important parameters for decision tree training is the minimal leaf size, that is, the minimal number of observations in a terminal node. You can find a detailed description of decision trees in Chapter 14; for now we appeal to your intuitive understanding. If the training set is composed of unique observations, a binary decision tree with leaf size one usually finds a perfect separation between the two classes in the training set. Clearly, the configuration of such a tree would vary greatly from one training set to another. On the other hand, if we allowed only one leaf per tree, it would contain the entire dataset. Such a tree would be very stable and quite useless. The optimal configuration is somewhere in between, at a point where both bias and variance are reasonably low. This behavior is shown in Figure 9.1. To estimate the tree bias and variance, we average over 50 trees grown at a fixed leaf size. Instead of generating a new training set for every tree, we use bootstrap aggregation (bagging) over an ensemble of trees. Bagging is described in Chapter 15. The minimum of the classification error is attained at a leaf size near 4% of the training set. We use Breiman's decomposition, but the results from James' decomposition in this case are very similar.

To demonstrate the dependence of the classification error on the size of the training set, we fix the leaf size at 4% of the training data and increase the number of training observations from 25 to 10 000 in 20 steps uniformly spaced on the logarithmic scale. We use 200 trees instead of 50 because the variance for small training sets is large and we need more trees to obtain a reliable estimate of the bias. The learning curve is shown in Figure 9.2. The bias shows little sensitivity to the size of the training set; this verifies our assumption that the leaf size equal to 4% of the training data is about optimal. The decrease in the classification error is entirely due to the decrease in variance. Again, we use Breiman's formalism, but the results are not sensitive to the choice of the decomposition.

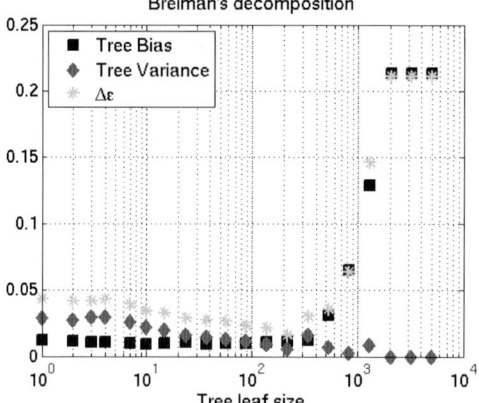

Figure 9.1 Bias, variance and classification error of decision tree versus leaf size. Here, $\Delta\epsilon$ is the difference between the tree error and the irreducible (Bayes) error (12%).

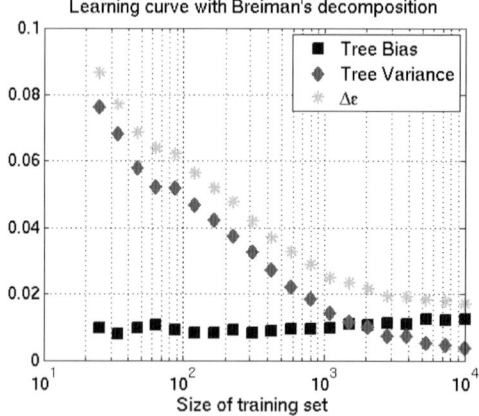

Figure 9.2 Bias, variance and classification error of decision tree versus size of the training set. The leaf size is set to 4% of the training data. Here, $\Delta\epsilon$ is the difference between the tree error and the irreducible (Bayes) error (12%).

We conclude this section by stating perhaps the obvious: to optimize the predictive power of a classifier, you need to carefully select its training parameters and use as many observations as possible for training. Selection of training parameters is discussed later in this book for specific classifiers. The second part of this recipe, using as many observations as possible for training, is tricky. In this section, we have presented an idealized scenario in which we have access to an unlimited amount of test data. In the real world, labeled data are limited and obtaining more data is often costly. We have to use these limited data efficiently. Solutions to this problem are discussed in the next two sections.

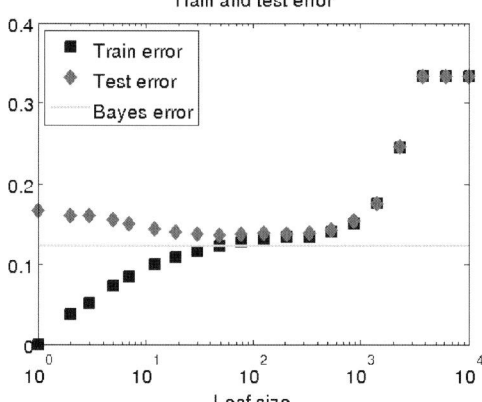

Figure 9.3 Estimates of the classification error obtained on training and test sets versus tree leaf size.

9.3
Training, Validating and Testing: The Optimal Splitting Rule

First and foremost, let us address the obvious question: why not use the same data for training and testing? Simply because the error estimated on the training data, often called *resubstitution error,* is optimistic. This phenomenon is called *overtraining*, and its effect is most pronounced for complex, flexible learners. When you build a model on training data, it attempts to learn the true joint distribution $P(x, y)$. The training data inevitably carries some amount of noise, and the model attempts to extract useful information about $P(x, y)$ from the noise component as well. The knowledge learned from this noise deteriorates the classifier accuracy on data not used for training.

We illustrate this issue by comparing the training and test errors in Figure 9.3, using the same dataset as in Figure 9.1. Again, we grow a decision tree varying its minimal leaf size. The minimal leaf size represents the model complexity. A tree with small leaves is a flexible, sensitive learner, and a tree with large leaves gives a robust, crude model. We observe two trends:

- The training error increases with leaf size; it is optimistic for small leaves and approaches the test error as the leaf size approaches the value maximizing the predictive power.
- The test error decreases with leaf size until the optimal value is reached. If we neglect the scale and focus on the left halves of the two curves, the test error looks almost like a mirror reflection of the training error about the Bayes error line.

The first trend can be observed for many classifiers. In general, flexible, complex classifiers such as ensembles or neural nets can be strongly affected by the noise found in the training data. Their training errors can be biased far below the true generalization error. In this case, a tree with leaf size 1 puts every training observation in a separate leaf. Not surprisingly, it separates the two classes perfectly in the training set, but its error on the test set is much larger. In contrast, simple, inflexible classifiers such as a linear discriminant or a decision split on one variable ("cut" in the physics jargon) are resistant to overtraining. In this case, a tree with large leaves has a simple structure with just a few decision splits and its training error is equal to its test error.

The second trend holds for some classifiers, but not for others. Ensembles, for instance, exhibit no sign of this "mirror effect". For ensembles, the test error is often minimized at zero training error. The relationship between the training and test errors has been studied extensively for ensemble learning and support vector machines; popular examples can be found in Schapire and Freund (1998) and Shawe-Taylor et al. (1996). Unfortunately, these theories suffer from two major drawbacks: applying them to real data is not straightforward, and the error bounds predicted by these theories are too loose to be practically useful. To the curious reader, we recommend a review of generalization bounds in Langford (2005). In practice, if you wish to obtain a relationship between the training and test errors, you would need to do this empirically, for a specific classifier applied to a specific problem.

We have established that the training error is usually not a good estimate of the generalization error. Therefore we need training data and *test data*. Obviously, we would like to learn the model from available labeled data as accurately as possible. At the same time, we would like to estimate its predictive power as accurately as possible. These two requirements are in conflict. As you have seen in the previous section, the predictive power of a model would improve if we provided more training data. On the other hand, the accuracy of the estimate of the predictive power would improve if we provided more test data. Suppose we have one set of labeled data and need to derive both training and test sets from it. How do we decide where to split?

Take a fixed classifier $\hat{y} = f(x)$ with generalization error ϵ. Success or failure of a prediction by this classifier can be modeled as a Bernoulli trial. The number of failures observed in a test set of size N then follows a binomial distribution,

$$P(M = m; N, \epsilon) = \binom{N}{m} \epsilon^m (1 - \epsilon)^{N-m},$$

where M is the count of failures ("misses") and $\binom{N}{m} = N!/m!/(N-m)!$ is the usual "N-choose-m" term. In practice, we choose N, observe m and need to find confidence bounds on ϵ. The unbiased estimator $\hat{\epsilon} = m/N$ has variance $\text{Var}(\hat{\epsilon}) = \epsilon(1 - \epsilon)/N$. If both ϵN and $(1 - \epsilon)N$ are sufficiently large, we can compute confidence bounds using a normal approximation, for instance: $\epsilon = m/N \pm \sqrt{m(N-m)/N^3}$ for the 1σ central interval. If either ϵN or $(1 - \epsilon)N$ is small, the confidence bounds need to be computed more rigorously. Estimation of confidence bounds has been discussed in the physics community quite extensively over the last decade, largely in the context of upper limit calculation for rare signal searches. We refer the reader to the materials of Phystat conferences for in-depth recipes. In practice, a simple-minded frequentist approach could suffice. Set the bounds at confidence level $1 - \alpha$ to $[\hat\epsilon_L, \hat\epsilon_U]$, where

$$P(M \leq m; N, \hat\epsilon_L) = 1 - \alpha/2$$
$$P(M \leq m; N, \hat\epsilon_U) = \alpha/2 \quad (9.8)$$

and $P(M \leq m; N, \epsilon) = \sum_{\tilde m=0}^{m} P(M = \tilde m; N, \epsilon)$ is the binomial cdf. To compute an upper-bound interval at confidence level $1 - \alpha$, set $\hat\epsilon_L = 0$ and find $\hat\epsilon_U$ from $P(M \leq m; N, \hat\epsilon_U) = \alpha$; similarly for lower-bound intervals. These bounds are known in the statistics literature as Clopper–Pearson intervals due to Clopper and Pearson (1934).

How can this help us decide where to split? Let us formalize the question. On one hand, we would like to have the error of the learned model, $\epsilon(N_{\text{train}})$, as low as possible. On the other hand, we would like to estimate ϵ as accurately as possible, that is, we would like to minimize the length of the confidence interval $[\hat\epsilon_L(\epsilon, N_{\text{test}}), \hat\epsilon_U(\epsilon, N_{\text{test}})]$. Above, we emphasize that ϵ depends on the size of the training set, and an estimate of ϵ depends on ϵ itself and the size of the test set.

For simplicity, let us define our task as minimizing the expected upper bound $E\hat\epsilon_U$, where the expectation is taken over an ensemble of training and test sets. Using the normal approximation, we can write for the 95% upper bound:

$$E\hat\epsilon_U = \epsilon(N_{\text{train}}) + 1.96\sqrt{\frac{\epsilon(N_{\text{train}})[1 - \epsilon(N_{\text{train}})]}{N_{\text{test}}}}.$$

This estimate is crude for two reasons. First, we use the normal approximation, and second, the expectation of the standard deviation $\sqrt{\hat\epsilon(1-\hat\epsilon)/N}$ is not equal to the standard deviation at the expected value $\sqrt{\epsilon(1-\epsilon)/N}$. Hence, we use this estimate for illustration only. Let us fix the size of available labeled data to N_{total} and set $N_{\text{test}} = N_{\text{total}} - N_{\text{train}}$. As N_{train} increases, the error $\epsilon(N_{\text{train}})$ decreases, but the error uncertainty

$$\sqrt{\frac{\epsilon(N_{\text{train}})[1 - \epsilon(N_{\text{train}})]}{N_{\text{test}}}}$$

can either increase or decrease because both numerator and denominator are decreasing functions of N_{train}. We plot the expected error uncertainty and 95% upper limit for $N_{\text{total}} = 1000$ and $N_{\text{total}} = 10\,000$ in Figure 9.4 using the same binary classification problem as above. Observe that the position of the optimal splitting point depends on the total size of the available data. If you had 10 000 observations, you could keep 20–60% of observations in the training set because the minimum is flat. If you had 1000 observations, you would have to retain about 50% of observations for training; shifting the training-test split away from 50-50 would lead to a less efficient model construction.

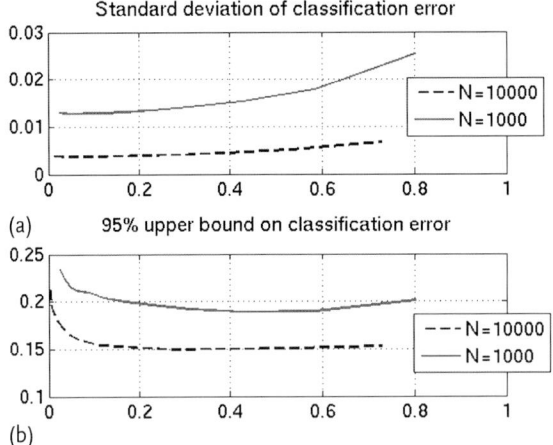

Figure 9.4 (a) Expected uncertainty; (b) 95% upper limit for classification error versus data fraction used for training. One set of curves is for the total data size $N_{\text{total}} = 1000$ and the other set of curves is for $N_{\text{total}} = 10\,000$.

The simulation study depicted in Figure 9.4 omits an important stage of a real-world analysis – optimizing the parameters of a classifier. We have omitted this step because we optimized the leaf size in the previous study shown in Figure 9.1. We could not have learned the optimal leaf size from the training set because the training error is usually less than the generalization error. Similarly, using the same test set for optimizing the classifier *and* estimating the generalization error would lead to an optimistic estimate. To understand this effect, perform a hypothetical experiment. Imagine that the generalization error shows no sensitivity to the leaf size at all: at any size, the error has the same distribution over test sets. Then, if you grow a number of trees at different leaf sizes and choose the one with the smallest test error, your estimate will lie below the true mean.

This is why often we need a third set, usually referred to as *validation data*. Since both validation and test sets are used to estimate the error of a trained classifier, it makes sense to keep them equal in size. We can find the optimal splitting rule by setting $N_{\text{total}} = N_{\text{train}} + 2N_{\text{VT}}$, where N_{VT} is the size of validation or test data, and computing the error bounds for $N_{\text{test}} = N_{\text{VT}}$.

Lectures and books on supervised learning offer heuristic advice such as "split into training-validation-test as 50-25-25". You can follow this advice blindly only if you have a lot of labeled data. In this case, your analysis should be similar to that shown in Figure 9.4 for $N_{\text{total}} = 10\,000$, and the result should have little sensitivity to the optimal split position. Note that "a lot of labeled data" refers not to the overall data size but to the size of the interesting data. For example, if you start with one million simulated background events and end up with just a few background events in the signal region, your measurement will suffer from a large statistical uncertainty associated with the background estimate. Allocating a larger fraction of the available data for testing could then reduce this uncertainty considerably.

We encourage you to search for the optimal splitting point whenever possible. If full Monte Carlo simulation is too slow, use simplified simulation or resampling techniques reviewed in the next section.

If you have a small labeled set to begin with, dividing it into independent pieces for training, validating and testing can be wasteful. Resampling techniques are widely used in statistical applications when data are limited. We discussed cross-validation and bootstrap in Chapter 4 in the context of density estimation. The same techniques apply to classification.

Splitting data into subsets for training and testing is a necessary condition for reliable estimation of the predictive performance. In some cases, a validation set is also necessary. We stress that this condition is not only necessary but also sufficient! The training-validation-test machinery, if applied correctly, protects analysts from obtaining biased estimates of the model accuracy. Extra layers of protection are not necessary. We have seen a physics analysis in which boosted decision trees were found to be the best classifier and immediately rejected because the distribution of classification scores in the training data differed from that in the test data. Sadly, the analysts did not realize that this behavior is most common for this classifier; yet tree ensembles have been applied with great success to many real-world problems.

9.4
Resampling Techniques: Cross-Validation and Bootstrap

Resampling effectively increases the amount of data without incurring the full cost of data simulation or collection. This trick comes at a price: datasets obtained by resampling are not independent. This lack of independence can affect the quality of estimates. Theory and application experience demonstrate that resampling does produce good estimates of various statistical quantities in certain problems. One of such problems, estimation of the predictive power by a classification algorithm, is described below.

9.4.1
Cross-Validation

Cross-validation works by splitting data into $\mathcal{K}$ disjoint subsets, or *folds*. (We use symbol $\mathcal{K}$ to distinguish from the number of classes denoted by K throughout the book.) Use $1 - 1/\mathcal{K}$ of data for training and hold out $1/\mathcal{K}$ of data for testing. Repeat this step $\mathcal{K}$ times, that is, use every observation once for testing and $\mathcal{K}$ times for training.

The number of folds can vary from 2 to N, where N is the number of available observations. As you learned in Chapter 4, $\mathcal{K} = N$ is called leave-one-out cross-validation. How many folds are optimal?

If you worry about the lack of independence among the resampled training sets, you may be tempted to enforce independence by setting $\mathcal{K} = 2$. This proves to be a poor choice for most problems. Cross-validation is used for small datasets. If plenty of data were available, cross-validation would not be needed in the first place. For small datasets, the learning curve shown in Figure 9.2 is far from saturation, and halving the size of the training set would lead to a larger generalization error. Experimentation with real datasets in Kohavi (1995) shows that estimates of the generalization error by two-fold cross-validation are indeed biased high.

What happens if we set $\mathcal{K} = N$? *Leave-one-out cross-validation* uses almost the entire available data, and there is no loss of performance due to a smaller size of the training set. However, the N classifiers are likely going to be very similar because they are constructed on essentially the same data. Leave-one-out validation, although unbiased, has high variance: it does not capture the variability of training sets drawn from the parent distribution $P(x, y)$. The severity of this shortcoming can vary across datasets and classifiers. As shown in Kohavi (1995), leave-one-out estimates of the generalization error can be comparable to those by other partitioning schemes in some cases. The present knowledge (and the present uncertainty) about leave-one-out cross-validation is summarized in Elisseeff and Pontil (2003). In practice, the question of whether leave-one-out is applicable may be moot because of the high computational cost of this method.

We use terms "bias" and "variance" in the two paragraphs above, but their meaning is different from that in Section 9.2. Earlier we discussed bias and variance of a classification algorithm for data drawn from a fixed distribution $P(x, y)$. Here, we talk of bias and variance of the error estimate for a fixed resampling scheme. For example, two-fold cross-validation gives an error estimate which is biased high due to the high variance of the classifier trained on a small dataset.

Two-fold and leave-one-out schemes represent the two extremes, each with its advantages and disadvantages. The most popular choice for $\mathcal{K}$ in supervised learning is 10. Ten-fold cross-validation avoids losing a significant fraction of the predictive power due to a smaller training size and obtains error estimates with sufficiently low variance. Application of 10-fold validation to real datasets is documented in Kohavi (1995). As a rule of thumb, we recommend using 10-fold validation by default.

Repeated cross-validation can likely reduce the variance of the error estimate even further. Repeating implies re-partitioning data for every cross-validation run. In practice, this amounts to running the respective software utility with a new seed for the random number generator.

Suppose we wish to estimate the classification error by $\mathcal{K}$-fold validation (without repetitions) for a dataset of size N. We could count observations in the held-out folds for which the predicted label disagrees with the true label and divide by the total sample size: $\hat{\epsilon} = N(\hat{y} \neq y)/N$. Alternatively, we could estimate the classification error $\hat{\epsilon}_k$ for each fold separately and take an average: $\hat{\epsilon} = \sum_{k=1}^{\mathcal{K}} N_k \hat{\epsilon}_k / N$, where N_k is the number of observations in fold k (this formula can be easily generalized for weighted observations). Which way is better?

Classification error is a linear statistic, and the two estimators give identical results. Physicists usually work with quantities other than classification error. For instance, you might be interested in estimating the background acceptance (false positive rate, FPR) at a fixed signal efficiency (true positive rate, TPR). For a nonlinear statistic, it is generally best to estimate the quantity of interest after combining predictions over all test folds. In other words, instead of computing FPR for each fold and then taking an average of $\mathcal{K}$ values, plot one Receiver Operating Characteristic (ROC) curve (to be described in Section 10.2) for all held-out predictions and estimate FPR using this curve. If you repeat $\mathcal{K}$-fold validation R times, compute R estimates $\hat{\epsilon}_r$, one per repetition, and then average over them: $\hat{\epsilon} = \sum_{r=1}^{R} \hat{\epsilon}_r / R$.

When you estimate a parameter by averaging, either explicitly as in $\hat{\epsilon} = \sum_{k=1}^{\mathcal{K}} N_k \hat{\epsilon}_k / N$ or implicitly as in $\hat{\epsilon} = N(\hat{y} \neq y)/N$, over several measurements, it is tempting to compute confidence intervals too. Sometimes analysts quote the standard deviation of the error estimate over the folds,

$$\sqrt{\frac{\sum_{k=1}^{\mathcal{K}} N_k (\hat{\epsilon}_k - \hat{\epsilon})^2}{N}} .$$

This estimate is illegitimate for two reasons: it assumes that the folds are independent although they are not, and it is based on the normal approximation known to be inaccurate for ϵ close to zero. You should try developing a more sound statistical procedure, before using this estimate as a last resort. If you wish to obtain confidence bounds on the generalization error, you can use the error bounds (9.8). If you are interested not in the generalization error but in some other quantity, you may be able to derive similar bounds under reasonable assumptions about the underlying distribution. For example, this quantity could be a function of the signal efficiency and background acceptance which follow a binomial distribution for a fixed sample size. If you wish to compare two classifiers by cross-validation, such a comparison should be based on one of the formal tests described in the next chapter.

9.4.2
Bootstrap

Bootstrap generates a replica of a dataset by sampling N out of N observations with replacement. On average, $(1 - 1/e) N$, or $0.6321 N$, observations in each replica are unique and the rest are duplicates of some observations from this unique set.

A *leave-one-out bootstrap* (LOOBS) estimate (Efron, 1983) of the generalization loss is obtained by taking

$$\hat{L}_{\text{loobs}} = \frac{1}{N} \sum_{n=1}^{N} \frac{1}{|B_{-n}|} \sum_{b \in B_{-n}} \ell(y_n, f_b(\boldsymbol{x}_n)), \tag{9.9}$$

where B_{-n} is a set of all bootstrap replicas that do not contain observation n, $|B_{-n}|$ is the number of such replicas, and f_b is a model learned on replica b. Substitute any loss function for $\ell(y, f)$; for example, use 0-1 loss for classification error. Replicas in B_{-n} are effectively obtained by sampling from the original set $\{\boldsymbol{x}_i\}_{i=1}^{N}$; $i \neq n$; with the nth observation removed; hence "leave one out" in the name.

The leave-one-out estimate is often larger than the generalization error. In contrast, the training loss

$$\hat{L}_{\text{train}} = \frac{1}{N} \sum_{n=1}^{N} \ell(y_n, f(x_n)) \tag{9.10}$$

usually underestimates the generalization loss. Here, f is the model learned on the full set $\{x_n\}_{n=1}^{N}$. An unbiased estimate of the generalization loss is obtained by averaging the optimistic and pessimistic estimates:

$$\hat{L} = (1-w)\hat{L}_{\text{train}} + w\hat{L}_{\text{loobs}}. \tag{9.11}$$

The weight w ranges from $1 - 1/e$ to 1. This estimator, proposed in Efron and Tibshirani (1997), is called *.632+ estimator*. The origin of .632 is obvious, and the plus sign is there for historical reasons. Originally, Efron (1983) proposed a *.632 estimator* in which w is fixed at the lower bound of the permissible range, $1 - 1/e$. The .632 estimator was shown in Kohavi (1995) to underestimate the generalization error for some datasets and classifiers by a large margin. Later the .632 estimate was upgraded to .632+ to correct this issue.

Efron and Tibshirani (1997) give a recipe for estimating w. Define

$$\gamma = \frac{1}{N^2} \sum_{i=1}^{N} \sum_{j=1}^{N} \ell(y_i, f(x_j)) \tag{9.12}$$

to be the average loss over all permutations of x_n and y_n. Most usually, $\hat{L}_{\text{train}} \leq \hat{L}_{\text{loobs}} < \gamma$ would hold. A violation of these inequalities would imply that the learned model is quite poor. For example, $\gamma < \hat{L}_{\text{train}}$ for classification error would imply that f predicts worse than a random coin toss on the training set. The relative overfitting rate is then given by

$$r = \frac{\hat{L}_{\text{loobs}} - \hat{L}_{\text{train}}}{\gamma - \hat{L}_{\text{train}}}. \tag{9.13}$$

For simple models r is close to 0, and for flexible models r is close to 1. The optimal weight w is then

$$w = \frac{e-1}{e-r}. \tag{9.14}$$

In practical applications, the number of bootstrap replicas varies from a few dozen to a few thousand. Bootstrap is usually more expensive than a single round of 10-fold validation. As shown in Kohavi (1995) and Efron (1983), bootstrap estimates typically have lower variance. This lower variance appears to be a virtue of multiple resampling runs, not of bootstrap *per se*. A comparative study in Kim (2009) suggests that repeated cross-validation can produce classification error estimates with variance as low as those from the .632 and .632+ methods.

9.4.3
Sampling with Stratification

In physics, analysts can sample with or without stratification. For example, to estimate the uncertainty of a parameter estimate, you could generate data by Monte Carlo. If the data are composed of several classes, you could fix the number of observations per class in every simulated dataset, that is, stratify by class. Alternatively, you could sample without stratification, that is, draw a number of observations for each class assuming a Poisson distribution.

Either sampling scheme can be preferred in various settings. It is important to remember that stratified sampling reduces the variance of an estimator in the presence of systematic effects. If the classification errors for two classes are substantially different, the uncertainty estimates obtained by sampling with and without stratification can dramatically differ as well. The same applies to resampling schemes such as cross-validation and bootstrap.

Let us illustrate this problem using a simple setup. Train a binary classifier and obtain the uncertainty of its error estimate using repeatedly simulated sets of test data. Suppose the model classifies observations of one class with error ϵ_1 and observations of the other class with error ϵ_2. The number of classification failures in each class, $N_k \hat{\epsilon}_k$, is modeled as a binomial random variable. The two binomial random variables, $N_1 \hat{\epsilon}_1$ and $N_2 \hat{\epsilon}_2$, are obtained on data not used for training the classifier and therefore independent. For simplicity, assume that the overall number of observations in every simulated experiment, $N_1 + N_2$, is fixed at N.

Suppose the class proportions in every simulation are kept at π_1 and $\pi_2 = 1 - \pi_1$. Estimate the error for every simulated experiment by putting

$$\hat{\epsilon} = \pi_1 \hat{\epsilon}_1 + \pi_2 \hat{\epsilon}_2. \tag{9.15}$$

We know that $\text{Var}[\hat{\epsilon}_k] = \epsilon_k(1-\epsilon_k)/N_k$ with $N_k = \pi_k N$. Taking the variance of both sides of (9.15), we obtain

$$\text{Var}[\hat{\epsilon}] = \frac{\pi_1 \epsilon_1(1-\epsilon_1) + \pi_2 \epsilon_2(1-\epsilon_2)}{N}. \tag{9.16}$$

Let us turn now to the case without stratification. Build a hierarchical model in which ϵ is a random variable with probability mass function $P(\epsilon = \epsilon_k) = \pi_k$ and the number of classification failures $N\hat{\epsilon}$ is a binomial random variable with rate ϵ. The variance of the error estimate is then

$$\text{Var}[\hat{\epsilon}] = E[\text{Var}[\hat{\epsilon}|\epsilon]] + \text{Var}[E[\hat{\epsilon}|\epsilon]]. \tag{9.17}$$

The within-class variance, $E[\text{Var}[\hat{\epsilon}|\epsilon]]$, is identical to (9.16). The second term, the between-class variance $\text{Var}[\epsilon]$, is trivially computed. The overall variance is then

$$\text{Var}[\hat{\epsilon}] = \frac{\pi_1 \epsilon_1(1-\epsilon_1) + \pi_2 \epsilon_2(1-\epsilon_2)}{N} + \pi_1 \pi_2 (\epsilon_1 - \epsilon_2)^2. \tag{9.18}$$

The variance reduction by stratification is most pronounced when the class fractions are comparable and their classification errors differ. The same conclusion generally applies to resampling techniques with and without stratification, although these techniques cannot be described by a simple model due to the lack of independence across data sets.

What sampling (or resampling) scheme should you choose? The answer depends on your objective. In a typical physics analysis, counts of signal and background candidates observed in the signal region obey the Poisson distribution. This calls for sampling without stratification if the objective is estimation of the uncertainty. On the other hand, cross-validation is usually undertaken to obtain an estimate of the predictive power of a model for a known number of signal and background observations in the training data. Stratification then reduces the variance of the estimate and is the recommended choice. Note that the (re)sampling method and the method for computing the variance of the estimate must be consistent. It would make no sense, for example, to sample with stratification and use an unstratified estimate of variance, $\hat{\epsilon}(1-\hat{\epsilon})/N$, for the simple problem above.

Beware of what your software tool does! Classifiers in the Statistics Toolbox in MATLAB by default use cross-validation stratified by class. You can choose between cross-validation with and without stratification using the `cvpartition` utility. The `bootstrp` function bootstraps data without stratification.

9.5
Data with Unbalanced Classes

In particle and astrophysics, training data are often simulated. The cost of this simulation often varies from one class to another. Some classes, for instance signal or hard-to-classify background, are intentionally produced in larger quantities. Consequently, class frequencies in the training data do not represent the true frequencies observed in the experiment.

Suppose we wish to minimize the error of classification into A and B, where A and B are two equally likely processes. For some reason, we generate 10 000 observations of class A and only 100 observations of class B. A classifier trained on this set would learn mostly how to minimize the classification error for class A. If we applied this classifier to a test set in which A and B were equally mixed (as in the real world), we would likely obtain a large error. If we trained a classifier on data with A and B mixed in equal proportion, we would obtain a model with low test error.

In this example, we know the expected probabilities of classes A and B before performing the experiment. In the experiment, the actual quantities of A and B may not be exactly equal but should be fairly close. We can therefore set the prior probabilities for the two classes equal: $P(A) = P(B) = 1/2$. Suppose $P(x|A)$ and $P(x|B)$ are the learned distributions for the two classes. Taking the assumed prior probabilities and these distributions, we apply the Bayes rule to estimate the posterior probabilities at x,

$$P(A|x) = \frac{P(x|A)P(A)}{P(x|A)P(A) + P(x|B)P(B)},$$

with a similar expression for $P(B|x)$.

We use the Bayesian terminology such as "prior" and "posterior", but in essence all we do is adjust the magnitudes of the class components in the data. Physicists usually describe this procedure as "weighting" and apply it routinely.

How can we mix the two classes in equal proportion if they are not represented equally in the training set? There are three major strategies to accomplish this:

1. If the classifier can adjust class probabilities for training, set the prior probabilities to uniform.
2. Learn the classifier on a replica of the training set in which the majority class is undersampled.
3. Learn the classifier on a replica of the training set in which the minority class is oversampled.

Let us discuss these strategies in detail.

9.5.1
Adjusting Prior Probabilities

Classifiers handle prior class probabilities in various ways. Section 11.1 (discriminant analysis), Section 12.1 (neural networks), Section 13.5.1 (support vector machines), Section 13.6.1 (probability density estimation), Section 13.6.3 (nearest neighbor rules), and Section 14.1 and 14.2 (decision trees). Here we consider three learners: decision tree, discriminant analysis, and nearest neighbor rule.

A classification tree reviewed in Chapter 14 constructs decision splits by minimizing a criterion such as the Gini index or cross-entropy. These criteria are functions of class probabilities in the tree nodes. A change in the class prior probabilities has a direct impact on the training process: all class probabilities passed to the split criterion function are multiplied by a set of constant factors, one factor per class. Posterior probabilities predicted by the tree incorporate these class priors.

Discriminant analysis reviewed in Chapter 11 essentially ignores prior probabilities for training. For simplicity, let us focus on quadratic discriminant analysis (QDA). For QDA, we assume that classes are drawn from multivariate normal distributions with unequal covariance matrices. To estimate the covariance matrix for a class, we use all observations from this class, no matter how probable they are with respect to the other classes. The prior class probability therefore does not affect the training process at all. A trained discriminant computes the pdf for each class k at point x, $P(x|k)$. The posterior probability of class k at x can be found using the Bayes rule: $P(k|x) \propto P(x|k)P(k)$. Thus, class prior probabilities are used by the discriminant for prediction but not for training.

One nearest neighbor (1-NN) rule is an example of a classifier insensitive to class priors. An unlabeled observation x has the same nearest neighbor in the training (labeled) data irrespective of how probable the classes are.

Can you use prior probabilities for your classification task? To answer this question, you need to understand the theory behind the chosen classifier and how the software implementation uses these probabilities. Treatment of class priors by decision trees, QDA, and 1-NN rule are, for the most part, unambiguous. In contrast, recipes for using prior probabilities by support vector machines or neural networks are less certain.

9.5.2
✄ Undersampling the Majority Class

Undersampling the majority class, also known as *one-sided sampling*, has been explored in several publications. Despite the obvious loss of information, undersampling has been shown to perform well on real-world data. Undersampling comes with an advantage: if the sampled fraction of the majority class is small, the classifier can be trained much faster on less data.

Drummond and Holte (2003) investigate the predictive performance of undersampling for various fractions of the majority class retained in the training set. In their analysis, observations from the majority class are deleted at random. They use repeated cross-validation to reduce the variance of the performance estimate. Their analysis therefore does not reflect the increase in the variance of a single classifier built on an undersampled set due to the data reduction.

Kubat and Matwin (1997) propose deleting only borderline observations from the majority class. Such borderline observations are identified as observations participating in *Tomek links*. Observations (x_1, y_1) and (x_2, y_2) form a Tomek link in a dataset if x_1 is the nearest neighbor of x_2, x_2 is the nearest neighbor of x_1, and $y_1 \neq y_2$. Wilson (1972) proposes deleting an observation from the training set if its label is different from that predicted by its k nearest neighbors; Batista *et al.* (2004) and Barandela *et al.* (2003) set $k = 3$ in their studies of Wilson's cleaning procedure. The fraction of the majority class removed by these data editing algorithms varies from one dataset to another and can be negligibly small for some.

Yen and Lee (2009) describe several undersampling algorithms based on clustering. After the labeled data are clustered, they select a fixed fraction of majority-class observations from every cluster, thus controlling the overall size of the majority class sample. They offer several modifications of this algorithm based on distance between majority-class and nearest minority-class observations.

The algorithms described so far grow one classifier on an undersampled version of the training set. A natural extension of this approach is to learn an ensemble of classifiers on many subsets of the training data with the undersampled majority class. We review ensemble learning in Chapter 15. Here, we mention a few algorithms tailored for learning on imbalanced data. If you are not familiar with ensemble learning, you may wish to read Chapter 15 first.

The `EasyEnsemble` algorithm proposed in Liu *et al.* (2006) forms a subset for every classifier in the ensemble by taking all observations from the minority class and sampling the same number of observations from the majority class at random with replacement. Chen *et al.* (2004) use essentially the same technique to generate data subsets for *balanced random forest* except, in the spirit of random forest, they bootstrap the minority class. The `BalanceCascade` algorithm, also proposed in Liu *et al.* (2006), uses the same algorithm to generate subsets. In addition, at every step `BalanceCascade` finds a threshold on the classifier output to match the desired false positive rate and removes observations of the majority class correctly classified by this threshold from the data available for future sampling. Ricamato *et al.* (2008) describe a modification of `EasyEnsemble` in which the majority class is sampled without replacement. To ensure that every observation from the majority class be used for training, they take the number of subsets for the ensemble to be the ratio of the majority and minority class size. Zhou and Liu (2006) describe a more involved sampling scheme for constructing an ensemble of neural networks with data editing based on a 1-NN rule. `EasyEnsemble` and `BalanceCascade` combine classifiers built on subsets of the training data by taking a simple average over their predictions.

Seiffert *et al.* (2008) propose a boosting algorithm based on the same sampling idea. `RUSBoost` (boosting by random undersampling) uses `AdaBoost.M2`, a multiclass version of the well-known `AdaBoost` algorithm, described in Section 15.1.6, modulo the sampling scheme. At every step, `RUSBoost` selects all observations from the minority class and selects the same number of observations from the majority class or classes at random using the observation weights as multinomial sampling probabilities. `RUSBoost` trains a weak classifier on the selected subset. Then `RUSBoost` applies (15.26) to reweight observations in the entire set, including observations not used for training this weak classifier. The weights for misclassified observations increase, and the weights for correctly classified observations decrease. The updated weights are then used in the next boosting iteration. The final prediction of the ensemble is computed by taking a weighted vote over the weak classifiers built on the data subsets. Note that Seiffert *et al.* (2008) apply the `AdaBoost.M2` algorithm even to problems with two classes because it is protected against premature termination. The two-class version of `AdaBoost`, also known as `AdaBoost.M1`, continues adding learners to the ensemble as long as the error from the last learner falls below 50%. Since the learner is trained on a fraction of the data and the error is evaluated on the entire set, this condition is often violated. As shown in Section 15.1.6, `AdaBoost.M2` avoids this problem by using an alternative definition of the error.

9.5.3
✂ Oversampling the Minority Class

An other way of mixing the two classes in equal proportion is by oversampling the minority class. Simple-minded oversampling increases the time spent on training a classifier and usually fails to produce a more accurate predictive model. A more promising approach is synthesizing new observations of the minority class.

A well-known method for generating synthetic data is SMOTE (synthetic minority oversampling technique), described in Chawla *et al.* (2002). This algorithm positions new observations randomly between a minority-class observation and its five same-class nearest neighbors. The position of a new observation is generated uniformly on a nearest-neighbor link. The number of used links depends on the oversampling factor. For example, if the minority class needs to be oversampled by a factor of two, SMOTE places one new observation on each of the two nearest-neighbor links randomly selected out of five. If the oversampling factor is ten, SMOTE places two new observations on each of the five links.

Several modifications of the SMOTE algorithm have been proposed. He *et al.* (2008) set the number of synthesized examples for every minority-class observation proportional to the number of majority-class cases among its nearest neighbors. Han *et al.* (2005) synthesize new minority-class examples mostly in the regions where the two classes overlap; again, such regions are found using nearest neighbor rules. Batista *et al.* (2004) explore SMOTE in conjunction with Tomek links.

The SMOTEBoost algorithm described in Chawla *et al.* (2003) deploys the well-known AdaBoost algorithm described in Sections 15.1.2 and 15.1.3 with one modification: at every boosting step, the minority-class data are oversampled to compensate the skew in the class distribution. The DataBoost-IM algorithm due to Guo and Viktor (2004) finds hard-to-classify observations and uses them as templates to produce synthetic examples for both classes. Since boosting increases weights of misclassified observations, these hard-to-classify cases are selected from observations with largest weights. The algorithm generates many synthetic examples for the minority class and a few for the majority class, maintaining the class balance for every weak learner.

9.5.4
Example: Classification of Forest Cover Type Data

We apply two algorithms to the forest cover type data hosted at the UCI repository (Frank and Asuncion, 2010). Britsch *et al.* (2010) describe a three-stage technique for learning on severely skewed data. They use this technique for reconstruction of $D^0 \to \pi^+ K^-$ decays in the simulated LHCb data and, to verify their findings on a substantially different learning problem, for classification of the forest cover types. As we do not have access to their $D^0 \to \pi^+ K^-$ data, we can only compare results for the forest cover types.

Britsch *et al.* (2010) focus on separating the minority class (signal) from the majority class (background) to extract a signal of high purity. They evaluate the predictive performance of their method by plotting ROC curves for very low values of the false positive rate (FPR). ROC curves are discussed in Section 10.2.

The terms "minority class" and "majority class" in this case are somewhat deceptive. Because Britsch *et al.* (2010) aim at obtaining a signal of high purity, signal becomes the majority class and background becomes the minority class in the relevant region of the input space. This is why Britsch *et al.* (2010) oversample background in their study.

Following Britsch *et al.* (2010), we approach learning on the forest cover type data as a binary classification problem aimed at separating type Cottonwood/Willow from all other types. We use only 10 integer variables out of 54 variables available for these data. We split the data evenly between training and test sets, which leads to a slightly different splitting than the one used in Britsch *et al.* (2010); it seems unlikely that this difference could have a noticeable effect on this comparison.

We apply two ensemble algorithms, `GentleBoost` described in Chapter 15 and `RUSBoost` described in this chapter, using decision tree as the weak learner. These algorithms, provided by the `fitensemble` function in the Statistics Toolbox of MATLAB, are conceptually simpler than the three-stage technique described in Britsch *et al.* (2010). For `GentleBoost`, we set the learning rate to 0.1 and the minimal leaf size in a tree to one half of the number of observations in the minority class. Instead of optimizing these parameters, we simply guessed their values based on experience. For `RUSBoost`, we use the default setting of `fitensemble` to grow an ensemble on majority and minority classes mixed in equal amounts for every tree. We reduce the minimal leaf size to grow trees with approximately as many leaves as for `GentleBoost`. In this example, `RUSBoost` is significantly faster than `GentleBoost` because every tree is grown on a much smaller set and therefore needs to consider fewer split candidates. Having trained the two classifiers, we compute their ROC curves using the `perfcurve` function. This function can compute confidence bounds for an ROC curve by bootstrap, as described in Section 10.2.

We include a MATLAB example below. As seen from Figure 9.5, the two ROC curves are very close to the best curve shown in Britsch *et al.* (2010). The technique described in Britsch *et al.* (2010) produces the best ROC curve when the training set includes as many background observations as possible. To obtain the best performance on the forest cover type data, they oversample background by a factor of 5 using `SMOTE`. In contrast, no oversampling is applied in our study. `GentleBoost` uses the entire data "as is". `RUSBoost` undersamples background by more than a factor of 200.

Contents
- Load data
- How many observations of each class do we have?
- GentleBoost
- RUSBoost
- ROC curves
- ROC with confidence bounds
- Plot the ROC curves.

Load data
Follow the paper "Classifying extremely imbalanced data sets" by Britsch *et al.* (2010). Use only 10 integer variables for the `covtype` data from the UCI reposi-

tory. The data are split evenly across training and test sets. We omit the MATLAB code loading the data in memory.

How Many Observations of Each Class Do We Have?

```
tabulate(Ytrain) % classes for training data
```

```
   Value    Count    Percent
     0     289132    99.53%
     1       1374     0.47%
```

```
tabulate(Ytest) % classes for test data
```

```
   Value    Count    Percent
     0     289133    99.53%
     1       1373     0.47%
```

GentleBoost

Grow an ensemble of decision trees by GentleBoost. Set the minimal leaf size to roughly one half of the minority class 1. Use a popular value of 0.1 for the learning rate. Balance the two classes by enforcing equal prior probabilities.

```
minleafForGentle = round(sum(Ytrain==1)/2)

minleafForGentle =
    687

t = ClassificationTree.template('minleaf',700);
gentle = fitensemble(Xtrain,Ytrain,'GentleBoost',100,t,...
    'prior','uniform','LearnRate',.1);
```

RUSBoost

Grow an ensemble of decision trees by RUSBoost. The size of the training set for every tree is twice the size of class 1 in the training data. Reduce the minimal leaf size proportionally to produce trees with the same number of leaves as GentleBoost.

```
minleafForRUS = round(minleafForGentle/numel(Ytrain)*2*sum(Ytrain==1))

minleafForRUS =
    6

t = ClassificationTree.template('minleaf',minleafForRUS);
rus = fitensemble(Xtrain,Ytrain,'RUSBoost',100,t);
```

ROC Curves

Compute predicted classification scores for the test data. The returned array of scores is of size N-by-2 for N observations in the test data and 2 classes. Class 1 (signal) is the second class, and that is why we pick the second column from the array of scores.

Focus on the range [0, 0.0025] for FPR (false positive rate, or background acceptance). Divide this range into 21 intervals. Compute TPR (true positive rate, or signal efficiency) vs FPR at the interval endpoints. Compare with the best curve in Britsch *et al.* (2010, Figure 6).

```
fprVals = linspace(0,2.5e-3,21);
[~,Sgentle] = predict(gentle,Xtest);
SgentleSignal = Sgentle(:,2);
[fprGentle,tprGentle,threGentle] = ...
    perfcurve(Ytest,SgentleSignal,1,'xvals',fprVals);

[~,Srus] = predict(rus,Xtest);
SrusSignal = Srus(:,2);
[fprRUS,tprRUS,threRUS] = perfcurve(Ytest,SrusSignal,1,'xvals',fprVals);

fprPaper = [5 7.5 17 95]*1e-5;
tprPaper = [0.3 0.42 0.64 0.88];
```

ROC With Confidence Bounds

Use 1000 bootstrap replicas to compute confidence bounds for the ROC obtained by RUSBoost. Relax the minimal threshold to select a representative sample for bootstrapping.

```
sum(SrusSignal>min(threRUS)) % number of observations
                             % used for the ROC curve

ans =
       2025

keep = SrusSignal>3;
sum(keep) % number of observations used for bootstrapping

ans =
       4450

SrusSignalKeep = SrusSignal(keep);
YtestKeep = Ytest(keep);
N0 = sum(Ytest==0);
N1 = sum(Ytest==1);
[fp,tp] = perfcurve(YtestKeep,SrusSignalKeep,1,'xcrit','FP',...
            'ycrit','TP','xvals',fprVals*N0,'nboot',1000);
fpr = fp/N0;
tpr = tp/N1;
```

Plot the ROC Curves (Figure 9.5)
```
figure;
plot(fprPaper,tprPaper,'kd','MarkerSize',8);
```

```
hold;
plot(fprGentle,tprGentle,'k-');
errorbar(fpr,tpr(:,1),tpr(:,2)-tpr(:,1),tpr(:,3)-tpr(:,1),'k*');
xlim([0 0.003]);
ylim([0.2 1]);
fprTicks = cellstr(num2str(fprVals'));
set(gca,'xticklabel',[fprTicks(1:4:21)' {'0.003'}]);
hold off;
legend('Best in paper','GentleBoost','RUSBoost with bounds',...
       'Location','SE');
xlabel('False Positive Rate');
ylabel('True Positive Rate');
```

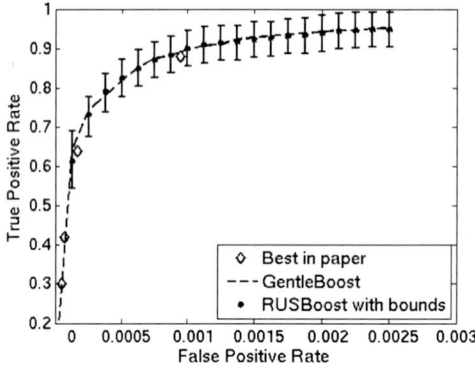

Figure 9.5 ROC curves obtained by the best approach in Britsch et al. (2010), GentleBoost and RUSBoost for the covtype data. The curve obtained by RUSBoost includes confidence bounds.

9.6
Learning with Cost

In Section 9.5, we set to minimize the classification error for balanced test data using an unbalanced training sample. Sometimes the analyst needs to face a different challenge. Suppose the two classes must be treated asymmetrically, that is, correctly predicting one class is more important than correctly predicting the other. In machine learning, this problem is addressed by using unequal misclassification costs. For instance, to obtain a signal of high purity at the expense of low signal efficiency, you could heavily penalize the learner for identifying background observations as signal. To do so, set the cost of misclassifying background as signal, c_{BS}, to a large value, and the cost of misclassifying signal as background, c_{SB}, to a small value. For K classes, form an asymmetric $K \times K$ matrix $\mathbf{C}$ in which element $c_{ij} \geq 0$ is the cost of classifying an observation of class i into class j. Set the diagonal elements of this matrix to zero.

Tools for learning on imbalanced data and learning with misclassification costs have a lot in common. For binary classification, unequal misclassification costs can be included by adjusting the prior class probabilities,

$$\tilde{\pi}_1 = \frac{c_{12}\pi_1}{c_{12}\pi_1 + c_{21}\pi_2}, \tag{9.19}$$

with a similar expression for $\tilde{\pi}_2$. For multiclass learning, the conversion is less straightforward and can be ambiguous in some problems. Yet in practice it is often possible to express the class asymmetry in terms of adjusted prior probabilities.

Sometimes adjusting the prior class probabilities does not improve the classification accuracy. An alternative approach is to apply post-training adjustments to the classifier predictions. Mix the classes in the training set in equal proportions and learn the classifier as usual. For binary classification, plot a ROC curve and find its optimal operating point. For two or more classes, predict into the class with the minimal cost of misclassification:

$$\hat{y}(x) = \arg\min_{y=1,\dots,K} \sum_{k=1}^{K} \hat{P}(k|x) c_{ky}, \tag{9.20}$$

where $\hat{P}(k|x)$ is the estimated posterior probability.

Assigning different weights to observations originating from different processes and inspecting the ROC curve are common approaches in physics analysis. Learning with cost, on the other hand, is unknown. This is why we refrain from describing this approach further.

9.7
Exercises

1. Label signal by +1 and background by -1. Assume that the pdfs for the signal and background, $P(x|Y = +1)$ and $P(x|Y = -1)$, are accurately known and positive over the entire domain $\mathcal{X}$. Define $s(\mathcal{X}) = S \int_{\mathcal{X}} P(x|Y = +1) dx$ and $b(\mathcal{X}) = B \int_{\mathcal{X}} P(x|Y = -1) dx$ to be the expected signal and background in a region $\mathcal{X}$ for unknown magnitudes S and B. The region $\mathcal{X}$ lies in the domain $\mathcal{X}$, that is, $\mathcal{X} \subset \mathcal{X}$. Let $Q(s(\mathcal{X}), b(\mathcal{X}))$ be a function with partial derivatives $\partial Q/\partial s > 0$ and $\partial Q/\partial b < 0$ (for example, $Q = s/\sqrt{s+b}$). Consider searching for the optimal region $\mathcal{X}^*$ maximizing the criterion Q: $\mathcal{X}^* = \arg\max_{\mathcal{X}} Q(s(\mathcal{X}), b(\mathcal{X}))$. Show that the optimal region $\mathcal{X}^*$ is defined by $P(x|Y = +1)/P(x|Y = -1) > C$ for some constant C.
2. Consider a modification of the above problem in which the background pdf $P(x|Y = -1)$ is known poorly. The expected background $b(\mathcal{X})$ then has a large uncertainty $\sigma_b(\mathcal{X})$ varying with the region $\mathcal{X}$. Replace $Q(s, b)$ with an appropriate optimization criterion to take the uncertainty information into account (for example, consider maximizing $Q(s(\mathcal{X}), b(\mathcal{X}) + \sigma_b(\mathcal{X}))$). Show that the optimal region $\mathcal{X}^*$ is not necessarily defined by $P(x|Y = +1)/P(x|Y = -1) > C$.
3. Consider estimating classification error, ϵ, by $\mathcal{K}$-fold cross-validation. Let $\hat{\epsilon}_k$ be the error estimate obtained for the kth fold with N_k observations. Assume that

the $\mathcal{K}$ estimates $\hat{\epsilon}_k$ are independent. Let $N = \sum_{k=1}^{\mathcal{K}} N_k$ be the total number of observations in the data. Consider two estimates of the classification error, $\hat{\epsilon}^{(1)} = \sum_{k=1}^{\mathcal{K}} N_k \hat{\epsilon}_k / N$ and $\hat{\epsilon}^{(2)} = \sum_{k=1}^{\mathcal{K}} \hat{\epsilon}_k / \mathcal{K}$. Show that both estimates are unbiased and the variance of the second estimate is not less than the variance of the first estimate.

4. In Section 9.4.3, we have compared sampling with and without stratification. Consider imbalanced data in which class 1 is under-represented in the training set, $\pi_1 \ll \pi_2$. Assume that in the real world the class proportions are close. How would you change the error computation (9.15) for such data? Using the new error definition, show that the variance values for the stratified and unstratified estimators could be very different.

References

Barandela, R., Sanchez, J., Garcia, V., and Rangel, E. (2003) Strategies for learning in class imbalance problems. *Pattern Recognit.*, **36**, 849–851.

Batista, G., Prati, R., and Monard, M. (2004) A study of the behavior of several methods for balancing machine learning training data. *ACM SIGKDD Explorations Newsletter – Special issue on learning from imbalanced datasets*, **6** (1).

Breiman, L. (1998) Arcing classifiers. *Ann. Stat.*, **26** (3), 801–849.

Britsch, M., Gagunashvili, N., and Schmelling, M. (2010) Classifying extremely imbalanced data sets, arXiv:1011.6224v1 [physics.data-an].

Chawla, N., Bowyer, K., Hall, L., and Kegelmeyer, W. (2002) SMOTE: Synthetic minority over-sampling technique. *J. Artif. Intell. Res.*, **16**, 321–357.

Chawla, N., Lazarevic, A., Hall, L., and Bowyer, K. (2003) SMOTEBoost: improving prediction of the minority class in boosting. *7th Eur. Conf. Princ. Pract. Knowl. Discov. Databases (PKDD'03)*, Lecture Notes on Computer Science, vol. 2838, pp. 107–119.

Chen, C., Liaw, A., and Breiman, L. (2004) Using random forest to learn imbalanced data. *Stat. Tech. Rep.* 666, University of California Berkeley.

Clopper, C. and Pearson, E. (1934) The use of confidence or fiducial limits illustrated in the case of the binomial. *Biometrika*, **26**, 404–413.

Domingos, P. (2000) A unified bias-variance decomposition for zero-one and squared loss. *Proc. 17th Natl. Conf. Artif. Intell. and 12th Conf. Innov. Appl. Artif. Intell.*

Drummond, C. and Holte, R. (2003) C4.5, class imbalance, and cost sensitivity: Why under-sampling beats over-sampling. *Int. Conf. Mach. Learn.*

Efron, B. (1983) Estimating the error rate of a prediction rule: Improvement on cross-validation. *J. Am. Stat. Assoc.*, **78** (382), 316–331.

Efron, B. and Tibshirani, R. (1997) Improvements on cross-validation: The .632+ bootstrap method. *J. Am. Stat. Assoc.*, **92** (438), 548–560.

Elisseeff, A. and Pontil, M. (2003) Leave-one-out error and stability of learning algorithms with applications. *Comput. Syst. Sci.*, **190**.

Frank, A. and Asuncion, A. (2010) UCI machine learning repository. http://archive.ics.uci.edu/ml (accessed 14 July 2013).

Friedman, J. (1997) On bias, variance, 0/1 loss, and the curse-of-dimensionality. *Data Min. Knowl. Discov.*, **1**, 55–77.

Guo, H. and Viktor, H. (2004) Learning from imbalanced data sets with boosting and data generation: the DataBoost-IM approach. *SIGKDD Explor.*, **6** (1), 30–39.

Han, H., Wang, W.Y., and Mao, B.H. (2005) Borderline-SMOTE: A new over-sampling method in imbalanced data sets learning. *Adv. Intell. Comput.*, **3644**, 878–887.

He, H., Bai, Y., Garcia, E., and Li, S. (2008) ADASYN: Adaptive synthetic sampling approach for imbalanced learning. *Int. Joint Conf. Neural Netw.*, IEEE, 3, pp. 1322–1328.

James, G. (2003) Variance and bias for general loss functions. *Mach. Learn.*, **51** (2), 115–135.

Kim, J.H. (2009) Estimating classification error rate: Repeated cross-validation, repeated hold-out and bootstrap. *Comput. Stat. Data Anal.*, **53** (11), 3735–3745.

Kohavi, R. (1995) A study of cross-validation and bootstrap for accuracy estimation and model selection. *Proc. 14th Int. Joint Conf. Artif. Intell.*, vol. 2, Morgan Kaufmann.

Kohavi, R. and Wolpert, D. (1996) Bias plus variance decomposition for zero-one loss functions. *ICML'96*, pp. 275–283.

Kubat, M. and Matwin, S. (1997) Addressing the curse of imbalanced training sets: One-sided selection. *Int. Conf. Mach. Learn.*, pp. 179–186.

Langford, J. (2005) Tutorial on practical prediction theory for classification. *J. Mach. Learn. Res.*, **6**, 273–306.

Liu, X.Y., Wu, J., and Zhou, Z.H. (2006) Exploratory under-sampling for class-imbalance learning. *Int. Conf. Data Min.*, pp. 965–969.

Ricamato, M.T., Marrocco, C., and Tortorella, F. (2008) MCS-based balancing techniques for skewed classes: An empirical comparison. *Int. Conf. Pattern Recognit.*, pp. 1–4.

Schapire, R. and Freund, Y. (1998) Boosting the margin: A new explanation for the effectiveness of voting methods. *Ann. Stat.*, **26** (5), 1651–1686.

Seiffert, C., Khoshgoftaar, T., Hulse, J.V., and Napolitano, A. (2008) RUSBoost: Improving classification performance when training data is skewed. *Int. Conf. Pattern Recognit.*, pp. 1–4.

Shawe-Taylor, J., Holloway, R., Bartlett, P., Williamson, R., and Anthony, M. (1996) A framework for structural risk minimisation. *Proc. 9th Annu. Conf. Comput. Learn. Theory*, ACM Press, pp. 68–76.

Tibshirani, R. (1996) Bias, variance and prediction error for classification rules, *Tech. Rep. 9602*, University of Toronto, Dept. of Statistics.

Valentini, G. and Dietterich, T. (2004) Bias-variance analysis of support vector machines for the development of SVM-based ensemble methods. *J. Mach. Learn. Res.*, **5**, 725–775.

Wilson, D. (1972) Asymptotic properties of nearest neighbor rules using edited data. *IEEE Trans. Syst. Man Cybern.*, **2** (3).

Yen, S.J. and Lee, Y.S. (2009) Cluster-based under-sampling approaches for imbalanced data distributions. *Expert Syst. Appl.*, **36**, 5718–5727.

Zhou, Z.H. and Liu, X.Y. (2006) Training cost-sensitive neural networks with methods addressing the class imbalance problem. *IEEE Trans. Knowl. Data Eng.*, **18** (1), 63–77.

10
Assessing Classifier Performance

Many measures of classification power are available. Their popularity varies from one field of application to another. In this chapter, we describe two popular measures, classification error and Receiver Operating Characteristic (ROC) curves. We also describe statistical tests for comparing two or more classification models.

The term "Receiver Operating Characteristic curve" may sound unfamiliar to some physicists. In effect, a ROC curve is background rejection rate plotted against signal efficiency. Such plots are widely used in analysis of physics data.

10.1
Classification Error and Other Measures of Predictive Power

The simplest estimate of the predictive power is classification error, or equivalently 0-1 loss (9.3) averaged over the joint distribution $P(x, y)$, as shown in (9.1). As a measure of the predictive power, classification error is often inadequate or crude. For example, if your goal is learning the class probabilities $P(y|x)$, a classification error estimate does little to help you assess the learned model. Even if the goal of your analysis is minimization of the 0-1 loss, classification error by itself does not tell the whole story. For instance, data are often re-analyzed when more become available. As you saw in Section 9.2, classification error can be decomposed into a sum of bias and variance terms. Learning with more data often leads to reduction of the variance term and does little to bias. If you design two classifiers with equal error rates, one with low bias and high variance and another one with high bias and low variance, only one of them can benefit from learning on a larger amount of data.

In the physics community, 0-1 loss is used less often than elsewhere. Examples of performance measures popular among physicists include ROC curves and specialized figures of merit such as precision, $s/\sqrt{s+b}$, for expected magnitudes of signal and background, s and b, respectively. The choice of an appropriate measure is tied to specifics of the analysis and is largely beyond the scope of this book.

In this chapter, we mostly discuss classification error and ROC curves. Although classification error is hardly a popular choice in physics analysis, it is the choice that is best understood, based on theory and empirical studies. When possible, we emphasize that an algorithm used for classification error can be used for a generalized measure of performance. ROC curves are relatively well studied and almost inevitably appear in every physics analysis with the exception of multiclass problems.

Assessing the power of a classification model is usually disentangled from building the optimal classification model. We discuss this issue in Section 9.1 and briefly repeat here. The workflow consists of two steps. First, build a model by minimizing an appropriate loss function. Second, measure the predictive power of this model using a metric other than the loss used for training. If the predictive performance is not satisfactory, select a new model and go back to the first step, and so on. In an ideal world, the metrics (loss functions) used for training a model and assessing its power would be identical. In practice, they differ. Until better tools are developed, the presented workflow will remain the most practical choice.

10.2
Receiver Operating Characteristic (ROC) and Other Curves

Receiver Operating Characteristic (ROC) curves are a popular tool for evaluation of binary classification models. Physicists are used to plotting background acceptance versus signal efficiency, or equivalently background and signal fractions identified as signal. The standard ROC curve is exactly that, but with the axes swapped and notation changed. *False positive rate* (FPR), or the fraction of background misclassified as signal, is plotted on the *x*-axis. *True positive rate* (TPR), or the correctly classified fraction of signal, is plotted on the *y*-axis. Various versions of this plot are encountered in analysis of physics data. We use the conventional notation and definitions borrowed from the statistics literature.

10.2.1
Empirical ROC curve

Suppose we have two classes, "positive" and "negative". Misclassified observations are called "false", and correctly classified observations are called "true". Here is a recipe for making an empirical ROC curve by applying a trained classifier to labeled data:

1. Compute soft scores predicted by this classifier for these data.
2. Sort the scores in ascending order. By convention, a high score indicates that this observation is likely from the positive class, and a low score indicates otherwise.
3. Remove duplicate entries from the sorted array of scores. The classifier can predict identical scores for distinct observations, but we only need to consider distinct score values.
4. Go through the sorted array of scores. For every score, count the number of positive and negative observations with scores above and below this value (in-

cluding the duplicates eliminated at the previous step). Observations from the positive class with scores above this threshold are *true positives* (TP), and observations from the positive class with scores below this threshold are *false negatives* (FN). Likewise, this score threshold splits observations from the negative class into *false positives* (FP) above the threshold and *true negatives* (TN) below the threshold. Record the values of TP and FP.

5. Divide the recorded TP counts by the total number of observations from the positive class to obtain TPR. Similarly, divide the recorded FP counts by the total number of negative observations to obtain FPR.
6. Plot the FPR value on the *x*-axis and the respective TPR value on the *y*-axis.

This recipe generalizes to weighted observations. Instead of adding counts, add weights below or above a given threshold and divide by the total weight in the respective class.

Note a subtle point: if we count observations above and below a certain threshold at step 4, we can include an observation located exactly at the threshold in either count. Various software implementations use various conventions. The perfcurve function available from the Statistics Toolbox in MATLAB counts an observation as a positive classification if its score is greater or equal to the threshold and negative otherwise.

We include a MATLAB script to plot ROC curves obtained by pseudolinear discriminant analysis (PLDA) described in Chapter 11 and 200 boosted stumps[1] described in Chapter 15 for ionosphere data available from the UCI repository (Frank and Asuncion, 2010). This is a binary classification problem aimed at separating "good" and "bad" radar returns. We use PLDA instead of conventional LDA because some variables in the ionosphere data have zero variance. The predicted scores are computed by 10-fold cross-validation. Boosted stumps outperform PLDA, and both classifiers perform better than random guess shown with a diagonal line in Figure 10.1.

```
% Load data
load ionosphere;

% For reproducibility, set the random number generator seed for data
% partitioning.
rng(1);

% Build a cross-validated pseudo linear discriminant
cvplda = ClassificationDiscriminant.fit(X,Y,...
    'DiscrimType','pseudoLinear','crossval','on');

% Grow a cross-validated ensemble of decision stumps boosted by
% AdaBoostM1
cvada = fitensemble(X,Y,'AdaBoostM1',200,'Tree','crossval','on');

% Compute predicted scores for data not used for training
```

---

1) A stump is a decision tree with two leaves.

10 Assessing Classifier Performance

```
[~,SfitLDA] = kfoldPredict(cvplda);
[~,SfitAda] = kfoldPredict(cvada);

% Obtain FPR and TPR choosing 'good returns' as the positive class
[fprLDA,tprLDA] = perfcurve(Y,SfitLDA(:,2),'g');
[fprAda,tprAda] = perfcurve(Y,SfitAda(:,2),'g');

% Plot the ROC curves
plot(fprLDA,tprLDA,'b-');
hold;
plot(fprAda,tprAda,'r-.');
line([0 1],[0 1],'color','c');
xlabel('False positive rate');
ylabel('True positive rate');
legend('Pseudo LDA','Boosted stumps','Fair coin toss','Location','SE');
```

The best ROC curve is the one closest to the upper left corner of the plot in Figure 10.1, furthest away from the diagonal line. If a ROC curve falls below the diagonal line or wiggles around it, the classifier is rather useless.

10.2.2
Other Performance Measures

A variety of ROC-like curves are used in different fields. We summarize popular measures of performance and various versions of their names in Table 10.1. In this book, we discuss only ROC and related quantities. Other measures are included in the table for readers willing to explore the literature on their own.

10.2.3
Optimal Operating Point

After computing a ROC curve, the analyst often wants to find an optimal operating point on the curve. The choice of the optimal operating point is usually driven by specifics of the experiment such as acceptable background levels or minimal signal efficiency. If your goal is to minimize the overall classification error, the op-

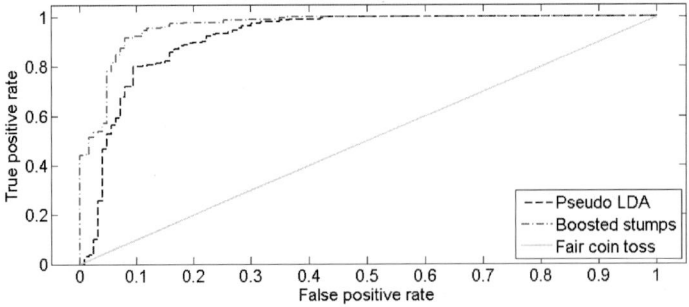

Figure 10.1 Cross-validated ROC curves for ionosphere data.

10.2 Receiver Operating Characteristic (ROC) and Other Curves

Table 10.1 Statistics used in ROC-like curves. P = positive, N = negative, T = true, F = false, and R = rate. ALL = TP + FN + FP + TN is the number of all observations.

Name	Definition
Rate of positive predictions	$\dfrac{TP + FP}{ALL}$
Rate of negative predictions	$\dfrac{TN + FN}{ALL}$
Accuracy	$\dfrac{TP + TN}{ALL}$
TPR, or sensitivity, or recall, or signal efficiency	$\dfrac{TP}{TP + FN}$
FPR, or fallout, or background acceptance	$\dfrac{FP}{TN + FP}$
TNR, or specificity	$\dfrac{TN}{TN + FP}$
FNR, or miss	$\dfrac{FN}{TP + FN}$
Positive predictive value, or precision, or signal purity	$\dfrac{TP}{TP + FP}$
Negative predictive value	$\dfrac{TN}{TN + FN}$

timal operating point can be found using a prescription in Provost and Fawcett (2001). Draw a line with slope N/P through the upper left corner of the ROC plot; here, P and N are observation counts in the positive and negative class, respectively. Slide this line down and to the right until it touches the ROC curve. The point of intersection is the one with the minimal classification error. Indeed, the number of misclassifications is equal to (borrowing notation from Table 10.1) FP + FN, or equivalently FPR × N + (1 − TPR) × P. Therefore, two points on the ROC curve, (FPR_1, TPR_1) and (FPR_2, TPR_2), give equal classification errors if $(TPR_2 − TPR_1)/(FPR_2 − FPR_1) = N/P$. The point with the minimal error must be closest to the point of perfect classification, (0, 1) in ROC coordinates. The rest follows.

For weighted data, P and N ought to be replaced by the respective weights. For instance, for unequal class prior probabilities π_P and π_N, the classification error is proportional to $\pi_N \cdot FP + \pi_P \cdot FN$ and the optimal slope becomes $(\pi_N N)/(\pi_P P)$. Although the ROC curve is invariant with respect to the prior probabilities, the location of the optimal operating point is not.

Because the optimal operating point may depend on the prior class probabilities, or equivalently the weights in which the class distributions are mixed in the data, a reliable estimation of this point may not be possible. In this case, the area under the curve, discussed in the next section, can be used.

10.2.4
Area Under Curve

Compare the two ROC curves in Figure 10.1. We have a winner: boosted stumps outperform PLDA everywhere. How would we determine which classifier is better if the two curves crossed? If we found optimal operating points for the two curves (having defined "optimality" in any way we like), we could compare the two classifiers at these operating points. Comparing points is easier than comparing curves. For example, if the optimal operating points are obtained as described in the previous section, one of them would be closer to the upper left corner of the ROC plot than the other.

Unfortunately, sometimes it is not clear how to select a good operating point. Consider, for example, a problem in which we know the distribution for the negative class (background) reasonably well but only have a crude guess about its magnitude. Since the choice of the optimal operating point depends on the class probabilities, this choice must be deferred until we obtain a better estimate of the background yield. How do we compare classifiers before that stage of the analysis is reached? It is tempting to summarize a ROC curve with a single number and base classifier comparison on that number. One popular measure of this kind is the *area under the ROC curve* (AUC).

AUC has a well-defined statistical interpretation. If you choose a pair of observations at random, one from the positive class and one from the negative class, AUC is the probability of obtaining a larger classification score for the positive observation. As shown in Mason and Graham (2002), AUC is related to the Wilcoxon rank sum test (also called Mann–Whitney U test) for equality of two independent univariate distributions. The test statistic, $U = P \times N \times AUC$, for positive and negative class sizes P and N, respectively, has been well studied.

AUC was advertised in Bradley (1997) as a powerful tool for estimating the predictive power and included in classic introductory papers on ROC curves such as Fawcett (2006) and others. Later AUC was criticized as an incoherent measure of the classifier performance in Hand (2009). This criticism was countered by Flach *et al.* (2011) who showed that AUC is a coherent measure of the predictive power as long as we make no assumptions about the relative misclassification costs of the two classes, or equivalently about the optimal operating point. While the reader may find this debate entertaining, we dare summarize it in a simple proposition: AUC is a poor measure of the predictive power if we have some knowledge of the optimal operating point. The more we know about the operating point, the less relevant AUC is.

10.2.5
Smooth ROC Curves

Empirical ROC curves are not smooth: the small steps observed in Figure 10.1 are due to the discreteness of the FPR and TPR statistics. An empirical ROC curve attains smoothness asymptotically in the limit of a large sample size. If the analyst can afford simulating or collecting many observations from both classes, a discrete ROC curve should suffice. Sometimes the analyst is not able to do so. In rare signal searches, for instance, signal selection requirements are typically optimized to obtain signal of high purity (low FPR). Simulated background is suppressed by a large factor producing just a few observations in the interesting region of the ROC curve. Smoothing can then let the analyst search for the optimal operating point *between* the few obtained ROC points.

If you wish to compute a smooth curve for a small sample, you can use a parametric, semi-parametric or nonparametric model. Before describing these models, let us formalize the problem. Let S be a random variable for the classification soft score, and let $F(s) = P(S \leq s | Y = -1)$ and $G(s) = P(S \leq s | Y = +1)$ be score cdfs for the negative and positive class, respectively. Using terminology from Table 10.1, F is the true negative rate (TNR) and G is the false negative rate (FNR). If p is FPR,

$$r(p) = 1 - G(F^{-1}(1-p)) \tag{10.1}$$

is TPR. The equation above defines a ROC curve.

To parametrize (10.1), we can assume specific forms for F and G such as normal distributions. There is no reason to expect a priori that the scores are normally distributed; however, reality conforms to this assumption in certain problems. We do not entertain this approach further because most physicists feel comfortable working with nonparametric models and such models, if properly constructed, produce good results for any pair of F and G.

A less restrictive approach is a semi-parametric binormal model. The ROC model (10.1) is invariant under a strictly monotone transformation of score S. If we postulate the existence of such a transformation converting both F and G to normal distributions, we can write

$$r(p) = \Phi(\mu + \sigma \Phi^{-1}(p)) \tag{10.2}$$

for the standard normal cdf Φ and unknown parameters μ and σ. The actual transformation converting F and G to this form does not need to be explicitly found. Hsieh and Turnbull (1996) describe a generalized least-squares procedure for fitting parameters μ and σ in (10.2). Cai and Moskowitz (2004) and Zhou and Lin (2008) propose fitting techniques based on profile likelihood.

A fully nonparametric technique amounts to approximating $F(s)$ and $G(s)$ by a sum of kernel functions,

$$\hat{F}(s) = \frac{1}{N_{-1}} \sum_{n=1}^{N_{-1}} K\left(\frac{s - s_n}{h_{-1}}\right), \tag{10.3}$$

with a similar expression for $\hat{G}(s)$. Here, $K((s-s_n)/h)$ is a cdf produced by a kernel pdf placed at s_n, N_{-1} is the size of the negative class, and h_{-1} is the kernel width for the negative class. The conventional wisdom predicts that selecting a functional form for the kernel is less important than finding a good bandwidth h.

By choosing a small bandwidth, we can make the smooth curve follow the empirical one very closely. But if the smooth curve is as jagged as the empirical one, there is no point in smoothing. On the other hand, if we choose a large bandwidth, the smooth curve will approximate the empirical one poorly. Put in a different way, there is a trade-off between bias and variance. Small bandwidths produce curves with small bias and large variance, whereas large bandwidths produce curves with large bias and small variance. The optimum is somewhere in between.

As shown in Chapter 5, the optimal bandwidth for density estimation scales as $O(N^{-1/5})$ for the number of fitted observations N. In contrast, Lloyd and Yong (1999) show that the optimal bandwidths for estimating $\hat{r}(p)$ at fixed p scale as $O(N_i^{-1/3})$ for $i \in \{-1, +1\}$. If scores for class i follow a normal distribution, they recommend estimating the bandwidth by putting

$$h_i = 1.58 \frac{\hat{\sigma}_i}{N_i^{1/3}} \tag{10.4}$$

for the normal kernel, where $\hat{\sigma}_i$ is the estimated standard deviation of the score distribution. A robust estimator of σ_i is obtained from quantiles of the empirical distribution, $\hat{\sigma} = (q_{0.75} - q_{0.25})/1.34$. If the scores do not follow a normal distribution, the formulas for estimation of h_i in Lloyd and Yong (1999) become noticeably more complex.

Hall and Hyndman (2003) minimize the mean integrated squared error (MISE)

$$E_{\hat{G},\hat{F},p}\{[\hat{r}(p) - r(p)]^2\} = \int_0^1 E_{\hat{G},\hat{F}}\{[\hat{G}(\hat{F}^{-1}(p)) - G(F^{-1}(p))]^2\}$$
$$\times F'(F^{-1}(p))dp \tag{10.5}$$

for the score pdf in the negative class $F'(s)$. Similarly, they conclude that the optimal bandwidths scale as $O(N_i^{-1/3})$ and provide a recipe for their empirical estimation. The MISE estimate is driven by high-density intervals in classification scores. In practice, one could be interested in poorly populated parts of the ROC curve, for example, the lower left corner. In that case, the MISE formula and recipe for computing the bandwidths need to be modified accordingly.

The advantage of ROC smoothing by kernel densities is its low coding cost. Utilities for kernel density estimation are available from many software suites. Although computing the optimal bandwidths may not be trivial, it is often possible to convert the score densities for both classes to roughly normal distributions by a simple monotone transformation. After conversion, the estimator (10.4) should be close enough to optimal for a good smoothing approximation.

We illustrate this approach by smoothing the ROC curve obtained by PLDA for the ionosphere data shown in Figure 10.1. We use the `ksdensity` function from the Statistics Toolbox in MATLAB for kernel smoothing. The scores returned by the `ClassificationDiscriminant` model are posterior class probabilities with distributions peaking near 0 and 1. Applying an inverse logit transformation, $\log(s/(1-s))$, we convert these scores to essentially the linear term $x^T \Sigma^{-1}(\mu_1 - \mu_2)$ modulo an additive constant. An explanation of this term can be found in Section 11.1. A test of normality due to Lilliefors (1967) applied to the score distributions shown in Figure 10.2 rejects the hypothesis of normality for the "good" class at the 5% level and barely fails to reject it for the "bad" class, returning a p-value of 7.9%. The estimated bandwidths, 0.41 and 1.69 for the "good" and "bad" classes, nevertheless lead to a good smoothed approximation in the well-populated region of the curve above FPR $= 0.2$, as seen in Figure 10.3. At FPR values below 0.2, the smoothed curve lies below the empirical one because TPR is overestimated. If we repeat the analysis after rejecting all observations with negative scores, the optimal bandwidth for the "bad" class decreases by roughly a factor of 3 leading to a much better agreement between the two curves, as seen in Figure 10.4.

```
% Find good and bad radar returns
isGood = strcmp(Y,'g');
isBad = strcmp(Y,'b');

% Count numbers of good and bad radar returns
Ngood = sum(isGood);
Nbad = sum(isBad);

% Recompute classification scores after applying inverse logit
% transformation. Select the second column of SfitLDA which corresponds to
% good radar returns ('g' class).
cvplda.ScoreTransform = 'invlogit';
[~,SfitLDA] = kfoldPredict(cvplda);
SfitLDA = SfitLDA(:,2);

% Compute empirical curve with score thresholds
[fprLDA,tprLDA,threLDA] = perfcurve(Y,SfitLDA,'g');

% Obtain quantiles for kernel bandwidth estimation
Qgood = quantile(SfitLDA(isGood),[0.25 0.75]);
Qbad = quantile(SfitLDA(isBad),[0.25 0.75]);

% Estimate optimal kernel bandwidth
Hgood = 1.58*(Qgood(2)-Qgood(1))/1.34/Ngood^(1/3);
```

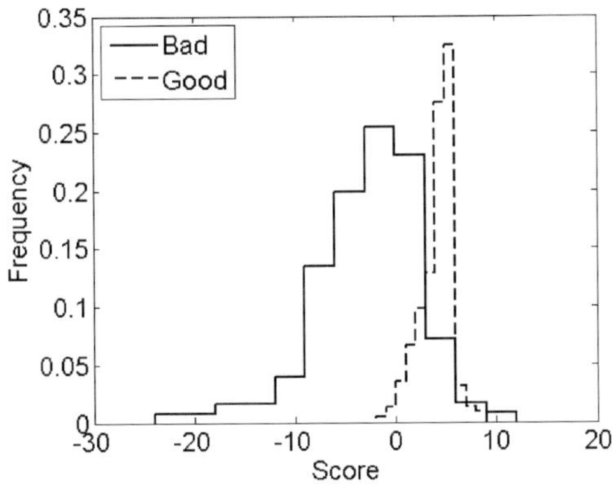

Figure 10.2 Score densities for the "bad" and "good" class obtained by PLDA with an inverse logit transformation applied.

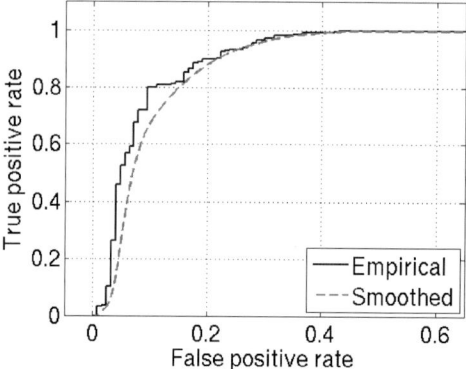

Figure 10.3 Empirical and smoothed ROC curves for ionosphere data.

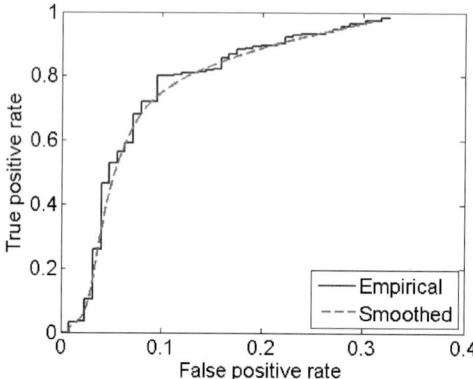

Figure 10.4 Empirical ROC curve and ROC curve smoothed in the low FPR region for ionosphere data.

```
Hbad = 1.58*(Qbad(2)-Qbad(1))/1.34/Nbad^(1/3);

% Compute smoothed TNR and FNR at 1000 points uniformly spaced from
% minimal to maximal score
thre = linspace(min(SfitLDA),max(SfitLDA),1000);
F = ksdensity(SfitLDA(isBad),thre,'width',Hbad,'function','cdf'); % TNR
G = ksdensity(SfitLDA(isGood),thre,'width',Hgood,...
    'function','cdf'); % FNR

% Plot empirical and smoothed ROC curve
figure;
plot(fprLDA,tprLDA);
hold;
plot(1-F,1-G,'r-');
legend('Empirical','Smoothed','Location','SE');
xlabel('False positive rate');
ylabel('True positive rate');
```

10.2.6
Confidence Bounds for ROC Curves

An introduction to ROC curves in Fawcett (2006) lists two most popular approaches for estimation of pointwise confidence intervals: *vertical averaging* (VA) and *threshold averaging* (TA). Several alternative approaches are reviewed in Mackassy and Provost (2005).

For VA, we fix FPR values and find confidence bounds for TPR. To use VA for horizontal averaging, swap the ROC axes, that is, fix TPR and find confidence bounds for FPR. The most common approaches to VA in the literature are:

1. Treat TPR as a binomial random variable and construct confidence bounds following one of the available prescriptions for the binomial distribution. If $\hat{r}$ is the measured TPR, the expected TPR can be set to $\hat{r}$ and the standard deviation can be estimated as $\hat{\sigma}_r = \sqrt{\hat{r}(1-\hat{r})/N_+}$ for the size of the positive class N_+ leading to 1σ confidence bands $\hat{r} \pm \hat{\sigma}_r$. This approximation works well for $\hat{r}$ far from 0 and 1. Alternatively, we can obtain 1σ confidence bounds on r as solutions to the likelihood-ratio equation $-2\log(L(r)/L(\hat{r})) = 1$ with $n_+ = \hat{r}N_+$ and $L(r) = \binom{N_+}{n_+} r^{n_+}(1-r)^{N_+ - n_+}$. Even better, we can find these confidence bounds by full frequentist construction.
2. Estimate the distribution of TPR empirically. Take the scores $\{s_n\}_{n=1}^N$ and respective class labels $\{y_n\}_{n=1}^N$ used to obtain the ROC curve. Resample or simulate the score-label pairs $\{(y_n, s_n)\}_{n=1}^N$ many times. Obtain a ROC curve for every simulation or resampling run. Record values of TPR at fixed FPR to form the empirical distribution. Set the confidence bounds to the respective quantiles of this distribution. For example, for 1σ bands use, roughly, 0.16 and 0.84 quantiles. Optionally smooth the empirical distribution by kernel density estimation or some other nonparametric (or semi-parametric) technique.
3. Sample TPR values, as described above, and fit the obtained sample to a parametric distribution, most usually normal.

For TA, we fix thresholds on the soft classification score and find bounds for TPR and FPR. The same approaches as for VA apply. Usually FPR and TPR are assumed independent, and two unrelated confidence intervals are found, one for FPR and one for TPR, at every point on the ROC curve. The assumption of independence is clearly violated, but in practice this issue is commonly ignored. Hilgers (1991) constructs independent confidence intervals for FPR and TPR and then forms a confidence rectangle. For example, to construct a 95% confidence rectangle, construct two 97.5% intervals since $(1 - 0.025)^2 \cong 0.95$. This technique obviously adds very little information. Constructing a bivariate confidence region for FPR and TPR would require estimation of the joint probability distribution. Similar to the univariate case, this could be done empirically but would require larger samples.

The binomial model is most appropriate for TA. Consider a continuous random variable. If its value is above some fixed threshold, count this event as success; otherwise count it as failure. If observations are drawn independently, the assumptions of the binomial model are satisfied. For VA, however, this threshold is found from the respective value of FPR. In this case, the binomial model fails to account for variance due to threshold estimation.

Mackassy et al. (2005) study frequentist coverage of several methods for computation of pointwise ROC confidence bounds. They report that intervals produced by the binomial model are too narrow and significantly undercover. Mackassy et al. (2005) emphasize this conclusion for VA and refer to a similar trend for TA. We could not verify their findings in our studies. A direct comparison of our results is not possible because Mackassy et al. (2005) do not specify if they use intervals based on the normal approximation $\hat{\sigma}_r = \sqrt{\hat{r}(1-\hat{r})/N_+}$, intervals obtained from the likelihood ratio, or intervals found by full frequentist construction. In our studies, frequentist (Clopper–Pearson) bounds exhibit sensible behavior. An example is shown in Figure 10.5. For this exercise, we use two bivariate normal distributions with equal covariance matrices. Linear discriminant reviewed in Chapter 11 attains Bayes error for these data. The ROC curve produced by the perfect classifier is shown in Figure 10.5a. To estimate coverage, we generate 10 000 datasets. Every dataset has 100 observations split evenly between the two classes. The soft classification score is set to the posterior probability of observing an instance of the positive class, $P(Y = +1|x)$. We divide the [0, 1] range for FPR into 30 equal intervals; similarly for the soft score threshold. For every dataset, we generate a ROC curve using scores predicted by the perfect classifier. We then find a set of TPR values for FPR at the interval midpoints and another set of TPR values for the chosen midpoint thresholds. We count how often the Clopper–Pearson binomial interval for the desired coverage level covers the true value. The result is shown in Figure 10.5b.

Computing confidence bounds by a nonparametric procedure of item 2 offers several advantages over the other two approaches:

1. We avoid making any assumptions about the distributions of FPR and TPR.

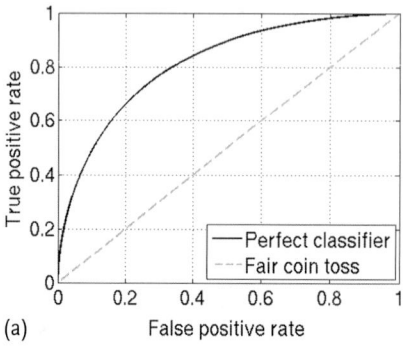
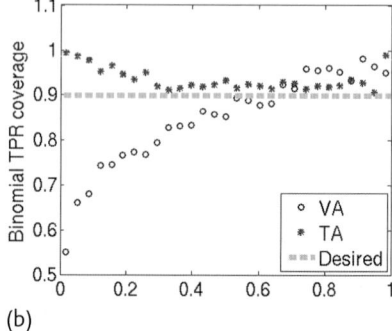

Figure 10.5 ROC curve obtained by linear discriminant (a) and TPR coverage obtained by the binomial model for the 90% confidence level (b). The horizontal axis in the (b) shows FPR for VA and $1 - P(Y = +1|x)$ for TA, where $P(Y = +1|x)$ is the posterior probability for the positive class.

2. We can apply this technique to any performance curve, not just ROC.
3. If we use raw quantiles of the empirical distribution (without smoothing and semi-parametric approximation), the obtained confidence bounds are guaranteed to be contained within the allowed parameter space. For example, the lower bounds for FPR and TPR are nonnegative, and the upper bounds do not exceed one.

If simulation of many ROC curves is costly, resampling can be used. A popular resampling approach is bootstrap reviewed in Section 9.4.2.

Empirical confidence bounds can be improved by kernel smoothing reviewed in Chapter 5. This is especially true for FPR and TPR close to 0 or 1. Choosing the optimal kernel width is not trivial. In the procedure by Hall *et al.* (2004), one has to compute six kernel widths. The two widths for estimation of the FPR and TPR cdfs scale as $O(N_i^{-1/3})$ for the number of observations in the respective class N_i, the two widths for the FPR and TPR pdfs scale as $O(N_i^{-1/5})$, and finally the two widths for smoothing the resampled distributions scale as $O(N_i^{-1/9})$. If the widths are properly chosen, coverage of the bootstrap intervals converges to the desired value at the rate of $O(N^{-2/3})$ for N observations in the sample. The empirical study in Mackassy *et al.* (2005) shows that bootstrap vertical bounds provide, on average, good coverage, especially if used with kernel smoothing. Bootstrap TA intervals overcover.

So far, discussion has been focused on pointwise confidence bounds. To estimate confidence bands for the entire ROC curve, draw a smooth curve through the endpoints of the pointwise bounds. Smoothing the ROC curve by kernels allows bound computation at an arbitrary point and therefore defines vertical bands unambiguously.

TA involves computing bounds for FPR and TPR at fixed thresholds. The inverse problem, estimating confidence bounds for threshold at a fixed value of FPR or TPR, amounts to computing bounds for the quantile of the score distribution. This computation can be carried out by nonparametric bootstrap. Bootstrap confidence bounds (without smoothing) for distribution quantiles converge to the desired frequentist coverage rather slowly. As shown in Hall and Martin (1989), the rate of this convergence is $O(N^{-1/2})$. One of the best alternatives to bootstrap is presented in Chen and Hall (1993). They define an empirical likelihood as a weighted sum over kernels placed at the observed points and offset by the quantile value. Treating these weights as nuisance parameters, they find a confidence interval for the quantile using a likelihood ratio. Their confidence intervals converge to the correct coverage at the rate of $O(N^{-2})$. Ho and Lee (2005) give a good review of confidence bound estimation techniques for quantiles in their introduction.

The `perfcurve` function in the Statistics Toolbox of MATLAB can compute bootstrap confidence bounds by VA or TA. The actual computation is carried out in the `bootci` (bootstrap confidence intervals) function. By default `bootci` computes BCa (bias-corrected percentile intervals with acceleration) bounds described in Efron and Tibshirani (1998).

10.2.6.1 Example: Computation of ROC Confidence Bounds for the Ionosphere Data

We include a snippet of MATLAB code for computing VA bounds by bootstrap and TA bounds by the binomial model. Both techniques are applied to the ionosphere data analyzed in Section 10.2.1. The perfcurve function by default computes 95%, or 1.96σ, intervals. The ROC bounds computed by the two methods are shown in Figures 10.6 and 10.7. Can we say conclusively that AdaBoost is more accurate than pseudo-LDA? For a dataset of this size, the advantage of one technique over the other is not obvious.

Contents

- Vertical averaging by bootstrap
- Threshold averaging using binomial intervals.

Vertical Averaging by Bootstrap Use 1000 bootstrap replicas to compute confidence bounds for Y at fixed X values, where X and Y are first and second output arguments from perfcurve. By default, perfcurve computes FPR for X and TPR for Y and therefore performs vertical averaging.

```
[fprLDA,tprLDAboot] = perfcurve(Y,SfitLDA(:,2),'g',...
    'xvals',0:0.01:0.4,'nboot',1000);
[fprAda,tprAda] = perfcurve(Y,SfitAda(:,2),'g',...
    'xvals',0:0.01:0.4);
```

Plot the ROC curve obtained by pseudo-LDA with error bars. Plot the ROC curve obtained by AdaBoost without error bars, to keep the plot clean.

```
figure;
errorbar(fprLDA,tprLDAboot(:,1),tprLDAboot(:,2)-tprLDAboot(:,1),...
    tprLDAboot(:,3)-tprLDAboot(:,1),'*');
hold;
```

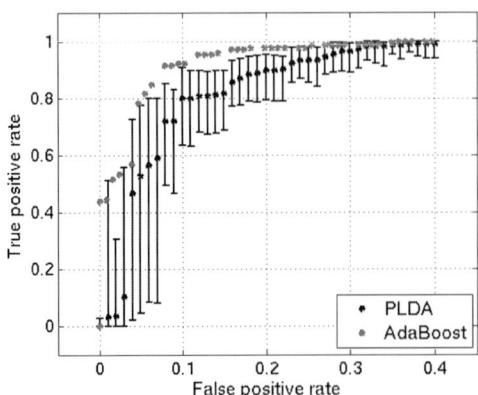

Figure 10.6 VA 95% bootstrap bounds for the ROC curve obtained by pseudo-LDA applied to the ionosphere data. The curve for AdaBoost is shown without errors.

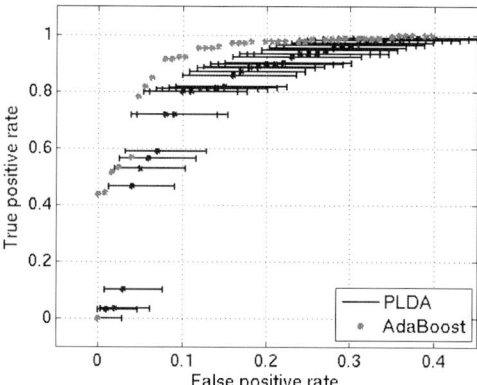

Figure 10.7 TA 95% binomial horizontal bounds for the ROC curve obtained by pseudo-LDA applied to the ionosphere data. Vertical bounds are not shown to avoid overloading the plot. The curve for AdaBoost is shown without errors.

```
plot(fprAda,tprAda,'r*');
hold off;
legend('PLDA','AdaBoost','Location','SE');
xlabel('False positive rate');
ylabel('True positive rate');
ylim([-0.1 1.1]);
xlim([-0.05 0.45]);
grid on;
```

Threshold Averaging Using Binomial Intervals Copy TPR and FPR arrays obtained earlier into TPR and FPR arrays with binomial intervals. The new `bino` arrays have 3 columns. The first column is for central values; we copy the FPR and TPR values obtained earlier into the first columns of the respective `bino` arrays. The second column is for lower bounds and the third column is for upper bounds.

```
tprLDAbino = zeros(size(tprLDAboot));
tprLDAbino(:,1) = tprLDAboot(:,1);
fprLDAbino = zeros(size(tprLDAboot));
fprLDAbino(:,1) = fprLDA;

N1 = sum(strcmp(Y,'g')); % Number of observations in the 'good' class
N0 = sum(strcmp(Y,'b')); % Number of observations in the 'bad' class
```

Compute Clopper–Pearson binomial confidence intervals (lower and upper bounds) using binofit.

```
[~,tprLDAbino(:,2:3)] = binofit(tprLDAbino(:,1)*N1,N1);
[~,fprLDAbino(:,2:3)] = binofit(fprLDAbino(:,1)*N0,N0);
```

Plot the ROC curve for pseudo-LDA with horizontal error bars only. Although we compute the vertical errors as well, we do not show them because the plot would look messy. To plot horizontal error bars, we use `herrorbar` utility by Jos van der Geest, downloadable from the MATLAB File Exchange

http://www.mathworks.com/matlabcentral/fileexchange/3963-herrorbar

```
figure;
hLDA = herrorbar(fprLDAbino(:,1),tprLDAbino(:,1),...
    fprLDAbino(:,2)-fprLDAbino(:,1),fprLDAbino(:,3)-fprLDAbino(:,1),'*');
hold;
hAda = plot(fprAda,tprAda,'r*');
hold off;
legend([hLDA(1) hAda],{'PLDA' 'AdaBoost'},'Location','SE');
xlabel('False positive rate');
ylabel('True positive rate');
ylim([-0.1 1.1]);
xlim([-0.05 0.45]);
grid on;
```

10.3
Testing Equivalence of Two Classification Models

Suppose two classification models are constructed and tested on the same data. One model is observed to have a smaller classification error than the other. If the ultimate goal of the analysis is classification accuracy, the analyst can choose the model with smaller error. Sometimes the decision is not so simple. The analyst could favor a model with larger error (called "inferior model" below) for several reasons:

1. The inferior model may be more interpretable. For instance, if boosted trees and linear discriminant give comparable classification errors, most practitioners would prefer the discriminant. Linearity is a simple concept, and boosted trees are a black box.
2. It may be easier to estimate systematic uncertainties for the inferior model. For example, naive Bayes models the data distribution as a product of marginal densities. If a parameter enters only one marginal density, estimating its effect on the entire model is straightforward. In contrast, boosted trees would require a more sophisticated approach.
3. The analyst may feel more comfortable using the inferior model due to prior experience with such models. Neural nets gained popularity among physicists decades ago, while boosted trees are a recent addition to the toolkit. Although boosted trees are conceptually simpler, or at least not more difficult, learning and practicing a new tool takes some time. If the new technique does not beat the old one, why not stick to the path well traveled?

How much accuracy would you be willing to sacrifice for the inferior model? One way to answer this question would be to set a limit on the difference between the two accuracy estimates. For example, if the relative loss in accuracy is 1%, you may consider it negligibly small to affect your physics analysis. Another way to answer the same question would be by performing a formal statistical test for equality of the two accuracy values. If the two values are statistically equal, choose the model favored for subjective reasons. If the two values are significantly different, choose the more accurate model. In this section, we describe a few recipes for performing such formal statistical tests.

We test the null hypothesis

H_0: the two models have equal accuracy

against its alternative

H_1: the two models do not have equal accuracy .

The accuracy above does not necessarily mean "fraction of observations correctly labeled by the classifier". Use any quantity you like as long as you can design and carry out a formal statistical test.

To test H_0 against H_1, define an appropriate statistic and model its distribution under H_0. Then look at the data and measure the value of this statistic. Compute the probability of observing this or a less likely value under H_0 and call this probability *p-value*. A small *p*-value indicates that H_0 should be rejected. Suppose we reject H_0 if and only if the *p*-value falls below α. In statistical terms, we perform an α-size test. The value of α must be chosen before looking at the data. Popular choices are 5 and 10%.

An α-size test must reject H_0 for $100\alpha\%$ of samples produced under H_0. Put differently, the *Type I error* of the test must equal the specified size, where the Type I error is the probability of rejecting H_0 if it is true. Sometimes it is hard to design a test satisfying this requirement. In that case, a test must be designed to ensure that the Type I error does not exceed α. If the Type I error is less than α, the test is called an α-*level test*.

Another important measure of the test quality is *Type II error*, the probability of not rejecting H_0 when H_1 is true. The *test power* is one minus Type II error, or equivalently the probability of rejecting H_0 when H_1 is true. A good test has low Type II error and large power.

Dieterich (1998) evaluates Type I error and power of five popular tests for comparing classifiers. These procedures make assumptions about the distribution of the test statistic and therefore are not guaranteed to produce Type I error equal to the desired size. Two of these five procedures need an independent test set, and the other three use resampling. Based on this study, Dieterich recommends two procedures to practitioners: McNemar's test if an independent set is available and 5×2 cross-validation paired t test otherwise. These two procedures are recommended because they produce the lowest Type I error among the considered alternatives.

To compare classifiers A and B by *McNemar's test*, count the number of observations in the test data misclassified by model A and correctly classified by model B as well as the number of observations correctly classified by model A and misclassified by model B. Denote these counts by n_{01} and n_{10}. The test statistic

$$T_{\text{McNemar}} = \frac{(|n_{01} - n_{10}| - 1)^2}{n_{01} + n_{10}} \tag{10.6}$$

is approximately distributed as χ^2 with 1 degree of freedom. To compute the p-value, estimate the upper tail of the χ_1^2 distribution above T_{McNemar}.

McNemar's test is two-sided and can only be used for testing the null hypothesis "the two models have equal accuracy" against the alternative "the two models do not have equal accuracy". In addition, McNemar's test is not recommended for values of $n_{01} + n_{10}$ below 20. Both shortcomings can be cured by the *binomial test*. Suppose we test the null hypothesis "model A is as accurate as model B" against the one-sided alternative "model A is more accurate than model B". The null hypothesis should be rejected for small values of n_{01} and large values of n_{10}. The test statistic,

$$T_{\text{binomial}} = \sum_{n=0}^{n_{01}} P(n; n_{01} + n_{10}, 0.5), \tag{10.7}$$

is the probability of observing n_{01} or fewer successes in $n_{01} + n_{10}$ independent drawings with success probability 0.5. This statistic is uniformly distributed under the null. Here, $P(n; N, p) = \binom{N}{n} p^n (1 - p)^{N-n}$ is the binomial probability mass function.

What if we do not have an independent test set? We could try resampling. First, let us refresh some basic statistics. Consider a normal random variable Z with mean μ and variance σ^2. Draw N independent observations $\{z_n\}_{n=1}^{N}$. The sample mean $\bar{z} = \sum_{n=1}^{N} z_n / N$ and the sample variance $S^2 = \sum_{n=1}^{N} (z_n - \bar{z})^2 / (N-1)$ are independent, and their ratio

$$T = \frac{\bar{z} - \mu}{S/\sqrt{N}} \tag{10.8}$$

has a *Student t distribution* with $N - 1$ degrees of freedom. If Z is not normal, the statistic (10.8) converges to the Student t distribution with $N - 1$ degrees of freedom as N approaches infinity. Note that we use the $\sqrt{N}$ factor in the denominator of (10.8) for $N - 1$ degrees of freedom.

Naively, we could use the same statistic for comparison of two classifiers on resampled data. Divide the available data into training and test sets. Train both models on the training set and estimate their classification errors, ϵ_A and ϵ_B, on the test set. Record $\delta = \epsilon_A - \epsilon_B$. Under H_0, models A and B have equal accuracy and the expectation of δ must be zero. Execute R times: split the data into training and test sets and compute $\delta_r = \epsilon_A^{(r)} - \epsilon_B^{(r)}$. Compute the sample mean $\bar{\delta} = \sum_{r=1}^{R} \delta_r / R$ and variance $S^2 = \sum_{r=1}^{R} (\delta_r - \bar{\delta})^2 / (R-1)$. Assume that statistic

$$T_{\text{resampled}} = \frac{\bar{\delta}}{S/\sqrt{R}} \tag{10.9}$$

has a Student t distribution with $R - 1$ degrees of freedom.

This recipe, of course, is not going to work because δ_rs are no longer independent. The δ_r values are usually positively correlated and the variance of $\bar{\delta}$ is larger than what it would be if they were independent. The distribution of the $T_{\text{resampled}}$ statistic has tails heavier than those for the Student t distribution with $R-1$ degrees of freedom. Our naive procedure would reject H_0 more often than in $100\alpha\%$ of trials.

The 5×2 *cross-validation paired t test* alleviates this problem by replacing the numerator in (10.9) with the difference observed in the first run. The test works as follows. In every run, the data are split in equal parts for training and testing. Both classifiers are trained on the first half and evaluated on the second half producing errors $\epsilon_{r,A}^{(1)}$ and $\epsilon_{r,B}^{(1)}$. Then the training and test sets are swapped, and we obtain errors $\epsilon_{r,A}^{(2)}$ and $\epsilon_{r,B}^{(2)}$. The difference in errors for this run is taken to be $\delta_r = (\delta_r^{(1)} + \delta_r^{(2)})/2$ with $\delta_r^{(1)} = \epsilon_{r,A}^{(1)} - \epsilon_{r,B}^{(1)}$ and $\delta_r^{(2)} = \epsilon_{r,A}^{(2)} - \epsilon_{r,B}^{(2)}$. The contribution to the S^2 estimate from run r is $s_r^2 = (\delta_r^{(1)} - \delta_r)^2 + (\delta_r^{(2)} - \delta_r)^2$. If all $\delta_r^{(k)}$'s are independent normal random variables with zero mean and variance σ^2, the ratio s_r^2/σ^2 is distributed as χ_1^2 (the subscript indicates degrees of freedom), and the sum of 5 independent χ_1^2 variables $\sum_{r=1}^{5} (s_r/\sigma)^2$ is distributed as χ_5^2. The ratio of the independent normal and χ_5^2 random variables

$$T_{5 \times 2\ t\ \text{paired}} = \frac{\delta_1^{(1)}}{\sqrt{\sum_{r=1}^{5} s_r^2/5}} \quad (10.10)$$

then has a Student t distribution with 5 degrees of freedom. The assumption of independence for $\delta_r^{(k)}$'s is clearly violated, but the resulting statistic (10.10) has been empirically shown to produce Type I error close to the desired α, as well as good rejection power.

The 5×2 paired t test suffers from two drawbacks. First, it halves the data and therefore tends to produce classifiers with noticeably larger variance than those trained on the full set. Second, there is something unsettling about choosing the first observed difference $\delta_1^{(1)}$. Why not choose the second or third?

Alpaydin (1999) observes that if all $\delta_r^{(k)}$'s are independent, their sum

$$\sum_{r=1}^{5} \sum_{k=1}^{2} \left(\frac{\delta_r^{(k)}}{\sigma} \right)^2$$

must have a χ_{10}^2 distribution. The ratio of the independent χ_{10}^2 and χ_5^2 variables

$$T_{5 \times 2\ F\ \text{paired}} = \frac{\sum_{r=1}^{5} \sum_{k=1}^{2} \left(\delta_r^{(k)} \right)^2 / 10}{\sum_{r=1}^{5} s_r^2 / 5} \quad (10.11)$$

then has an $F_{10,5}$ *distribution* with 10 and 5 degrees of freedom. Of course, $\delta_r^{(k)}$'s are not independent, and the statistic (10.11) may not have the specified distribution. Nevertheless, the proposed $T_{5 \times 2\ F\ \text{paired}}$ test has been shown to produce lower Type

I error and higher power than $T_{5\times 2\ t\ \text{paired}}$ on several real-world datasets. This comparison hints that the Type I error for the 5 × 2 CV F test is consistently below α. In other words, the test is conservative.

The statistic (10.9) can be obtained by cross-validation. Set the number of runs to the number of folds $\mathcal{K}$. Estimate every δ_k; $k = 1, \ldots, \mathcal{K}$; by training both classifiers on all data with the kth fold excluded and then evaluating their accuracy on the kth fold. As shown in Dietterich (1998), the *10-fold CV test* provides good rejection power. Unfortunately, its Type I error is well above the desired α. Repeated cross-validation

$$T_{\text{repeated CV}} = \frac{\bar{\delta}}{S/\sqrt{\mathcal{K}R}}$$

with

$$\bar{\delta} = \frac{\sum_{r=1}^{R} \sum_{k=1}^{\mathcal{K}} \delta_r^{(k)}}{\mathcal{K}R}$$

and

$$S^2 = \frac{\sum_{r=1}^{R} \sum_{k=1}^{\mathcal{K}} \left(\delta_r^{(k)} - \bar{\delta}\right)^2}{\mathcal{K}R - 1}$$

would increase the Type I error even further. Bouckaert (2003) considers statistic

$$T_f = \frac{\bar{\delta}}{S/\sqrt{f+1}}, \tag{10.12}$$

where $\bar{\delta}$ and S are computed in the usual way and f is unknown. Assuming that T_f has a Student t distribution with f degrees of freedom, he calibrates f experimentally on simulated data to match the desired test size α. For 10 repetitions of 10-fold cross-validation ($R = 10, \mathcal{K} = 10$), the optimal f is far below the naive $\mathcal{K}R - 1$ estimate. Details on the exact choice of f are a little fuzzy. Bouckaert (2003) states in one place that the optimal f remains constant at 11 for $\mathcal{K} = 10$ folds in a broad range of R; yet in conclusion he recommends setting f to 10. Subsequently, Bouckaert and Frank (2004) give a general formula for f implying the optimal value of 8 for 10 folds. Of course, the t distributions for 8 and 11 degrees of freedom are reasonably close. When we compare two classification models using the 10 × 10 CV test (10 repetitions of 10-fold cross-validation), we set f to 10.

The 10-fold CV test has good statistical properties. This test compares classifiers on training sets close to the full data size and is known to have good rejection power. In addition, Bouckaert (2003) and Bouckaert and Frank (2004) show that the 10 × 10 CV test is more replicable than the 5 × 2 CV paired t test. Note that neither the 5 × 2 nor the 10 × 10 test is deterministic: the outcome depends on an arbitrary data partition. "Replicability" is defined by the probability that applications of the same test procedure to the same data produce the same outcome. The disadvantage of the 10×10 CV test is its high computational cost for complex classifiers and large datasets.

All tests discussed in this section have been developed for comparing classification errors. They should be equally applicable to comparing other measures of performance ϵ_A and ϵ_B. The distributions of ϵ_A and ϵ_B measured over independent identical experiments must satisfy the Central Limit Theorem.

The t statistic for testing $H_0: \epsilon_A = \epsilon_B$ against $H_1: \epsilon_A \neq \epsilon_B$ obtained from either 5×2 paired t or 10×10 procedure can range from $-\infty$ to $+\infty$. Either a low or high value of this statistic would indicate that H_0 should be rejected. Suppose we observe value t_0. To obtain the p-value for the hypothesis test, compute the tail below t_0 if t_0 is negative and above t_0 if t_0 is positive; then multiply the tail area by 2. In contrast, the 5×2 paired F test is one-sided: the F statistic is always positive, and the p-value is obtained from the upper tail.

The hypothesis test described above cannot be used to choose one classification model over the other – it only tells us if the difference in their errors is significant or not. In practice, a more interesting question could be: Is model A more accurate than model B? To answer this question, we must test $H_0: \epsilon_A \geq \epsilon_B$ against $H_1: \epsilon_A < \epsilon_B$. The relationship between two-sided and one-sided uniformly most powerful (UMP) tests captured in the Karlin–Rubin theorem suggests that the t statistic (10.10) or (10.12) can be used here as well. The associated p-value is obtained from the upper tail of the t distribution, without multiplication by 2. The paired F test is not applicable in this situation.

In summary, we recommend the 10×10 CV test if the computational burden is manageable and 5×2 CV paired F test otherwise for comparing classification errors or other measures of performance. If you can afford an independent test set and if you want to test the equality of classification errors, you can use McNemar's test or the binomial test.

10.4
Comparing Several Classifiers

In the previous section, we have described statistical tests for comparing two classification models. Here, we discuss a more general problem of comparing several classifiers.

Suppose we have L models with classification errors (or other measures of predictive performance), $\{\epsilon_l\}_{l=1}^{L}$. To test their equivalence, we define hypotheses

$$H_0: \epsilon_1 = \ldots = \epsilon_L$$

and

$$H_1: \exists i, j: \epsilon_i \neq \epsilon_j.$$

Demsar (2006) and Pizarro *et al.* (2002) review a handful of applicable techniques. Unfortunately, all these techniques require repeated measures of the predictive performance over independent sets of data.[2] Most usually, the analyst does not have access to independent datasets.

Even if we could carry out a formal test, the obtained conclusion might not be of practical interest. If H_0 is rejected, all we can say is that at least two classifiers are significantly different. The test does not tell us what what pair of classifiers has triggered the rejection of H_0.

A more interesting question would be: Is one classifier significantly better than any other? To answer this question, we could run $L(L-1)/2$ pairwise tests. This approach faces two problems:

1. Type I error could greatly exceed the desired test size due to multiple rounds of comparison.
2. There may be no clear winner in all pairings.

Below we discuss steps to address these problems.

Let α be the desired test size, that is, the probability of committing at least one Type I error in all individual tests. Let us estimate β, the size of each individual test, to account for the test multiplicity.

Consider T independent tests, each of size β. If H_0 is true, the probability of not rejecting H_0 in each test is $1-\beta$, and the probability of not rejecting H_0 in any test is therefore $(1-\beta)^T$. The probability of rejecting H_0 in T tests is then $1-(1-\beta)^T$, and we would like this probability to equal α. We obtain the *Sidak correction*, $\beta_{\text{Sidak}} = 1-(1-\alpha)^{1/T}$.

The pairwise comparisons are not independent because every classifier participates in $L-1$ tests. Boole's inequality implies that if the size of each individual test is β, the size of T tests is at most $T\beta$; this result holds if the tests are not independent. We obtain the *Bonferroni correction*, $\beta_{\text{Bonf}} = \alpha/T$. Incidentally, this correction can be obtained by expanding the Sidak correction up to the first order in α. The Bonferroni correction is more conservative, $\beta_{\text{Bonf}} < \beta_{\text{Sidak}}$, because it does not use the assumption of independence.

If we order the compared classifiers by preference, we can choose the most preferred one among the best candidates in the absence of a clear winner. This ordering is subjective and must be done before testing. Yildiz and Alpaydin (2006) sort classifiers in descending order by simplicity (simple models are placed at the beginning of the list) and conduct pairwise one-sided tests $H_0: \epsilon_i \leq \epsilon_j$ against $H_1: \epsilon_i > \epsilon_j$ for sorted indices $1 \leq i < j \leq L$. In every test, H_0 is rejected if simple model i produces a significantly larger classification error than complex model j. If H_0 is rejected, classifier j is declared the winner; otherwise classifier i wins. The size for every pairwise test is set to $\alpha/(L(L-1)/2)$ by the Bonferroni correction. Because the comparison is one-sided, Yildiz and Alpaydin (2006) use the 5×2 paired t test; they could use the corrected 10×10 test (10.12) with equal success. If two or more classifiers win all rounds, the simplest among them becomes the ultimate winner. An obvious weakness of this scheme is the lack of a well-defined

---

2) Although Pizarro *et al.* (2002) apply these techniques to resampled data, they never compare the actual Type I error to the desired test size α.

pretest sorting criterion. We should prefer simple over complex, but what is simpler – neural net or boosted decision trees? You have to exercise your judgment. Consider this ordering as an opportunity for promoting your favorite classification technique.

The test procedures described here are not your only choice. Devise a procedure of your own to meet the needs of your analysis. For example, you may wish to choose a simple model such as linear discriminant as the baseline for comparison. Then you can ask: Are there models, among the remaining $L-1$ classifiers, performing significantly better than the baseline? To answer this question, you can run $L-1$ pairwise tests, each comparing the respective classifier with the discriminant. The size of each test would be set to $\alpha/(L-1)$ by the Bonferroni correction.

Beware: To prevent Type I error from exceeding the desired size, you must design the test procedure under the assumptions of the null hypothesis. Consider, for instance, the test described in the previous paragraph. You may be tempted to preselect $\tilde{L} < L-1$ most accurate classifiers and run pairwise tests, each with size $\alpha/\tilde{L}$, against the baseline model. By doing so, you may obtain a large Type I error. Under the null hypothesis, all classifiers have equal accuracy. Selecting a few largest values and comparing them to the baseline would increase the number of rejections, even if all values were drawn from the same distribution. The test would need to account for the bias introduced by this preselection.

10.5
Exercises

1. Consider estimating FPR, p, at a fixed threshold imposed on the soft classification score. Consider estimation by $\mathcal{K}$-fold cross-validation. Let $\hat{p}_k$ be the FPR estimate in the kth fold with N_k^- observations from the negative class. Assume that the $\mathcal{K}$ estimates $\hat{p}_k$ are independent. Let $N^- = \sum_{k=1}^{\mathcal{K}} N_k^-$ be the total number of negative-class observations in the data. Consider two FPR estimates: the first estimate $\hat{p}^{(1)}$ is obtained by plotting one ROC curve for all cross-validated scores and counting negative-class observations above the chosen threshold, and the second estimate is set to

$$\hat{p}^{(2)} = \sum_{k=1}^{\mathcal{K}} N_k^- \hat{p}_k / N^- .$$

Are the two estimates equivalent? Instead of estimating FPR at a fixed score threshold, now estimate FPR at a fixed value of TPR. Are the two FPR estimates still equivalent? (Hint: How do $\hat{p}^{(1)}$ and $\hat{p}^{(2)}$ account for the variance in the score threshold estimate?)

2. Consider the empirical ROC curves in Figure 10.1. Due to finite statistics, each curve is made of vertical and horizontal steps. Suppose we move left to right, that is, in the direction of increasing FPR. If the next distinct classification score is for an observation of the negative class, make a horizontal step. If the next

distinct classification score is for an observation of the positive class, make a vertical step. If one observation of the positive class and one observation of the negative class had equal scores, how would the empirical ROC curve look in the neighborhood of the respective point?

3. In Section 10.4, we have suggested performing $L(L-1)/2$ pairwise tests to find the best classifier out of L learners. Consider alternative strategies such as, for instance, a single-elimination tournament. Discuss their pros and cons versus the proposed scheme.

References

Alpaydin, E. (1999) Combined 5 × 2 CV F test for comparing supervised classification learning algorithms. *Neural Comput.*, **11** (8), 1885–1992.

Bouckaert, R. (2003) Choosing between two learning algorithms based on calibrated tests. *Int. Conf. Mach. Learn.*, pp. 51–58.

Bouckaert, R. and Frank, E. (2004) Evaluating the replicability of significance tests for comparing learning algorithms. *Adv. Knowl. Discov. Data Min., 8th Pacific-Asia Conf.*, pp. 3–12.

Bradley, A. (1997) The use of the area under the ROC curve in the evaluation of machine learning algorithms. *Pattern Recognit.*, **30** (7), 1145–1159.

Cai, T. and Moskowitz, C. (2004) Semi-parametric estimation of the binormal ROC curve for a continuous diagnostic test. *Biostatistics*, **5** (4), 573–586.

Chen, S. and Hall, P. (1993) Smoothed empirical likelihood confidence intervals for quantiles. *Ann. Stat.*, **21** (3), 1166–1181.

Demsar, J. (2006) Statistical comparisons of classifiers over multiple data sets. *J. Mach. Learn. Res.*, **7**, 1–30.

Dietterich, T. (1998) Approximate statistical tests for comparing supervised classification learning algorithms. *Neural Comput.*, **10** (7), 1895–1923.

Efron, B. and Tibshirani, R. (1998) *An Introduction to the Bootstrap, Monographs on Statistics and Applied Probability*, vol. 57, Chapman & Hall/CRC.

Fawcett, T. (2006) An introduction to ROC analysis. *Pattern Recognit. Lett.*, **27**, 861–874.

Flach, P., Hernandez-Orallo, J., and Ferri, C. (2011) A coherent interpretation of AUC as a measure of aggregated classification performance. *Proc. 28th Int. Conf. Mach. Learn.*, ACM, New York, NY, USA, ICML '11, pp. 657–664.

Frank, A. and Asuncion, A. (2010) UCI machine learning repository. http://archive.ics.uci.edu/ml (accessed 15 July 2013).

Hall, P. and Hyndman, R. (2003) Improved methods for bandwidth selection when estimating ROC curves. *Stat. Probab. Lett.*, **64** (2), 181–189.

Hall, P., Hyndman, R., and Fan., Y. (2004) Nonparametric confidence intervals for receiver operating characteristic curves. *Biometrika*, **91**, 743–750.

Hall, P. and Martin, M. (1989) A note on the accuracy of bootstrap percentile method confidence intervals for a quantile. *Stat. Probab. Lett.*, **8** (3), 197–200.

Hand, D. (2009) Measuring classifier performance: a coherent alternative to the area under the ROC curve. *Mach. Learn.*, **77**, 103–123.

Hilgers, R. (1991) Distribution-free confidence bounds for ROC. *Methods Inf. Med.*, **30**, 96–101.

Ho, Y. and Lee, S. (2005) Iterated smoothed bootstrap confidence intervals for population quantiles. *Ann. Stat.*, **33** (1), 437–462.

Hsieh, F. and Turnbull, B. (1996) Nonparametric and semiparametric estimation of the receiver operating characteristic curve. *Ann. Stat.*, **24** (1), 25–40.

Lilliefors, H. (1967) On the Kolmogorov–Smirnov test for normality with mean and variance unknown. *J. Am. Stat. Assoc.*, **62**, 399–402.

Lloyd, C. and Yong, Z. (1999) Kernel estimators of the ROC curve are better than empirical. *Stat. Probab. Lett.*, **44**, 221–228.

Mackassy, S. and Provost, F. (2005) Confidence bands for ROC curves: Methods and an empirical study. *Proc. 22nd Int. Conf. Mach. Learn.*, vol. 119, ACM, pp. 537–544.

Mackassy, S., Provost, F., and Rosset, S. (2005) Pointwise ROC confidence bounds: An empirical evaluation. *Proc. Workshop on ROC Anal. Mach. Learn. (ROCML-2005) at ICML-2005.*

Mason, S. and Graham, N. (2002) Areas beneath the relative operating characteristics (ROC) and relative operating levels (ROL) curves: Statistical significance and interpretation. *Q. J. R. Meteorol. Soc.*, **128**, 2145–2166.

Pizarro, J., Guerrero, E., and Galindo, P. (2002) Multiple comparison procedures applied to model selection. *Neurocomputing*, **48**, 155–173.

Provost, F. and Fawcett, T. (2001) Robust classification for imprecise environments. *Mach. Learn.*, **42**, 203–231.

Yildiz, O. and Alpaydin, E. (2006) Ordering and finding the best of $K > 2$ supervised learning algorithms. *IEEE Trans. Pattern Anal. Mach. Intell.*, **28** (3).

Zhou, X.H. and Lin, H. (2008) Semi-parametric maximum likelihood estimates for ROC curves of continuous-scale tests. *Stat. Med.*, **27** (25), 5271–5290.

11
Linear and Quadratic Discriminant Analysis, Logistic Regression, and Partial Least Squares Regression

In this chapter, we review, for the most part, linear methods for classification. The only exception is quadratic discriminant analysis, a straightforward generalization of a linear technique. These methods are best known for their simplicity. A linear decision boundary is easy to understand and visualize, even in many dimensions. An example of such a boundary is shown in Figure 11.1 for Fisher iris data.

Because of their high interpretability, linear methods are often the first choice for data analysis. They can be the only choice if the analyst seeks to discover linear relationships between variables and classes. If the analysis goal is maximization of the predictive power and the data do not have a linear structure, nonparametric nonlinear methods should be favored over simple interpretable techniques.

Linear discriminant analysis (LDA), also known as Fisher discriminant, has been a very popular technique in particle and astrophysics. Quadratic discriminant analysis (QDA) is its closest cousin.

11.1
Discriminant Analysis

Suppose we observe a sample drawn from a multivariate normal distribution $N(\boldsymbol{\mu}, \boldsymbol{\Sigma})$ with mean vector $\boldsymbol{\mu}$ and covariance matrix $\boldsymbol{\Sigma}$. The data are D-dimensional, and vectors, unless otherwise noted, are column-oriented. The multivariate density is then

$$P(\boldsymbol{x}) = \frac{1}{\sqrt{(2\pi)^D |\boldsymbol{\Sigma}|}} \exp\left[-\frac{1}{2}(\boldsymbol{x} - \boldsymbol{\mu})^\mathsf{T} \boldsymbol{\Sigma}^{-1} (\boldsymbol{x} - \boldsymbol{\mu})\right], \tag{11.1}$$

where $|\boldsymbol{\Sigma}|$ is the determinant of $\boldsymbol{\Sigma}$.

Suppose we observe a sample of data drawn from two classes, each described by a multivariate normal density

$$P(\boldsymbol{x}|k) = \frac{1}{\sqrt{(2\pi)^D |\boldsymbol{\Sigma}_k|}} \exp\left[-\frac{1}{2}(\boldsymbol{x} - \boldsymbol{\mu}_k)^\mathsf{T} \boldsymbol{\Sigma}_k^{-1} (\boldsymbol{x} - \boldsymbol{\mu}_k)\right] \tag{11.2}$$

Statistical Analysis Techniques in Particle Physics, First Edition. Ilya Narsky and Frank C. Porter.
©2014 WILEY-VCH Verlag GmbH & Co. KGaA. Published 2014 by WILEY-VCH Verlag GmbH & Co. KGaA.

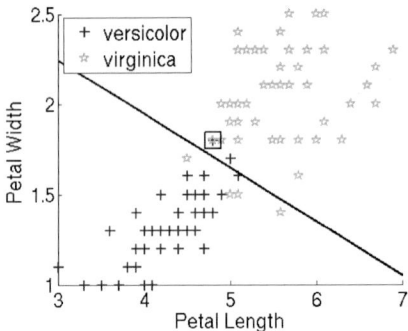

Figure 11.1 Class boundary obtained by linear discriminant analysis for Fisher iris data. The square covers one observation of class versicolor and two observations of class virginica.

for classes $k = 1, 2$. Recall that Bayes rule gives

$$P(k|x) = \frac{\pi_k P(x|k)}{P(x)} \tag{11.3}$$

for the posterior probability $P(k|x)$ of observing an instance of class k at point x. The unconditional probability $P(x)$ in the denominator does not depend on k. The prior class probability π_k was introduced in Chapter 9; we discuss its role in the discriminant analysis below.

Let us take a natural logarithm of the posterior odds:

$$\begin{aligned}\log \frac{P(k=1|x)}{P(k=2|x)} = &\log \frac{\pi_1}{\pi_2} - \frac{1}{2} \log \frac{|\Sigma_1|}{|\Sigma_2|} \\&+ x^T \left(\Sigma_1^{-1} \mu_1 - \Sigma_2^{-1} \mu_2 \right) \\&- \frac{1}{2} x^T \left(\Sigma_1^{-1} - \Sigma_2^{-1} \right) x \\&- \frac{1}{2} \left(\mu_1^T \Sigma_1^{-1} \mu_1 - \mu_2^T \Sigma_2^{-1} \mu_2 \right) .\end{aligned} \tag{11.4}$$

The hyperplane separating the two classes is obtained by equating this log-ratio to zero. This is a quadratic function of x, hence *quadratic discriminant analysis*. If the two classes have the same covariance matrix $\Sigma_1 = \Sigma_2$, the quadratic term disappears and we obtain *linear discriminant analysis*.

Fisher (1936) originally derived discriminant analysis in a different fashion. He searched for a direction q maximizing separation between two classes,

$$S(q) = \frac{\left[q^T (\mu_1 - \mu_2) \right]^2}{q^T \Sigma q} . \tag{11.5}$$

This separation is maximized at $q = \Sigma^{-1}(\mu_1 - \mu_2)$. The terms not depending on x in (11.4) do not change the orientation of the hyperplane separating the two distributions – they only shift the boundary closer to one class and further away from the other. The formulation by Fisher is therefore equivalent to (11.4) for LDA.

To use this formalism for $K > 2$ classes, choose one class, for example the last one, for normalization. Compute posterior odds $\log[P(k|\mathbf{x})/P(K|\mathbf{x})]$ for $k = 1, \ldots, K-1$. The logarithm of the posterior odds is additive, that is,

$$\log[P(i|\mathbf{x})/P(j|\mathbf{x})] = \log[P(i|\mathbf{x})/P(K|\mathbf{x})] - \log[P(j|\mathbf{x})/P(K|\mathbf{x})]$$

for classes i and j. The computed $K-1$ log-ratios give complete information about the hyperplanes of separation. If we need to compute the posterior probabilities, we require that $\sum_{k=1}^{K} P(k|\mathbf{x}) = 1$ and obtain estimates of $P(k|\mathbf{x})$ from the log-ratios. The same trick is used in other multiclass models such as multinomial logistic regression. For prediction on new data, the class label is assigned by choosing the class with the largest posterior probability.

In this formulation, the class prior probabilities merely shift the boundaries between the classes without changing their orientations (for LDA) or their shapes (for QDA). They are not used to estimate the class means or covariance matrices; hence, they can be applied after training. Alternatively, we could ignore the prior probabilities and classify observations by imposing thresholds on the computed log-ratios. These thresholds would be optimized using some physics-driven criteria. Physicists often follow the second approach.

As discussed in Chapter 9, classifying into the class with the largest posterior probability $P(y|\mathbf{x})$ minimizes the classification error. If the posterior probabilities are accurately modeled, this classifier is optimal. If classes indeed have multivariate normal densities, QDA is the optimal classifier. If classes indeed have multivariate normal densities with equal covariance matrices, LDA is the optimal classifier. Most usually, we need to estimate the covariance matrices empirically.

LDA is seemingly simple, but this simplicity may be deceiving. Subtleties in LDA implementation can change its result dramatically. Let us review them now.

11.1.1
Estimating the Covariance Matrix

Under the LDA assumptions, classes have equal covariance matrices and different means. Take the training data with known class labels. Let $\mathbf{M}$ be an $N \times K$ class membership matrix for N observations and K classes: $m_{nk} = 1$ if observation n is from class k and 0 otherwise. First, estimate the mean for each class in turn,

$$\hat{\boldsymbol{\mu}}_k = \frac{\sum_{n=1}^{N} m_{nk} \mathbf{x}_n}{\sum_{n=1}^{N} m_{nk}}. \tag{11.6}$$

Then compute the pooled-in covariance matrix. For example, use a maximum likelihood estimate,

$$\hat{\boldsymbol{\Sigma}}_{\text{ML}} = \frac{1}{N} \sum_{k=1}^{K} \sum_{n=1}^{N} m_{nk} (\mathbf{x}_n - \hat{\boldsymbol{\mu}}_k)(\mathbf{x}_n - \hat{\boldsymbol{\mu}}_k)^{\mathsf{T}}. \tag{11.7}$$

Vectors $\mathbf{x}_n$ and $\hat{\boldsymbol{\mu}}_k$ are $D \times 1$ (column-oriented), and $(\mathbf{x}_n - \hat{\boldsymbol{\mu}}_k)(\mathbf{x}_n - \hat{\boldsymbol{\mu}}_k)^{\mathsf{T}}$ is therefore a symmetric $D \times D$ matrix. This maximal likelihood estimator is biased. To remove the bias, apply a small correction:

$$\hat{\Sigma} = \frac{N}{N-K} \hat{\Sigma}_{ML} . \tag{11.8}$$

Elementary statistics textbooks derive a similar correction for a univariate normal distribution and include a formula for an unbiased estimate, $S^2 = \sum_{n=1}^{N}(x_n - \bar{x})^2/(N-1)$, of the variance, σ^2. The statistic $(N-1)S^2/\sigma^2$ is distributed as χ^2 with $N-1$ degrees of freedom. We use $N-K$ instead of $N-1$ because we have K classes. Think of it as losing one degree of freedom per linear constraint. In this case, there are K linear constraints for K class means.

In physics analysis, datasets are usually large, $N \gg K$. For unweighted data, this correction can be safely neglected. For weighted data, the problem is a bit more involved. The weighted class means are given by

$$\hat{\mu}_k = \frac{\sum_{n=1}^{N} m_{nk} w_n x_n}{\sum_{n=1}^{N} m_{nk} w_n} . \tag{11.9}$$

The maximum likelihood estimate (11.7) generalizes to

$$\hat{\Sigma}_{ML} = \sum_{k=1}^{K} \sum_{n=1}^{N} m_{nk} w_n (x_n - \hat{\mu}_k)(x_n - \hat{\mu}_k)^T . \tag{11.10}$$

Above, we assume that the weights are normalized to sum to one: $\sum_{n=1}^{N} w_n = 1$. The unbiased estimate is then

$$\hat{\Sigma} = \frac{\hat{\Sigma}_{ML}}{1 - \sum_{k=1}^{K} \frac{W_k^{(2)}}{W_k}} , \tag{11.11}$$

where $W_k = \sum_{n=1}^{N} m_{nk} w_n$ is the sum of weights in class k and $W_k^{(2)} = \sum_{n=1}^{N} m_{nk} w_n^2$ is the sum of squared weights in class k. For class-free data $K = 1$, this simplifies to $\hat{\Sigma} = \hat{\Sigma}_{ML}/(1 - \sum_{n=1}^{N} w_n^2)$. If all weights are set to $1/N$, (11.11) simplifies to (11.8). In this case, the corrective term $\sum_{n=1}^{N} w_n^2$ attains minimum, and the denominator in (11.11) is close to 1. If the weights are highly nonuniform, the denominator in (11.11) can get close to zero.

For LDA with two classes, this bias correction is, for the most part, irrelevant. Multiplying all elements of the covariance matrix by factor a is equivalent to multiplying $x^T \Sigma^{-1}(\mu_1 - \mu_2)$ by $1/a$. This multiplication does not change the orientation of the hyperplane separating the two classes, but it does change the posterior class probabilities at point x. Instead of using the predicted posterior probabilities directly, physicists often inspect the ROC curve and select a threshold on classification scores (in this case, posterior probabilities) by optimizing some function of true positive and false positive rates. If we fix the false positive rate, multiply the log-ratio at any point x by the same factor and measure the true positive rate, we will obtain the same value as we would without multiplication.

Unfortunately, this safety mechanism fails for QDA, multiclass LDA, and even LDA with two classes if the covariance matrix is estimated as a weighted combination of the individual covariance matrices, as described in Section 11.1.3. We refrain from recommending the unbiased estimate over the maximum likelihood estimate or the other way around. We merely point out this issue. If you work with highly nonuniform weights, you should investigate the stability of your analysis procedure with respect to weighting.

11.1.2
Verifying Discriminant Analysis Assumptions

The key assumptions for discriminant analysis are multivariate normality (for QDA and LDA) and equality of the class covariance matrices (for LDA). You can verify these assumptions numerically.

Two popular tests of normality proposed in Mardia (1970) are based on multivariate skewness and kurtosis. The sample kurtosis is readily expressed in matrix notation

$$\hat{\kappa} = \frac{1}{N} \sum_{n=1}^{N} \left[(x_n - \hat{\mu})^\mathsf{T} \hat{\Sigma}^{-1} (x_n - \hat{\mu}) \right]^2, \tag{11.12}$$

where $\hat{\mu}$ and $\hat{\Sigma}$ are the usual estimates of the mean and covariance matrix. Asymptotically, $\hat{\kappa}$ has a normal distribution with mean $D(D+2)$ and variance $8D(D+2)/N$ for the sample size N and dimensionality D. A large observed value indicates a distribution with tails heavier than normal, and a small value points to a distribution with tails lighter than normal.

A less rigorous but more instructive procedure is to inspect a *quantile-quantile (QQ) plot* of the squared Mahalanobis distance (Healy, 1968). If X is a random vector drawn from a multivariate normal distribution with mean μ and covariance Σ, its squared Mahalanobis distance $(X - \mu)^\mathsf{T} \Sigma^{-1} (X - \mu)$ has a χ^2 distribution with D degrees of freedom. We can plot quantiles of the observed Mahalanobis distance versus quantiles of the χ_D^2 distribution. A departure from a straight line would indicate the lack of normality, and extreme points would be considered as candidates for outliers. In practice we know neither μ nor Σ and must substitute their estimates. As soon as we do, we, strictly speaking, can no longer use the χ_D^2 distribution, although it remains a reasonable approximation for large N. No statistical test is associated with this approach, but visual inspection often proves fruitful.

Equality of the class covariance matrices can be verified by a Bartlett multivariate test described in popular textbooks such as Andersen (2003). Formally, we test hypothesis $H_0: \Sigma_1 = \ldots = \Sigma_K$ against $H_1:$ at least two Σ's are different. The test statistic,

$$-2 \log V = (N - K) \log |\hat{\Sigma}| - \sum_{k=1}^{K} (n_k - 1) \log |\hat{\Sigma}_k|, \tag{11.13}$$

resembles a log-likelihood ratio for the pooled-in unbiased estimate $\hat{\Sigma}$ and unbiased class estimates $\hat{\Sigma}_k$. Here, n_k is the number of observations in class k. This formula would need to be modified for weighted data. When the sizes of all classes are comparable and large, $-2\log V$ can be approximated by a χ^2 distribution with $(K-1)D(D+1)/2$ degrees of freedom (see, for example, Box, 1949). For small samples, the exact distribution can be found in Gupta and Tang (1984). H_0 should be rejected if the observed value of $-2\log V$ is large. Intuitively, $-2\log V$ measures the lack of uniformity of the covariance matrices across the classes. The pooled-in estimate is simply a sum over class estimates $(N-K)\hat{\Sigma} = \sum_{k=1}^{K}(n_k-1)\hat{\Sigma}_k$. If the sum of several positive numbers is fixed, their product (equivalently, the sum of their logs) is maximal when the numbers are equal. This test is based on the same idea. The Bartlett test is sensitive to outliers and should not be used in their presence.

The tests described here are mostly of theoretical value. Practitioners often apply discriminant analysis when its assumptions do not hold. The ultimate test of any classification model is its performance. If discriminant analysis gives a satisfactory predictive power for nonnormal samples, don't let the rigor of theory stand in your way. Likewise, you can verify that QDA improves over LDA by comparing the accuracies of the two models using one of the techniques reviewed in Chapter 10.

11.1.3
Applying LDA When LDA Assumptions Are Invalid

Under the LDA assumptions, all classes have multivariate normal distributions with different means and the same covariance matrix. The maximum likelihood estimate (11.10) is equal to the weighted average of the covariance matrix estimates per class:

$$\hat{\Sigma}_{ML} = \sum_{k=1}^{K} W_k \hat{\Sigma}_k . \tag{11.14}$$

Above, $W_k = \sum_{n=1}^{N} m_{nk} w_n$ is the sum of weights in class k. As usual, we take $\sum_{n=1}^{N} w_n = 1$.

In practice, physicists apply LDA when none of the LDA conditions holds. The class densities are not normal and the covariance matrices are not equal. In these circumstances, you can still apply LDA and obtain some separation between the classes. But there is no theoretical justification for the pooled-in covariance matrix estimate (11.14). It is tempting to see how far we can get by experimenting with the covariance matrix estimate.

Let us illustrate this problem on a hypothetical example. Suppose we have two classes with means

$$\mu_1 = \begin{pmatrix} 1/2 \\ 0 \end{pmatrix} \quad \mu_2 = \begin{pmatrix} -1/2 \\ 0 \end{pmatrix} \tag{11.15}$$

and covariance matrices

$$\Sigma_1 = \begin{pmatrix} 2 & 1 \\ 1 & 2 \end{pmatrix} \quad \Sigma_2 = \begin{pmatrix} 2 & -1 \\ -1 & 2 \end{pmatrix}. \tag{11.16}$$

The two covariance matrices are inverse to each other, up to some constant. If we set the covariance matrix for LDA to $(\Sigma_1 + \Sigma_2)/2$, the predicted line of optimal separation is orthogonal to the first coordinate axis and the optimal classification is given by x_1. If we set the covariance matrix for LDA to Σ_1, the optimal classification is $2x_1 - x_2$. If we set the covariance matrix for LDA to Σ_2, the optimal classification is $2x_1 + x_2$. If we had to estimate the pooled-in covariance matrix on a training set, we would face the same problem. If classes 1 and 2 were represented equally in the training data, we would obtain $(\Sigma_1 + \Sigma_2)/2$. If the training set were composed mostly of observations of class 1, we would obtain Σ_1. Similarly for Σ_2.

Let us plot ROC curves for the three pooled-in matrix estimates. As explained in the previous chapter, a ROC curve is a plot of true positive rate (TPR) versus false positive rate (FPR), or accepted signal versus accepted background. Take class 1 to be signal and class 2 to be background. The three curves are shown in Figure 11.2. Your choice of the optimal curve (and therefore the optimal covariance matrix) would be defined by the specifics of your analysis. If you were mostly concerned with background suppression, you would be interested in the lower left corner of the plot and choose Σ_2 as your estimate. If you goal were to retain as much signal as possible at a modest background rejection rate, you would focus on the upper right corner of the plot and choose Σ_1. If you wanted the best overall quality of separation measured by the area under the ROC curve, you would choose $(\Sigma_1 + \Sigma_2)/2$. It is your analysis, so take your pick!

If we took this logic to the extreme, we could search for the best a in the linear combination $\hat{\Sigma} = a\hat{\Sigma}_1 + (1-a)\hat{\Sigma}_2$ by minimizing some criterion, perhaps FPR at fixed TPR. If you engage in such optimization, you should ask yourself if LDA is the right tool. At this point, you might want to give up the beloved linearity and switch to a more flexible technique such as QDA. In this example, QDA beats LDA at any FPR, no matter what covariance matrix estimate you choose.

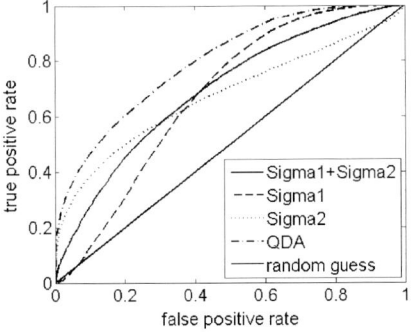

Figure 11.2 ROC curves for LDA with three estimates of the covariance matrix and QDA.

11.1.4
Numerical Implementation

First we estimate the class means $\hat{\boldsymbol{\mu}}_k$; $k = 1, \ldots, K$; and center the training matrix **X** by subtracting the estimated means from respective rows: if row n of matrix **X** is an observation of class k, this row is centered by subtracting $\hat{\boldsymbol{\mu}}_k$. To avoid computing and inverting a covariance matrix, we can take a convenient decomposition of the centered matrix **X**. The motivation is accuracy: we don't want to propagate measurement errors in **X** into $\mathbf{X}^\mathsf{T}\mathbf{X}$ by squaring them. We reviewed SVD in Section 8.3.2. Another good choice is QR decomposition $\mathbf{X} = \mathbf{QR}$, where **Q** is orthogonal of size $N \times D$ and **R** is upper triangular of size $D \times D$ for N observations and D variables. "Upper triangular" means that all elements below the main diagonal are set to zero. It is easy to see that $\mathbf{X}^\mathsf{T}\mathbf{X} = \mathbf{R}^\mathsf{T}\mathbf{R}$. The term $(x - \mu)^\mathsf{T}(\mathbf{X}^\mathsf{T}\mathbf{X})^{-1}(x - \mu)$ then reduces to $((x - \mu)^\mathsf{T}\mathbf{R}^{-1})((x - \mu)^\mathsf{T}\mathbf{R}^{-1})^\mathsf{T}$. The row vector $(x - \mu)^\mathsf{T}\mathbf{R}^{-1}$ of length D can be easily computed thanks to the special structure of **R**.

So far, we have ignored observation weights and bias corrections. To include them, we need to multiply every row in **X** by the corresponding weight. Let $\mathbf{diag}(a)$ be a diagonal matrix with vector a on its main diagonal and zeros elsewhere. Since **Q** is orthogonal, square roots of the weights in the QR decomposition $\mathbf{diag}(\sqrt{w})\mathbf{X} = \mathbf{QR}$ are absorbed in **R**. The weighted maximum likelihood estimate of the covariance matrix is then given by $\mathbf{R}^\mathsf{T}\mathbf{R}$. The bias correction applies to all weights and can be accounted for by multiplying all elements in w by the same factor.

Discriminant analysis is implemented in many software suites. One example is the `ClassificationDiscriminant` class in the Statistics Toolbox of MATLAB.

11.1.5
Regularized Discriminant Analysis

If the rank of the observable matrix **X** is less than the data dimensionality, matrix **R** from the QR decomposition, or similarly the covariance matrix estimate $\hat{\boldsymbol{\Sigma}}$, cannot be inverted. Discriminant analysis, the way it has been described above, cannot be applied to wide data ($D > N$). Physicists usually work with very tall data ($D \ll N$) but occasionally can run into the same problem. A tall matrix **X** is rank-deficient if any variable (column) is a linear combination of other variables (columns). The simplest cause of this rank deficiency is collinearity between two variables. In statistical terms, collinearity means that the magnitude of the linear correlation is close to 1. We can always identify such pairs of variables and exclude them. Testing all linear combinations of variables is much harder. Even if we have no reason to believe that one of the input variables is a linear combination of the others, measurement and computational errors can lead to a singular matrix **X**.

How can we invert a noninvertible covariance matrix? One way is to use *pseudoinverse*. If we take thin SVD of the covariance matrix $\boldsymbol{\Sigma} = \mathbf{USV}^\mathsf{T}$, we can compute its inverse as $\boldsymbol{\Sigma}^{-1} = \mathbf{VS}^{-1}\mathbf{U}^\mathsf{T}$. Inverting **U** and **V** amounts to transposing them. The trick is inverting **S**, the diagonal matrix of singular values. If all diagonal elements of **S** were different from zero within numerical accuracy, inverting **S** would amount to taking the inverse of every diagonal element. If some diagonal elements of **S** are zeros, we can set the respective elements in $\mathbf{S}^{-1}$ to zero too. This operation is called pseudo-inversion and is denoted by superscript $^+$: $\mathbf{\Sigma}^+ = \mathbf{V}\mathbf{S}^+\mathbf{U}^\mathsf{T}$. In practice, the equality of the diagonal elements to zero is tested by an appropriate numerical criterion such as, roughly, comparison with the change in the largest element of **S** induced by a one-bit change to its floating point representation.

Another option is to regularize the covariance matrix by adding positive values to its main diagonal. Since diagonal elements of a covariance matrix are positive, this addition increases their magnitude with respect to off-diagonal elements. If we make the diagonal elements sufficiently large, the matrix gets sufficiently close to a diagonal one and becomes invertible. Several regularization schemes for discriminant analysis have been proposed in Friedman (1988); Guo et al. (2007); Witten et al. (2009); Qiao et al. (2008). Here, we mention only one:

$$\tilde{\mathbf{\Sigma}} = (1-\gamma)\hat{\mathbf{\Sigma}} + \gamma \, \mathbf{diag}(\hat{\mathbf{\Sigma}}) \,. \tag{11.17}$$

The regularized covariance matrix estimate $\tilde{\mathbf{\Sigma}}$ is a weighted average of the plain estimate $\hat{\mathbf{\Sigma}}$ and its main diagonal $\mathbf{diag}(\hat{\mathbf{\Sigma}})$. Now $\gamma = 0$ gives the standard estimate without regularization, and $\gamma = 1$ gives the diagonal version of the covariance matrix estimate. We can choose the best value of γ in the range $[0, 1]$ by optimizing an appropriate criterion, for example, by minimizing classification error for the discriminant analysis model on test data.

Earlier we used a QR decomposition of training data **X**. For regularization of tall data $N > D$, a more convenient choice is EVD. Compute the covariance matrix $\mathbf{\Sigma} = \mathbf{X}^\mathsf{T}\mathbf{X}$ after centering **X**. (We drop the hat notation for the remainder of this section.) Take the diagonal part of the covariance matrix (variance for each variable), $\mathbf{\Omega} = \mathbf{diag}(\mathbf{\Sigma})$. Take EVD of the correlation matrix $\mathbf{\Omega}^{-1/2}\mathbf{\Sigma}\mathbf{\Omega}^{-1/2} = \mathbf{V}\mathbf{\Lambda}\mathbf{V}^\mathsf{T}$. A matrix **R**, similar to the one obtained from the QR decomposition, can be obtained by regularizing $\mathbf{\Lambda}$,

$$\tilde{\mathbf{R}} = [(1-\gamma)\mathbf{\Lambda} + \gamma \mathbf{I}_{D\times D}]^{1/2} \mathbf{V}^\mathsf{T} \mathbf{\Omega}^{1/2} \,. \tag{11.18}$$

The regularized covariance matrix can be now computed as $\tilde{\mathbf{\Sigma}} = \tilde{\mathbf{R}}^\mathsf{T}\tilde{\mathbf{R}}$. For prediction on new data, we can use $\tilde{\mathbf{R}}^{-1}$ in place of $\mathbf{R}^{-1}$ with $\tilde{\mathbf{R}}^{-1}$ trivially obtained from (11.18). This regularization scheme is too slow and memory-inefficient for wide data $D \gg N$. Instead, the trick described in Guo et al. (2007) should be used.

11.1.6
LDA for Variable Transformation

Fisher discriminant (11.5) takes a particularly simple form for two classes. Let us modify it a bit. Take $\boldsymbol{\mu}_C = W_1\boldsymbol{\mu}_1 + W_2\boldsymbol{\mu}_2$ to be the pooled-in mean and $\mathbf{\Sigma}_B = \sum_{k=1}^{2} W_k(\boldsymbol{\mu}_k - \boldsymbol{\mu}_C)(\boldsymbol{\mu}_k - \boldsymbol{\mu}_C)^\mathsf{T}$ to be the between-class covariance matrix.

Maximizing (11.5) is then equivalent to maximizing

$$S(q) = \frac{q^\mathsf{T} \Sigma_\mathrm{B} q}{q^\mathsf{T} \Sigma_\mathrm{W} q}, \tag{11.19}$$

where Σ_W is the within-class covariance matrix; earlier we used Σ without a subscript for this quantity. Above, W_1 and W_2 are class weights obtained by summing weights of all observations in the respective class. The choice of factors W_1 and W_2 is not essential – maximizing the criterion above is equivalent to maximizing (11.5) as long as the mean point μ_C lies on a line connecting μ_1 and μ_2.

This problem looks remarkably similar to the principal component decomposition (8.3). In PCA we search for orthogonal eigenvectors q satisfying $\Sigma q = \lambda q$ and obtain them through eigenvalue decomposition (EVD) of the covariance matrix Σ. Here, we search for q satisfying $\Sigma_\mathrm{B} q = \lambda \Sigma_\mathrm{W} q$ and obtain them through generalized eigenvalue decomposition (GEVD) of Σ_B and Σ_W. For binary classification, Σ_B is computed using two independent vectors with one linear constraint and therefore has rank 1. The one and only eigenvector defines the optimal classification boundary between the two classes. Solving the generalized eigenvalue problem is equivalent to the formalism described in Section 11.1.

For $K > 2$ classes, the problem is more involved. Now Σ_B is of rank $K - 1$, and the $K - 1$ eigenvectors are Σ_W-orthogonal: $q_i^\mathsf{T} \Sigma_\mathrm{W} q_j = 0$ for two eigenvectors, q_i and q_j, with $i \neq j$. LDA thus defines a nonorthogonal transformation from D to $K - 1$ variables and can be used as a tool for dimensionality reduction. Here, we assume $K \leq D$.

Several versions of this technique are available. For good reviews, we recommend Ye and Xiong (2006) and Zhang et al. (2010).

Classification and variable transformation are two different perspectives on LDA. For classification we find $K(K-1)/2$ linear boundaries between K classes, and for variable transformation we find $K - 1$ most informative directions in space.

Examples of LDA application for variable transformation in physics analysis are rare. Physics analysis is typically focused on separation of signal and background formulated as binary classification. Background is often composed of observations (events) produced by different processes. This multiclass structure can be exploited by LDA to derive new variables potentially useful for understanding the individual background components. Unfortunately, LDA has not been really explored in this capacity.

The LDA variable transformation can be computed by GEVD. Function `eig` in MATLAB is capable of performing both EVD and GEVD. Once you estimate between-class and within-class covariance matrices, call

```
[TransCoeffs,Lambda] = eig(BetweenSigma,WithinSigma,'chol');
```

This call returns a $D \times D$ matrix of transformation coefficients `TransCoeffs` and a $D \times D$ diagonal matrix of eigenvalues `Lambda`. Option `'chol'` tells the function to use the Cholesky decomposition of the symmetric positive semidefinite covariance matrices `BetweenSigma` and `WithinSigma`. Function `eig` does not know that the rank of BetweenSigma is at most $K - 1$, and you have to remove zero eigenvalues from Lambda and the corresponding columns from TransCoeffs manually. Then you can multiply an $N \times D$ matrix of new observations by the coefficients, X*TransCoeffs, to transform this matrix to the reduced space.

11.2
Logistic Regression

As seen in (11.4), LDA uses a simple linear model for the log of the posterior odds. Suppose we have two classes, $Y \in \{0, 1\}$, and let $p = P(Y = 1|x)$ be the posterior probability for class 1 at point x. We now write the same linear model as

$$\log \frac{p}{1 - p} = \beta_0 + \boldsymbol{\beta}^T x , \qquad (11.20)$$

where $\boldsymbol{\beta}$ is a column vector of coefficients and β_0 is an intercept.

This simple change of notation introduces a new interpretation of the model and a new set of algorithmic tools. We no longer require that the two classes have multivariate normal densities with equal covariance matrices. We have thus obtained a model which is still simple yet considerably more flexible. Now we can deploy the machinery of linear regression. This machinery provides a range of attractive tools such as formulas for computing the regression coefficients and their covariance matrix, well-known approaches to treatment of categorical variables, tests for significance of variables in the model, methods for selecting the optimal number of variables, and so on. Regression is largely beyond the scope of this book, and we are not going to review this machinery in detail. Good reviews of logistic regression can be found in Agresti (2002); Hastie et al. (2008); McCullagh and Nelder (1983), and Hosmer and Lemeshow (1989).

Despite its simplicity and certain advantages over LDA, logistic regression is virtually unknown in the world of particle and astrophysics. Statisticians and practitioners in other fields use logistic regression a lot.

11.2.1
Binomial Logistic Regression: Theory and Numerical Implementation

Usually, we collect observations into a matrix **X** of size $N \times D$ with one observation per row and one variable per column. Vector x_n represents the nth observation (transposed row) from the collected data. We label the two classes 0 and 1 and collect the observed class labels in a vector **y**. We then set $p(x)$ in (11.20) to the probability of observing class 1 at point x.

We can treat the class label Y as a Bernoulli random variable with a probability mass function

$$P_{Y|X=x}(y|x) = p^y (1 - p)^{1-y} . \qquad (11.21)$$

If the observations are independent, the likelihood is a simple product,

$$L(\beta_0, \boldsymbol{\beta}; \mathbf{y}) = \prod_{n=1}^{N} [p(\mathbf{x}_n; \beta_0, \boldsymbol{\beta})]^{y_n} [1 - p(\mathbf{x}_n; \beta_0, \boldsymbol{\beta})]^{1-y_n}. \quad (11.22)$$

We can now obtain the optimal estimates of β_0 and $\boldsymbol{\beta}$ by maximizing the likelihood.

Before we do this, let us address a minor technical point – the intercept term β_0. Instead of writing it down every time, we define vector $\tilde{\mathbf{x}} \equiv \{1, x_1, \ldots, x_D\}$ with $D+1$ elements and the first element set to 1. Similarly we define $\tilde{\boldsymbol{\beta}} \equiv \{\beta_0, \beta_1, \ldots, \beta_D\}$. We set $\tilde{\mathbf{x}}$ and $\tilde{\boldsymbol{\beta}}$ to be column vectors. Now we can replace $\beta_0 + \boldsymbol{\beta}^T \mathbf{x}$ with $\tilde{\boldsymbol{\beta}}^T \tilde{\mathbf{x}}$. We also define a new matrix $\tilde{\mathbf{X}}$ by setting it equal to $\mathbf{X}$ and adding the leftmost column filled with ones.

Taking the log of (11.22) and plugging (11.20) in, we obtain

$$\log L(\tilde{\boldsymbol{\beta}}; \mathbf{y}) = \mathbf{y}^T \tilde{\mathbf{X}} \tilde{\boldsymbol{\beta}} - \sum_{n=1}^{N} \log \left[1 + \exp\left(\tilde{\boldsymbol{\beta}}^T \tilde{\mathbf{x}}_n\right)\right]. \quad (11.23)$$

The gradient of $\log L$ in the $(D+1)$-dimensional space $\tilde{\boldsymbol{\beta}}$ is then

$$\nabla_{\tilde{\boldsymbol{\beta}}} \log L(\tilde{\boldsymbol{\beta}}; \mathbf{y}) = \tilde{\mathbf{X}}^T (\mathbf{y} - \mathbf{p}), \quad (11.24)$$

where $\mathbf{p}$ is a column vector with elements equal to $p(\tilde{\mathbf{x}}_n; \tilde{\boldsymbol{\beta}})$. The Hessian (matrix of second derivatives) is readily computed as well:

$$H_{\tilde{\boldsymbol{\beta}}} \log L(\tilde{\boldsymbol{\beta}}; \mathbf{y}) = -\tilde{\mathbf{X}}^T \, \mathrm{diag}\left[\mathbf{p}(\mathbf{1}_{N \times 1} - \mathbf{p})^T\right] \tilde{\mathbf{X}}. \quad (11.25)$$

Above, $\mathbf{diag}(\mathbf{A})$ is a matrix with the main diagonal equal to that of $\mathbf{A}$ and all other elements set to zero, and $\mathbf{1}_{N \times 1}$ is a column vector with N elements, all set to 1.

Since we know both gradient and Hessian, numerical optimization is easy. This problem can be solved by one of the minimization utilities found in the CERN math libraries. Alternatively, you could use one of many implementations such as the GeneralizedLinearModel class available from the Statistics Toolbox in MAT-LAB. Such implementations often use the iterative weighted least-squares algorithm described in McCullagh and Nelder (1983).

The estimated coefficients $\hat{\tilde{\boldsymbol{\beta}}}$ maximizing the likelihood (11.22) are unique and finite if the two classes overlap. If you can draw a hyperplane perfectly separating the two classes in the training set, the coefficients along the direction of perfect separation must go to infinity. Unless software can detect the perfect separation, you get coefficients with arbitrarily large, unstable values. In this case, you might prefer a different linear classifier. Linear discriminant and partial least-squares reviewed in this chapter, as well as linear support vector machines reviewed in Chapter 13, could be a better choice.

To include observation weights $\mathbf{w}$, replace (11.23) with

$$\log L(\tilde{\boldsymbol{\beta}}; \mathbf{y}) = \mathbf{y}^T \, \mathrm{diag}(\mathbf{w}) \tilde{\mathbf{X}} \tilde{\boldsymbol{\beta}} - \sum_{n=1}^{N} w_n \log \left[1 + \exp\left(\tilde{\boldsymbol{\beta}}^T \tilde{\mathbf{x}}_n\right)\right], \quad (11.26)$$

where $\mathbf{diag}(\mathbf{w})$ is an $N \times N$ matrix with weights on the main diagonal.

11.2.2
Properties of the Binomial Model

The covariance matrix and confidence intervals for the regression coefficients are estimated using the Hessian matrix $\mathbf{H} = -H_{\tilde{\beta}} \log L(\tilde{\boldsymbol{\beta}}; \mathbf{y})$ at the minimum of the negative log-likelihood. We can translate these estimates to confidence intervals for the class probabilities. For the binary logit, we have $\text{Var}(\hat{\boldsymbol{\beta}}^T \tilde{x}) = \tilde{x}^T \mathbf{H}^{-1} \tilde{x}$, where the variance $\text{Var}(\hat{\boldsymbol{\beta}}^T \tilde{x})$ is taken over coefficient estimates $\hat{\boldsymbol{\beta}}$ at fixed $\tilde{x}$. Since $p(\tilde{x})$ is a monotone function of $\hat{\boldsymbol{\beta}}^T \tilde{x}$, we immediately obtain a confidence interval for $p(\tilde{x})$ by varying the estimate $\hat{\boldsymbol{\beta}}^T \tilde{x}$ by one standard deviation in either direction. Most classification models do not allow for simple computation of the uncertainty of the posterior probability estimate. Logistic regression is a nice exception.

Logistic regression can be extended for ordinal classification. For K ordered classes, we model the cumulative probability

$$\log \frac{P(Y \leq k | \mathbf{x})}{P(Y > k | \mathbf{x})} = \beta_{0k} + \boldsymbol{\beta}^T \mathbf{x} \tag{11.27}$$

for $k = 1, \ldots, K-1$ and $P(Y \leq K | \mathbf{x}) = 1$. The intercepts for the $K-1$ logit models are different, but the slopes are constrained to be the same. In the statistics literature, this formalism is known as the *proportional odds model*. The optimal coefficients are found by maximizing the likelihood formed as a product of probabilities (11.30) with $p_k(\mathbf{x}) = P(Y \leq k | \mathbf{x}) - P(Y \leq k-1 | \mathbf{x})$.

Using the likelihood function (11.23), we can test for significance of individual terms in the logit model. Suppose we would like to test if variable d is useful for classification. Fit the logit model with and without this variable and take a difference $2 \log L_0 - 2 \log L_1$ of the fitted likelihood values without and with variable d. If the variable is continuous, this difference is asymptotically distributed as χ^2 with one degree of freedom. If the variable is categorical with C levels, it is encoded by $C-1$ dummy variables. To test for significance, we need to take all these dummy variables out and therefore set the number of degrees of freedom for the χ^2 distribution to $C-1$. Computing the upper tail of the χ^2 cdf above the observed log-likelihood difference, we obtain a p-value for testing the null hypothesis (variable d is not significant for classification) against its alternative (variable d is significant). A low p-value indicates usefulness of the variable in question.

This technique could be used for variable ranking and selection, discussed in detail in Chapter 18. If the decision to keep or discard a variable is made by comparing the p-value of the respective test with a prespecified level (for instance, 5%), this p-value needs to be calibrated to account for multiple testing. This problem was discussed in Chapter 10 and will be discussed again in Chapter 18.

11.2.3
Verifying Model Assumptions

A popular measure of goodness of fit for logistic regression is described in Hosmer and Lemeshow (1989). It is based on comparing the observed distribution of the predicted class probabilities with the known class labels. Let $p(x)$ be the predicted probability of observing $Y = 1$ at x. We can group the predicted probabilities into B bins between 0 and 1 using, for example, an equal number of observations per bin. Let n_b be the number of observations, $\bar{y}_b$ the average true response, and $\bar{p}_b$ the average predicted probability in bin b. If the data are adequately modeled by logistic regression, the Hosmer–Lemeshow statistic

$$C = \sum_{b=1}^{B} \frac{n_b(\bar{y}_b - \bar{p}_b)^2}{\bar{p}_b(1 - \bar{p}_b)} \tag{11.28}$$

is asymptotically distributed as χ^2 with $B - 2$ degrees of freedom. The number of bins is usually set to 10.

Copas (1989) describes a simpler test based on the statistic

$$S = \sum_{n=1}^{N} (y_n - p(x_n))^2 . \tag{11.29}$$

It has the advantage of not using a somewhat arbitrary choice of binning. Hosmer et al. (1997) show that the test by Copas performs as well as more complex procedures on simulated and real-world datasets. If the data are adequately modeled by logistic regression, the test statistic asymptotically follows a normal distribution with estimated mean and variance. Refer to Hosmer et al. (1997) for details on their estimation.

11.2.4
Logistic Regression with Multiple Classes

So far, we have reviewed logistic regression with two classes, also known as the binomial logit model. Let us generalize our analysis to multiclass problems. The probability mass function for categorical variable Y with K outcomes is

$$P_{Y|X=x}(y_1, \ldots, y_K|x) = \prod_{k=1}^{K} [p_k(x)]^{y_k} , \tag{11.30}$$

where $p = \{p_k\}_{k=1}^{K}$ is a vector of K multinomial probabilities. Only one of the K elements in $y_1, \ldots, y_K$ can be set to 1; the rest are set to 0. The probabilities p_k of observing Y in state k sum to one. Now we use the same trick as the one we used for LDA: choose class K for reference and assume linear models for $K - 1$ odds:

$$\log \frac{P(k|\tilde{x})}{P(K|\tilde{x})} = \tilde{\beta}_k^T \tilde{x} . \tag{11.31}$$

We use "tilde" quantities implying that the intercept term is included.

To simplify the notation further, let us introduce an $N \times (K - 1)$ matrix $\mathbf{Y}$ and a $(D+1) \times (K-1)$ matrix $\tilde{\mathbf{B}}$. The matrix $\mathbf{Y}$ encodes nominal response with K levels, as discussed in Section 7.1. Every row of $\mathbf{Y}$ has one in kth position if the observation is of class k and zeros everywhere else. If the observation is of class K, the entire row is filled with zeros. The matrix $\tilde{\mathbf{B}}$ is composed of $K-1$ column vectors $\tilde{\boldsymbol{\beta}}_k$. Now we can write the log-likelihood as

$$\log L(\tilde{\mathbf{B}}; \mathbf{Y}) = \sum_{n=1}^{N} \sum_{k=1}^{K-1} y_{nk} \tilde{\boldsymbol{\beta}}_k^\top \tilde{\mathbf{x}}_n - \sum_{n=1}^{N} \log\left[1 + \sum_{k=1}^{K-1} \exp\left(\tilde{\boldsymbol{\beta}}_k^\top \tilde{\mathbf{x}}_n\right)\right], \quad (11.32)$$

where y_{nk} is the element of $\mathbf{Y}$ in the nth row and kth column. Using this formula, you can derive expressions for the gradient and Hessian.

We can fit the multinomial logistic model in the same way we fit its binomial counterpart, except now we have $K-1$ times as many unknown coefficients. Many software implementations such as, for instance, the `mnrfit` function available from the Statistics Toolbox in MATLAB are capable of carrying out this task. If your implementation can work with two classes only, you can fit $\tilde{\boldsymbol{\beta}}_k$ separately for every $k = 1, \ldots, K-1$. In that case, choose the class with most observations for reference. The estimates obtained by these $K-1$ independent fits are different from those obtained by optimizing the likelihood (11.32) but should be fairly close if the reference class dominates by a large margin.

11.3
Classification by Linear Regression

In the previous section, you saw how logistic regression can be applied to classification. Here, we describe a simpler model, linear regression of multivariate response. Formally, we put

$$\mathbf{Y} = \boldsymbol{\beta}_0 + \mathbf{B}^\top \mathbf{X} \quad (11.33)$$

for response $\mathbf{Y}$, variables $\mathbf{X}$, intercept terms $\boldsymbol{\beta}_0$, and a matrix of regression coefficients $\mathbf{B}$. Above, $\mathbf{X}$ and $\mathbf{Y}$ are vector random variables. As in the previous section, we can incorporate the intercepts in the definitions of $\mathbf{X}$ and $\mathbf{B}$ to write

$$\mathbf{Y} = \tilde{\mathbf{B}}^\top \tilde{\mathbf{X}}. \quad (11.34)$$

What is the size of the response vector $\mathbf{Y}$ for K classes? As you saw in Section 7.1, categorical response with K levels can be reduced to $K-1$ binary classification problems. We could adopt the same approach here and encode $\mathbf{Y}$ as a vector of length $K-1$. For example, we could represent every class k but K (the last one) by putting one in the kth position and zeros elsewhere. The last class would be then encoded as all zeros. This approach could work in principle. But imagine how this model would predict the class label based on the predicted response vector $\hat{\mathbf{y}}$. If we assigned an observation to the class with the largest element in $\hat{\mathbf{y}}$, we would never assign any observation to class K. We could compute some loss between the predicted $\hat{\mathbf{y}}$ and the class template ($\mathbf{y}$ used to encode this class) and assign an observation to the class with minimal loss; in particular, computing the quadratic error would be appropriate. But then we would have to make an extra, not entirely trivial step. Instead of facing these difficulties, we prefer to encode $\mathbf{y}$ as a vector with K elements. Matrix $\tilde{\mathbf{B}}$ then has size $(D + 1) \times K$ for D input variables. Every class k is represented by one in kth position and zeros elsewhere. An observation is assigned to the class with the largest element in the predicted vector $\hat{\mathbf{y}}$. This setup is similar to the "one versus all" strategy discussed in Chapter 16.

In the statistics literature, this method is called *regression of an indicator matrix*. Regression of an indicator matrix is a special case of multivariate regression, that is, regression with several response variables per observation. You can find a deeper description in Hastie *et al.* (2008) and other textbooks.

Suppose class labels for the training data are collected into an $N \times K$ matrix $\mathbf{Y}$, and variables are collected into an $N \times D$ matrix $\mathbf{X}$. As usual, we augment $\mathbf{X}$ by including the leftmost column of ones to account for the intercept terms; the augmented matrix is $\tilde{\mathbf{X}}$. Coefficients $\tilde{\mathbf{B}}$ can be found by minimizing the squared error in the training set, $\|\mathbf{Y} - \tilde{\mathbf{X}}\tilde{\mathbf{B}}\|_F^2$, where $\|\mathbf{A}\|_F^2 = \sum_{n=1}^{N} \sum_{k=1}^{K} a_{nk}^2$ is the squared Frobenius norm of an $N \times K$ matrix $\mathbf{A}$. The analytical solution is given by

$$\tilde{\mathbf{B}} = (\tilde{\mathbf{X}}^T \tilde{\mathbf{X}})^{-1} \tilde{\mathbf{X}}^T \mathbf{Y}. \tag{11.35}$$

As shown in Section 13.6.2, this solution can be modified to accommodate weighted observations. Details of the numerical implementation can be found in textbooks such as Moler (2008), and implementations can be found in software suites such as MATLAB.

Linear regression offers the same advantages as logistic regression: simplicity, easy treatment of categorical variables, standard errors for regression coefficients, and others. In addition, it is faster than logistic regression, although this advantage in speed may not matter in practice if datasets are moderate in size. Elements of the predicted vector $\hat{\mathbf{y}}$ can lie outside the $[0, 1]$ interval and therefore cannot be interpreted as class probabilities.

The technique we describe in the next section takes linear regression further. It provides regression of an indicator matrix as well. In addition, this technique can be used for dimensionality reduction and variable transformation.

11.4
✄ Partial Least Squares Regression

Partial least squares (PLS) regression resembles linear transformations described in Chapter 8 and linear classification methods discussed in this chapter.

Similar to ICA, the PLS technique searches for a nonorthogonal transformation to a new basis, one component at a time. The search can be stopped when a predefined number of PLS components is found. If the number of components is kept far below the number of variables, a low-dimensional representation of the data is obtained.

Similar to LDA, PLS searches for components that best predict the observed response. In textbooks, PLS is described as a multivariate regression method. In the statistics literature, "multivariate regression" implies several response variables per observation (one-dimensional response leads to multiple regression). As you saw in Section 11.3, classification with K classes can be reduced to multivariate regression with K response variables. Such regression can be carried out by PLS.

In a sense, PLS takes the best of the two approaches. PLS can provide a simple compact representation of high-dimensional data with large classification accuracy.

Introduced by Wold (1966) for research in economics, PLS has become popular in chemometrics and medical imaging. It is most useful when the number of variables is large but can be applied in a low-dimensional setting as well. Perhaps, it is about time PLS found its way to physics analysis.

Suppose X and Y are random vectors with D and K elements, respectively. Assume that the two random vectors have been centered to zero means and scaled to unit variance. We search for directions w and c such that the product of $T = w^T X$ and $U = c^T Y$ has maximal expectation $E(TU)$. Since T and U have been centered, $E(TU)$ is their covariance. We require that vectors w and c have unit norms to prevent $E(TU)$ from taking arbitrarily large values.

In practice, we need to replace random vectors X and Y with their observed values **X** and **Y**. Here, **X** is of size $N \times D$ and **Y** is of size $N \times K$ for N observations, D variables and K classes. Let us find the first PLS component. Set $\mathbf{X}_1 = \mathbf{X}$ and $\mathbf{Y}_1 = \mathbf{Y}$ and search for unit vectors w_1 and c_1 maximizing $b_1 = t_1^T u_1$, the product of $t_1 = \mathbf{X}_1 w_1$ and $u_1 = \mathbf{Y}_1 c_1$. These two vectors are found by an appropriate iterative procedure. Vector t_1 is stored as the first column of the *score* matrix **T**. Vector $p_1 = \mathbf{X}_1^T t_1$ is stored as the first column of the *loading* matrix **P**. Vector $b_1 c_1$ is stored as the first column of the *weight* matrix **C**.

To find the second component, partial out the effect of the first component. This step is similar to Gram–Schmidt orthogonalization (deflation) in ICA. Take $\mathbf{X}_2 = \mathbf{X}_1 - t_1 p_1^T = (\mathbf{I}_{N \times N} - t_1 t_1^T) \mathbf{X}_1$ and $\mathbf{Y}_2 = \mathbf{Y}_1 - t_1 c_1^T$. Then search for a new pair, w_2 and c_2, maximizing $t_2^T u_2$, and so on.

If we stop after L components are found, we can estimate the original centered and scaled **X** and **Y** by

$$\hat{\mathbf{X}} = \mathbf{T}\mathbf{P}^T \qquad (11.36)$$

$$\hat{\mathbf{Y}} = \mathbf{T}\mathbf{C}^T . \qquad (11.37)$$

Above, **T** is of size $N \times L$, **P** is of size $D \times L$, and **C** is of size $K \times L$. The loading matrix **P** and the weight matrix **C** save the learned model. The score matrix **T** provides a representation of the observed data **X** and **Y** in the PLS space.

By construction, the score matrix **T** is orthogonal, but the loading matrix **P** is not. This is an important distinction from PCA where the loadings are orthogonal. Both for PCA and PLS, the directions of the optimal components in the space spanned by x are given by columns of the loading matrix. PCA performs a rotation, and PLS performs a more general linear transformation.

To predict the class vector y from an observed vector x, take

$$\hat{y} = \mathbf{C}\mathbf{P}^+ x , \qquad (11.38)$$

where $\mathbf{P}^+$ is the pseudo-inverse of $\mathbf{P}$. To compute $\mathbf{P}^+$, take its SVD, $\mathbf{P} = \mathbf{USV}^\mathsf{T}$, and set $\mathbf{P}^+ = \mathbf{VS}^+\mathbf{U}^\mathsf{T}$. The pseudo-inverse of the diagonal matrix $\mathbf{S}$ is obtained by taking the inverse of every nonzero diagonal element and taking zero for every zero diagonal element. To use this prediction formula, both x and $\hat{y}$ need to be corrected for centering and scaling.

How many PLS components should be chosen? One way of answering this question is to assess the importance of every PLS component by measuring the amount of variance it explains. The total observed variance in $\mathbf{X}$ is proportional to the trace of the covariance matrix $\operatorname{tr}(\mathbf{X}^\mathsf{T}\mathbf{X})$; similarly for $\mathbf{Y}$. The variance explained by the PLS model is $\operatorname{tr}(\mathbf{P}^\mathsf{T}\mathbf{P})$. The fraction of variance explained by the lth PLS component is therefore proportional to the squared magnitude of the lth column of the loading matrix $\sum_{d=1}^{D} p_{dl}^2$. PLS is a predictive model, and we are mostly interested in variance in $\mathbf{Y}$. By a similar argument, take the fraction of variance explained by the lth PLS component to be $\sum_{k=1}^{K} c_{kl}^2 / \operatorname{tr}(\mathbf{Y}^\mathsf{T}\mathbf{Y})$. The magnitude of the lth column of the weight matrix $\mathbf{C}$ happens to be b_l, the covariance between t_l and u_l. Plot the fraction of explained variance versus the component index. Often this plot can give you a good idea for the optimal number of components. The elbow effect mentioned in Section 8.3.4 could be put to work for PLS, too.

Although PLS is a simple linear model, it is prone to overtraining if a large number of components is used. If your primary concern is the predictive power, choose the number of PLS components by maximizing the classification accuracy of the PLS model using cross-validated data or an independent test set.

If $\mathbf{W} = (\mathbf{P}^\mathsf{T})^{-1}$ is the $D \times L$ inverse of the transposed matrix of loadings (sometimes called the matrix of PLS weights), use

$$\hat{i}_d = \sqrt{D \sum_{l=1}^{L} \tilde{w}_{dl}^2 \sum_{k=1}^{K} \tilde{c}_{kl}^2} \tag{11.39}$$

as a measure of importance for variable d; $d = 1, \ldots, D$; in a standardized input matrix $\mathbf{X}$. Here, $\tilde{w}_{dl}^2 = w_{dl}^2 / \sum_{i=1}^{D} w_{il}^2$ and $\tilde{c}_{kl}^2 = c_{kl}^2 / \sum_{i=1}^{K} \sum_{j=1}^{L} c_{ij}^2$ are normalized weight matrices. In other words, assign large importance values to variables with large contributions to most important PLS components. The contribution of variable d to PLS component l is measured by the respective weight w_{dl}. The sign of w_{dl} should not matter; hence, square the weights. The sum $\sum_{k=1}^{K} \tilde{c}_{kl}^2$ measures the importance of PLS component l for class separation. Since we assume $\mathbf{X}$ to be standardized, all PLS scores (columns of $\mathbf{T}$) have unit variance; if this were not the case, an additional normalization term in (11.39) would have to be included.

One of the difficulties in describing and interpreting PLS is its diversity. Several versions of PLS exist. Here, we have described essentially the NIPALS algorithm due to Wold (1966) focused on modeling relations between variables X and Y. This technique is similar to canonical correlation analysis (CCA) in spirit. A computationally-efficient PLS algorithm due to Bookstein (1994) obtains loadings and weights by taking SVD of $\mathbf{X}^\mathsf{T}\mathbf{Y}$. Barker and Rayens (2003) propose a modified PLS algorithm for classification and show its connection to LDA and CCA. Rosipal and

Kramer (2006) and Abdi (2010) provide good reviews of PLS basics and its incarnations.

Function `plsregress` in the Statistics Toolbox of MATLAB implements the SIMPLS algorithm described in de Jong (1993). Implementations of PLS are available from R and other statistical packages.

11.5
Example: Linear Models for MAGIC Telescope Data

To illustrate application of linear classifiers to real-world data, we include a MATLAB example.

Contents
- Load data
- Standardize variables
- Train a linear discriminant and select 4 best predictors
- Train LDA, logistic regression (LR) and partial least squares (PLS)
- Estimate classification errors using test data
- Test the models for equivalence
- Verify LDA assumptions (Figure 11.3)
- Verify LR assumptions
- Estimate variable importance
- Visualize the classification models (Figures 11.4 and 11.5).

Load Data

We use the MAGIC telescope data from the UCI repository http://archive.ics.uci.edu/ml/datasets/MAGIC+Gamma+Telescope. These are simulated data for detection of high energy gamma particles in a ground-based atmospheric Cherenkov gamma telescope using the imaging technique. The goal is to separate gamma particles from hadron background.

The analysis described in Bock *et al.* (2004), applies a number of powerful classifiers such as random forest and neural net. The power of these classifiers is evaluated by measuring true positive rates (TPR) at false positive rates (FPR) 0.01, 0.02, 0.05, 0.1, and 0.2. Here, we present an example illustrating properties of three linear techniques for binary classification. These linear techniques are not meant to improve the classification accuracy obtained in the publication. We use a reduced set of variables. As seen below, the optimal boundary between the two classes is far from linear.

We split the data into training and test sets in the proportion 2 : 1 and code class labels as a logical vector (true for gamma and false for hadron).

```
load MagicTelescope;
Ytmp = false(numel(Ytrain),1);
Ytmp(strcmp(Ytrain,'Gamma')) = true;
```

```
Ytrain = Ytmp;
Ytmp = false(numel(Ytest),1);
Ytmp(strcmp(Ytest,'Gamma')) = true;
Ytest = Ytmp;
```

Standardize Variables

In these data, variables have substantially different standard deviations. These standard deviations do not carry information useful for separation of the two classes. In this situation, it is recommended to standardize variables by subtracting the mean and dividing by the standard deviation. This is especially true if you plan to judge the importance of variables by the magnitudes of the respective coefficients in the linear model.

For some implementations, standardizing variables is not necessary. For example, `ClassificationDiscriminant.fit` standardizes variables separately for every class. Even if you perform the overall standardization by the `zscore` function, it is unlikely to affect the analysis outcome. In contrast, estimates of variable importance from `plsregress` are sensitive to the standardization.

We use function `zscore` to standardize the training data and then use the computed means and standard deviations to standardize the test data.

```
[Xtrain,mu,sigma] = zscore(Xtrain);
Xtest = bsxfun(@rdivide,bsxfun(@minus,Xtest,mu),sigma);
```

Train a Linear Discriminant and Select 4 Best Predictors

There are 10 variables in the MAGIC data. In physics analysis, linear techniques are usually applied to data with lower dimensionality. We reduce the number of variables to 4 by selecting variables most important for the linear discriminant analysis (LDA). To select the most important variables for LDA, we look at the `DeltaPredictor` property of the discriminant object. This property shows thresholds (on some normalized scale) at which variables are eliminated from the LDA model, one threshold per variable. A large threshold indicates an important variable.

```
VarNames = {'fLength' 'fWidth' 'fSize' 'fConc' 'fConc1' ...
    'fAsym' 'fM3Long' 'fM3Trans' 'fAlpha' 'fDist'};
LDA = ClassificationDiscriminant.fit(Xtrain,Ytrain,...
        'PredictorNames',VarNames);
[~,sorted] = sort(LDA.DeltaPredictor,'descend');
D = 4;
keepVarIdx = sorted(1:D); % indices of variables kept in the model

keepVarNames = VarNames(keepVarIdx) % names of kept variables
Xtrain = Xtrain(:,keepVarIdx); % keep only selected variables
                                % for training
Xtest = Xtest(:,keepVarIdx); % keep only selected variables for test

keepVarNames =
    'fLength'  'fAsym'  'fConc'  'fAlpha'
```

Train LDA, logistic regression (LR) and partial least squares (PLS)

To train LDA, we call the `ClassificationDiscriminant.fit` method which returns an object of class `ClassificationDiscriminant`. We then copy LDA coefficients from the respective property of this object to a new variable.

```
LDA = ClassificationDiscriminant.fit(Xtrain,Ytrain);
LDAcoeffs = LDA.Coeffs(1,2).Linear;
```

To train LR, we call the `GeneralizedLinearModel.fit` method returning an object of class `GeneralizedLinearModel`. We then copy the fitted coefficients from the respective property of this object to a new variable.

```
glm = GeneralizedLinearModel.fit(Xtrain,Ytrain,'distribution','binomial');
LRcoeffs = glm.Coefficients.Estimate;
```

To train PLS, we call the `plsregress` function. It returns

- X loadings (XplsAxes, a matrix of size 4-by-4 for 4 input variables and 4 PLS components, one column per component)
- Y loadings (YplsAxes, a row vector with 4 elements for one column in Y and 4 PLS components)
- X coordinates in the PLS frame (XplsCoord, a matrix of size N-by-4 for N observations in the training data and 4 PLS components)
- Y coordinates in the PLS frame (YplsCoord, a matrix of size N-by-4 for N observations in the training data and 4 PLS components)
- PLS coefficients (PLScoeffs, a vector with 5 elements in which the first element is for the intercept)
- percentage of variance explained by the PLS model (PLSpctvar, a matrix of size 2-by-4 in which the first row contains the percentage of variance explained in X by every PLS component and the second row contains the percentage of variance explained in Y by the same PLS component)
- various PLS statistics (PLSstat, a struct with several fields one of which is the matrix of PLS weights W)

```
[XplsAxes,YplsAxes,XplsCoord,YplsCoord,PLScoeffs,PLSpctvar,~,PLSstat] = ...
    plsregress(Xtrain,Ytrain);
```

Estimate Classification Errors Using Test Data

We use the three models to compute predictions for the test data and compare these predictions with the true class labels in Ytest.

To compute the predictions by LDA, we call the `predict` method of the discriminant object. The first output of this method is a vector of the predicted class labels, and the second output is an N-by-2 matrix of the estimated posterior class probabilities for N observations in the test data and two classes. Since the "true" class is second in LDA.ClassNames, we retain the second column in this matrix only (for binary classification, one of the two scores can be obtained from the other).

```
LDA.ClassNames
[YfitLDA,PfitLDA] = predict(LDA,Xtest);
PfitLDA = PfitLDA(:,2);

ans =
     0
     1
```

To compute the binomial probabilities predicted by LR, we use the predict method of the glm object. We then assign every observation to the "true" class if the predicted probability is above 0.5 and "false" class otherwise (the same procedure is used for LDA in the call to predict).

```
PfitLR = predict(glm,Xtest);
YfitLR = PfitLR>0.5;
```

To compute the predictions by PLS, we multiply Xtest by the vector of PLS coefficients. Since the first coefficient is for the intercept, we add a column of ones to Xtest. Then we assign class labels to observations using the same recipe as for LR.

```
SfitPLS = [ones(size(Xtest,1),1) Xtest]*PLScoeffs;
YfitPLS = SfitPLS>0.5;
```

Classification error for every model is estimated as a fraction of misclassified observations in the test data.

```
Ntest = numel(Ytest);
errLDA = sum(YfitLDA~=Ytest)/Ntest
errLR = sum(YfitLR~=Ytest)/Ntest
errPLS = sum(YfitPLS~=Ytest)/Ntest

errLDA =
    0.2155
errLR =
    0.2044
errPLS =
    0.2162
```

Test the Models for Equivalence

We run McNemar's test on every pair of models. This test is described in Section 10.3. Although LR offers only a marginal 5% relative error reduction over LDA, this improvement is highly significant. LDA and PLS have comparable errors, and both are outperformed by LR. Note: We use the mcnemary function not included in the official MATLAB distribution. You can code this function yourself with little effort. Refer to Section 10.3 for detail.

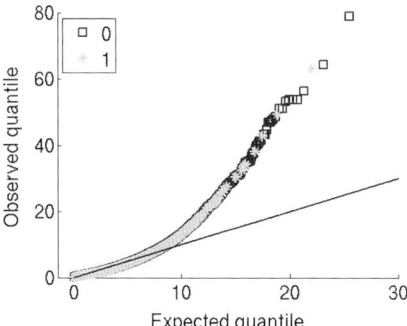

Figure 11.3 QQ plot for linear discriminant applied to the MAGIC telescope data.

```
[~,pLRvsLDA] = mcnemary(Ytest,YfitLDA,YfitLR)

pLRvsLDA =
   4.8992e-08

[~,pLRvsPLS] = mcnemary(Ytest,YfitPLS,YfitLR)

pLRvsPLS =
   1.1957e-08

[~,pLDAvsPLS] = mcnemary(Ytest,YfitPLS,YfitLDA)

pLDAvsPLS =
    0.6171
```

Verify LDA Assumptions

For every training observation, we compute the squared Mahalanobis distance to its true class. We then compute the expected quantiles assuming that the squared Mahalanobis distance has a χ^2 distribution with 4 degrees of freedom. As seen in Figure 11.3, the empirical quantiles in the QQ-plot deviate up from the expected values indicating that the data in Xtrain have tails heavier than normal. The assumption of multivariate normality is clearly no good, and LDA cannot be the optimal classifier for these data.

```
mah = mahal(LDA,Xtrain,'ClassLabels',Ytrain);
Ntrain = numel(Ytrain);
expQ = chi2inv(((1:Ntrain)-0.5)/Ntrain,D); % expected quantiles
[mah,sorted] = sort(mah); % sorted observed quantiles
figure;
gscatter(expQ,mah,Ytrain(sorted),'bg','s*',10,'off'); % plot by class
legend('0','1','Location','NW');
xlabel('Expected quantile');
ylabel('Observed quantile');
line([0 30],[0 30],'color','k');
```

Verify LR Assumptions

We use the Hosmer–Lemeshow test to estimate the goodness of fit for the LR model. We divide the fitted binomial probabilities into 10 bins with an equal number of observations per bin. We then compute the Hosmer–Lemeshow statistic and the associated p-value assuming a χ^2 distribution with 8 degrees of freedom. The p-value is low suggesting that the optimal separation between the two classes is not linear.

```
PfitLR = predict(glm,Xtrain);
edges = [0, quantile(PfitLR,0.1:0.1:0.9), Inf]; % 11 bin edges
Nbin = zeros(10,1);
Pbin = zeros(10,1);
Ybin = zeros(10,1);
for n=1:10
    inThisBin = PfitLR>=edges(n) & PfitLR<edges(n+1);
    Nbin(n) = sum(inThisBin);
    Pbin(n) = mean(PfitLR(inThisBin));
    Ybin(n) = mean(Ytrain(inThisBin));
end
HL = sum( Nbin.*(Ybin-Pbin).^2./Pbin./(1-Pbin) ); % Hosmer-Lemeshow
                                                  % statistic
pHL = 1 - chi2cdf(HL,8) % p-value for Hosmer-Lemeshown test

pHL =
    1.2434e-04
```

Estimate Variable Importance

As noted above, the `DeltaPredictor` property of the discriminant object can be used to judge the relative importance of every variable in the model. This property stores magnitudes of the standardized LDA coefficients (that is, coefficients computed after standardizing variables for every class separately).

```
VarImpLDA = LDA.DeltaPredictor

VarImpLDA =
    0.7827  0.6374  0.3736  0.2232
```

For LR, we simply look at the magnitudes of the coefficients.

```
VarImpLR = abs(LRcoeffs(2:end))'

VarImpLR =
    1.1968  1.2289  0.5919  0.2923
```

For PLS, we estimate variable importance as described in the PLS section. The first PLS component explains 29% of variance in the class label, far more than the other three:

```
YplsAxes

YplsAxes =
   29.1008  7.4754  4.2798  1.0913
```

The importance of every variable is therefore determined mostly by its contribution to the first PLS component. The first two variables have largest (in magnitude) coefficients in the first component (first column of XplsAxes):

```
XplsAxes(:,1)

ans =
  -84.4776
  -72.3538
   32.0900
  -46.1461
```

Here are the exact estimates of variable importance from PLS:

```
W = PLSstat.W;
Wtilde2 = bsxfun(@rdivide,W.^2,sum(W.^2,1));
Ctilde2 = YplsAxes.^2/sum(YplsAxes.^2);
VarImpPLS = sqrt(D * sum(bsxfun(@times,Wtilde2,Ctilde2),2) )'

VarImpPLS =
    1.6034  1.0314  0.4045  0.4488
```

All models agree that the first two variables, fLength and fAsym, are more important for prediction than the other two.

Visualize the Classification Models

We make a scatter plot of the two most important variables in Figure 11.4 and superimpose the first and second PLS axes. We do not plot the LDA or LR axes because they are very close to the first PLS axis: the angle between the projection of the first PLS component and the projection of the LDA (LR) axis on the (fLength,fAsym) plane is 3 (5) degrees. The angle between the first PLS axis and the LDA (LR) axis is much larger in the full 4D space, but we focus on the two most significant variables only. To plot the axes, we compute tangent of the angle between the plotted line and the horizontal axis and multiply it by the abscissa limits to get the ordinate limits.

```
figure;
h = gscatter(Xtrain(:,1),Xtrain(:,2),Ytrain,[],'s*',[],'off');
Xlims = [-1 2.5];
PLStan = XplsAxes(2,:)./XplsAxes(1,:);
h1 = line(Xlims,PLStan(1)*Xlims,'color','k','LineStyle','-');
h2 = line(Xlims,PLStan(2)*Xlims+2,'color','k','LineStyle','-');
legend([h(1) h(2) h1 h2],...
    {'Class 0' 'Class 1' '1st PLS axis' '2nd PLS axis'},...
```

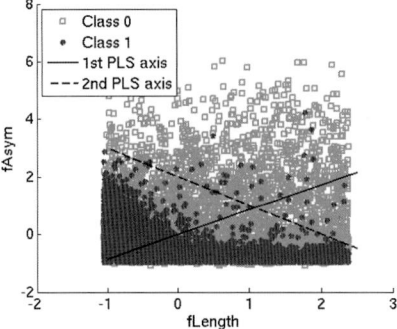

Figure 11.4 Class distributions in two most important variables with the first two PLS axes superimposed.

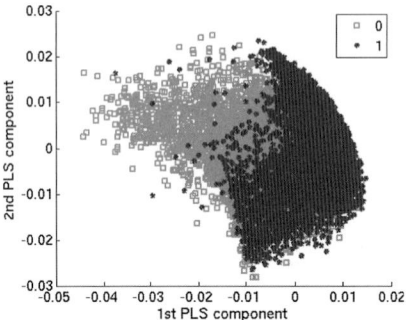

Figure 11.5 Class distributions in the 1st and 2nd PLS components.

```
    'Location','NW');
xlabel('fLength');
ylabel('fAsym');
```

We also make a scatter plot of the first two coordinates in the PLS reference frame in Figure 11.5. The line of optimal separation in this plot coincides with the horizontal axis. Because PLS, unlike LDA and LR, computes more than one linear component, we can explore the data in the transformed coordinates. The usefulness of this exploration would vary from one analysis to another.

```
figure;
gscatter(XplsCoord(:,1),XplsCoord(:,2),Ytrain,[],'s*');
xlabel('1st PLS component');
ylabel('2nd PLS component');
```

11.6
Choosing a Linear Classifier for Your Analysis

In this chapter, we have reviewed, for the most part, methods for linear classification. The only exception has been quadratic discriminant analysis. Here, we give a brief summary of their properties.

The most attractive feature of a linear model is its simplicity. A linear decision boundary can be easily visualized in a multivariate space. Contributions of individual variables to the classifier prediction can be trivially estimated. Inclusion of systematic measurement uncertainties would be straightforward as well, relative to powerful nonparametric models such as neural networks or tree ensembles, both discussed in the later chapters.

Training a linear classifier on tall data with up to hundreds of variables, typical for physics analysis, takes little time. Prediction is equally fast.

The biggest flaw of linear classification is low accuracy. The optimal decision boundary for real-world data is seldom linear. LDA is based on the most restrictive set of assumptions about the class distributions and for that reason tends to be least accurate. Regularized discriminant is a powerful tool for analysis of data with few observations and hundreds or more variables; however, such datasets are atypical in particle and astrophysics. For tall datasets, logistic regression can provide a noticeable improvement over LDA, especially for data in which the class distributions are not multivariate normal and the optimal decision boundary is close to linear. Both linear discriminant and logistic regression provide estimates of the class posterior probabilities, accurate under the model assumptions. These model assumptions can be validated using formal statistical tests.

Regression techniques such as linear regression of an indicator matrix and PLS can be used for linear classification as well. PLS regression, in addition, can reveal hidden relations among variables. Conceptually, PLS regression is not simple and may require some practicing. The response predicted by these regression techniques is not related to the class posterior probabilities in an obvious way and can lie outside the $[0, 1]$ range. If estimates of the posterior probabilities are desired, this limitation can represent a serious problem.

Categorical variables can be included in any linear model using dummy variables described in Section 7.1. Dummy variables manifestly violate the assumption of multivariate normality required for discriminant analysis.

11.7
Exercises

1. Prove that the estimate (11.8) of the pooled-in covariance matrix is unbiased.
2. Prove that the weighted estimate (11.11) of the pooled-in covariance matrix is unbiased. Treat the weights w_n as constant factors.
3. Adapt (11.12) to estimate the pooled-in kurtosis for several samples drawn from distributions with different means and equal covariance matrices.

4. Let $\mathbf{X} = \mathbf{U}\mathbf{S}\mathbf{V}^\mathsf{T}$ be SVD of $\mathbf{X}$. Assuming that the variables in $\mathbf{X}$ are centered, write down an estimate of the covariance matrix $\mathbf{\Sigma}$.
5. Verify (11.24) and (11.25).
6. Think of a binary classification problem in which the optimal decision boundary is linear and the two overlapping class distributions are *not* normal. Show empirically that for these data logistic regression gives a lower generalization error than linear discriminant.
7. Consider linear regression $Y = \beta_0 + \boldsymbol{\beta}^\mathsf{T} \mathbf{X}$ with scalar response $Y \in \{-1, +1\}$. Assume that all observation weights in the training set are equal and the covariance matrix for the input variables in $\mathbf{X}$ is not singular. Prove that the coefficient estimates $\hat{\boldsymbol{\beta}}$ obtained by this linear regression model and coefficient estimates $\hat{\boldsymbol{\Sigma}}^{-1}(\hat{\boldsymbol{\mu}}_1 - \hat{\boldsymbol{\mu}}_2)$ obtained by linear discriminant are proportional. How do the coefficient estimates change for each model if $Y = -1$ is replaced by $Y = 0$?
8. Consider the same linear regression as in the previous problem. Prove that the intercept estimate $\hat{\beta}_0$ and the constant term in the linear discriminant model $\log(\pi_1/\pi_2) - (\hat{\boldsymbol{\mu}}_1^\mathsf{T}\hat{\boldsymbol{\Sigma}}^{-1}\hat{\boldsymbol{\mu}}_1 - \hat{\boldsymbol{\mu}}_2^\mathsf{T}\hat{\boldsymbol{\Sigma}}^{-1}\hat{\boldsymbol{\mu}}_2)/2$ are proportional, with the same proportionality factor as the coefficients, if the two classes are mixed in equal proportion in the training set with $\pi_1 = \pi_2$. (Hint: Eliminate the intercept term from the linear regression model by centering the training set and express the posterior log-odds for linear discriminant using $(\mathbf{x} - (\hat{\boldsymbol{\mu}}_1 + \hat{\boldsymbol{\mu}}_2)/2)^\mathsf{T} \hat{\boldsymbol{\Sigma}}^{-1} (\hat{\boldsymbol{\mu}}_1 - \hat{\boldsymbol{\mu}}_2)$.)

References

Abdi, H. (2010) Partial least squares regression and projection on latent structure regression (PLS regression). *Computational Statistics*, vol. 2, pp. 97–106, Wiley Interdisciplinary Reviews.

Agresti, A. (2002) *Categorical Data Analysis*, Probability and Statistics, John Wiley & Sons, 2nd edn.

Andersen, T. (2003) Testing hypotheses of equality of covariance matrices and equality of mean vectors and covariance matrices, in *An Introduction to Multivariate Statistical Analysis*, John Wiley & Sons, Wiley Series in Probability and Statistics, 3rd edn.

Barker, M. and Rayens, W. (2003) Partial least squares for discrimination. *J. Chemometr.*, **17**, 166–173.

Bock, R.K., Chilingarian, A., Gaug, M., Hakl, F., Hengstebeck, T., Jirina, M., Klaschka, J., Kotrc, E., Savicky, P., Towers, S., Vaiciulis, A., Wittek, W. (2004) Methods for multidimensional event classification: A case study using images from a Cherenkov gamma-ray telescope. *Nucl. Instrum. Methods A*, **516**, 511–528.

Bookstein, F. (1994) Partial least squares: a dose-response model for measurement in the behavioral and brain sciences. *Psycoloquy*, **5**.

Box, G. (1949) A general distribution theory for a class of likelihood criteria. *Biometrika*, **36** (3/4), 317–346.

Copas, J. (1989) Unweighted sum of squares test for proportions. *J. R. Stat. Soc.*, **38** (1), 71–80.

de Jong, S. (1993) SIMPLS: An alternative approach to partial least squares regression. *Chemom. Intell. Lab. Syst.*, **18**, 251–263.

Fisher, R. (1936) The use of multiple measurements in taxonomic problems. *Ann. Eugen.*, **7**, 179–188.

Friedman, J. (1988) Regularized discriminant analysis, *SLAC-PUB 4389*, Stanford Linear Accelerator.

Guo, Y., Hastie, T., and Tibshirani, R. (2007) Regularized linear discriminant analysis and its application in microarrays. *Biostatistics*, **8** (1), 86–100.

Gupta, A. and Tang, J. (1984) Distribution of likelihood ratio statistic for testing equality of covariance matrices of multivariate Gaussian models. *Biometrika*, **71** (3), 555–559.

Hastie, T., Tibshirani, R., and Friedman, J. (2008) Linear methods for classification, in *The Elements of Statistical Learning*, Springer, pp. 119–128, 2nd edn.

Healy, M. (1968) Multivariate normal plotting. *J. R. Stat. Soc.*, **17** (2), 157–161.

Hosmer, D., Hosmer, T., Cessie, S.L., and Lemeshow, S. (1997) A comparison of goodness-of-fit tests for the logistic regression model. *Stat. Med.*, **16**, 965–980.

Hosmer, D. and Lemeshow, S. (1989) *Applied Logistic Regression*, John Wiley & Sons.

Mardia, K. (1970) Measures of multivariate skewness and kurtosis with applications. *Biometrika*, **57**, 519–530.

McCullagh, P. and Nelder, J. (1983) *Generalized Linear Models*, Chapman & Hall.

Moler, C. (2008) *Numerical Computing with MATLAB*, SIAM. http://www.mathworks.com/moler/ (accessed 22 July 2013).

Qiao, Z., Zhou, L., and Huang, J. (2008) Sparse linear discriminant analysis with applications to high dimensional low sample size data. *IAENG Int. J. Appl. Math.*, **39**, 48–60.

Rosipal, R. and Kramer, N. (2006) Overview and recent advances in partial least squares, in *Subspace, Latent Structure and Feature Selection: Statistical and Optimization Perspectives Workshop (SLSFS 2005)*, Lecture Notes in Computer Science, vol. 3940, Springer, Berlin, pp. 34–51.

Witten, D., Hastie, T., and Tibshirani, R. (2009) A penalized matrix decomposition, with applications to sparse principal components and canonical correlation analysis. *Biostatistics*, **10** (3), 515–534.

Wold, H. (1966) Estimation of principal components and related models by iterative least squares, in *Multivariate Analysis* (ed. P. Krishnaiah), Academic Press, New York, pp. 391–420.

Ye, J. and Xiong, T. (2006) Computational and theoretical analysis of null space and orthogonal linear discriminant analysis. *J. Mach. Learn. Res.*, **7**, 1183–1204.

Zhang, L.H., Liao, L.Z., and NG, M. (2010) Fast algorithms for the generalized Foley-Sammon discriminant analysis. *SIAM J. Matrix Anal. Appl.*, **31**, 1584–1605.

12
Neural Networks

The genesis of the neural network is in attempts to model biological intelligence, including the notion where the firing of a neuron occurs once a number of summed "inputs" crosses some threshold. Now with the biological motivation removed, the technology of neural networks applied to classification problems has enjoyed great popularity and success. There is a vast literature on this subject (we note the treatments in Bishop (2006); Haykin (2006); MacKay (2003)). In this chapter we introduce neural networks with physics applications in mind.

As with other classifiers to be discussed, the neural network is a nonlinear transformation on a set of input variables $x = x_1, \ldots, x_D$ to a set of output variables o, where the transformation depends on a set of adjustable parameters. The adjustable parameters are tuned on a training set such that the performance of the network on a classification problem is optimized according to a desired criterion.

The method of neural networks is one of the earliest supervised learning methods to be used in particle physics. It has enjoyed extensive use in diverse contexts, such as particle identification, event reconstruction, and optimization of physics analyses. Most of this usage has been of the classification ilk, and the neural net is a popular tool for distinguishing signal from background.

Neural networks can also be applied to regression (curve fitting) problems. Indeed, Cybenko (1989) demonstrates that any continuous function of D variables with compact support can be approximated arbitrarily well with feed-forward networks with one hidden layer such as we discuss below. While it is important to mention this application, our discussion in this chapter will focus on the classification problem. Neural networks have been used as well for regression in particle physics, though not as extensively as for classification. For example, Lees *et al.* (2013) uses a feed-forward neural network to parameterize the efficiency over a Dalitz plot.

12.1
Perceptrons

A *perceptron* is a model for a neuron, charged with selecting between two classes, C_1 and C_2, based on the inputs. It first takes a weighted linear combination of the

Statistical Analysis Techniques in Particle Physics, First Edition. Ilya Narsky and Frank C. Porter.
©2014 WILEY-VCH Verlag GmbH & Co. KGaA. Published 2014 by WILEY-VCH Verlag GmbH & Co. KGaA.

inputs, with the addition of a constant term b (called the "bias"):

$$u = \mathbf{w}^\mathsf{T}\mathbf{x} + b . \tag{12.1}$$

In general, $\mathbf{x}$ could be replaced by some fixed nonlinear transformation on $\mathbf{x}$. The perceptron assigns scalar output o according to the signum function:

$$o(u) = \begin{cases} +1, & u \geq 0 \\ -1, & u < 0 . \end{cases} \tag{12.2}$$

That is, the perceptron assigns an observation with $u \geq 0$ to C_1, and with $u < 0$ to C_2. The "weights" w and bias b are adjusted to optimize the discrimination between the two classes.

Notice that the trained perceptron defines a hyperplane in the input space:

$$\mathbf{w}^\mathsf{T}\mathbf{x} + b = 0 , \tag{12.3}$$

giving the boundary between the two output classifications. In fact, the perceptron can find a solution to any linearly separable classification problem. Two classes are *linearly separable* if it is possible to draw a hyperplane such that the two classes are separated.

It will be convenient for further development to incorporate the bias term into our vector notation. Hence, we modify our input vector, adding a dimension at index zero: $\tilde{\mathbf{x}} = \{x_0 = 1, x_1, \ldots, x_D\}$. Likewise, the weight vector is modified by adding component $w_0 = b$. Then we may write simply $u = \tilde{\mathbf{w}}^\mathsf{T}\tilde{\mathbf{x}}$.

Various learning algorithms and criteria could be imagined for the perceptron. The *perceptron criterion* provides a suitable approach, which we now develop. Suppose we have a training dataset of size N, that is, we have the vectors (observations) $\mathbf{x}_1, \ldots, \mathbf{x}_N$, with known classification, $y_i = +1$ if observation i is in class 1 and $y_i = -1$ if the observation is in class 2. For example, class 1 could be signal and class 2 could be background. The quantity y_i is called the *target coding* (as well as true label and known label). Letting $u_i = \tilde{\mathbf{w}}^\mathsf{T}\tilde{\mathbf{x}}_i$, the quantity $u_i y_i$ is greater than zero if the classification is correct, and negative if it is incorrect.

The perceptron criterion assigns a weighted penalty for any misclassifications. Let $\mathcal{M}(\tilde{\mathbf{w}})$ be the set of all misclassified observations. Form an error function (perceptron criterion) according to:

$$e(\tilde{\mathbf{w}}) = -\sum_{i \in \mathcal{M}} u_i y_i = -\sum_{i \in \mathcal{M}} \tilde{\mathbf{w}}^\mathsf{T}\tilde{\mathbf{x}}_i y_i . \tag{12.4}$$

This is a nonnegative quantity. It would simply be the number of misclassifications if $u_i y_i$ were replaced by -1, but it is better for gradient search algorithms to have a continuous function of the weights. The perceptron learns on our training sample by an iterative updating of the weights in the direction of reducing the error function. A simple algorithm that is guaranteed to converge if the training set is linearly separable is the *stochastic gradient descent*. Imagine that we have randomized the order of the training vectors. Loop through this set one at a time, with

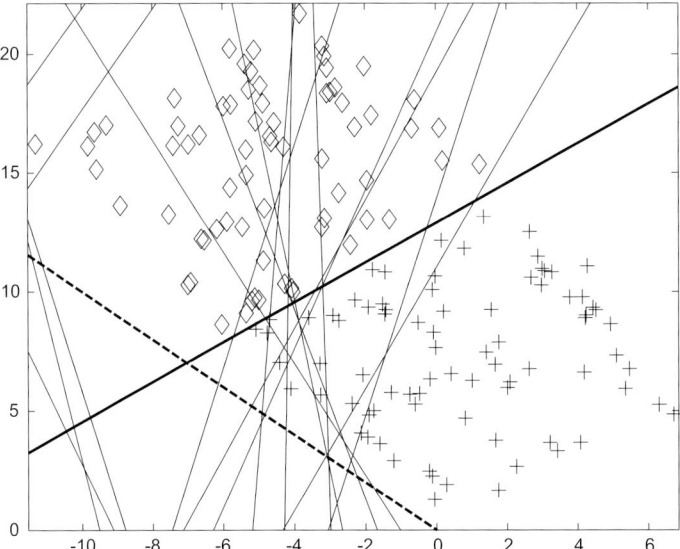

Figure 12.1 Demonstration of the training of the perceptron algorithm on a linearly separable dataset (diamonds and plusses). The dashed line corresponds to the starting weight vector. The heavy solid line shows the final solution, and the lighter solid lines show the intermediate solutions after each iteration through the training data.

index $n = 1, \ldots, N$. If we denote the weights at learning step k by $\tilde{w}(k)$, then the weights at step $k+1$ are obtained by

$$\tilde{w}(k+1) = \tilde{w}(k) - \nabla e(\tilde{w}(k)) = \begin{cases} \tilde{w}(k), & i \notin \mathcal{M}, \\ \tilde{w}(k) + \tilde{x}_i y_i, & i \in \mathcal{M}. \end{cases} \quad (12.5)$$

If vector i is already correctly classified, the weight vector does not change. Otherwise the weights are adjusted in the direction to reduce the contribution from the x_i to $e(w)$.

Figure 12.1 demonstrates the training of the perceptron on a linearly separable dataset of two-dimensional vectors. Twenty-two loops through the training set were made before obtaining convergence, a typical number for this dataset. The solution for the weights is not unique, other stochastic gradient attempts will give different results, all successfully separating the two classes.

In neural networks, the discontinuous signum of the perceptron is smoothed out to a differentiable function such as the sigmoid (or *logistic sigmoid*):

$$\sigma(u) = \frac{1}{1 + e^{-u}}. \quad (12.6)$$

Besides getting continuity, there is a somewhat intuitive rationale for this particular choice. Express the posterior probability to be in class 1 as a sigmoid:

$$P(C_1|x) = \frac{P(x|C_1)P(C_1)}{P(x|C_1 \cup C_2)} = \sigma(u). \quad (12.7)$$

The inverse is the *logit* function:

$$u = \log \frac{\sigma}{1-\sigma} = \log \frac{P(C_1|x)}{P(C_2|x)}. \qquad (12.8)$$

That is, u is the log of the ratio of the posterior probabilities for the two classes (*log odds*). We will make some use of this interpretation below; in any event the sigmoid is a convenient continuous approximation to the signum.

The simple perceptron is not especially suited to the nonlinear case, and we can do much better with other algorithms. In the context of the neural network idea, we may extend our perceptron by combination into a *multi-layer perceptron*, or *feed-forward neural network*. This extension permits useful application in nonlinear classification problems.

12.2
The Feed-Forward Neural Network

A feed-forward neural net consists of layers – a set of inputs, an output layer, and any number of "hidden" layers in between. Each layer has a number of units or nodes that take inputs from the next lower layer and provide outputs to the next higher layer. It is not required that each node in a layer have a connection with every node in its input layer or with every node in its output layer. It is also permissible to construct networks with connections that skip layers. The essential requirement for a feed-forward architecture is that there be no circular references, so that the outputs are deterministic functions of the inputs. However, the most commonly used network consists of complete connections and no skipped layers. A possible such net architecture with one hidden layer is illustrated in Figure 12.2. We adopt the convention (Bishop, 2006) that this is a two-layer network, omitting the inputs from the count in order to emphasize the layers of computation.

We label the nodes of the network by the layer, $\ell = 1, \ldots, L$, and the nodes within a layer ℓ by $j = 1, \ldots, n_\ell$, where n_ℓ is the number of nodes in layer ℓ. Considering a given node in the network, we label its inputs by $z_i, i = 0, \ldots, Z$, where Z is the number of inputs and $z_0 \equiv 1$ is the bias. Where necessary, we add indices to describe the node, that is $z_{ij\ell}$ describes input i to node j in layer ℓ, but we may omit the extra indices when discussing a generic node. For example, for a node in the first layer, $Z = D$ and $z_i = x_i, i = 1, \ldots, D$ are the inputs to the network. We will treat every node as receiving all of the variables from the preceding layer; if any are to be omitted that is logically equivalent to fixing the corresponding weights to be zero.

At each node of the neural network, the inputs to that node, including a bias input, are summed with linear weighting. Thus, at our given node, we form the quantity

$$a = \sum_{i=0}^{Z} w_i z_i. \qquad (12.9)$$

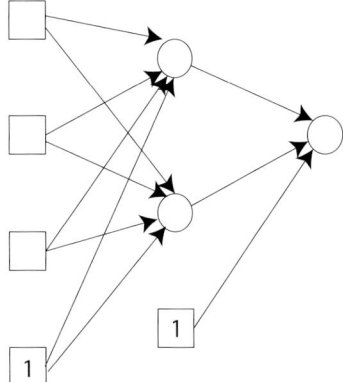

Figure 12.2 Illustration of a simple neural net architecture with one hidden layer. The open squares represent the inputs; open squares with the number 1 represent the bias inputs, and the circles represent the nodes where the computation resides.

This quantity is known as an *activation*; the node's output is computed from it. Where required, we will denote our weights with three indices, $w_{ij\ell}$.

The output of the node is obtained by transforming the activation with a nonlinear, differentiable, *activation function*, $h(a)$. There are various choices of activation function for the hidden layers, including the sigmoid. A nice choice for hidden layers is

$$h(a) = \tanh(a), \tag{12.10}$$

because its derivative is readily evaluated from the value of h:

$$h'(a) = 1 - h(a)^2. \tag{12.11}$$

The derivative of a sigmoid is similarly easy to compute. For the output layer of the neural net, a sigmoid is usually used if there is only one output. This is the case if there are only two classes (e.g., "signal" and "background") to distinguish. In the case of multiple classes a *softmax* generalization of the sigmoid is appropriate:

$$o_k = \sigma(a_k) = \frac{e^{a_k}}{\sum_{i=1}^{K} e^{a_i}}, \tag{12.12}$$

where o_k is the output corresponding to class k, and K is the number of classes. This form enables the interpretation of outputs as posterior probabilities.

Thus, consider the two-layer network illustrated in Figure 12.2. There are $D = 3$ inputs, x_1, x_2, x_3, and $n_1 = 2$ nodes in the hidden layer, layer $\ell = 1$. Then the activations in the hidden nodes are

$$a_{j1} = \sum_{i=0}^{3} w_{ij1} x_i, \quad j = 1, 2. \tag{12.13}$$

The inputs to the single node in the output layer are then obtained by letting the activation function operate on these activations:

$$z_j = h(a_{j1}) = \tanh(a_{j1}), \quad j = 1, 2. \tag{12.14}$$

The activation in the output layer node is

$$a = \sum_{i=0}^{2} w_{i12} z_i. \tag{12.15}$$

Finally, the output is

$$o = \sigma(a)$$

$$= \left\{ 1 + \exp\left[-w_{012} - \sum_{j=1}^{2} w_{j12} \tanh\left(w_{0j1} + \sum_{i=1}^{3} w_{ij1} x_i \right) \right] \right\}^{-1}, \tag{12.16}$$

where we have written the output explicitly as it expands to a function of the inputs in the second line. There are a total of eleven weights to determine to completely specify this function. Note that this just corresponds to the number of lines in the diagram: we may think of the weights as being associated with the lines connecting the nodes.

12.3
Backpropagation

The weights are determined by training the network to perform the desired task. This entails optimizing the classification performance. Suppose we have a single binary target variable, y, which can take on values 0 (for background perhaps) and 1 (e.g., signal). The network output is taken to be a sigmoid:

$$o(x, \tilde{w}) = \frac{1}{1 + e^{-a}}, \tag{12.17}$$

where $a = a(x, \tilde{w})$ is the activation in the output node. Taking a cue from (12.7), we may interpret $o(x, \tilde{w})$ as the conditional probability $P(Y = 1|x)$, and $1 - o$ to be $P(Y = 0|x)$. That is,

$$P(y|x) = o(x, \tilde{w})^y [1 - o(x, \tilde{w})]^{1-y}. \tag{12.18}$$

Then we may formulate a likelihood function over a training sample $x_1, \ldots, x_N$, and define an error function as the negative log likelihood:

$$e(\tilde{w}) = -\sum_{n=1}^{N} [y_n \log o_n(x_n, \tilde{w}) + (1 - y_n) \log(1 - o_n(x_n, \tilde{w}))]. \tag{12.19}$$

This is called the *cross-entropy* error function. Other error functions, or *loss functions*, may be used, but this one is known to work well. It is not difficult to generalize this to multiclass problems. Neural network predictions can similarly be interpreted as posterior class probabilities for multiclass problems, including nets minimizing the quadratic loss.

Our problem is to find those weights that minimize the loss function. This is a multivariate minimization problem. Using our knowledge of the network architecture helps us to reduce the computing burden, compared with general-purpose minimization tools. Let us denote the contribution to our error function from training observation n by e_n. This will be a function of the network weights. We can find those weights where the error is minimized (for the training dataset) by following the gradient of the error function with respect to each weight.

We may carry out the error minimization program as follows. Consider the contribution of training observation n to the error using the cross-entropy, for example,

$$e_n(\tilde{w}) = -\{y_n \log o_n(x_n, \tilde{w}) + (1 - y_n) \log[1 - o_n(x_n, \tilde{w})]\} \:. \tag{12.20}$$

Take the partial derivative of this with respect to the weight for the ith input to node j in layer ℓ:

$$\frac{\partial e_n}{\partial w_{ij\ell}} = \frac{\partial e_n}{\partial a_{j\ell}} \frac{\partial a_{j\ell}}{\partial w_{ij\ell}}$$

$$= e_{nj\ell} z_{ij\ell} \:, \tag{12.21}$$

where we have defined

$$e_{nj\ell} \equiv \frac{\partial e_n}{\partial a_{j\ell}} \:, \tag{12.22}$$

and have used the fact that (defining $n_0 = D$ for the first layer inputs)

$$a_{j\ell} = \sum_{i=0}^{n_{\ell}-1} w_{ij\ell} z_{ij\ell} \:. \tag{12.23}$$

To evaluate all of the derivatives, we start at the output node. If we use the cross-entropy, for example, we obtain

$$e_{n1L} = o_n - y_n \:. \tag{12.24}$$

This result is also obtained if we use an error function given by one-half the sum of the squared deviations between the output and the target value.

Now move back to layer $\ell = L - 1$. We wish to evaluate, for $j = 1, \ldots, n_\ell$:

$$e_{nj\ell} = \frac{\partial e_n}{\partial a_{j\ell}} = \sum_{q=1}^{n_{\ell+1}} \frac{\partial e_n}{\partial a_{q,\ell+1}} \frac{\partial a_{q,\ell+1}}{\partial a_{j\ell}}$$

$$= \sum_{q=1}^{n_{\ell+1}} e_{nq,\ell+1} \frac{\partial a_{q,\ell+1}}{\partial a_{j\ell}} \:. \tag{12.25}$$

In the case of a single output node, the sum only has one term, but we have written the formula generally, so that it may be applied to all hidden layers. We may evaluate the derivatives according to

$$\frac{\partial a_{q,\ell+1}}{\partial a_{j\ell}} = \sum_{i=1}^{n_\ell} w_{iq,\ell+1} \frac{\partial z_{iq,\ell+1}}{\partial a_{j\ell}}$$

$$= \sum_{i=1}^{n_\ell} w_{iq,\ell+1} \frac{\partial h(a_{i\ell})}{\partial a_{j\ell}}$$

$$= w_{jq,\ell+1} h'(a_{j\ell}) . \tag{12.26}$$

Thus, we may write

$$e_{nj\ell} = h'(a_{j\ell}) \sum_{q=1}^{n_{\ell+1}} w_{qj,\ell+1} e_{nq,\ell+1} . \tag{12.27}$$

This is *backpropagation*: starting at the output layer, we work our way back through the network recursively evaluating derivatives using the chain rule. If the activation function is chosen to be the tanh function for the hidden layers, then we have the computationally easy job of evaluating

$$e_{nj\ell} = [1 - h(a_{j\ell})^2] \sum_{q=1}^{n_{\ell+1}} w_{qj,\ell+1} e_{nq,\ell+1} . \tag{12.28}$$

We now have a fast method for estimating the gradients that will be used in searching for the optimal weights. If we know the vector of weights at iteration k, $\tilde{w}^{(k)}$, we obtain the weights for the next iteration according to

$$\tilde{w}^{(k+1)} = \tilde{w}^{(k)} - \eta \nabla e(\tilde{w}^{(k)}) . \tag{12.29}$$

The positive parameter η is called the *learning rate*. It should be adjusted for a suitable compromise between small values implying slow learning (high computation) and large values with fast learning, but potential instability. The form of the iteration given here assumes that the error determined over the entire training dataset is used to compute the step to the next iteration. This is known as *batch learning* (see Section 6.3). A variant is to use one training event at a time, replacing the e in (12.29) with e_n. Then different training events are used at each iteration, either by systematically looping over them, or by choosing at random. This approach is known as *online learning*. The online method has some advantages, besides applicability to real "online" situations with evolving datasets. It provides the opportunity for the algorithm to extract itself from local minima, and likewise can more efficiently handle situations with redundant data. However, the online advantages come at the cost of accuracy, since each evaluation of the gradient is based on only a single event. The batch process lends itself more readily to parallel computation as well, since different observations may be treated simultaneously in computing a new loss.

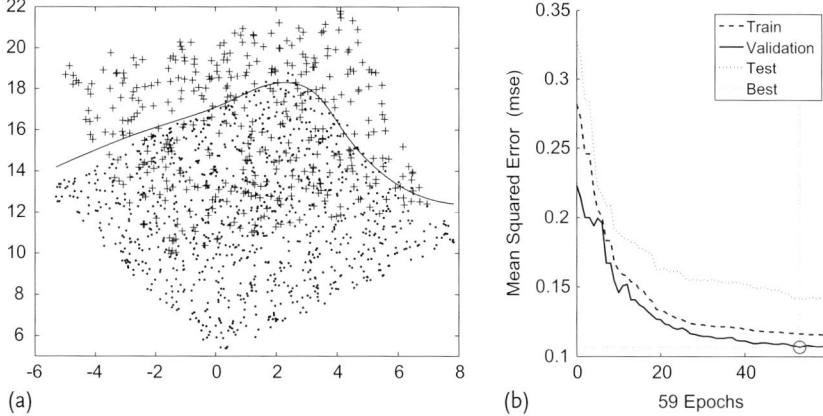

Figure 12.3 Example of a neural network classification on a two-class problem with a two-dimensional input space. (a) The dataset with different symbols indicating the two classes. The line shows the boundary where the network output is equal for the two classes. (b) The network performance (MSE) as a function of training epoch.

Figure 12.3 provides a simple example of a feed-forward neural network with backpropagation in action. The dataset consists of two classes sampled according to pdfs described by two variables. The pdfs used are simply uniform in the populated regions, but with a large overlap, as seen in the figure. The network is a two-layer network with 10 nodes in the hidden layer. The MATLAB Neural Network Toolbox is used to perform the training. The available data is divided into training, validation, and test samples in a 70 : 15 : 15 split. The optimization criterion is minimization of the squared error. In this example, 59 batch iterations are performed before the training is stopped. Training is stopped when no improvement is achieved in six successive attempts on the validation set, as seen in Figure 12.3b. Note that the performance on the test dataset is somewhat poorer on the test dataset than on the datasets used for training and validation.

A curve is drawn on Figure 12.3a showing where the network output is equal for the two classes. This is the nominal boundary between the two classifications: above the curve events will be classified as "pluses", below as "dots". We say "nominal" because other choices could in principle be made besides the line of equality, based on some additional criterion. It may seem puzzling that the neural net has chosen to approximately follow the boundary of one of the regions. In fact, this is correct behavior, for the given task of minimizing squared error. The reason is that the dot class has a higher density of points than the plus class. Whether this is desirable behavior for a particular problem depends on the details of the problem and what figure of merit the analyst really wants to optimize.

The supervised learning approach to the classification problem corresponds in general to solving an ill-posed problem, and steps must be taken to ensure sensible results. Loosely, the problem is that the training data is noisy, and this can result in erratic behavior in learning from it. Thus, we typically have to add some assumptions, such as "smoothness" of the underlying model. This process is known as *regularization*, and there are many approaches that have been proposed. Note that when we used our validation sample to decide when to stop training in our example above, this helped us to avoid over-training on the noise in the training set.

We have already encountered the difficulty inherent in an ill-posed problem when we discussed unfolding (Chapter 5). Thus, we have already seen some techniques for regularization in Section 5.14.1. These same ideas may be applied in the regularization of neural networks. We may take our network loss function, and augment it with a term that becomes large if things start to go wild. Following the discussion in Section 5.14.1, Tikhonov regularization can be used. We could also simply adjust our loss function by adding a term proportional to the sum of the squares of the weights:

$$e(\tilde{w}) \rightarrow e(\tilde{w}) + \frac{\lambda}{2}\tilde{w}^\top \tilde{w}. \tag{12.30}$$

Thus, the weights are prevented from blowing up. This method is known as *weight decay*, and λ is the regularization coefficient. Small values of λ permit the network to follow complicated behavior; large values restrict it to a "smooth" classifier.

12.4
Bayes Neural Networks

We have already introduced the possible interpretation of the network output as a posterior probability for an event to be in a given class. In this case, the cross-entropy corresponds to the negative log-likelihood of the likelihood function. The regularization in (12.30) looks like the negative logarithm of a Gaussian with independent random variables $\tilde{W}_i$, each distributed according to variance $1/\lambda$. Thus, the error function may be interpreted as the negative logarithm of a likelihood function. Development of this idea is most natural in a Bayesian context. In this case, the Gaussian distribution in the weights corresponds to an assumed prior distribution for the weights. Regularization is hence achieved via imposing a prior distribution in the Bayesian approach to neural networks.

The output of the Bayesian methodology should be the posterior distribution (or class posterior probability) for an observation with input vector x to be in, say, class 1 (we will assume a two-class problem for discussion). We need to construct this distribution. The weights are additional random variables, and our network output for observation n depends on them, that is, $o_n = o_n(x, \tilde{w})$. The (negative log) likelihood function we work with is

$$-\log L = -\sum_{n=1}^{N}\{y_n \log o_n(x_n, \tilde{w}) + (1 - y_n)\log[1 - o_n(x_n, \tilde{w})]\} + \frac{\lambda}{2}\tilde{w}^\top \tilde{w}. \tag{12.31}$$

The maximum posterior distribution corresponds to maximizing (up to symmetries, which are considered equivalent) this function over w, for a given value of λ, for example, using the backpropagation techniques discussed above. We will use $\hat{w}$ to denote the location of this maximum.

The parameter λ, entering the prior distribution for the weights, is so far taken as fixed and known. However, it is clear that some values will work better than others. For example, setting $\lambda = 0$ eliminates regularization, and very large λ corresponds to weights clustered about zero, almost independent of the training data. Thus, we would like to optimize further over λ. In the Bayesian context, we could incorporate a prior distribution over λ into our likelihood and then maximize over all parameters. However, the problem is usually simplified to make it more tractable. Rather than simultaneously optimizing over all of the parameters, we form a marginal likelihood for λ by integrating over the posterior distribution for the weights:

$$L(\lambda; x) = \int L(\lambda, w; x) p(w|\lambda) dw . \tag{12.32}$$

This computation may be accomplished, approximately, with a technique known as the *Laplace approximation*. Our distribution $p(w|\lambda)$ is Gaussian, but what about $L(\lambda, w; x)$? The Laplace approximation consists in approximating this with a Gaussian as well. Thus, considering the posterior distribution according to (12.31), we center our Gaussian on $w = \hat{w}$, and estimate the inverse of the covariance matrix according to

$$\Sigma^{-1} = -\nabla\nabla \log L = \lambda + H . \tag{12.33}$$

Here, H is the Hessian matrix of second derivatives of our cross-entropy error with respect to the weights. This may be evaluated with backpropagation or other methods.

The approximate marginalized log-likelihood for λ is then

$$\log L(\lambda; x) \approx -e(\hat{w}) - \frac{1}{2} \log |\Sigma^{-1}| + \frac{n_w}{2} \log \lambda , \tag{12.34}$$

where n_w is the number of weight parameters (i.e., H and Σ are $n_w \times n_w$ matrices). We then maximize this with respect to λ, with the result

$$\hat{\lambda} = \frac{1}{\hat{w}^T \hat{w}} \sum_{i=1}^{n_w} \frac{\lambda_i}{\hat{\lambda} + \lambda_i} , \tag{12.35}$$

where $\{\lambda_i\}$ are the eigenvalues of the Hessian matrix. Note that this is approximate unless one then iterates with the new value for λ to obtain new weights and hence a new Hessian matrix.

To make our Bayesian network predictive for any given input x, we must marginalize our posterior over the weights:

$$p(y|x) = \int p(y|x, w) p(w) dw , \tag{12.36}$$

where $p(w)$ is the posterior distribution for the weights as determined using the training data. The simplest approximation to performing this step is to suppose that the posterior in the weights is very narrow, approximate it as $p(w) = \delta(w - \hat{w})$, and obtain

$$p(y|x) = p(y|x, \hat{w}). \tag{12.37}$$

We can do better than this by using the Laplace approximation for the posterior over the weights in the form of a Gaussian, as described for example in Bishop (2006). The effect of including the uncertainty in w is to spread out the predictive distribution, which of course makes sense in terms of the degree of belief, or confidence, in the predictions.

12.5
Genetic Algorithms

In this chapter, we look at neural networks with a fixed structure. However, it is natural to ask how to choose an optimal structure, or topology, for the network, and the answer tends to be empirical. "Optimal" itself has various components, including how well it does the job, and how quickly it learns. The goal in particle physics is often not minimization of classification error, but how sensitive we are to new phenomena, or how precisely we can make a measurement. While classification error is often a reasonable surrogate, with a suitably optimized cut placed on the output of the so-trained network, this is not guaranteed, and other figures of merit may not map as well onto the backpropagation technology. We mention here another, computer-based, approach to this problem, that of *genetic algorithms*. A general reference is Haupt and Haupt (1998).

The notion behind genetic algorithms is the observation that living organisms can undergo mutations in their genetic codes, apparently randomly, and that sometimes these mutations can lead to a "fitter" individual. In application to neural networks, one starts with a "population" of networks. Mutations may be implemented with random variations on weights and/or randomly altering the link and node structure. The population evolves from one "generation" to the next either asexually, or sexually in a process called *crossover*. The probability for a given individual network to influence the population in the next generation is based on its "fitness", where fitness is defined in terms of the performance on the task the network is to perform.

The use of genetic algorithms has been investigated, with promising results, for particle physics applications. In particular Whiteson and Whiteson (2009) looks at the application of genetic algorithms to measurement of the top quark mass and searching for the Higgs boson using the NEAT (Neural Networks through Augmenting Technologies) method (Stanley and Miikkulainen, 2002), as well as rtNEAT (for real-time NEAT) (Stanley *et al.*, 2005).

12.6
Exercises

1. Generate a dataset of two-dimensional vectors that is not linearly separable and try the perceptron algorithm on it. A contrived but simple and visual choice is to generate data uniformly in two half-annular rings. The rings may be arranged to be completely separable, but not linearly separable.
2. Demonstrate (12.24) for an error given by the cross-entropy.
3. Demonstrate (12.35) for the estimation of regularization parameter λ.
4. Design a feed-forward network with backpropagation to solve the dataset in the first problem. You should divide your dataset into training, validation, and test datasets and apply methodology similar to that for Figure 12.3. Make a plot of your test dataset with a line indicating where the network output is equal for the two classes.

References

Bishop, C.M. (2006) *Pattern Recognit. Mach. Learn.*, Springer.

Cybenko, G. (1989) Approximation by superpositions of a sigmoidal function. *Math. Controls Signals Syst.*, **2**, 303–314.

Haupt, R.L. and Haupt, S.E. (1998) *Practical Genetic Algorithms*, John Wiley & Sons.

Haykin, S. (2006) *Neural Networks and Learning Machines*, 3rd edn, Prentice Hall.

Lees, J.P. *et al.* (2013) Search for direct CP violation in singly Cabibbo-suppressed $D^{\pm} \to K^{+} K^{-} \pi^{\pm}$ decays. *Phys. Rev. D*, **87**, 052010 (12 pp).

MacKay, D.J.C. (2003) *Information Theory, Inference, and Learning Algorithms*, Cambridge University Press.

Stanley, K.O., Bryant, B.D., and Miikkulainen, R. (2005) Real-time neuroevolution in the nero video game. *IEEE Trans. Evol. Comput.*, **9**, 653–668.

Stanley, K.O. and Miikkulainen, R. (2002) Evolving neural networks through augmenting technologies. *Evol. Comput.*, **10**, 99–127.

Whiteson, S. and Whiteson, D. (2009) Machine learning for event selection in high energy physics. *Eng. Appl. Artif. Intell.*, **22**, 1203–1217.

13
Local Learning and Kernel Expansion

This chapter combines two subjects. One is empirical local methods such as nearest neighbor (NN) rules. The other one is kernel methods such as support vector machines (SVM). Algorithms of the first kind are simple and intuitive. Algorithms of the second kind are founded on a rigorous theory developed mostly in the 1980s and 1990s. This theory is anything but intuitive, at least to people without a strong background in math. Despite this striking distance between the two types of algorithms, they have one thing in common. At the core of every algorithm in this chapter is the idea of a *pairwise similarity* measure: assign a new observation to the class whose members (observations) have the highest pairwise similarity to the new point. A simple example of such similarity is the inverse of the Euclidean distance. More generally, a measure of similarity is defined by the chosen kernel function.

We introduce the kernel formalism first and then describe empirical local methods, pointing out why they do not conform to the strict theory.

Following a popular interpretation, we describe local learning as mapping the original variables into a high-dimensional feature space and then searching for a linear class boundary in that space. We illustrate this approach on a hypothetical dataset. Two classes are shown in Figure 13.1a with crosses and stars. Clearly, we cannot find a straight line separating the two classes. We plot pairwise distances in Figure 13.1b. The three crosses are grouped together forming the dark 3×3 upper left quadrant with small pairwise distances. We could classify a new observation as a cross if its mean distance to the three crosses is less than a certain cutoff. To compute this cutoff, we take the average of the mean pairwise distance within the crosses (excluding zero distance to self) and the mean pairwise distance between the crosses and stars. The linear class boundary in the distance space maps onto an ellipse in the original two-dimensional space. We obtain a simple linear boundary at the expense of increasing the data dimensionality from 2 to 7.

Although a powerful and popular tool, local methods are not without flaws. Kernel methods based on a strict theoretical foundation do not allow for computation of the posterior class probabilities. For instance, a soft score computed by a two-class SVM can be used to measure the confidence of classification; yet it cannot be converted to an estimate of the class posterior probabilities in a straightforward way. In addition, local methods pose a serious computational challenge for data with many observations. Kernel algorithms typically require computing

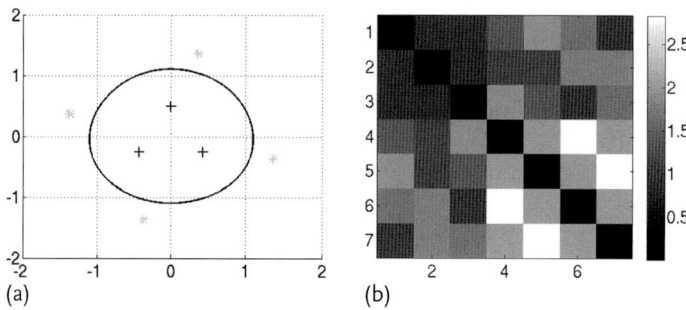

Figure 13.1 (a) Class boundary obtained by linear separation in the distance space. (b) Squared distances between points; the three crosses take the upper left quadrant of the distance matrix.

$N(N-1)/2$ pairwise similarities for N observations in the training set and storing them in memory. Empirical methods such as NN rules typically require computing N similarities between a test observation (query point) and every observation in the training set. Speeding up local methods and reducing their memory footprint has been a subject of ongoing research. Computational tricks for SVM and NN are discussed in this chapter as well.

13.1
From Input Variables to the Feature Space

Suppose we have a set of training data with N observations, $\{x_n\}_{n=1}^{N}$. Every observation is described by D input variables. As usual, we collect N transposed column vectors x_n in a matrix $\mathbf{X}$ of size $N \times D$.

Let us introduce a measure of similarity, $G(x, x')$, between vectors x and x'. We call G a *kernel function* and require that it be continuous and symmetric. We discuss specific forms of kernel functions later in this chapter. Examples are the Gaussian kernel, $G(x, x') = \exp(-\|x - x'\|^2 / 2)$, or a dot product, $G(x, x') = x^T x'$.

We construct a *Gram matrix*, $\mathbf{G}$, of size $N \times N$ by setting elements of this matrix to the kernel function evaluated at the respective observations, $g_{ij} = G(x_i, x_j)$. If $\mathbf{G}$ is positive semidefinite, we can take eigenvalue decomposition (EVD),

$$\mathbf{G} = \mathbf{V} \mathbf{\Lambda} \mathbf{V}^T, \tag{13.1}$$

where $\mathbf{\Lambda}$ is a diagonal matrix of nonnegative eigenvalues, and matrix $\mathbf{V} = (v_1, \ldots, v_N)$ is composed of orthogonal column vectors v_n. Define a vector-valued function ϕ such that

$$\phi(x_n) = v_n. \tag{13.2}$$

This function maps every observation x_n from the original D-dimensional space to the N-dimensional feature space. In physics analysis, data are typically tall ($N \gg D$), and the introduced mapping increases the dimensionality by orders of magnitude.

Define now a weighted dot product in the space spanned by vectors $\boldsymbol{\phi}$:

$$\langle \boldsymbol{\phi}(\boldsymbol{x}_i), \boldsymbol{\phi}(\boldsymbol{x}_j)\rangle = \sum_{n=1}^{N} \lambda_n \phi_n(\boldsymbol{x}_i)\phi_n(\boldsymbol{x}_j). \tag{13.3}$$

Above, λ_n is the nth diagonal element of $\boldsymbol{\Lambda}$ and ϕ_n is the nth element of vector $\boldsymbol{\phi}$. Elements of the Gram matrix $\mathbf{G}$ can be then expressed as dot products in the feature space, $g_{ij} = \langle \boldsymbol{\phi}(\boldsymbol{x}_i), \boldsymbol{\phi}(\boldsymbol{x}_j)\rangle$. Let $\boldsymbol{\Phi} = (\boldsymbol{\phi}(\boldsymbol{x}_1), \ldots, \boldsymbol{\phi}(\boldsymbol{x}_N))$ be an $N \times N$ matrix of features. The Gram matrix is then simply a matrix product $\mathbf{G} = \boldsymbol{\Phi}^\top \boldsymbol{\Phi}$ in the feature space, where the product is weighted by the eigenvalues.

Instead of defining a weighted dot product, we could set

$$\boldsymbol{\Phi} = \boldsymbol{\Lambda}^{1/2} \mathbf{V}^\top \tag{13.4}$$

to absorb λ_n in the definition of $\boldsymbol{\phi}(\boldsymbol{x}_n)$. No matter what convention we choose, we can interpret $\mathbf{G}$ as a matrix product in the feature space and replace $\boldsymbol{\Phi}^\top \boldsymbol{\Phi}$ by $\mathbf{G}$ in all formulas in this chapter.

The requirement of positive semidefiniteness for matrix $\mathbf{G}$ is essential. If we allowed negative eigenvalues λ_n, we would have to allow vectors $\boldsymbol{\phi}$ with negative squared norms $\boldsymbol{\phi}^\top \boldsymbol{\phi}$ in the feature space. We don't know how to work in this geometry.

Above, we have shown how to map a specific dataset onto a high-dimensional feature space. If we wanted to follow this route in practice, we would have to perform EVD of the Gram matrix to ensure its positive semidefiniteness every time. Decomposing a large square matrix can be time consuming and prone to numerical errors. Fortunately, this is seldom necessary. Most popular kernel functions are known to accept a dot product representation in the feature space. This fact is captured in

Theorem 13.1. *Mercer's Theorem: Let $\mathcal{G}$ be an integral operator such that*

$$\mathcal{G}(\phi) = \int_{\mathcal{X}} G(\boldsymbol{x}, \boldsymbol{x}')\phi(\boldsymbol{x}')d\boldsymbol{x}' \tag{13.5}$$

for kernel function $G(\boldsymbol{x}, \boldsymbol{x}')$. Require that G be positive semidefinite, that is,

$$\int_{\mathcal{X}}\int_{\mathcal{X}} G(\boldsymbol{x}, \boldsymbol{x}')\phi(\boldsymbol{x})\phi(\boldsymbol{x}')d\boldsymbol{x}d\boldsymbol{x}' \geq 0 \tag{13.6}$$

for any function ϕ. Let ϕ_n; $n = 1, \ldots, \infty$; be the orthonormal eigenfunctions for this operator: $\mathcal{G}(\phi_n) = \lambda_n \phi_n$ and $\int_{\mathcal{X}} \phi_i(\boldsymbol{x})\phi_j(\boldsymbol{x})d\boldsymbol{x} = \delta_{ij}$. Then

$$G(\boldsymbol{x}, \boldsymbol{x}') = \sum_{n=1}^{\infty} \lambda_n \phi_n(\boldsymbol{x})\phi_n(\boldsymbol{x}'). \tag{13.7}$$

We state this theorem rather loosely. For rigor, refer to Mercer (1909) and Scholkopf and Smola (2002).

Again, we define a vector-valued function $\boldsymbol{\phi}(x)$ with components $\{\phi_n(x)\}_{n=1}^{\infty}$ mapping x onto a high-dimensional feature space. The kernel function $G(x, x')$ can be then interpreted as a weighted dot product in this space, $G(x, x') = \langle \boldsymbol{\phi}(x), \boldsymbol{\phi}(x') \rangle$.

Working in a space of infinite dimensionality, as suggested by (13.7), does not sound appealing. Fortunately, we don't have to. If the set of nonzero eigenvalues λ_n is finite, the dimensionality is also finite. Otherwise the sum in (13.7) is known to converge uniformly to the function value $G(x, x')$. In other words, for any $\delta > 0$ we can find N_0 such that for any $N > N_0$ we have $|G(x, x') - \sum_{n=1}^{N} \lambda_n \phi_n(x) \phi_n(x')| < \delta$. The feature space can be thought of as finite-dimensional with accuracy δ.

This dry theory has a powerful implication for real-world data analysis known as

Proposition 13.2. *The Kernel Trick:*
If a numeric algorithm can be applied to a dot product $x^T x'$, it can also be applied to any positive semidefinite kernel $G(x, x')$.

In particular, any algorithm designed to work on the covariance matrix $\mathbf{X}^T\mathbf{X}$ can be used on the Gram matrix $\mathbf{G} = \boldsymbol{\Phi}^T \boldsymbol{\Phi}$. Note that we do not need to know $\boldsymbol{\Phi}$. All we need is a kernel function $G(x, x')$. We can then form the Gram matrix $\mathbf{G}$ and substitute it for $\mathbf{X}^T\mathbf{X}$. Then we can put "kernel" in front of the algorithm name and use it in our analysis. The workflow is not entirely that simple as inevitably there are various computational issues to consider. This is however the essence of the kernel analysis. We illustrate this idea on a simple example in Section 13.1.1.

The empirical mapping in (13.4) is not unique. We could construct an alternative map for a finite set $\{x_n\}_{n=1}^N$ by putting

$$\boldsymbol{\Phi} = \mathbf{V}\Lambda^{1/2}\mathbf{V}^T, \tag{13.8}$$

or equivalently, $\boldsymbol{\Phi} = \mathbf{G}^{1/2}$. In this representation, $\boldsymbol{\Phi}$ is symmetric and commutes with $\mathbf{G}$.

Yet another mapping could be obtained by putting

$$\boldsymbol{\Phi} = \mathbf{G}. \tag{13.9}$$

This mapping requires a special dot product in the feature space to ensure that $\langle \boldsymbol{\phi}(x_i), \boldsymbol{\phi}(x_j) \rangle = g_{ij}$. Let

$$\langle \boldsymbol{\phi}_i, \boldsymbol{\phi}_j \rangle = \boldsymbol{\phi}_i^T \mathbf{M} \boldsymbol{\phi}_j \tag{13.10}$$

for some matrix $\mathbf{M}$ and $\boldsymbol{\phi}_i = \boldsymbol{\phi}(x_i)$. This equality implies that $\mathbf{GMG} = \mathbf{G}$ and can be satisfied, for example, by $\mathbf{M} = \mathbf{G}^{-1}$ or, since $\mathbf{G}$ is most usually singular, by $\mathbf{M} = \mathbf{G}^+$, where $\mathbf{G}^+$ is the pseudo-inverse. In this representation, the high-dimensional features are computed explicitly.

One of the three mappings can be chosen for a particular problem for pedagogical or intuitive reasons. All three mappings are equivalent as long as the algorithm is based on a properly defined product in the feature space $\langle \boldsymbol{\phi}(x_i), \boldsymbol{\phi}(x_j) \rangle$.

Above, we have explained the kernel formalism in terms of mapping into a high-dimensional feature space. In doing so, we have followed the view adopted in the SVM community. The algorithmic machinery described in this chapter could be stripped of all talk of high-dimensional features and introduced in alternative ways, as in fact it has been introduced in various books and papers. As shown in Section 13.2, the representation chosen here allows you to think of the kernel approximation as a linear function fitting the target relation in a high-dimensional space.

13.1.1
Kernel Regression

In Section 11.3 we discussed classification by linear regression. For simplicity, let us assume either binary classification or multiple regression. In either case, the modeled response Y is a scalar variable. As usual, we put the training data in a matrix $\mathbf{X}$ of size $N \times D$ for N observations and D input variables. Centering the input variables (columns) in $\mathbf{X}$ to zero mean, we effectively drop the intercept term from the model. Scaling the input variables to unit variance, we improve the numerical accuracy of matrix operations on $\mathbf{X}$. Let $\mathbf{y}$ be a column vector of N response values in the training data and let $\boldsymbol{\beta}$ be a column vector of D linear coefficients. Our objective is to minimize the squared error $(\mathbf{y} - \mathbf{X}\boldsymbol{\beta})^\mathsf{T}(\mathbf{y} - \mathbf{X}\boldsymbol{\beta})$. The analytic least-squares solution to the system of linear equations $\mathbf{y} = \mathbf{X}\boldsymbol{\beta}$ is given by Seber and Lee (2003):

$$\hat{\boldsymbol{\beta}} = (\mathbf{X}^\mathsf{T}\mathbf{X})^{-1}\mathbf{X}^\mathsf{T}\mathbf{y}. \tag{13.11}$$

The scalar prediction for a new (centered and scaled) observation at x is then

$$f(x) = x^\mathsf{T}(\mathbf{X}^\mathsf{T}\mathbf{X})^{-1}\mathbf{X}^\mathsf{T}\mathbf{y}. \tag{13.12}$$

This solution is mostly of theoretical value. Direct inversion of the covariance matrix $\mathbf{X}^\mathsf{T}\mathbf{X}$ is slow and suffers from numerical errors, especially in high dimensions. A faster and more accurate approach is to solve $\mathbf{y} = \mathbf{X}\boldsymbol{\beta}$ by computing QR decomposition of $\mathbf{X}$ through a sequence of Householder reflections, as described in Moler (2008) and other textbooks. This technique is provided, for instance, by the backslash \ operator in MATLAB. Other approaches are discussed in Seber and Lee (2003).

The analytic solution (13.12) opens a door to "kernelization" of the algorithm. Replacing x with ϕ and $\mathbf{X}$ with $\boldsymbol{\Phi}$, we obtain

$$f(x) = \phi^\mathsf{T}\mathbf{G}^{-1}\boldsymbol{\Phi}^\mathsf{T}\mathbf{y}. \tag{13.13}$$

To simplify this expression, we need to move $\boldsymbol{\Phi}^\mathsf{T}$ to the left of $\mathbf{G}^{-1}$. We can easily do so if we choose the representation (13.8) in which $\boldsymbol{\Phi}$ commutes with $\mathbf{G}$. We obtain

$$f(x) = \mathbf{g}^\mathsf{T}\mathbf{G}^{-1}\mathbf{y}, \tag{13.14}$$

where $g = \boldsymbol{\Phi}\phi$ is a vector with N elements $G(x_n, x)$. Instead of inverting $\mathbf{G}$ directly, we can solve $\mathbf{y} = \mathbf{G}\boldsymbol{\beta}$ for $\boldsymbol{\beta}$ by QR decomposition and then set $f(x) = \boldsymbol{\beta}^\mathsf{T} g$.

Should we center and scale the data in the feature space, as we did for the original variables in $\mathbf{X}$? The linear equations in $\mathbf{y} = \mathbf{G}\boldsymbol{\beta}$ are similar to those in $\mathbf{y} = \mathbf{X}\boldsymbol{\beta}$. We could exploit this analogy and apply column-wise centering and scaling to the Gram matrix $\mathbf{G}$. In practice, there is little incentive to do so. The intercept term is not as important in the high-dimensional space as it is for low-dimensional regression because the model has enough fitting flexibility already. Besides, centering or scaling would remove symmetry from $\mathbf{G}$, and decomposing a nonsymmetric matrix is generally harder than a symmetric one.

Kernel regression is hardly ever carried out in the form (13.14). The rank of $\mathbf{G}$ is typically far below the number of features N; that is, only a small fraction of features is useful. A model trained on abundant and useless features can be numerically unstable and overfit, often dramatically. To prevent overfitting, we need to regularize the solution (13.14).

13.2
Regularization

In Section 9.3 we discussed overfitting and how to avoid it. A common recipe is to obtain an unbiased estimate of the generalization error by cross-validation or using an independent validation set. Model parameters are then found by minimizing this unbiased estimate. This procedure is effective as long as the model does indeed have parameters capable of controlling its flexibility. Take a neural network with backpropagation or an ensemble of decision trees, for instance. At first such a model obtains a rough initial approximation to the distribution $P(x, y)$. Then the model learns the distribution gradually while accumulating training cycles. Overfitting can be prevented by stopping after an appropriate number of learning cycles or by reducing the learning rate of the model, or a combination of both. The incremental nature of the fitting process in this case allows for high precision control.

But what about models without a similar defense mechanism? For example, a one-dimensional cubic spline must pass through every fitted point on the (x, y) plane. The cubic spline in Figure 13.2 most likely overfits and would predict poorly on new data. To improve its accuracy, we would need to smoothen the curve. Smoothing means regularization.

A common way of regularizing splines and similar functions $f(x)$ is by including a penalty term in the minimized objective,

$$\Omega(f) = \lambda \int_x (\mathcal{D}f)^2 dx \tag{13.15}$$

with $\lambda \geq 0$. This definition holds for multivariate x. Here, $\mathcal{D}$ is a linear differential operator. In the case of one-dimensional splines, we can take $\mathcal{D} = \frac{d^2}{dx^2}$ and find the

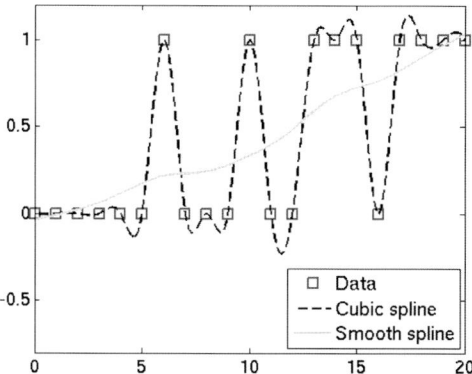

Figure 13.2 Smoothing spline approximations to empirical data. The case without regularization ($\lambda = 0$), or equivalently cubic spline, is shown with a dashed line, and the heavily regularized approximation ($\lambda = 0.9$) is shown with a solid line.

best fit $f(x)$ by minimizing

$$(1 - \lambda) \sum_{n=1}^{N} (y_n - f(x_n))^2 + \lambda \int_{\mathcal{X}} \left(\frac{d^2 f}{dx^2} \right)^2 dx . \tag{13.16}$$

The second term enforces smoothness by penalizing functions $f(x)$ with large second derivatives. The regularized spline in Figure 13.2 likely shows a much better generalization accuracy.

The goal of classification is minimization of the expected loss (9.1). In Chapter 9 we described a general workflow for such minimization: train a classifier on one set of data and monitor the empirical loss (9.2) on a test set to prevent overtraining. Similar to spline regularization, a classifier can be regularized by including a penalty term in the empirical loss minimized on the training data,

$$\tilde{L}_{\text{train}} = \sum_{n=1}^{N} \ell(y_n, f(\boldsymbol{x}_n)) + \Omega(f) . \tag{13.17}$$

For simplicity, we focus on binary classification and therefore require f to be scalar. To monitor the performance of this classifier on the test data, we use the empirical loss (9.2) without the penalty term $\Omega(f)$.

Let us take a short trip into the theory of differential equations. Operator $\mathcal{D}^*$ is adjoint to $\mathcal{D}$ if for any two scalar differentiable functions, $u(\boldsymbol{x})$ and $v(\boldsymbol{x})$, satisfying appropriate boundary conditions we can write

$$\int_{\mathcal{X}} u \mathcal{D} v \, d\boldsymbol{x} = \int_{\mathcal{X}} v \mathcal{D}^* u \, d\boldsymbol{x} . \tag{13.18}$$

Applying this rule to the regularization term (13.15), we obtain

$$\Omega(f) = \lambda \int_{\mathcal{X}} f \mathcal{D}^* \mathcal{D} f \, d\boldsymbol{x} . \tag{13.19}$$

A function $G(x', x)$ is called the *Green's function* for the differential operator $\mathcal{D}^*\mathcal{D}$ if

$$\mathcal{D}^*\mathcal{D}G(x', x) = \delta(x - x'). \tag{13.20}$$

Since the operator $\mathcal{D}^*\mathcal{D}$ is self-adjoint, the Green's function must be symmetric. Using the properties of the Dirac delta function, we have

$$\int_{\mathcal{X}} G(x', x)\mathcal{D}^*\mathcal{D}G(x'', x)dx = G(x', x'') \tag{13.21}$$

for x' and x'' arbitrarily chosen from domain $\mathcal{X}$. Our short trip ends here and we get back to the task at hand – regularizing the kernel methods.

As it turns out, the minimizer $f(x)$ of (13.17) under some fairly broad conditions can be expressed as a kernel expansion over the training points $\{x_n\}_{n=1}^N$. This proposition, first proven for quadratic loss functions $\ell(y, f)$ in Kimeldorf and Wahba (1971) and later generalized to other forms of loss in Cox and O'Sullivan (1990), is now known as the *Representer Theorem*. The connection between differential operators and kernels was established in Girosi et al. (1993) and Smola et al. (1998). We summarize these findings in

Proposition 13.3. *Kernel Expansion:*
A minimizer of the loss (13.17) admits a representation of the form

$$f(x) = \sum_{n=1}^N \alpha_n G(x_n, x), \tag{13.22}$$

where G is the Green's function for the operator $\mathcal{D}^*\mathcal{D}$.

In matrix notation, model estimates of the observed response vector **y** are given by $\hat{\mathbf{y}} = \mathbf{G}\boldsymbol{\alpha}$, where **G** is the Gram matrix and $\boldsymbol{\alpha}$ is a column vector of the expansion coefficients. To minimize (13.17), we need to find the optimal set of coefficients $\boldsymbol{\alpha}$ for the known kernel function G.

The theory behind this proposition is a bit more involved than shown here. In particular, the regularized loss (13.17) can be minimized by kernels other than the Green's function for $\mathcal{D}^*\mathcal{D}$. We recommend Scholkopf and Smola (2002) for a detailed review.

An example of a multivariate differential operator is

$$\mathcal{D}^*\mathcal{D} = \sum_{m=0}^{\infty} (-1)^m \frac{\sigma^{2m}}{m!2^m} \nabla^{2m}. \tag{13.23}$$

This operator gives rise to the normal kernel with radial symmetry,

$$G(x, x') = \exp\left(-\frac{\|x - x'\|^2}{2\sigma^2}\right). \tag{13.24}$$

The radial symmetry is induced by the translational and rotational invariance of the operator (13.23). In this case, the Green's function depends only on the Euclidean distance between the two points: $G(x, x') = G(\|x - x'\|)$. Kernels satisfying this property are called *radial basis functions* (RBF).

How should one choose the kernel function for a specific classification problem? On the basis of this discussion, you could be led to believe that you need to choose the regularization operator first and then obtain the respective kernel. In practice, the explicit form of the regularization operator does not matter much. To see why this is the case, take the expansion (13.22) and plug it in (13.19). Using the property (13.21) of the Green's function, it is easy to show that

$$\Omega(f) = \lambda \boldsymbol{\alpha}^\mathsf{T} \mathbf{G} \boldsymbol{\alpha} . \tag{13.25}$$

To find coefficients $\boldsymbol{\alpha}$ minimizing the regularized loss (13.17), you can work with the Gram matrix directly, without ever taking the integral in (13.15). From the empirical perspective, you do not even need to know that the regularization operator for the chosen kernel exists. If you can find coefficients $\boldsymbol{\alpha}$ minimizing (13.17) with the regularization term (13.25), you can measure the error of the obtained model by cross-validation or using an independent test set. If the error is low, you can consider this as empirical evidence indicating the lack of overtraining.

The kernel expansion (13.22) is sometimes written in a different form. If $G(x, x')$ is a Mercer kernel, $1 + G(x, x')$ is a Mercer kernel as well. The expansion (13.22) is then equivalent to

$$f(x) = \sum_{n=1}^{N} \alpha_n G(x_n, x) + b , \tag{13.26}$$

where $b = -\sum_{n=1}^{N} \alpha_n$ is a bias term. Poggio *et al.* (2001) investigate the relation between the two forms of kernel expansion in an article with a succinct title (b). Positive semidefinite kernels such as the homogeneous polynomial kernel $(x^\mathsf{T} x')^p$ do need the bias term. Although positive definite kernels such as the Gaussian kernel do not require a bias term, including this term does not hurt.

In summary, learning by kernels is carried out in four steps:

1. Choose a kernel function $G(x', x)$, or alternatively, choose a regularization operator and set the kernel to its Green's function.
2. Choose a specific form of the classification loss $\ell(y, f(x))$.
3. Assume a value of the regularization parameter λ.
4. Find the optimal set of linear coefficients $\boldsymbol{\alpha}$ by minimizing the regularized loss (13.17) with the penalty term (13.25) using the expansion (13.22).

Steps 3 and 4 are repeated until the optimal value of λ is found by monitoring the chosen classification loss on a dataset not used for training. An unbiased estimate of the generalization error is then obtained by applying the optimal model to an independent test set.

We illustrate how this workflow can be applied to extend the kernel regression example in Section 13.1.1.

13.2.1
Kernel Ridge Regression

Linear regression finds the least-squares solution to the system of linear equations $\mathbf{X}\boldsymbol{\beta} = \mathbf{y}$. If the data $\mathbf{X}$ have many variables, the solution can be numerically unstable and have a large test error. This problem can be cured by shrinking the magnitude of the coefficient vector. A popular approach described in Hoerl and Kennard (2000) is solving

$$\min_{\boldsymbol{\beta}}[(\mathbf{y} - \mathbf{X}\boldsymbol{\beta})^\top(\mathbf{y} - \mathbf{X}\boldsymbol{\beta}) + \lambda \boldsymbol{\beta}^\top \boldsymbol{\beta}]. \tag{13.27}$$

The $\lambda \boldsymbol{\beta}^\top \boldsymbol{\beta}$ term is called *ridge penalty*, and the method is called *ridge regression*. The idea of penalizing on the magnitude of $\boldsymbol{\beta}$, although an interesting subject in its own right, is not discussed here. The curious reader can find this topic described in various textbooks such as, for instance, Hastie *et al.* (2008).

Similar to standard linear regression, this minimization problem can be solved by differentiating over each component in $\boldsymbol{\beta}$ and setting the derivatives to zero. The obtained model is a simple adjustment to the one in (13.12):

$$f(\mathbf{x}) = \mathbf{x}^\top \left(\mathbf{X}^\top \mathbf{X} + \lambda \mathbf{I}_{D \times D}\right)^{-1} \mathbf{X}^\top \mathbf{y}, \tag{13.28}$$

where $\mathbf{I}_{D \times D}$ is an identity matrix. Adding positive values to the diagonal of the covariance matrix raises its rank and removes the numerical instability.

Using the kernel trick, we could "kernelize" (13.28) right away. Substituting $\boldsymbol{\phi}$ for $\mathbf{x}$ and $\boldsymbol{\Phi}$ for $\mathbf{X}$ and following the same steps as in Section 13.1.1, we obtain a similar adjustment to the kernel regression model (13.14):

$$f(\mathbf{x}) = \mathbf{g}^\top (\mathbf{G} + \lambda \mathbf{I}_{N \times N})^{-1} \mathbf{y}. \tag{13.29}$$

Alternatively, we could derive (13.29) by minimizing (13.17). The objective is to solve

$$\min_{\boldsymbol{\alpha}}[(\mathbf{y} - \mathbf{G}\boldsymbol{\alpha})^\top(\mathbf{y} - \mathbf{G}\boldsymbol{\alpha}) + \lambda \boldsymbol{\alpha}^\top \mathbf{G}\boldsymbol{\alpha}]. \tag{13.30}$$

Setting the derivative over each component of $\boldsymbol{\alpha}$ to zero, we obtain

$$\mathbf{G}(\mathbf{G}\boldsymbol{\alpha} + \lambda \boldsymbol{\alpha} - \mathbf{y}) = \mathbf{0}_{N \times 1}. \tag{13.31}$$

This equality is satisfied if $\mathbf{G}\boldsymbol{\alpha} + \lambda \boldsymbol{\alpha} = \mathbf{y}$, and the rest follows. Note that $\mathbf{G}\boldsymbol{\alpha} + \lambda \boldsymbol{\alpha} = \mathbf{y}$ is sufficient but not necessary for (13.31) to hold. Since the rank of $\mathbf{G}$ can be (much) less than N, the equality (13.31) can be satisfied by any $\mathbf{G}\boldsymbol{\alpha} + \lambda \boldsymbol{\alpha} - \mathbf{y}$ in the null space of $\mathbf{G}$. This derivation is therefore not strict.

Another way of deriving (13.29) is by setting G to the Green's function of operator $\mathcal{D}$ (instead of the Green's function of operator $\mathcal{D}^*\mathcal{D}$) in (13.19). Then $G(\mathbf{x}', \mathbf{x}'')$ on the right-hand side of (13.21) is replaced by $\delta(\mathbf{x}' - \mathbf{x}'')$, and the regularization term turns into $\lambda \boldsymbol{a}^T \boldsymbol{a}$. The minimization problem then becomes identical to the one in (13.27) for linear regression. It can be solved using the feature map (13.8).

As mentioned in Section 13.1.1, kernel regression is of little practical use for tall ($N \gg D$) datasets typical for physics analysis. In contrast, kernel ridge regression usually offers high accuracy for a carefully chosen kernel and can be very effective on small and mid-size datasets. Using kernel ridge regression for large data can be difficult. To obtain $\boldsymbol{g}$ in (13.29), you need to compute the kernel distance from the test observation to every observation in the training set. This operation can be slow and normally requires loading the entire training set into memory.

Being a well-recognized tool in other fields, kernel ridge regression is virtually unknown in the world of particle and astrophysics. A similar technique reviewed in Section 13.4 and known as "radial basis functions" has been used in a variety of physics analyses.

13.2.1.1 Example: Kernel Ridge Regression for the MAGIC Telescope Data

We apply kernel ridge regression to the MAGIC telescope data described in Bock *et al.* (2004) using MATLAB. Most MATLAB examples in this book require tools provided by the Statistics Toolbox and occasionally by the Optimization Toolbox. An implementation of kernel ridge regression only needs the backslash operator available from the base MATLAB.

Contents
- Prepare data
- Standardize variables
- Choose the kernel
- Compute the Gram matrix
- Map the test data into the feature space
- Compute the predicted response
- Compute HIACC

Prepare Data

We use the MAGIC telescope data from the UCI repository http://archive.ics.uci.edu/ml/datasets/MAGIC+Gamma+Telescope These are simulated data for detection of high energy gamma particles in a ground-based atmospheric Cherenkov gamma telescope using the imaging technique. The goal is to separate gamma particles from hadron background.

The analysis described in Bock *et al.* (2004), applies a number of powerful classifiers such as random forest, neural net and others. The power of these classifiers is evaluated by measuring true positive rates (TPR) at false positive rates (FPR) 0.01, 0.02, 0.05, 0.1, and 0.2. One such measure is HIACC (high accuracy) defined as the mean TPR for FPR values 0.1 and 0.2.

We split the data into training and test sets in the proportion 2 : 1 and code class labels as a numeric vector: $+1$ for gamma and -1 for hadron.

```
load MagicTelescope;
Ntrain = size(Xtrain,1);
Ntest = size(Xtest,1);
Ytrain = 2*strcmp('Gamma',Ytrain)-1;
Ytest = 2*strcmp('Gamma',Ytest)-1;
```

Standardize Variables

In these data, variables have substantially different standard deviations. These standard deviations do not carry information useful for separation of the two classes. We therefore standardize each variable by subtracting the mean and dividing by the standard deviation. To standardize, we pass the training data to the `zscore` function, which returns the estimated means and standard deviations. We then use these means and standard deviations to standardize the test data.

```
[Xtrain,mu,sigma] = zscore(Xtrain);
Xtest = bsxfun(@minus,Xtest,mu);
Xtest = bsxfun(@rdivide,Xtest,sigma);
```

Choose the Kernel

We use the standard normal kernel. Often this kernel is a good choice for standardized data in a low-dimensional setting (in this case, we have 10 variables). In this example, we are not making any effort to find the kernel with the best classification accuracy. It is possible that the predictive power of kernel ridge regression for this dataset could be improved by selecting a different kernel.

We define `kernelfun`, a function handle to compute pairwise kernel distances for observations collected in two matrices, X1 and X2, of the same size. In each matrix, observations are arranged in rows and variables are arranged in columns. `kernelfun` returns a column vector with one element per observation in X1 and X2. The first element of this vector is the kernel distance between the first observation (row) in X1 and the first observation (row) in X2, the second element of this vector is the kernel distance between the second observation (row) in X1 and the second observation (row) in X2 and so on. We use this function handle later to fill the Gram matrix and map the test data into the feature space.

```
sigma = 1;
kernelfun = @(x1,x2) exp(-sum((x1-x2).^2,2)/2/sigma^2);
```

Compute the Gram Matrix

We compute the Gram matrix for the training data. First, we create a square Ntrain-by-Ntrain matrix filled with zeros. Then we loop over training observations. For every training observation represented by vector q, we compute the kernel distance between this observation and all other observations in the training set. To do so, we replicate this observation using `repmat` to create a square Ntrain-by-Ntrain matrix in which all rows are set to q. This computation could be sped up using the symmetry of the Gram matrix. The loop however runs pretty fast for the chosen training data (12 000 observations in 10 dimensions), and we do not attempt to reduce the run time.

13.2 Regularization

```
Gtrain = zeros(Ntrain);
for n=1:Ntrain
    q = Xtrain(n,:);
    Gtrain(n,:) = kernelfun(Xtrain,repmat(q,Ntrain,1));
end
```

Map the Test Data into the Feature Space

We apply `kernelfun` to compute an Ntest-by-Ntrain feature matrix for the test data.

```
Gtest = zeros(Ntest,Ntrain);
for n=1:Ntest
    q = Xtest(n,:);
    Gtest(n,:) = kernelfun(Xtrain,repmat(q,Ntrain,1));
end
```

Compute the Predicted Response

By quick experimentation on cross-validated data (not shown here) we find that the value of the regularization parameter $\lambda = 1$ is about optimal. To find the regression coefficients `alpha`, we obtain a regularized least-squares solution to the linear system of equations `(Gtrain+lambda*I)*alpha=Ytrain` using the backslash operator. The output of `eye(Ntrain)` is I, an Ntrain-by-Ntrain identity matrix. We use the obtained coefficients to compute `Yfit`, the predicted response. The `Yfit` values can range from $-\infty$ to $+\infty$. We could assign class labels by thresholding the response at zero. Computation of a ROC curve does not require that we assign class labels, and we choose not to assign them.

```
alpha = (Gtrain+eye(Ntrain))\Ytrain;
Yfit = Gtest*alpha;
```

Compute HIACC

Following the paper by Bock *et al.* (2004), we estimate the mean signal efficiency at the background acceptance 0.1 and 0.2. We find TPR obtained by kernel regression, `tprkr`, at the specified values of FPR using the `perfcurve` function available from the Statistics Toolbox. The first two inputs to `perfcurve` are the true class labels and computed classification scores for the test data. We set the third input to +1, the label for the gamma class (signal). By default, `perfcurve` returns the computed FPR and TPR values as the first and second output arguments. We pass the `'xvals'` option to `perfcurve` to choose the desired values for FPR. Because the FPR values do not need to be computed, we place the tilde sign, ~, to ignore the first output from `perfcurve`.

```
[~,tprkr] = perfcurve(Ytest,Yfit,1,'xvals',[0.1 0.2])

tprkr =
    0.7651
    0.9086
```

The best HIACC value across all classification algorithms in the study by Bock et al. (2004), 0.852, is obtained by random forest. The second best classifier in their study, a neural net, returns 0.839. The value obtained by kernel ridge regression comes close.

```
accKR = mean(tprkr)

accKR =
    0.8369
```

We compute 95% confidence bounds for the two values of FPR by vertical averaging. To do so, we run the `perfcurve` function with the 'nboot' parameter set to 1000 bootstrap repetitions. We do not show this computation in this example to save space. To estimate the 95% lower bound on the mean signal efficiency, we average the two lower bounds; similarly for the upper bound (this estimate is not quite accurate but should suffice for this exercise). The confidence interval for the mean signal efficiency contains the HIACC values obtained by the random forest and neural net. The value computed by kernel ridge regression is statistically equal to the two best values quoted in the paper.

13.3
Making and Choosing Kernels

We list popular kernels in Table 13.1.

Use the following rules to construct new kernels satisfying the Mercer's Theorem:

1. If G_1 and G_2 are proper kernels, $G = a_1 G_1 + a_2 G_2$ with $a_1 > 0$ and $a_2 > 0$ is a proper kernel.
2. If G_1 and G_2 are proper kernels, $G = G_1 G_2$ is a proper kernel.
3. If $h(x)$ is a real-valued function, $G(x, x') = h(x)h(x')$ is a proper kernel.

Table 13.1 Popular kernels. If variables are standardized to zero mean and unit variance, the h_d values for all variables in the Gaussian product kernel are usually set equal. The Gaussian and Gaussian product kernels then become identical. The value of p in the polynomial kernels can be set to a positive integer.

Gaussian	$\exp\left(-\frac{\|x-x'\|^2}{2\sigma^2}\right)$
Gaussian product	$\exp\left(-\sum_{d=1}^{D} \frac{(x_d - x'_d)^2}{2h_d^2}\right)$
Polynomial	$(x^T x')^p$
Inhomogeneous polynomial	$(1 + x^T x')^p$
Sigmoid	$\tanh(a x^T x' + b)$

4. If **A** is a square positive semidefinite matrix, $G(x, x') = x^T A x'$ is a proper kernel.
5. If $h(z)$ is a real-valued function with a Taylor series expansion $h(z) = \sum_{n=1}^{\infty} a_n z^n$, then $G(x, x') = h(x^T x')$ is a proper kernel if and only if all coefficients a_n are nonnegative.

Interestingly, the sigmoid kernel violates rule 5 and is therefore improper.

In practice, the optimal kernel for your classification problem can be chosen by trial and error. Try the linear kernel ($p = 1$). If the class separation is poor, try the Gaussian kernel and a few polynomial kernels with low values of p. If the accuracy is still poor, try the sigmoid and customized kernels.

13.4
Radial Basis Functions

As established in Section 13.2, if the regularization operator is invariant under translations and rotations, the induced kernel depends only on the Euclidean distance between the two points. The kernel expansion in (13.22) then simplifies to

$$f(x) = \sum_{n=1}^{N} a_n G(\|x - x_n\|). \tag{13.32}$$

The objective is minimization of the quadratic loss. Popular kernel choices are normal $G(r) = \exp(-r^2/2)$, exponential $G(r) = \exp(-r)$ and thin plate spline $G(r) = r^2 \log(r)$. Here we use r to denote the magnitude $\|r\|$.

Stated this way, the RBF formalism is equivalent to kernel ridge regression described in Section 13.2.1. The next step however puts a new spin on the known equation. We can think of x_n in (13.32) as centers of the kernel expansion. We can question if we need to use *all* available points as the expansion centers. If we did not use all, we would likely lose some accuracy, but we could considerably reduce the CPU and memory consumption.

Thinking in terms of mapping onto a high-dimensional feature space, we could interpret this step as dimensionality reduction. To perform kernel ridge regression, we transform low-dimensional data into a high-dimensional space full of potentially useless features. Then we shrink the coefficient vector to reduce sensitivity to these useless features. Instead of doing and partially undoing the same work, we could try reducing the number of features before mapping the data. Effectively, we would implement an alternative way of regularization.

Suppose we select M centers for expansion out of N available points. The vector of coefficients $\boldsymbol{a}$ then has only M elements, and an equivalent of the Gram matrix, $\tilde{G}$, has N rows and M columns. The regularization term in (13.25) becomes $\boldsymbol{a}^T C \boldsymbol{a}$, where the symmetric matrix **C** of size $M \times M$ is obtained by taking rows of $\tilde{G}$ for the selected expansion centers. Substituting, with obvious modifications, the new matrix $\tilde{\mathbf{G}}$ and penalty term into (13.31), we obtain

$$(\tilde{\mathbf{G}}^T\tilde{\mathbf{G}} + \lambda\mathbf{C})\alpha = \tilde{\mathbf{G}}^T\mathbf{y}. \tag{13.33}$$

If a small set of expansion centers is used, the regularization term is typically not needed. In this case, the equation simplifies to the well-known linear regression formula $\tilde{\mathbf{G}}\alpha = \mathbf{y}$.

How can we select the expansion centers efficiently? A simple-minded strategy is to specify M and choose M points out of N at random. This strategy works reasonably well on some datasets. A more enlightened approach is to divide the data into M distance-based clusters and then choose one representative observation from every cluster. Various clustering algorithms such as, for instance, k-means can be used for this task.

The most accurate approach is training an RBF network. Such a network can be depicted as a neural net with an input layer, a hidden layer, and an output layer. The input layer has D units, one for each coordinate of the D-dimensional data. The hidden layer is composed of M expansion centers. The output layer for binary classification is one node carrying the simulated scalar response of the network. The output from the mth hidden node with coordinates $\mathbf{x}_m$ is given by the kernel function $G(\|\mathbf{x} - \mathbf{x}_m\|)$, and the weight for the link between the mth hidden node and the output node equals the expansion coefficient α_m. RBF networks are provided in many software suites, for example, the Neural Network Toolbox in MATLAB. Their descriptions and implementations vary. Full training of an RBF network can amount to updating the parameters of the kernel function for all hidden nodes at once or for each node separately, adjusting the link weights, and iteratively expanding the hidden layer with new nodes. Due to this flexibility, an RBF network typically attains higher accuracy than simple-minded selection of cluster centers followed by linear regression. In practice, this improvement in accuracy occurs at the expense of long training times and can be minor.

13.4.1
Example: RBF Classification for the MAGIC Telescope Data

Using the same MAGIC telescope dataset, we show how to perform binary classification on a set of kernel expansion centers preselected by the k-means algorithm.

Contents
- Find radial basis function (RBF) centers
- Map the training data into the feature space
- Map the test data into the feature space
- Compute the predicted response
- Compute HIACC.

Just like in the kernel ridge regression example, we use the MAGIC telescope data from the UCI repository. We split the data into training and test sets in the proportion 2 : 1 and code class labels as a numeric vector: $+1$ for gamma and -1 for hadron. The training data have $N_{train} = 12\,742$ observations and $D = 10$ input variables. We then standardize the variables and choose the kernel following the same steps as in the kernel ridge regression example.

Find Radial Basis Function (RBF) Centers

To find the kernel expansion centers, we cluster the data using the kmeans function provided in the Statistics Toolbox. By quick experimentation (not shown here) we find that $M = 500$ clusters give a reasonable balance between speed and accuracy of the classification model.

The kmeans function searches for cluster centers iteratively. At the first iteration, it assigns observations to clusters at random. At every next iteration, kmeans evaluates cluster fitness and moves observations across clusters if the move improves the fitness measure. We set the maximal number of iterations to 1000 to limit the search time. The output of kmeans can depend on the initial random cluster assignment. By setting the 'replicates' parameter to 10, we instruct kmeans to run 10 rounds of the initial random partitioning and select the one producing the most fit cluster assignment. We set the random generator seed by executing the rng function prior to calling kmeans for reproducibility.

```
M = 500;
options = statset('MaxIter',1000);
rng(1);
[~,C] = kmeans(Xtrain,M,'options',options,'replicates',10);
```

The kmeans function returns C, an M-by-D array of cluster centers. These centers may not coincide with the points in the dataset. To position every expansion center exactly at a training point, we find the nearest neighbor to every cluster center in the training set using the knnsearch function provided in the Statistics Toolbox. This function returns indices of observations in Xtrain closest to the cluster centers in C. To shift the cluster centers to the found positions, we assign the found points to C.

```
idxNearestX = knnsearch(Xtrain,C);
C = Xtrain(idxNearestX,:);
```

Map the Training Data into the Feature Space

We compute an Ntrain-by-M matrix of features for the training data. This computation is similar to the one in the kernel ridge regression example.

```
Gtrain = zeros(Ntrain,M);
for m=1:M
    q = C(m,:);
    Gtrain(:,m) = kernelfun(Xtrain,repmat(q,Ntrain,1));
end
```

We then compute the condition number of the feature matrix. The condition number is defined as the ratio of the largest over smallest singular value (by magnitude). In the kernel ridge regression example, the condition number of the full Ntrain-by-Ntrain matrix of features is at the order of 10^{28}. Here, the condition number is at the order of 10^3. The new matrix with 500 features likely does not need to be regularized. We assume that the optimal regularization parameter λ is zero.

```
cond(Gtrain)

ans =
    2.5416e+03
```

Map the Test Data into the Feature Space
In a similar fashion, we fill an Ntest-by-M matrix of features for the test data.

```
Gtest = zeros(Ntest,M);
for m=1:M
    q = C(m,:);
    Gtest(:,m) = kernelfun(Xtest,repmat(q,Ntest,1));
end
```

Compute the Predicted Response
In the absence of regularization, the optimal coefficients of the kernel expansion can be found in the same way as linear regression coefficients, by using the backslash operator. We multiply the test feature matrix by the estimated coefficients to obtain the predicted response.

```
alpha = Gtrain\Ytrain;
Yfit = Gtest*alpha;
```

Compute HIACC
Following the paper by Bock *et al.* (2004), we estimate the mean signal efficiency at the background acceptance 0.1 and 0.2, just like we did in the kernel ridge regression example. The obtained HIACC value is noticeably lower than 0.837 obtained by kernel ridge regression.

```
[~,tprkr] = perfcurve(Ytest,Yfit,1,'xvals',[0.1 0.2])
accKR = mean(tprkr)

tprkr =
    0.7118
    0.8968
accKR =
    0.8043
```

The modest loss in accuracy comes with more than 25-fold reduction in the CPU and memory requirements for the predictive model.

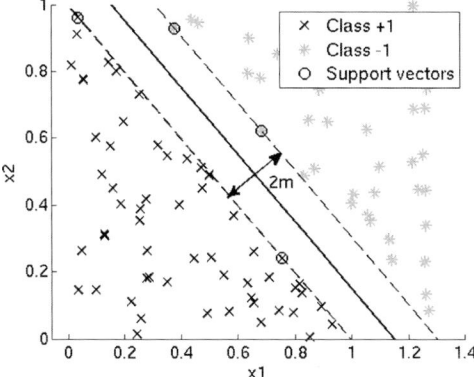

Figure 13.3 SVM classification for linearly separable data. The decision boundary $\boldsymbol{\beta}^T x + \beta_0 = 0$ is shown with a solid line. The class boundaries, $\boldsymbol{\beta}^T x + \beta_0 = -1$ and $\boldsymbol{\beta}^T x + \beta_0 = +1$, are shown with dashed lines. The distance between the class boundaries is $2m = 2/\|\boldsymbol{\beta}\|$.

```
Ntrain/M

ans =
    25.4840
```

13.5
Support Vector Machines (SVM)

Suppose we have multivariate data with two classes, $+1$ and -1. If we can draw a hyperplane such that all observations of one class lie on one side and all observations of the other class lie on the opposite side, the classes are said to be *linearly separable*. We draw two parallel hyperplanes through edge points of each class, as shown in Figure 13.3, and the decision boundary halfway between them. Define the optimal decision boundary as the one maximizing the distance between the two parallel hyperplanes. How can we find this boundary?

Formally, set the soft classification score to $f(x) = \boldsymbol{\beta}^T x + \beta_0$ and put the decision boundary at $f(x) = 0$. Observations with positive scores are then classified into class $+1$, and observations with negative scores are classified into -1. Multiplying all elements of $\boldsymbol{\beta}$ and β_0 by the same factor has no effect on the generalization error. Therefore we can scale $\boldsymbol{\beta}$ and β_0 to satisfy $y(\boldsymbol{\beta}^T x + \beta_0) = 1$ at the plane passing through the edge points of each class. The *classification margin*, $y f(x)$, is an important concept in the SVM theory and beyond. We will discuss this concept again in Section 15.1.4 in the context of ensemble learning.

By a simple geometric argument, the distance between the two planes is then $2/\|\boldsymbol{\beta}\|$. Since we have put the two planes at $y f(x) = 1$, the *hinge loss*,

$$\sum_{n=1}^{N} [1 - y_n f(x_n)]_+ , \qquad (13.34)$$

in the linearly separable training set equals zero. Here, $[a]_+$ equals zero if a is negative and a otherwise. Finding the optimal decision boundary is thus equivalent to solving the *primal problem*:

$$\text{minimize } \boldsymbol{\beta}^T \boldsymbol{\beta}$$
$$\text{subject to } \forall n: y_n \left(\boldsymbol{\beta}^T \boldsymbol{x}_n + \beta_0\right) \geq 1 . \tag{13.35}$$

It is convenient to solve this problem by Lagrange multipliers. The objective is to maximize the Lagrangian function,

$$L(\boldsymbol{\beta}, \beta_0, \boldsymbol{\alpha}) = \frac{1}{2} \boldsymbol{\beta}^T \boldsymbol{\beta} - \sum_{n=1}^{N} \alpha_n \left[y_n \left(\boldsymbol{\beta}^T \boldsymbol{x}_n + \beta_0\right) - 1 \right], \tag{13.36}$$

for N nonnegative multipliers α_n (see Appendix A.2 for details). Differentiating the Lagrangian over $\boldsymbol{\beta}$ and β_0 gives $D + 1$ linear equations:

$$\sum_{n=1}^{N} \alpha_n y_n \boldsymbol{x}_n = \boldsymbol{\beta}$$

$$\sum_{n=1}^{N} \alpha_n y_n = 0 . \tag{13.37}$$

Plugging $\boldsymbol{\beta}$ into (13.36), we obtain the *dual* optimization problem:

$$\text{maximize } L(\boldsymbol{\alpha}) = \sum_{n=1}^{N} \alpha_n - \frac{1}{2} \sum_{i=1}^{N} \sum_{j=1}^{N} \alpha_i \alpha_j y_i y_j \boldsymbol{x}_i^T \boldsymbol{x}_j$$
$$\text{subject to } \sum_{n=1}^{N} \alpha_n y_n = 0 \tag{13.38}$$
$$\forall n: \alpha_n \geq 0 .$$

The *bias term* β_0 has been eliminated from the Lagrangian by the second constraint in (13.37). After we obtain the optimal coefficients $\boldsymbol{\alpha}$, we can compute $\boldsymbol{\beta}$. Then we can find β_0 explicitly from the margin constraint $y(\boldsymbol{\beta}^T \boldsymbol{x} + \beta_0) = 1$ at the edge points for each class. Finally, we can write down the classification score

$$f(\boldsymbol{x}) = \sum_{n=1}^{N} \alpha_n y_n \boldsymbol{x}_n^T \boldsymbol{x} + \beta_0 \tag{13.39}$$

as the expansion seen in (13.26) with a simple dot product kernel. In matrix form, we have $\boldsymbol{\beta} = \mathbf{X}^T \text{diag}(\boldsymbol{y}) \boldsymbol{\alpha}$.

As seen in Figure 13.3, the hyperplane of optimal separation is defined by the edge points in each class. We would expect that only coefficients α_n for these edge points be of sizable magnitude. Our expectation is supported by the Karush–Kuhn–

Tucker (KKT) complementarity condition (see Appendix A.2) for the optimization problem (13.38):

$$\forall n: \alpha_n \left[y_n \left(\boldsymbol{\beta}^\mathsf{T} \boldsymbol{x}_n + \beta_0 \right) - 1 \right] = 0. \tag{13.40}$$

This implies that coefficient α_n must be zero unless observation $\boldsymbol{x}_n$ lies exactly on one of the two bounding hyperplanes. Observations $\boldsymbol{x}_n$ for which coefficients α_n are strictly positive are called *support vectors*.

Unlike any other method reviewed in this chapter, a support vector machine can find a *sparse* solution, that is, a solution in which support vectors form a vanishingly small fraction of the training set. Such a solution offers a great computational advantage – low memory storage and high speed of prediction for new data. A sparse solution is not necessarily the most accurate one. For instance, classes may be linearly separable in the training data but not in the test data. It is up to the analyst to ensure that a sparse solution does not lead to a significant loss of accuracy.

What if the data are not linearly separable? We modify the margin constraint (13.35) by introducing nonnegative slack variables ξ_n,

$$y_n \left(\boldsymbol{\beta}^\mathsf{T} \boldsymbol{x}_n + \beta_0 \right) \geq 1 - \xi_n \tag{13.41}$$

for all n in the training set. Observations with $\xi_n > 1$ lie on the wrong side of the decision boundary. To separate the two classes perfectly, we need the slack variables to be as small as possible. To minimize their magnitude, we introduce a penalty term. The 1-norm optimization problem

$$\text{minimize} \quad \frac{1}{2} \boldsymbol{\beta}^\mathsf{T} \boldsymbol{\beta} + C \sum_{n=1}^{N} \xi_n \tag{13.42}$$

$$\text{subject to} \quad \forall n: y_n \left(\boldsymbol{\beta}^\mathsf{T} \boldsymbol{x}_n + \beta_0 \right) \geq 1 - \xi_n$$

$$\forall n: \xi_n \geq 0$$

leads to the Lagrangian formulation

$$\text{maximize} \quad L(\boldsymbol{\alpha}) = \sum_{n=1}^{N} \alpha_n - \frac{1}{2} \sum_{i=1}^{N} \sum_{j=1}^{N} \alpha_i \alpha_j y_i y_j \boldsymbol{x}_i^\mathsf{T} \boldsymbol{x}_j \tag{13.43}$$

$$\text{subject to} \quad \sum_{n=1}^{N} \alpha_n y_n = 0$$

$$\forall n: C \geq \alpha_n \geq 0$$

with KKT complementarity conditions

$$\forall n: \alpha_n \left[y_n \left(\boldsymbol{\beta}^\mathsf{T} \boldsymbol{x}_n + \beta_0 \right) - 1 + \xi_n \right] = 0$$

$$\forall n: \xi_n (C - \alpha_n) = 0. \tag{13.44}$$

An observation violates the KKT conditions either if it is easily classified $y_n f(x_n) > 1$ and has a nonzero coefficient α_n or if it lies in the uncertain region $y_n f(x_n) < 1$ and has a coefficient less than C. Observations violating the KKT conditions play an important role in SVM training procedures described in Section 13.5.3. One of the convergence criteria used for SVM is the absence of KKT violators in the training set.

An alternative approach based on the 2-norm penalty, $C \sum_{n=1}^{N} \xi_n^2$, is discussed in Cristianini and Shawe-Taylor (2000) and Scholkopf and Smola (2002).

The *box constraint C* is a positive parameter that remains constant during optimization. As the box constraint approaches $+\infty$, the contribution due to the slack variables vanishes and the problem becomes identical to the one for linearly separable data. In contrast, by setting $C = 0$ we would allow slack variables to be arbitrarily large and might not be able to separate the classes. Effectively, the box constraint controls how many support vectors are used. Its optimal value can be found by minimizing the classification error of the SVM model by cross-validation or using an independent validation set.

Even if the classes are linearly separable, the box constraint can play an important role. In real-world data, each class can have only one support vector at $C = +\infty$. If the support vector is an outlier, the generalization error of the SVM model can be poor. Setting C to a finite value, we increase the number of support vectors in each class to obtain a more robust solution.

So far, we have discussed the case of linear separation. Using the kernel trick described in Section 13.1, we can replace the dot product $x^T x'$ in the SVM formalism with a function $G(x, x')$ satisfying the conditions of the Mercer's Theorem to obtain

$$f(x) = \sum_{n=1}^{N} \alpha_n y_n G(x_n, x) + \beta_0. \tag{13.45}$$

A hyperplane of separation in the feature space then induces a nonlinear decision boundary in the space of original variables.

The SVM formalism is built on a substantial body of theoretical work. Various papers relate the geometric margin (distance between the bounding hyperplanes) to the generalization error of the SVM model. In Section 15.1.4 we show how such a bound is used to construct new boosting algorithms. Since this book is aimed at practitioners, we omit these theoretical results here. You can find a good summary in Cristianini and Shawe-Taylor (2000).

13.5.1
SVM with Weighted Data

Inclusion of data weights in the SVM formalism is not straightforward. Here, we consider a simpler case of including class prior probabilities. Using prior probability values might be desired, for instance, for learning on data with class imbalance, as discussed in Section 9.5.

If the classes are separable, an observation is classified confidently unless this observation lies between the two hyperplanes defined by the support vectors. For two classes with equal prior probabilities, an observation positioned exactly halfway between the class boundaries can be assigned to either class. If the classes have unequal prior probabilities, it would make sense to shift the hyperplane of uncertain classification (solid line in Figure 13.3) toward the class with the smaller prior probability. The distance between the shifted hyperplane of uncertain classification and the class boundary would then equal the prior probability for the respective class.

If the classes are inseparable, the SVM optimization problem must be modified. Lee et al. (2004) and Bach et al. (2006) use a cost-sensitive definition of the hinge loss (13.34) to derive the respective primal and dual problems. For binary classification $y \in \{-1, +1\}$, misclassification costs can be unambiguously converted to prior probabilities using (9.19). Applying this conversion, we include the prior probabilities in the hinge loss (13.34) to obtain

$$\sum_{n=1}^{N} \gamma_{y_n}[1 - y_n f(x_n)]_+ \tag{13.46}$$

with $\gamma_{y_n} = N\pi_{y_n}/N_{y_n}$. Here, N_{+1} and N_{-1} are observation counts for class +1 and −1, respectively, with constraint $N_{-1} + N_{+1} = N$, and π_{+1} and π_{-1} are the prior probabilities for class +1 and −1, respectively, with constraint $\pi_{-1} + \pi_{+1} = 1$. The standard SVM problem with $\gamma_{-1} = \gamma_{+1} = 1$ has been discussed earlier. The primal problem (13.42) for the modified hinge loss transforms to

$$\text{minimize} \quad \frac{1}{2}\beta^T\beta + C \sum_{n=1}^{N} \gamma_{y_n} \xi_n \tag{13.47}$$

subject to the same constraints as earlier. This primal leads to the dual formulation (13.43) with a modified second constraint

$$\forall n: C\gamma_{y_n} \geq \alpha_n \geq 0. \tag{13.48}$$

If the classes are perfectly separable in the training set, the slack variables ξ_n are optimized to zero, obliterating any effect of the prior probability correction.

How can we incorporate observations weights in the two schemes for inclusion of the class prior probability? Since observations with zero expansion coefficients α_n have no effect on the separation boundary, their weights are irrelevant. If the classes are separable, the shift for the hyperplane of uncertain classification could be determined using the weights for the support vector observations only, ignoring the class prior probabilities. If the classes are inseparable, the constraint (13.48) could be modified to

$$\forall n: NCw_n \geq \alpha_n \geq 0, \tag{13.49}$$

allowing large coefficients α_n for observations with large weights w_n. Here, we assume that the weights w_n sum to one. These recipes appear sensible, but we are not aware of any publications describing them. Popular software implementations of SVM do not support weighted observations.

13.5.2
SVM with Probabilistic Outputs

As noted earlier, the soft score predicted by an SVM classifier represents signed distance to the decision boundary. For many practical problems such as plotting a ROC curve, the soft score would suffice.

Sometimes it may be desirable to obtain estimates of the class posterior probabilities at point x as well. Transformation of the soft score to the class posterior probability for SVM is ambiguous. Various methods have been proposed.

One simple approach is to estimate the posterior probabilities using cross-validation or an independent test set. We describe this approach for binary SVM. Choose a specific value for the SVM score, f_0. Compute scores for the test data. Set the posterior probability for the positive class to the number of positive observations with scores above f_0 divided by the total number of observations (originating from both classes) with scores above f_0. This approach has one caveat: the probability estimates in the tails of the score distribution can be rather inaccurate.

Platt (1999) observes that the posterior probability ratio often follows an exponential distribution. He proposes a sigmoid transformation,

$$p(f) = \frac{1}{1 + \exp(a f + b)}, \tag{13.50}$$

where $p = P(Y = +1|x)$ is the probability of observing the positive class at x and f is the soft score. Parameters a and b are found by maximizing the log-likelihood,

$$\log L = \sum_{n=1}^{N} [\tilde{y}_n \log p_n + (1 - \tilde{y}_n) \log(1 - p_n)], \tag{13.51}$$

over N observations in the test data with $p_n = p(f(x_n))$. Here, $\tilde{y} \in \{0, 1\}$ is the class label obtained from the original class label $y \in \{-1, +1\}$ using $\tilde{y} = (1 + y)/2$. Details of the fitting procedure can be found in Platt (1999).

13.5.3
✂ Numerical Implementation

Consider solving (13.43) in which $x_i^T x_j$ is replaced, thanks to the kernel trick, by an element of the positive semidefinite Gram matrix $g_{ij} = G(x_i, x_j)$. Define matrix Q of size $N \times N$ with elements $q_{ij} = y_i y_j g_{ij}$. Then express (13.43) in a vectorized form:

$$\text{minimize} \quad L(\alpha) = \frac{1}{2} \alpha^T Q \alpha - \alpha^T \mathbf{1}_{N \times 1} \tag{13.52}$$

$$\text{subject to} \quad \alpha^T y = 0 \tag{13.53}$$

$$\forall n: 0 \leq \alpha_n \leq C. \tag{13.54}$$

This is a *quadratic programming* (QP) problem with $2N + 1$ linear constraints, two (lower and upper) for each α_n and the constraint (13.53) induced by the bias term β_0. Matrix $\mathbf{Q}$ has size $N \times N$ for N observations.

For small datasets, this problem can be solved by well-known QP algorithms such as those described in Nocedal and Wright (2006). Such algorithms factorize $\mathbf{Q}$ to an appropriate form. This factorization is expensive to compute and must be kept in the computer memory until a solution to the QP problem is found. For large N, this approach is infeasible. A lot of effort in the last two decades has been spent on designing efficient algorithms for training SVM on large data. Some of the proposed algorithms have earned recognition from practitioners and are now in broad use. We describe them here. If you have no desire to look under the hood of SVM engines, you might want to skip this section.

A large-scale QP problem can be solved by interior-point methods if a convenient factorization of the matrix $\mathbf{Q}$ can be found. Application of interior-point methods to SVM at present remains unpopular. Below we describe a more popular approach to solving large-scale SVM problems known as *working set methods*, as well as its two extensions, sequential minimal optimization (SMO) and iterative single data algorithm (ISDA).

13.5.3.1 Chunking and Working Set Methods

Split the training data in two disjoint subsets, $\mathcal{M}$ with size M and $\mathcal{P}$ with size $P = N - M$. Without loss of generality, assume that the M observations are placed first in the data followed by the P observations. Split the coefficient vector $\boldsymbol{\alpha}$ in two subvectors, $\boldsymbol{\alpha}_M$ and $\boldsymbol{\alpha}_P$; similarly for the vector of class labels $\mathbf{y}$. Then split the matrix $\mathbf{Q}$ in four blocks. Decompose the problem (13.52) with constraints (13.53) and (13.54) as follows:

$$\text{minimize} \quad L(\boldsymbol{\alpha}_M) = \frac{1}{2}\boldsymbol{\alpha}_M^T \mathbf{Q}_{MM} \boldsymbol{\alpha}_M + \boldsymbol{\alpha}_M^T \mathbf{r} + R \tag{13.55}$$

$$\text{subject to} \quad \boldsymbol{\alpha}_M^T \mathbf{y}_M + z = 0 \tag{13.56}$$

$$\forall m; 1 \leq m \leq M : 0 \leq \alpha_m \leq C \tag{13.57}$$

$$\forall p; 1 \leq p \leq P : 0 \leq \alpha_p \leq C \tag{13.58}$$

with

$$\mathbf{r} = \mathbf{Q}_{MP} \boldsymbol{\alpha}_P - \mathbf{1}_{M \times 1} \tag{13.59}$$

$$R = \frac{1}{2}\boldsymbol{\alpha}_P^T \mathbf{Q}_{PP} \boldsymbol{\alpha}_P - \boldsymbol{\alpha}_P^T \mathbf{1}_{P \times 1} \tag{13.60}$$

$$z = \boldsymbol{\alpha}_P^T \mathbf{y}_P . \tag{13.61}$$

Now optimize $\boldsymbol{\alpha}_M$ keeping $\boldsymbol{\alpha}_P$ fixed. If M is small, the reduced problem can be managed by any QP solver. The M observations form a *working set*.

Suppose we minimize the reduced Lagrangian (13.55) for specific subsets $\mathcal{M}$ and $\mathcal{P}$. If we move an observation from $\mathcal{M}$ to $\mathcal{P}$ and re-optimize, we obtain the same solution for the remaining coefficients in $\mathcal{M}$. If we move an observation violating the KKT conditions (13.44) from $\mathcal{P}$ to $\mathcal{M}$ and re-optimize, we obtain a new solution with a lower value of the objective function $L(\boldsymbol{\alpha}_\mathcal{M})$. These two propositions are proved in Osuna et al. (1997). We can therefore minimize the Lagrangian (13.52) through a sequence of sub-problems, each obtained by moving a subset of KKT violators from $\mathcal{P}$ to $\mathcal{M}$ and possibly moving some observations from $\mathcal{M}$ to $\mathcal{P}$ to reduce the problem size.

A simple algorithm of this kind is *chunking*. Initialize $\boldsymbol{\alpha}_\mathcal{M}$ and $\boldsymbol{\alpha}_\mathcal{P}$ to some values. Select M observations for the initial set $\mathcal{M}$ at random. Minimize the reduced Lagrangian (13.55) over $\boldsymbol{\alpha}_\mathcal{M}$ keeping $\boldsymbol{\alpha}_\mathcal{P}$ fixed. Compare the obtained coefficients $\boldsymbol{\alpha}_\mathcal{M}$ to 0 and C with some tolerance. Keep observations with $0 < \alpha_m < C$ in $\mathcal{M}$ and move the rest to $\mathcal{P}$. Find a desired number of observations in $\mathcal{P}$ that violate the KKT conditions most and move them to $\mathcal{M}$. Minimize the reduced Lagrangian (13.55) again. Continue until convergence. In this approach, the size of the working set can grow large over iterations and become comparable to N.

A more sophisticated strategy is to fix the size of the working set in advance. Subsets $\mathcal{M}$ and $\mathcal{P}$ then trade equal numbers of observations at every step. Various heuristics can be used to determine what observations should be exchanged. Joachims (1999) describes a set of such heuristics used in SVMlight, a well-known software package for the working set method.

In the following sections, we describe two popular SVM solvers that can be viewed as extreme implementations of the working-set idea. Sequential minimal optimization (SMO) maintains a working set of size 2, and iterative single data algorithm (ISDA) is based on the coordinate descent algorithm optimizing one coefficient at a time.

13.5.3.2 Sequential Minimal Optimization (SMO)

The SMO algorithm proposed in Platt (1998) operates on a working set of size 2. We give here an outline of this algorithm. Choose a numeric tolerance δ on the KKT conditions (13.44) to be used as a measure of convergence. Define a vector of residuals,

$$r = Q\alpha + \beta_0 y - 1_{N \times 1}. \tag{13.62}$$

Find all points violating the KKT conditions by more than δ. As seen from (13.44), a point violates the KKT conditions if

$$(\alpha_n > 0 \text{ and } r_n > \delta) \text{ or } (\alpha_n < C \text{ and } r_n < -\delta). \tag{13.63}$$

Loop over the found KKT violators. For every point i, find point j maximizing $|r_i - r_j|$. Given the two KKT violators, i and j, minimize the reduced Lagrangian (13.55) with respect to α_i and α_j. The linear constraint (13.56) implies that this minimization must conserve the sum $y_i \alpha_i + y_j \alpha_j$. Since the two coefficients are bound by the box constraints (13.57) and the linear constraint above, the solution (α_1, α_2)

must lie on a line segment connecting the adjacent sides of a square with vertices $(0, 0)$ and (C, C) and parallel to the diagonal line of this square. After the new values of α_i and α_j are obtained, compute a new value for the bias term β_0 to satisfy the KKT conditions for points i and j. We can now see the rationale behind maximizing $|r_i - r_j|$: since the residuals for both points move much closer to zero in the update, optimizing this pair would typically lead to a large decrease in the full Lagrangian (13.52). Continue the outer loop past index i to process all known KKT violators.

Now find a subset of the violators with $0 < \alpha_n < C$. Loop over these unbound violators repeatedly, in the same manner as above, until their residuals drop below δ in magnitude. Focusing on unbound violators accelerates the optimization: bound points ($\alpha_n = 0$ or $\alpha_n = C$) are likely to remain bound after the update and therefore have less chance of improving the objective. Then take another pass through the entire set to find new violators, and so on. Optimization is continued until no KKT violators are left in the training set.

Modern SVM implementations do not use the original Platt's algorithm. Two points for the next update are chosen by various heuristics other than maximizing $|r_i - r_j|$. The two-loop structure of the original SMO is not necessarily used. The term "SMO" broadly applies to any algorithm based on the two-point update. At present, such two-point algorithms are the most popular way of solving large-scale SVM problems for classification. LIBSVM, one the better known software packages, is described in Fan *et al.* (2005).

13.5.3.3 Iterative Single Data Algorithm (ISDA), Also Known as Kernel AdaTron

As discussed in Section 13.2, the kernel expansion can be used with or without a bias term. Dropping the bias term in (13.45), we eliminate the linear constraint (13.53). Minimizing the Lagrangian (13.52) is then equivalent to solving $\mathbf{Q}\boldsymbol{\alpha} = \mathbf{1}_{N \times 1}$ subject to the box constraints (13.54).

Iterative methods for solving symmetric systems of linear equations are well known. The Gauss–Seidel (GS) method is a coordinate descent algorithm. At every step, it updates one coefficient only. To update α_n, GS uses the current approximation to $\boldsymbol{\alpha}$ and $\mathbf{q}_n$, the nth row of matrix $\mathbf{Q}$. Let $\mathbf{r} = \mathbf{Q}\boldsymbol{\alpha} - \mathbf{1}_{N \times 1}$ be a vector of residuals. At step t we have

$$\mathbf{q}_n \boldsymbol{\alpha}^{(t)} - 1 = r_n^{(t)}. \tag{13.64}$$

Now set $\boldsymbol{\alpha}^{(t+1)} = \boldsymbol{\alpha}^{(t)}$ and override the nth element by putting

$$\alpha_n^{(t+1)} = \alpha_n^{(t)} - \frac{r_n^{(t)}}{q_{nn}}, \tag{13.65}$$

where q_{nn} is the nth diagonal element of $\mathbf{Q}$. The new approximation then gives a perfect solution for the nth point, $\mathbf{q}_n \boldsymbol{\alpha}^{(t+1)} = 1$. Of course, all other points may have nonzero residuals after the update. GS continues updating one coefficient at a time until some convergence criterion is satisfied.

The update (13.65) does not respect the box constraints (13.54). Capping the updated value at the bounds of the allowed interval, we replace (13.65) with

$$\alpha_n^{(t+1)} = \min\left(\max\left(\alpha_n^{(t)} - \frac{r_n^{(t)}}{q_{nn}}, 0\right), C\right). \tag{13.66}$$

The described GS update is the key element of iterative single data algorithms (ISDA) for SVM, also known as Kernel AdaTron. ISDA methods are reviewed in Huang *et al.* (2006). MATLAB functions and Windows executables for ISDA are available from http://www.learning-from-data.com/download.htm.

To achieve a reasonable convergence rate for large data, ISDA applies various heuristics. Here is an outline of the algorithm. Split the training data in two subsets, the analyzed set and its complementary. Put the entire data in the analyzed set. Take a pass through points in the analyzed set to find the KKT violator with largest $|r_n|$. Update α_n using (13.66). Update residuals for all other points in the analyzed set using the new value of α_n. Find the next worst violator and update; repeat this procedure for a predefined number of steps. Shrink the analyzed set by moving points satisfying both $\alpha_n = 0$ and $r_n > -\delta$ from the analyzed set to the complementary set. Continue updating and shrinking the analyzed set in the same manner until no KKT violators are left. Then move KKT violators from the complementary set to the analyzed set and repeat.

We have described SMO as an algorithm for SVM with a bias term and ISDA as an algorithm for SVM without a bias term. In reality, the distinction is not so clear-cut. A version of SMO for bias-free SVM, and a version of ISDA for SVM with bias are also available. Refer to Huang *et al.* (2006) for details.

Although less popular than SMO, the ISDA approach is simpler and can perform well on large data. The run time comparison does not provide a clear winner. SMO algorithms tend to converge faster but require more computations per iteration.

13.5.3.4 Caching

The working set decomposition schemes described in Sections 13.5.3.1–13.5.3.3 can greatly reduce the computing time for large-scale SVM problems. Memory management presents another problem. Since matrix **Q** takes at least $N(N-1)/2$ elements for storage, the dataset does not need to be big to saturate your memory resource. For instance, a decent modern desktop with 8 GB of RAM can accommodate up to 32 000 observations if **Q** is stored in double precision.

Modern SVM implementations cache a subset of matrix **Q** in memory. This subset should be formed of entries used most often during optimization. For instance, ISDA repeatedly loops over the analyzed set composed of KKT violators with nonzero coefficients α_n. Caching the square block of matrix **Q** for the respective rows and columns can significantly speed up SVM training.

If the cache is not large enough to keep the relevant subset of **Q**, a caching strategy can be deployed. A well-known strategy is *first in first out*, also known as *queue*. To add a new observation to a full cache, remove the observation with the earliest entry time into the cache and put the new one in. The caching strategy can have a noticeable impact on the execution time.

13.5.4
✂ Multiclass Extensions

So far, we have discussed SVM for binary classification. What if we need to separate 3 or more classes?

Like any multiclass problem, SVM analysis of data with more than two classes can be reduced to a set of binary classification problems. A framework for this reduction is laid out in Chapter 16. One-versus-all (OVA) and one-versus-one (OVO) strategies are used most often. OVA induces K binary classifiers by separating each class from all others. OVO amounts to training $K(K-1)/2$ binary learners, one for each pair of classes.

Alternatively, you can build a single multiclass machine. Such algorithms are described in Vapnik (1998); Weston and Watkins (1999); Lee *et al.* (2004); Hsu and Lin (2002); Crammer and Singer (2001); Tsochantaridis *et al.* (2005); Bordes *et al.* (2007); Platt *et al.* (2000), and Tibshirani and Hastie (2007). Each paper includes an empirical study, often demonstrating a *slight* improvement in classification accuracy over OVA and OVO. Dedicated comparative studies of various multiclass SVM extensions can be found in Hsu and Lin (2002) and Statnikov *et al.* (2005). None of these studies convincingly shows superiority of the advanced extensions over OVO and OVA. Rifkin and Klautau (2004) argue that classification accuracy obtained by these single-machine methods can be reproduced in the OVA approach by fine-tuning the base binary classifiers. Likewise, there is no clear winner in sparsity (a sparse SVM model is one with a small number of support vectors). The OVO strategy is consistently one the fastest options for training.

We do not recommend investing your time in learning these single-machine algorithms. To apply SVM to multiclass data, use one of the approaches described in Chapter 16.

13.6
Empirical Local Methods

In Sections 13.1 and 13.2, we have reviewed the theory behind kernel methods. In practice, not every kernel method subscribes to this theory. Here, we review several popular heuristic algorithms.

Kernel ridge regression, radial basis functions and support vector machines (all reviewed earlier in this chapter) are best suited for binary classification. Extending them to multiclass problems is possible but not trivial. The empirical methods discussed in this section can be easily used for an arbitrary number of classes. They are also conceptually simple. A good review of these methods can be found in Atkeson *et al.* (1997).

13.6.1
Classification by Probability Density Estimation

Probability density estimation can be used as a tool for regression or classification. A well-known example is the Nadaraya–Watson regression estimator described in Nadaraya (1964) and Watson (1964). We re-derive this estimator for classification. Although trivial, this derivation escaped the popular textbooks we know of.

As discussed in Section 9.1, the classification error is minimized by predicting into the class with the largest posterior probability $P(y|x)$. The Bayes Rule gives

$$P(y|x) = \frac{P(x|y)P(y)}{P(x)} \tag{13.67}$$

for class label y and vector of input variables x. The prior probability $P(y)$, if not specified by the analyst, can be set to the fraction of observations of class y in the training data. The marginal distribution $P(x)$ in the denominator is obtained by averaging the joint distribution $P(x, y)$ over the set of possible labels, $P(x) = \sum_{y \in \mathcal{Y}} P(x|y)P(y)$. To estimate $P(y|x)$, we need to estimate $P(x|y)$. In other words, we need to estimate the multivariate pdf of x for every class.

A probability density estimator of the form

$$P(x|y) = \frac{A \sum_{n=1}^{N} I(y = y_n) G(x, x_n)}{\sum_{n=1}^{N} I(y = y_n)} \tag{13.68}$$

has been introduced in Chapter 5. Above, $I(y = y_n)$ is an indicator function equal to 1 when y and y_n coincide and 0 otherwise. The multivariate kernel G is a product of one-dimensional kernels G_1,

$$G(x, x') = \prod_{d=1}^{D} G_1 \left(\frac{x_d - x'_d}{h_d} \right), \tag{13.69}$$

and A is a normalization constant such that the kernel $G(x, x_n)$ in (13.68) integrates to one. The constant A does not need to be estimated – the reason for this will soon be clear. The kernel width h_d in (13.69) is set for every dimension separately. Its value can be found by minimizing the asymptotic mean integrated squared error (AMISE). If the data $\{x_n\}_{n=1}^{N}$ are drawn from a D-dimensional normal density, the optimal width for a normal univariate kernel $G_1(x) = \exp(-x^2/2)$ is approximately

$$h_d^* = \sigma_d N^{-\frac{1}{D+4}}, \tag{13.70}$$

where σ_d is the standard deviation for variable d. For kernels based on the Euclidean distance, it is common to standardize data; after standardization we can set $\sigma_d = 1$. The kernels (13.69) and (13.24) are then equivalent with $\sigma = N^{-\frac{1}{D+4}}$. If the true multivariate density is not normal, the width estimate (13.70) is suboptimal. Yet it is known to provide a good approximation to a sufficiently smooth density.

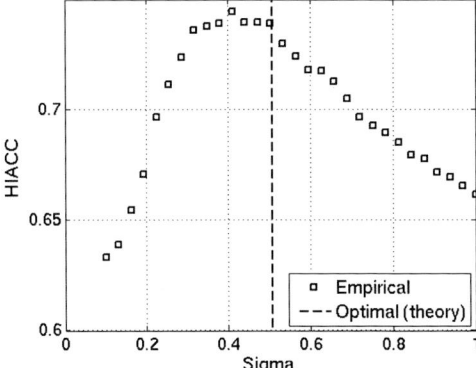

Figure 13.4 HIACC value for the MAGIC telescope data versus kernel width.

If the prior probability $P(y)$ is specified by the observation count in the respective class, $P(y) = \sum_{n=1}^{N} I(y = y_n)/N$, the joint probability is given by $P(x|y)P(y) = A \sum_{n=1}^{N} I(y = y_n) G(x, x_n)/N$. Plugging all the pieces into (13.67), we obtain

$$P(y|x) = \frac{\sum_{n=1}^{N} I(y = y_n) G(x, x_n)}{\sum_{n=1}^{N} G(x, x_n)}. \tag{13.71}$$

The normalization constant A cancels out in the numerator and denominator.

We can see now why the estimator (13.71) does not conform to the kernel theory. The coefficients α_n in (13.22) are completely defined by the training data. Here, they depend on the test point x.

The estimator (13.71) is usually less accurate than those obtained by kernel ridge regression or SVM. This estimator however can be computed quickly: if you follow the prescription (13.70) for setting the optimal kernel width, you eliminate the training step. Since the kernel width is the only optimizable parameter in this model, its optimization is more important than for other techniques reviewed in this chapter. Carrying out such an optimization increases the training time dramatically and typically produces a model still inferior to the more flexible techniques.

We plot the value of the HIACC criterion used to analyze the MAGIC telescope data against the kernel width in Figure 13.4. The width maximizing HIACC is close to the optimal theoretical value for density approximation. The best value of HIACC is well below the value obtained by any other technique in this chapter.

13.6.2
Locally Weighted Regression

In Chapter 9, we have defined the goal of classification as minimizing the expected loss (9.1), where the expectation is taken over the joint distribution $P(x, y)$. Alternatively, we could minimize the local expected loss at x,

$$L(x) = \frac{\sum_{y \in \mathcal{Y}} \int_{\mathcal{X}} \ell(y, f(x')) G(x, x') P(x', y) dx'}{\sum_{y \in \mathcal{Y}} \int_{\mathcal{X}} G(x, x') P(x', y) dx'}, \tag{13.72}$$

where $\ell(y, f)$ could be one of the loss functions in Table 9.1. Vapnik and Bottou (1993) demonstrate important theoretical properties of this approach such as consistency. The denominator is a constant completely defined by the unknown distribution $P(x', y)$ and assumed form of $G(x, x')$. In practice, we only need to minimize the numerator approximated by the empirical loss at x,

$$\hat{L}(x) = \frac{1}{N} \sum_{n=1}^{N} \ell(y_n, f(x_n)) G(x, x_n). \tag{13.73}$$

This approach amounts to building a model in the neighborhood of x by downweighting contributions from observations far from the point of interest. In this case, we do not need to compute the Gram matrix for the training data. If x is low-dimensional, regularization is typically not needed.

Let us take the familiar quadratic loss $\ell(y, f) = (y - f)^2$ and compute g, a vector with N elements for N observations in the training data. The nth element of this vector is the kernel distance between x and the respective training point, $g_n = G(x, x_n)$. Let $W = \text{diag}(g)$ be a diagonal $N \times N$ matrix with g on the main diagonal and zeros elsewhere. Do not confuse W with the Gram matrix G; the latter has plenty of nonzero off-diagonal elements! Assume a linear model $f(x) = \tilde{\beta}^T \tilde{x}$. We put the tilde sign above β and x indicating that the intercept term has been included, just like we did in Section 11.3. Now rewrite the empirical loss as

$$\hat{L}(x) = \frac{1}{N} \sum_{n=1}^{N} g_n \left(y_n - \tilde{\beta}^T \tilde{x}_n\right)^2. \tag{13.74}$$

The least-squares solution to this problem is identical to the one for weighted data. If y is a column vector of the observed response values and X is an $N \times (D + 1)$ matrix for N observations and D variables plus the intercept term, we solve the **locally weighted regression** (LWR) problem,

$$W^{1/2} y = W^{1/2} \tilde{X} \tilde{\beta}. \tag{13.75}$$

The analytical solution is given by

$$\tilde{\beta} = (\tilde{X}^T W \tilde{X})^{-1} \tilde{X}^T W y. \tag{13.76}$$

This solution can be used for regression or binary classification. Generalizing to a multiclass problem is easy. As shown in Section 11.3, for K classes the response is modeled by a vector with K elements. The multivariate regression model $f = \tilde{B}^T \tilde{x}$ is then solved analytically by putting

$$\tilde{B} = (\tilde{X}^T W \tilde{X})^{-1} \tilde{X}^T W Y \tag{13.77}$$

for an $N \times K$ response matrix Y and $(D + 1) \times K$ matrix of regression coefficients $\tilde{B}$.

LWR tends to be more accurate than classification by density estimation. It is less clear how LWR compares to kernel ridge regression. Either technique can be superior in various settings. Optimization of the expansion coefficients for kernel ridge regression is done once in a high-dimensional feature space. In contrast, the LWR coefficients are computed for every test point in the low-dimensional space of the original variables.

13.6.2.1 Example: LWR for the MAGIC Telescope Data

We demonstrate LWR using the MAGIC telescope dataset. The data are preprocessed in the same way as in the two previous examples for kernel ridge regression and RBF expansion. The HIACC criterion, the mean true positive rate at false positive rate values 0.1 and 0.2, is then computed in the same way as in the previous examples. We include code related to LWR only. The obtained HIACC value, 0.817, is worse than the one obtained by kernel ridge regression and better than the one obtained by the RBF expansion with 500 centers. In either case, the difference is not statistically significant.

Contents
- Include intercept in the linear model
- Fit the LWR model and compute its predictions.

Include Intercept in the Linear Model

We include an extra column, both in the training and test data. All elements in this column are set to 1. This is a standard trick for including an intercept term in a linear model. Later we estimate a vector of linear coefficients for the locally weighted data. The first element in this vector corresponds to the column filled with ones and represents the intercept term.

```
Xtrain = [ones(Ntrain,1) Xtrain];
Xtest = [ones(Ntest,1) Xtest];
```

Fit the LWR Model and Compute its Predictions

For every observation in the test data, we compute kernel distance to all observations in the training set. We collect square roots of the computed distances in a vector of weights, w. We then apply these weights to the training data to obtain Z, the weighted version of **X**, and v, the weighted version of **y**. To compute Z, we could take diag(w)*Xtrain, where diag(w) creates an Ntrain-by-Ntrain diagonal matrix. Instead, we use a more efficient bsxfun function to multiply every row in Xtrain by the respective element in w. To compute v, we perform elementwise multiplication of two vectors using the .* operator. Similar to the other examples in this chapter, we find the least-squares solution using the backslash operator.

```
Yfit = zeros(Ntest,1);
for n=1:Ntest
    q = Xtest(n,:);
    w = sqrt(kernelfun(Xtrain,repmat(q,Ntrain,1)));
    Z = bsxfun(@times,Xtrain,w);
    v = Ytrain.*w;
```

```
        b = Z\v;
        Yfit(n) = q*b;
    end
```

13.6.3
Nearest Neighbors and Fuzzy Rules

Nearest neighbor classification is one of the oldest and simplest algorithms. To classify a new observation, find its nearest neighbor in the training data. Then assign the class label of this nearest neighbor to the new observation. We have described 1-NN, the one nearest neighbor rule. A straightforward extension is the k-NN rule: Take k nearest neighbors and classify a new observation into the label most popular among them. We show classification boundaries obtained by 1-NN and 11-NN rules in Figure 13.5. Increasing the number of nearest neighbors smooths the class boundary.

Although we use unscaled variables for k-NN in Figure 13.5, it is common to scale input data before searching for nearest neighbors. The decision, to scale or not to scale, is ultimately yours. Obviously, you need to scale if you want all variables to be treated on equal footing.

If the training data are not weighted, the most popular class among nearest neighbors is determined by counting. If an even number of neighbors is used for two classes, we can get a tie. What happens then depends on the software implementation. The tie could be broken at random, but this would make the prediction nondeterministic. A better solution is to break the tie in favor of the class most popular in the entire data. The best solution is to avoid the ambiguity by using an odd number of nearest neighbors. If the training data are weighted, the most popular class among nearest neighbors is determined by summing weights. In this case, classification ties are rare.

Another way of breaking ties is by using the proximity information. Consider a 2-NN rule. Take a test observation and find its two nearest neighbors in the training

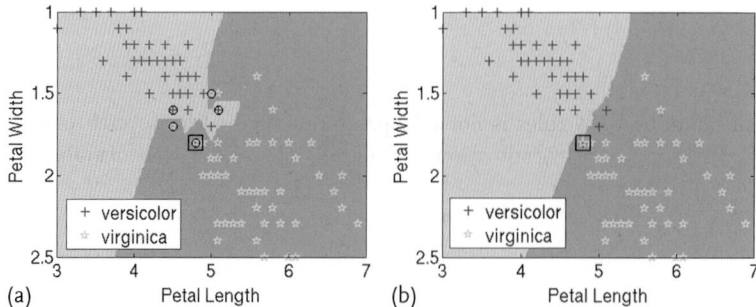

Figure 13.5 Classification of Fisher iris data by 1-NN (a) and 11-NN rules (b). Observations with nearest neighbors from the opposite class are circled in (a). The square covers one observation of class versicolor and two observations of class virginica. The k-NN rules are applied to unscaled data using the Euclidean distance.

set. Suppose this test observation is much closer to one of the two nearest neighbors than to the other. It would seem natural to treat the two neighbors as unequal for classification. For instance, if the nearest neighbor is of class A and the second nearest is of class B, it would make sense to assign the new observation to class A. Given that the second neighbor is of class B, we could lower the confidence of prediction into class A accordingly. To break ties by proximity, k-NN classification can use distance-based weighting. The weighting function, $G(x, x')$, for observations at x and x' is typically based on the Euclidean distance. Popular choices are $1/\|x - x'\|$ and $1/\|x - x'\|^2$. These choices are motivated by empirical research for k-NN classification.

If two or more observations in the training data are positioned at the same point, we may not be able to find exactly as many nearest neighbors as desired. The Fisher iris set depicted in Figure 13.5 has one observation of class versicolor and two observations of class virginica in one location shown with a square. If this point is nearest to a test observation, the minimal number of nearest neighbors for this observation is three. For classification, one neighbor out of three can be chosen at random or by a heuristic rule such as "use the neighbor with the lowest index in the training set". Alternatively, we could use the minimal unambiguous number of neighbors exceeding the desired number. For the 1-NN rule, we would use three neighbors for the test observation above and one neighbor for other test observations. The knnsearch function and the ClassificationKNN class, both available from the Statistics Toolbox in MATLAB, use the minimal unambiguous number of nearest neighbors if you set the IncludeTies parameter to 'on'.

Let us write down a formula for computing the class posterior probabilities by a k-NN rule:

$$P(k|x) = \frac{\sum_{m \in \mathcal{M}(x)} w_m G(x, x_m) I(y_m = k)}{\sum_{m \in \mathcal{M}(x)} w_m G(x, x_m)}. \tag{13.78}$$

As usual, $I(y_m = k)$ is 1 if the true class of the mth observation in the training data is k and 0 otherwise. This formula is similar to (13.71) for classification by probability density estimation, with two modifications. First, the summation is taken over the set $\mathcal{M}(x)$ of points identified as nearest neighbors of point x, whereas the summation in (13.71) is taken over the entire training set. Second, this formula includes observation weights. We could have included observation weights in (13.71), but chose not to do so for simplicity.

Classification by k-NN rules is not a proper kernel method for several reasons. First, just like the density-estimation classifier (13.71), the k-NN formula uses terms with $G(x, x_m)$ in the denominator. Second, if observation i is the nearest neighbor of observation j, the converse is not necessarily true. If we attempted to form a Gram matrix, it would not be necessarily symmetric and positive semidefinite. Third, the weighting function G may not be a proper kernel. Classification by k-NN is an empirical technique without a strict theoretical foundation.

We apply k-NN to the MAGIC telescope data. The optimal number of nearest neighbors is found by cross-validation. Then an unbiased estimate of the predictive power is estimated using an independent test set. We obtain a HIACC value of

0.73 for 50 nearest neighbors without distance weighting and 0.78 for 20 nearest neighbors with the squared inverse weighting. These values are significantly worse than those obtained by kernel ridge regression or LWR. Although k-NN classification is typically less accurate than more sophisticated techniques, it is often faster because it avoids complex fitting.

Coding a brute-force implementation of the 1-NN search is easy. Store the entire training data in memory. To find the nearest neighbor of a test observation (query point), compute distance to every point in the training set and choose the minimal one. The brute-force algorithm thus requires $O(DN)$ storage and $O(DN)$ query time for N training observations in D dimensions.

Friedman et al. (1977) propose a structure called *kd-tree* to achieve the expected $O(N)$ storage and $O(\log N)$ query time at fixed D. Similar to binary decision trees discussed in Chapter 14, a kd-tree recursively partitions the input space into boxes. Of course, the criterion used for this partitioning is different from that for a decision tree. Terminal nodes of a kd-tree are called *buckets*. To find the nearest neighbor to a query point in the training set, follow the tree from the root down to the bucket on which the query point lands. Then find the minimal distance between the query point and the training points in this bucket. Let this minimal distance be r. Then travel up to the root of the tree. In every branch (nonterminal) node, test if the sphere of radius r centered at the query point intersects the boundaries of the complementary partition. If there is no intersection, proceed to the parent node. Otherwise search the branch descending from the complementary node for nearest neighbors.

kd-trees can lead to substantial reduction of the query time for low-dimensional data. The query time depends on a factor exponential in D and quickly grows in high dimensions. We recommend using kd-trees for 10 dimensions or less as a rule of thumb.

An overview of other algorithms for the exact nearest neighbor search can be found in Arya et al. (1998).

Instead of searching for the exact nearest neighbor, could we settle for an approximate one? Suppose r is the distance between a query point and its exact nearest neighbor. Call a point that lies within $(1 + \epsilon)r$ of the query point its $(1 + \epsilon)$-approximate nearest neighbor (ANN). Arya et al. (1998) propose an algorithm to search for such approximate neighbors. Their algorithm can make use of either an improved version of the kd-tree or a new structure, *balanced box decomposition* tree. Arya et al. (1998) test their algorithm for various distributions in 16 dimensions. The query time is reduced by a factor of 10–50 for $\epsilon = 3$ relative to the exact search for most datasets. For highly clustered data, a significant reduction is observed for values $\epsilon = 1$ or smaller. In these experiments, the ANN algorithm typically finds the nearest neighbor with accuracy significantly better than specified. For example, Arya et al. (1998) report that for $\epsilon = 3$ the ratio of the approximate over exact nearest neighbor distance to the query point is at most 1.1.

The approaches described above make no use of the class information. Aha et al. (1991) propose an *IB3 algorithm* aimed at reducing the memory footprint and query time for classification. IB3 retains a subset of training data called *concept description* (CD) and discards the rest. Ideally, the CD is much smaller than the full training set and provides the same classification accuracy. IB3 starts with an empty CD set and processes training observations sequentially. We leave out details of filling out the empty set in the first iterations. Suppose the CD already contains some observations. Some of these observations are *acceptable,* and some are not. An observation is acceptable if its accuracy determined from all its nearest neighbors inspected by the algorithm so far is significantly larger than the fraction of observations of this class in the data. Given a set of observations in the CD, proceed as follows. For every new training observation, find its nearest acceptable neighbor in the CD. If this nearest neighbor has the same class as this observation, discard this observation. If the class labels are not the same, add this training observation to the CD. Update measures of acceptability for all unacceptable observations in the CD which are closer to this observation than its nearest acceptable neighbor. Remove observations whose measures of acceptability drop below a certain threshold from the CD. Continue until all training observations are processed.

Because IB3 processes training data sequentially, it is well suited for online learning. This algorithm can be used in the batch mode as well. In the latter case, observations in the training set should be shuffled at random before learning begins. It is best to try several rounds of shuffling to verify the stability of the algorithm.

Aha et al. (1991) test IB3 on real-world datasets with a few hundred observations in the training data, 7 to 21 variables, and up to 22 classes. The fraction of the training data retained in the CD in their experiments varies from 7 to 30%. IB3 gives accuracy comparable to that of the C4.5 decision tree described in Chapter 14.

Fuzzy rules are another heuristic local method. Developed in the engineering community, they exhibit characteristics of classification by density estimation discussed in Section 13.6.1 and nearest neighbor rules discussed here. The main distinction between fuzzy rules and the other two methods is the choice of a multivariate kernel function, or *membership function.* For instance, one popular choice for univariate analysis is a triangular function with support $|x - x'| < L$,

$$G_1(x, x') = \begin{cases} 0 & \text{if } |x - x'| \geq L \\ 1 - |x - x'|/L & \text{if } |x - x'| < L \end{cases}. \tag{13.79}$$

The membership function in D dimensions can be then defined as one of

$$G(x, x') = \min_{d=1,\dots,D} G_1\left(x_d, x'_d\right)$$

$$G(x, x') = \sum_{d=1}^{D} G_1\left(x_d, x'_d\right)$$

$$G(x, x') = \prod_{d=1}^{D} G_1\left(x_d, x'_d\right) . \qquad (13.80)$$

A review of fuzzy networks can be found in Abe (2001).

13.7
Kernel Methods: The Good, the Bad and the Curse of Dimensionality

Kernel methods are a powerful classification tool. Their application is limited by the number of observations in the training set, or equivalently by the dimensionality in the mapped feature space. For small to medium-size datasets, various kernel classifiers can be trained within reasonable time. In our experience, kernel ridge regression tends to provide superb accuracy for the carefully chosen kernel function and regularization parameter. Optimization is trickier for large datasets. An RBF method with a thoughtful scheme for selecting the expansion centers can reduce the required CPU and memory dramatically. SVM can find a sparse solution with an excellent predictive power and low memory requirements. Working-set algorithms such as SMO and ISDA are the most popular approach to solving large-scale SVM problems.

In this chapter, we have only considered continuous variables with smooth densities. Variables with outliers, multiple modes, or semi-discrete values present serious challenges to the kernel algorithms.

If data contain nominal variables, the discussed kernel methods cannot be applied without special preprocessing steps. If all variables in the data are nominal, you can use Hamming distance, defined as the number of unequal values between the two observations. *Value difference metric* (VDM) described in Stanfill and Waltz (1986) is a more accurate approach for computing pairwise distances in all-nominal data. An extension of this method, *heterogeneous value difference metric* (HVDM), described in Wilson and Martinez (1997) can be used for a mixture of continuous and nominal variables.

Most physics analyses apply classification tools to low-dimensional data. Application of advanced algorithms to dozens or hundreds of inputs is an emerging trend. We mention the curse of dimensionality in this book in a few places. We discuss it here in the context of local methods.

To demonstrate the curse of dimensionality in classification, we use two classes in a cube bounded between -1 and $+1$ in each dimension. Each observation is represented by a D-dimensional vector $x = \{x_d\}_{d=1}^{D}$. Observations with $x_1 > 0$ are labeled as class A, and observations with $x_1 \leq 0$ are labeled as class B. We

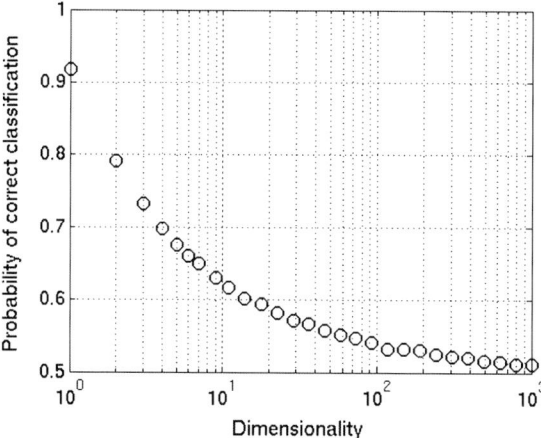

Figure 13.6 Probability of correct classification by the 1-NN rule for points uniformly distributed in a hypercube bounded between −1 and +1 in each dimension.

run 10 000 simulated experiments. In every experiment, we generate three observations, two of class A and one of class B. We treat the first observation from class A and the observation from class B as training data and classify the second observation from class A using the one nearest neighbor (1-NN) rule. We plot the classification accuracy against the number of dimensions in Figure 13.6. The curve descends to 0.5 indicating that the predictive power of the 1-NN rule is poor.

Earlier we explained the power of kernel methods through their ability to map data into a high-dimensional space. Now we are arguing that classifying data in high dimensions can be rather inaccurate. How can we solve this controversy?

The key ingredient of a kernel method is its numeric algorithm. Finding the optimal plane of separation in high-dimensional data can be easy with the right tool. In the chosen example, the linear boundary could be found by a linear SVM, linear regression or discriminant analysis in combination with straightforward variable selection. Algorithms that allow for fitting coefficients a_n in the kernel expansion (13.22) tend to be more powerful than simple-minded techniques such as nearest neighbor rules or classification by probability density estimation. Keeping this in mind, we state here a modified version of

Conjecture 13.4. *The curse of dimensionality: The kernel expansion (13.22) with simple-minded estimation of the coefficients a_n tends to break down in high dimensions.*

13.8
Exercises

1. The $1/\|x - x'\|$ kernel is positive semidefinite (one way to demonstrate this is by observing that electrostatic energy of an arbitrary continuous charge distribution must be nonnegative). Using this kernel in practice could be difficult

for datasets with pairs of observations separated by zero or small distance. How would you modify the kernel for practical computations? Would the kernel remain positive semidefinite after these modifications?
2. Compute the ridge regression coefficients in (13.28) for large λ. Assume that λ is much greater than the magnitude of any element in $\mathbf{X}^T\mathbf{X}$.
3. In Section 13.4.1, we set the RBF expansion centers to observations closest to the cluster centroids found by k-means. Consider using the cluster centroids for the expansion centers. Assume that G is a positive semidefinite kernel. Are matrices $\tilde{\mathbf{G}}^T\tilde{\mathbf{G}}$ and $\mathbf{C}$ in (13.33) guaranteed to be positive semidefinite?
4. Prove that the distance between the two hyperplanes defined by $y(\boldsymbol{\beta}^T\mathbf{x} + \beta_0) = 1$ for $y \in \{-1, +1\}$ in Figure 13.3 equals $2/\|\boldsymbol{\beta}\|$.
5. In Section 11.1 we introduced linear discriminant analysis (LDA) as a technique for separating two classes with multivariate normal distributions and equal covariance matrices. Show that the LDA model (11.4) is equivalent to the SVM model (13.45) with kernel $G(\mathbf{x}, \mathbf{x}') = \mathbf{x}^T\mathbf{A}\mathbf{x}'$ and all coefficients α_n equal. Express $\mathbf{A}$, α, and β_0 in terms of the class means and covariance matrices.
6. Modify the estimator (13.71) for the case when the class prior probabilities $P(y)$ are not set by observation counts in the training set, that is, $P(y) \neq \sum_{n=1}^{N} I(y = y_n)/N$.
7. Explain why multiplying both sides of (13.75) by $\mathbf{W}^{-1/2}$ and taking $\tilde{\boldsymbol{\beta}}$ to be the solution to $\mathbf{y} = \tilde{\mathbf{X}}\tilde{\boldsymbol{\beta}}$ is not equivalent to solving (13.76).

References

Abe, S. (2001) *Pattern Classification: Neuro-fuzzy Methods and their Comparison*, Springer.

Aha, D., Kibler, D., and Albert, M. (1991) Instance-based learning algorithms. *Mach. Learn.*, 6, 37–66.

Arya, S., Mount, D., Netanyahu, N., Silverman, R., and Wu, A. (1998) An optimal algorithm for approximate nearest neighbor searching fixed dimensions. *J. ACM*, 45 (6), 891–923.

Atkeson, C., Moore, A., and Schaal, S. (1997) Locally weighted learning. *Artif. Intell. Rev.*, 11, 11–73.

Bach, F., Heckerman, D., and Horvitz, E. (2006) Considering cost asymmetry in learning classifiers. *J. Mach. Learn. Res.*, 7, 1713–1741.

Bock, R., Chilingarian, A., Gaug, M., Hakl, F., Hengstebeck, T., Jirina, M., Klaschka, J., Kotrc, E., Savicky, P., Towers, S., Vaiciulis, A., and Wittek, W. (2004) Methods for multidimensional event classification: a case study using images from a Cherenkov gamma-ray telescope. *Nucl. Instrum. Methods A*, 516, 511–528.

Bordes, A., Bottou, L., Gallinari, P., and Weston, J. (2007) Solving Multiclass Support Vector Machines with LaRank, Proceedings of the 24th International Conference on Machine Learning (ICML), 20–24 June 2007, Corvallis, ACM, New York.

Cox, D. and O'Sullivan, F. (1990) Asymptotic analysis of penalized likelihood and related estimators. *Ann. Stat.*, 18 (4), 1676–1695.

Crammer, K. and Singer, Y. (2001) On the algorithmic implementation of multiclass kernel-based vector machines. *J. Mach. Learn. Res.*, 2, 265–292.

Cristianini, N. and Shawe-Taylor, J. (2000) *Support Vector Machines and other kernel-based learning methods*, Cambridge University Press.

Fan, R.E., Chen, P.H., and Lin, C.J. (2005) Working set selection using second order information for training support vector machines. *J. Mach. Learn. Res.*, **6**, 1889–1918.

Friedman, J., Bentley, J., and Finkel, R. (1977) An algorithm for finding best matches in logarithmic expected time. *ACM Trans. Math. Softw.*, **3**, 209–226.

Girosi, F., Jones, M., and Poggio, T. (1993) Priors, stabilizers and basis functions: from regularization to radial, tensor and additive splines, *A.I. Memo 1430*, MIT.

Hastie, T., Tibshirani, R., and Friedman, J. (2008) Linear methods for regression, *The Elements of Statistical Learning*, Springer, pp. 61–67, 2nd edn.

Hoerl, A. and Kennard, R. (2000) Ridge regression: Biased estimation for nonorthogonal problems. *Technometrics*, **42** (1), 80–86.

Hsu, C.W. and Lin, C.J. (2002) A comparison of methods for multiclass support vector machines. *IEEE Trans. Neural Netw.*, **13** (2), 415–425.

Huang, T.M., Kecman, V., and Kopriva, I. (2006) *Kernel Based Algorithms for Mining Huge Datasets, Studies in Computational Intelligence*, vol. 17, Springer.

Joachims, T. (1999) Making large-scale svm learning practical, in *Advances in Kernel Methods – Support Vector Learning* (eds B. Schölkopf, C. Burges, and A. Smola), MIT Press.

Kimeldorf, G. and Wahba, G. (1971) Some results on Tchebycheffian spline functions. *J. Math. Anal. Appl.*, **33** (1), 82–96.

Lee, Y., Lin, Y., and Wahba, G. (2004) Multicategory support vector machines: Theory and application to the classification of microarray data and satellite radiance data. *J. Am. Stat. Assoc.*, **99** (465), 67–82.

Mercer, J. (1909) Functions of positive and negative type, and their connection with the theory of integral equations. *Philos. Trans. R. Soc. A*, **209**, 415–446.

Moler, C. (2008) *Numerical Computing with MATLAB*, SIAM. http://www.mathworks.com/moler/ (accessed 24 July 2013).

Nadaraya, E. (1964) On estimating regression. *Theory Probab. Appl.*, **9** (1), 141–142.

Nocedal, J. and Wright, S. (2006) *Numerical Optimization*, Springer Series in Operations Research, Springer, 2nd edn.

Osuna, E., Freund, R., and Girosi, F. (1997) An improved training algorithm for support vector machines. *IEEE Workshop Neural Netw. Signal Process.*, pp. 276–285.

Platt, J. (1998) Sequential minimal optimization: A fast algorithm for training support vector machines, *Tech. Rep. MSR-TR-98-14*, Microsoft Research.

Platt, J. (1999) Probabilistic outputs for support vector machines and comparisons to regularized likelihood methods, in *Advances in Large Margin Classifiers*, MIT Press, pp. 61–74.

Platt, J., Cristianini, N., and Shawe-Taylor, J. (2000) Large margin DAGs for multiclass classification. *Adv. Neural Inf. Process. Syst.*, **12** (3), 547–553.

Poggio, T., Mukherjee, S., Rifkin, R., Rakhlin, A., and Verri, A. (2001) b, AI Memo 2011-011, MIT.

Rifkin, R. and Klautau, A. (2004) In defense of one-vs-all classification. *J. Mach. Learn. Res.*, **5**, 101–141.

Scholkopf, B. and Smola, A. (2002) *Learning with Kernels: Support Vector Machines, Regularization, Optimization and Beyond*, Adaptive Computation and Machine Learning, The MIT Press.

Seber, G. and Lee, A. (2003) *Linear Regression Analysis*, Wiley Series in Probability and Statistics, Wiley-Interscience, 2nd edn.

Smola, A., Scholkopf, B., and Muller, K.R. (1998) The connection between regularization operators and support vector kernels. *Neural Netw.*, **11**, 637–649.

Stanfill, C. and Waltz, D. (1986) Toward memory-based reasoning. *Commun. ACM*, **29**, 1213–1228.

Statnikov, A., Aliferis, C., Tsamardinos, I., Hardin, D., and Levy, S. (2005) A comprehensive evaluation of multicategory classification methods for microarray gene expression cancer diagnosis. *Bioinformatics*, **21** (5), 631–643.

Tibshirani, R. and Hastie, T. (2007) Margin trees for high-dimensional classification. *J. Mach. Learn. Res.*, **8**, 637–652.

Tsochantaridis, I., Joachims, T., Hofmann, T., and Altun, Y. (2005) Large margin methods for structured and interdependent output variables. *J. Mach. Learn. Res.*, **6**, 1453–1484.

Vapnik, V. (1998) *Statistical Learning Theory*, John Wiley & Sons.

Vapnik, V. and Bottou, L. (1993) Local algorithms for pattern recognition and dependencies estimation. *Neural Comput.*, **5** (6), 893–909.

Watson, G. (1964) Smooth regression analysis. *Indian J. Stat. A*, **26** (4), 359–372.

Weston, J. and Watkins, C. (1999) Support vector machines for multi-class pattern recognition, in *Eur. Symp. Artif. Neural Netw.*, vol. 4, pp. 219–224.

Wilson, D. and Martinez, T. (1997) Improved heterogeneous distance functions. *J. Artif. Intell. Res.*, **6**, 1–34.

14
Decision Trees

Decision tree is one of the oldest tools for supervised learning. A decision tree typically partitions the observable space $\mathcal{X}$ into disjoint boxes. Specifics of this partitioning can vary greatly across tree algorithms. An example of such partitioning obtained by the CART algorithm is shown in Figure 14.1. First, the tree splits the data at `PetalWidth`=1.75. All but one observation with `PetalWidth` values above this cut are from class virginica – one cross is hiding under the square. Observations with `PetalWidth` values below the cut are split further at `PetalLength`=4.95. Finally, the node with `PetalWidth` below 1.75 and `PetalLength` below 4.95 is split in two producing a leaf with most versicolor points and a small leaf with one observation of class virginica.

Trees gained popularity in the 1970s, although related research originated at least as early as the late 1950s. Many types of decision trees have been proposed in the last four decades. We only mention a few popular implementations.

The *CART* (classification and regression tree) algorithm is described in Breiman *et al.* (1984). Its commercial version is sold by Salford Systems. The `tree` package in R and the `ClassificationTree` class available from the Statistics Toolbox in MATLAB follow the book closely.

The ID3 package, introduced in Quinlan (1986), evolved into *C4.5* described in Quinlan (1993). The code for C4.5 is available on diskette with a copy of the book. The J48 package in Weka is an open-source Java implementation of the C4.5 algorithm. RuleQuest sells a commercial package, C5.0, and distributes a limited version of the C source code without charge.

Kass (1980), laying his work on earlier research on automatic interaction detection, proposes *CHAID* (chi-squared automatic interaction detection). A commercial version is included in the Statistica suite sold by StatSoft. A recent R-Forge project under the same name provides a free version.

Below we explain the CART algorithm and compare it, when appropriate, to C4.5. Kohavi and Quinlan (2002) offer a comparative summary of the two frameworks. A more complete list of tree algorithms can be found in Rokach and Maimon (2008).

Decision tree is a well-known tool, and we anticipate that most readers would be somewhat familiar with the subject. In physics analysis, decision tree is usually applied in its most basic form. After reviewing this basic functionality, we introduce advanced material such as tree pruning, surrogate splits and estimation of

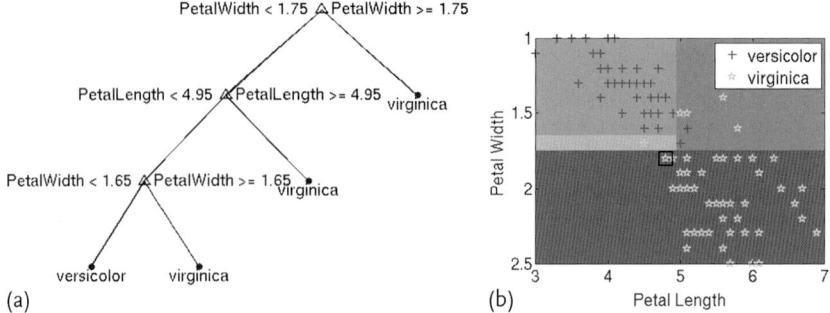

Figure 14.1 A decision tree with 4 leaves grown on Fisher iris data: (a) Tree structure, and; (b) node rectangles in the space of input variables. The square covers one observation of class versicolor and two observations of class virginica.

variable importance. We also discuss splits on categorical variables and variables with missing values; these subjects can be omitted at first reading.

14.1
Growing Trees

Take a training set of N observations $\pmb{x}_n$, each described by D input variables, with class labels y_n. Place the entire set in the root node of a tree. Partition the root node into left and right subsets, L and R, by assigning every observation to one of the two subsets. Apply the same partitioning procedure to each of the two new nodes. Continue splitting the nodes until a stopping criterion is satisfied.

What are the optimal subsets L and R? Assume that all variables are continuous. Search for splitting rules of the form

$$\pmb{x} \in L \quad \text{iff} \quad x_d < s$$
$$\pmb{x} \in R \quad \text{iff} \quad x_d \geq s$$

for $d = 1, \ldots, D$ and all possible values of s. Then select the rule maximizing a certain split criterion. Use this optimal rule to divide the node in two subsets.

What are the possible values of s? If a continuous variable takes M distinct values in this node, we can divide this node along this variable in at most $M-1$ ways. The exact scheme for dividing $\{v_m\}_{m=1}^M$ values into $M-1$ subsets is an implementation choice. For instance, we could place splits between every two adjacent distinct values, $s = (v_m + v_{m+1})/2;\ m = 1, \ldots, M-1$.

What could be a good split criterion? This question is key to successful tree construction. Before we discuss criteria for evaluating the splitting rules, let us introduce some notation.

Let t_0 be the parent node. The left and right child nodes are then denoted by t_L and t_R. Let N_0, N_L and N_R be the number of training observations in the parent, left, and right node, respectively. The probability of a node is defined by a fraction of training observations landing on this node, $P(t) = N_t/N$; $t \in \{0, L, R\}$. Here and below, N is the total number of observations in the training set. (All formulas in this section use observation counts but can easily accommodate observation weights.) The root node contains all observations, and its probability is one.

For simplicity, consider the case of two classes, A and B. If node t contains N_A observations of class A and N_B observations of class B, we estimate the class posterior probabilities by putting $P(A|t) = N_A/N_t$, and similarly for B. The sum of posterior probabilities equals one. A tree node predicts into the class with the largest posterior, and therefore its training error is $\epsilon(t) = \min_{y \in \{A,B\}} P(y|t)$.

If a tree node contains observations of one class only, it is called *pure* and its training error is zero. We prefer a tree with pure nodes because it would confidently separate the classes. At the same time, we do not want nodes with low probabilities $P(t)$ because a leafy tree would likely overfit the data. Growing a tree with big pure nodes may not be possible because the class distributions overlap. To choose the best split, we need a measure of impurity.

Let us define *node impurity* as a function of class probabilities,

$$i(t) = \phi(P(A|t), P(B|t)), \tag{14.1}$$

and set bounds on this function, $0 \le \phi(p,q) \le 1/2$. The impurity is maximal when the two classes are mixed in equal proportion and minimal when the node is pure. Given the imposed bounds, we require $\phi(0,1) = \phi(1,0) = 0$ and $\phi(1/2, 1/2) = 1/2$. A two-class impurity measure would be naturally symmetric, $\phi(p,q) = \phi(q,p)$.

A good decision split should minimize the impurity. Above, we have defined impurity for one node. A binary split produces two. Intuitively, we should minimize the average impurity for the two children. The two nodes can differ in size, and for averaging it would be natural to weight their impurities by the node probabilities. Introducing the weighted node impurity,

$$I(t) = P(t)i(t), \tag{14.2}$$

we write down the average impurity after the split as $I(t_L) + I(t_R)$. Finally, we define *impurity gain*,

$$\Delta I = I(t_0) - I(t_L) - I(t_R). \tag{14.3}$$

The best splitting rule can be found by maximizing the impurity gain ΔI over all possible splits for all variables. If there are no splits with positive gain, the node cannot be split and becomes a *leaf*, or *terminal* node. Split nodes are called *branch* nodes.

The last missing piece is a specific form of the impurity function ϕ. A common goal of classification is minimization of the classification error. It is tempting to choose the best split as the one attaining the minimal misclassification. Why not use the node error $\epsilon(t)$ for the node impurity $i(t)$?

As it turns out, the classification error is not a good impurity measure. Consider an example from Breiman et al. (1984). Imagine a node with 400 observations of

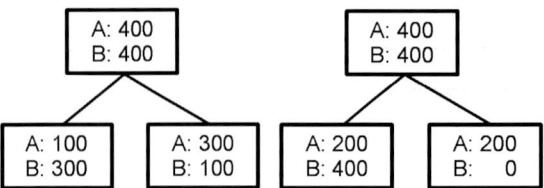

Figure 14.2 Two splits producing equal classification errors.

class A and 400 observations of class B. Suppose we find two splits shown in Figure 14.2. Either split produces 25% error. The first split gives two impure nodes, and in the second case only one node is impure. It appears that the second split is more likely to produce a tree with big pure nodes. Yet the chosen impurity measure treats the two splits equally.

This problem is caused by linearity of the node error. Indeed, if A is the minority class in the left child t_L, the weighted node error $P(A|t_L)P(t_L)$ equals $N_A^{(L)}/N$, where $N_A^{(L)}$ is the number of observations from class A in t_L. Similarly, if B is the minority class in the right child t_R, its weighted node error is $N_B^{(R)}/N$. The average error for the two child nodes is then N_{mis}/N, where $N_{\text{mis}} = N_A^{(L)} + N_B^{(R)}$ is the total number of misclassifications in the two nodes. We can move misclassified observations between the nodes without changing the average error. To penalize the tree for creating impure nodes, we need a more aggressive, nonlinear impurity measure. This nonlinear measure would have to be a concave function of $P(A|t)$.

One simple choice is a quadratic function called *Gini diversity index*:

$$\phi(p,q) = 1 - p^2 - q^2. \tag{14.4}$$

Another popular choice is *deviance*, or *cross-entropy*:

$$\phi(p,q) = -\frac{p \log_2 p + q \log_2 q}{2}. \tag{14.5}$$

As usual, for two classes with posterior probabilities p and q, we have $p + q = 1$. Classification error, Gini index, and cross-entropy are shown in Figure 14.3.

Many other measures of impurity have been proposed. A good fraction of them are reviewed in Rokach and Maimon (2008). None of the proposed measures universally outperforms others in classification accuracy. In practice, choosing an appropriate stopping rule and pruning level is usually more important than choosing an impurity measure. Stopping rules are reviewed in the next section. Pruning is discussed in Section 14.4.

The Gini and cross-entropy criteria were introduced in research predating decision trees. Breiman *et al.* (1984) apply these impurity measures to tree construction. Quinlan (1986) motivates the use of cross-entropy for decision splits by a heuristic argument. Quinlan and Rivest (1989) derive the cross-entropy splitting criterion from principles rooted in information theory.

The two impurity criteria can produce splits in which both child nodes are dominated by the same class. Consider, for example, a tree grown on the MAGIC telescope data described in Bock *et al.* (2004). We apply `ClassificationTree` from

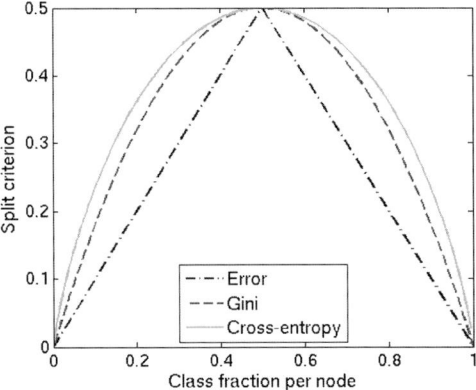

Figure 14.3 Impurity measure versus posterior probability for one class in binary classification.

the Statistics Toolbox in MATLAB to two thirds of the data randomly selected for training. The tree, shown in Figure 14.4, finds splits by utilizing the Gini index and is not allowed to produce leaf nodes with fewer than 1500 observations. We count the tree nodes top to bottom and left to right: the root node is assigned index 1, the left and right nodes at the level next to the root are assigned indices 2 and 3, respectively, the four nodes at the next level are assigned indices 4–7, and so on. The root node is split using the fAlpha variable. This splits sends most showers produced by high-energy gamma rays left (to node 2) and most hadronic showers right (to node 3). Splitting node 2 further appears at first sight unproductive: all showers in the branch off node 2 are classified as gammas. Is the tree wrong?

Let us take a closer look at node 2. This node contains 6209 gamma and 1348 hadronic showers, and its probability is 0.593. The Gini index weighted by the node probability equals 0.174. The left child has 4144 gamma and 481 hadronic showers; the respective numbers for the right child are 2065 and 867. The weighted Gini indices for the left and right child nodes are 0.0676 and 0.0958. The impurity gain due to this split is therefore 0.0104. This split divides the distribution of fAlpha in two intervals. In the first interval, hadronic and gamma showers are mixed in proportion 1 : 9, and in the second interval this proportion is close to 3 : 7. The

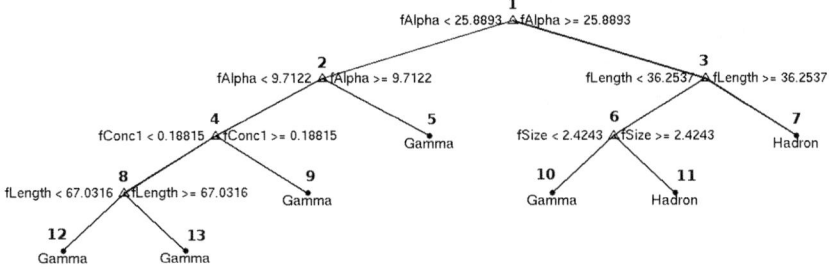

Figure 14.4 MATLAB decision tree for the MAGIC telescope data. A leaf node is required to have at least 1500 observations.

tree classifies observations landing on node 4 as gammas more confidently than observations landing on node 5.

14.2
Predicting by Decision Trees

Put an observation down the tree. Find the leaf node on which this observation lands. Assign this observation to the class with the largest posterior probability in this node. The posterior probability for class k in node t is estimated by

$$\hat{P}(k|t) = \frac{\frac{\pi_k N_k(t)}{N_k}}{\sum_{i=1}^{K} \frac{\pi_i N_i(t)}{N_i}} . \tag{14.6}$$

Above, π_k is the prior probability for class k, $N_k(t)$ is the observation count for class k in node t, and N_k is the total number of training observations in class k. When the prior probabilities are derived from the training data, $\pi_k = N_k/N$ for the data size N, (14.6) takes a particularly simple form. This formula easily generalizes to weighted observations.

14.3
Stopping Rules

If we split data recursively using the Gini index or cross-entropy, we can grow a deep tree with small leaves. For example, a split sending one or more observations of the minority class into a pure child node is guaranteed to produce a positive impurity gain for two classes. Most usually, such a split can be found for data with many continuous variables. A leafy tree can attain a low training error, but its generalization (test) error is often large. To avoid growing deep and inaccurate trees, we need to impose a stopping criterion.

Here is a list of stopping criteria found in the literature. A tree node cannot be split if:

1. The node is pure, that is, all observations originate from one class.
2. The size of this node is less than the allowed minimal size of a parent node.
3. Any split with a positive impurity gain produces child nodes with sizes less than the allowed minimal size of a child node.
4. The maximal tree depth has been reached.
5. The maximal number of leaf nodes has been reached.
6. The largest possible impurity gain is below a certain threshold.

"Node size" above means "number of observations" in this node.

Alternatively, we could grow a leafy tree and then remove branches with poor classification accuracy. Reducing a fully grown tree is called *pruning*. A popular belief is that trees produced by pruning tend to be more accurate than trees produced by applying a conservative stopping rule. We discuss pruning in the next section.

14.4 Pruning Trees

A tree is pruned by replacing its branch with the node at the root of this branch. Take the tree in Figure 14.4. Let us prune nodes 12 and 13, the two gamma leaf nodes at the bottom of the tree produced by a cut on fLength. To prune them, eliminate these two nodes from the tree and turn node 8, their parent, into a leaf. If we wanted to prune again, we could eliminate the new leaf and its gamma sibling (node 9) and turn node 4 into a leaf. Then we could prune away nodes 4 and 5 and turn node 3 into a leaf. The resulting tree is shown in Figure 14.5.

How do we decide if a branch can be pruned? For example, is it a good idea to eliminate nodes 12 and 13 from the tree in Figure 14.4? This question can be answered in a number of ways.

Pruning is a form of regularization. A numerical model is often regularized by modifying its objective to include a term penalizing for complexity. This approach to pruning trees is taken in Breiman *et al.* (1984). Before describing this approach, we introduce a new term, *risk*, and define the goal of learning a tree as *risk minimization*. In Section 14.1, we have discussed several specific forms of risk such as classification error and various impurity measures. Here, we use the term "risk" to include them both. Risk is similar to classification loss introduced in Section 9.1.

Let $r(t)$ be a measure of risk associated with node t. For example, we could set the node risk $r(t)$ to the classification error in this node weighted by the node probability: $r(t) = P(t)\epsilon(t)$. The risk for tree T is a sum of risk values over its leaf nodes, $L(T)$:

$$r(T) = \sum_{t \in L(T)} r(t). \tag{14.7}$$

The risk estimate obtained by using training, or resubstitution, data is biased low and most usually approaches zero as the tree depth grows. In contrast, the risk estimated using test data attains minimum at some optimal tree depth and increases

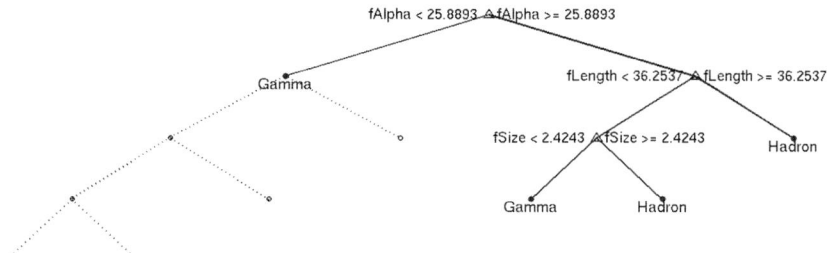

Figure 14.5 Tree shown in Figure 14.4 with 3 leaf nodes pruned away.

as the tree grows deeper. To mimic the behavior of the risk estimate on test data, we add a penalty term to the resubstitution risk. If the tree complexity is measured by counting its leaf nodes, a penalized form of risk is given by

$$\tilde{r}(T) = r(T) + \alpha |L(T)|, \qquad (14.8)$$

where $|L(T)|$ is the number of leaf nodes in tree T and $\alpha \geq 0$ is the penalty coefficient. An example for the MAGIC telescope data is shown in Figure 14.6. The optimal tree $T^*(\alpha)$ is obtained by minimizing the penalized risk at fixed α.

If we set $\alpha = 0$, we do not impose any penalty on the tree complexity. The optimal tree $T^*(0)$ is either identical to the full tree or can be obtained from the full tree by removing splits that do not lower the tree risk. As we increase α in (14.8), we reduce the size of the optimal tree $T^*(\alpha)$. Eventually, for sufficiently large α, the optimal tree is reduced to the root node. The *optimal pruning sequence* is a set of trees ordered by parameter α. If $\alpha_1 \leq \alpha_2$, optimal tree $T^*(\alpha_1)$ contains $T^*(\alpha_2)$ as a subtree. A rigorous proof of this proposition can be found in Breiman *et al.* (1984).

Let us now construct the optimal pruning sequence for the tree in Figure 14.4. At the zeroth pruning level, we put $\alpha_0 = 0$. At this level, we can only prune away splits that do not improve the training error. If child nodes produced by a split are dominated by different classes, this split does improve the training error. Child nodes sharing the majority class give the same training error as their parent. We can prune away nodes 12 and 13, and consequently nodes 9 and 5, as shown in Figure 14.5, because these four leaves are dominated by the gamma class. In Section 14.1, we argued that using the Gini index or cross-entropy as the splitting criterion produces better trees. Now we are removing splits obtained by minimizing the Gini index because they do not improve the training error. Are we contradicting ourselves?

There is one simple reason why using an impurity measure for splitting and the training error for pruning is sensible. Nodes split by an impurity criterion could

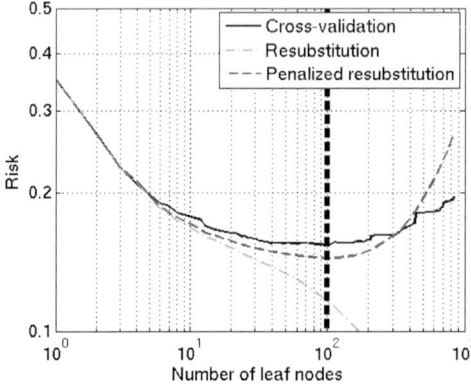

Figure 14.6 Cross-validated, training, and penalized training error versus number of leaf nodes for the MAGIC telescope data. The plots are obtained for $\alpha = 2.7 \times 10^{-4}$. The vertical line indicates the minimum of the cross-validated error.

be split further into nodes dominated by different classes. This did not happen for the tree in Figure 14.4 because splitting was stopped by the requirement on the minimal leaf size (1500 observations). When we split node 2, for instance, we did not know that all descendants of this node, namely nodes 4, 5, 8, 9, 12 and 13, would have gamma as the majority class. Now we have been made aware of this fact, and we are attempting to undo the unnecessary work.

We have obtained the optimal tree $T^*(\alpha_0)$ at the zeroth pruning level. Let us now find the smallest value of α for the first pruning level, α_1. Suppose a branch T is pruned and replaced by its root node t. Pruning is allowed if the penalized risk for the branch exceeds that of its root. The critical value α^* at which the two risk values are equal is found from

$$r(t) + \alpha^* = r(T) + \alpha^* |L(T)|. \tag{14.9}$$

For α^*, we obtain

$$\alpha^* = \frac{r(t) - r(T)}{|L(T)| - 1}. \tag{14.10}$$

The tree in Figure 14.5 has 3 branch nodes: 1, 3, and 6. We show their risk values in Table 14.1. Node 6 exhibits the lowest risk improvement due to branching. Substituting the numbers for node 6 into (14.10), we obtain $\alpha_1 = 8.8 \times 10^{-3}$. To prune the tree to the first level, $T^*(\alpha_1)$, we eliminate nodes 10 and 11. Node 6 becomes a leaf. At the second pruning level, we eliminate nodes 6 and 7. At the third level, we prune away nodes 2 and 3 reducing the tree to its root.

We have constructed the optimal pruning sequence. But how do we know to what level the tree should be pruned?

One approach is to choose the optimal pruning level by minimizing the classification error measured on an independent test set. This is the best course of action if there are enough data for training and testing. If the amount of data is limited, we can use cross-validation. Here, we have to deal with a subtle problem: the optimal pruning sequence can vary across the folds.

Suppose we use 10-fold cross-validation. We grow 10 trees, $\{T_k\}_{k=1}^{10}$, each on nine tenths of the data, and construct the optimal pruning sequence for each tree. There is no guarantee that every T_k tree is going to have as many pruning levels or the same $\{\alpha_m\}_{m=0}^{M}$ sequence as T, the tree grown on the entire data. If the number

Table 14.1 Risk analysis of branch nodes in Figure 14.4 at the zeroth pruning level. Node risk is the node error multiplied by the node probability. Branch risk is the sum over risk values for all leaves descending from this node.

Node	Node risk $r(t)$	Number of descendant leaves $|L(T)|$	Branch risk $r(T)$	$\frac{r(t)-r(T)}{|L(T)|-1}$
1	0.3516	4	0.2175	0.0447
3	0.1611	3	0.1118	0.0247
6	0.1056	2	0.0968	0.0088

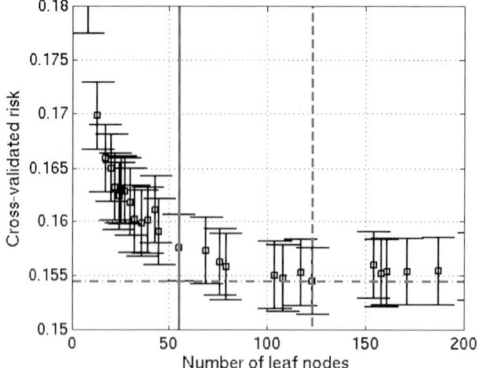

Figure 14.7 Cross-validated error with 1σ confidence bounds versus number of leaf nodes for the MAGIC telescope data. The vertical dash and horizontal dash-dot lines indicate the minimum of the cross-validation error. The vertical solid line shows the minimal tree size at which the lower 1σ bound stays below the minimal error.

of folds is sufficiently large, we can assume that every T_k is reasonably close to T. For simplicity, let us assume that the maximal pruning level for every T_k and for tree T is M. Compute the optimal $\{\alpha_m\}_{m=0}^{M}$ sequence for tree T. Suppose $T^*(\alpha_m)$ is the optimal tree at the mth pruning level. Because α_m is chosen as the *minimal* α allowing removal of some leaf nodes at pruning level m, the trees in the folds are not necessarily brought to the mth pruning level by α_m. That is, $T_k^*(\alpha_m)$ may be pruned to level $m-1$, not m.

To work around this issue, Breiman et al. (1984) propose a heuristic correction. Note that $T^*(\alpha)$ is the optimal tree at the mth pruning level for any α in the interval $\alpha_m \leq \alpha < \alpha_{m+1}$. Instead of pruning every fold to α_m, we can prune to the geometric average, $\alpha'_m = \sqrt{\alpha_m \alpha_{m+1}}$. Although $T_k^*(\alpha'_m)$ is not guaranteed to be pruned to the mth level, it is pruned to the desired level more often than $T_k^*(\alpha_m)$. To compute the cross-validated error at the mth pruning level, follow the usual procedure discussed in Section 9.4.1: Apply the pruned tree $T_k^*(\alpha'_m)$ in every fold to the one tenth of the data not used for training this tree, count misclassified observations across all folds, and divide the total number of misclassifications by the number of observations in the entire data.

As seen from Figure 14.6, the minimum of the cross-validation error is rather flat. We could make the tree substantially smaller at the expense of losing some, but not much accuracy. One way to accomplish this is by using a *one standard error rule* described in Breiman et al. (1984). As discussed in Section 9.3, the number of misclassifications in a fixed-size sample can be modeled by a binomial random variable. Confidence bounds on the classification error are then obtained using the estimated variance. The one standard error rule states that we can prune the tree to the level at which the lower risk bound is less or equal to the minimal risk. This technique is illustrated in Figure 14.7. The smallest cross-validated error, 0.1545, for the MAGIC telescope data is obtained by a tree with 123 leaves. The same tree pruned to 55 leaves gives cross-validated error 0.1576 ± 0.0031, consistent with the smallest possible value. The lower 1σ bound for the tree pruned to the next level (45 leaves) is above 0.1545. Therefore the tree must be pruned to 55 leaves.

Above, we have described the pruning procedure adopted in Breiman et al. (1984). C4.5 uses a different approach. Start from the bottom of the tree. Choose a specific branch node. Compute a 75% upper limit on the classification error for this node assuming a binomial distribution. Compute two similar upper limits – one for the aggregated prediction from all leaves descending from this node and another one for its most probable sub-branch. If the most probable sub-branch gives the smallest upper limit, replace the node by this sub-branch. If the node itself gives the smallest upper limit, prune away all children. If the sum of all descendant leaves gives the smallest upper limit, do not prune. If the node is pruned, continue to the top of the tree. Refer to Quinlan (1993) for further detail.

We conclude the section on pruning trees with a MATLAB example.

14.4.1
Example: Pruning a Classification Tree

Contents
- Grow a deep tree
- Find the optimal pruning level by cross-validation.

We use the MAGIC telescope data from the UCI repository http://archive.ics.uci.edu/ml/datasets/MAGIC+Gamma+Telescope These are simulated data for detection of high energy gamma particles in a ground-based atmospheric Cherenkov gamma telescope using the imaging technique. The goal is to separate gamma particles from hadron background.

We split the data into training and test sets in the proportion 2 : 1 and code class labels as a numeric vector: $+1$ for gamma and -1 for hadron. In this example, we use only the training set. Matrix Xtrain has about 13 000 rows (observations) and 10 columns (variables). Vector Ytrain holds class labels.

```
load MagicTelescope;
Ytrain = 2*strcmp('Gamma',Ytrain)-1;
size(Xtrain)

ans =
        12742          10
```

Grow a Deep Tree
We use the ClassificationTree class available from the Statistics Toolbox in MATLAB. By default, ClassificationTree.fit uses at least 10 observations for branch nodes and imposes no restriction on the leaf size.

```
tree = ClassificationTree.fit(Xtrain,Ytrain);
```

The tree is deep as evidenced by the number of nodes.

```
tree.NumNodes

ans =
        1651
```

The tree is pruned by default. The PruneList property of the tree object shows pruning levels for tree nodes. If the pruning level of a node is L, this node can be removed when the tree is pruned to level L + 1. The pruning levels range from 0 (no pruning) to the maximal level at which the tree is reduced to its root.

```
min(tree.PruneList)

ans =
     0

max(tree.PruneList)

ans =
    67
```

Find the Optimal Pruning Level by Cross-Validation

We search for the optimal pruning level using 10-fold cross-validation. The result could be sensitive to how exactly the data are partitioned in 10 folds. We set the seed for the random number generator to let the reader reproduce our results exactly.

The cvLoss method of the tree object returns a cross-validated estimate of the classification error E, its standard deviation SE based on the binomial approximation, number of leaf nodes at each pruning level nLeaf, and the best pruning level BestMin chosen by the "minimal risk" rule. The quantities E, SE, and nLeaf are vectors with max(tree.PruneList)+1 elements. By default, cvLoss uses 10-fold cross-validation.

```
rng(1);
[E,SE,nLeaf,BestMin] = cvLoss(tree,'subtrees','all','treesize','min');
```

Reset the random seed to partition the data into the same cross-validation folds. Then obtain the best pruning level using the "one standard error" rule. There is no need to recompute E, SE and nLeaf because they are fixed by the data partition.

```
rng(1);
[~,~,~,BestSE] = cvLoss(tree,'subtrees','all','treesize','se');
```

Show the pruning level, number of leaves and cross-validated error obtained by the "minimal risk" rule. Offset the level by 1 because the levels are counted from zero.

```
BestMin
```

```
BestMin =
    40
```

```
nLeaf(BestMin+1)
```

```
ans =
   123
```

```
E(BestMin+1)
```

```
ans =
   0.1545
```

14.5
Trees for Multiple Classes

So far, we have focused on trees for binary classification. What if we have 3 or more classes?

The impurity criteria (14.4) and (14.5) are easily generalized to an arbitrary number of classes. The optimal binary split is found, as before, by maximizing the impurity gain (14.3).

We could also use the formalism developed in Section 14.1 for binary classification by partitioning the available classes in two groups, or superclasses. Suppose we have K classes. Partition them in two mutually exclusive subsets, C and $\bar{C}$, and set p and q in (14.4) to be their probabilities. Compute the impurity gain as if computing it for two classes. Find the optimal split by maximizing the impurity gain over all possible class partitions and all possible splits. Trying all possible class partitions can be computationally demanding. Fortunately, there is a simple shortcut. Breiman *et al.* (1984) prove that maximizing the two-group Gini index over all class partitions is equivalent to maximizing the *twoing* criterion,

$$\Delta I_{\text{twoing}} = \frac{P(t_L)P(t_R)}{4}\left[\sum_{k=1}^{K}|P(k|t_L) - P(k|t_R)|\right]^2. \qquad (14.11)$$

This criterion does not require looping over class partitions. Its computational cost is the same as for the Gini index.

The twoing criterion is useful for finding groups of similar classes near the top of a tree. At the bottom of the tree, it isolates individual classes. Consider a hypothetical example. Take four classes, A, B, C and D, with 100 observations in each. Set variables as shown in Table 14.2.

Variable 1 separates class A from the rest, variable 2 separates classes A and B from classes C and D, and variable 3 separates class D from the rest. Trees grown on these data by using the Gini index and twoing criterion are shown in Figure 14.8.

The first split in the Gini tree produces a pure node with class A and an impure node with the three other classes. The first split in the twoing tree produces two impure nodes, first with classes A and B and second with classes C and D. The twoing

Table 14.2 Variables for the four-class problem.

Class	x1	x2	x3
A	0	0	0
B	1	0	0
C	1	1	0
D	1	1	1

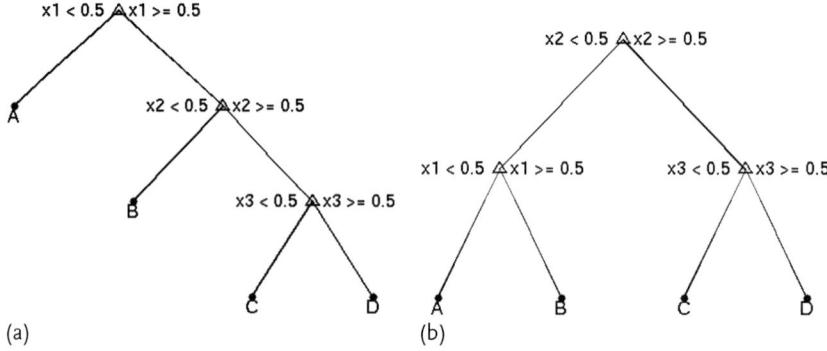

(a) (b)

Figure 14.8 Tree grown by using the Gini index (a) and twoing criterion (b) for the data in Table 14.2.

criterion thus prefers a split partitioning the four classes in two large groups, and the Gini index prefers a split producing a smaller pure node. Both trees achieve perfect class separation, but the twoing tree is more balanced. The twoing criterion is most useful when you have many classes. If we had 3 classes in the example above, the Gini and twoing trees would be identical.

14.6
✂ Splits on Categorical Variables

As mentioned in Section 7.1, categorical variables can be ordinal and nominal. A decision tree treats ordinal variables in the same way as it treats numeric (continuous) variables. Decision boundaries are orthogonal to the coordinate axes and therefore insensitive to the distance metric in the multivariate space.

Nominal variables can present a challenge. C nominal categories can be partitioned in two subsets in 2^C ways. Eliminating redundant and useless partitions, we reduce this number to $2^{C-1} - 1$. The number of possible splits grows exponentially with the number of categories.

For binary classification, we can use a computational shortcut. Take category c and count observations from both classes in this category, $N_A^{(c)}$ and $N_B^{(c)}$. Define probability of class A in category c, $P_A^{(c)} = N_A^{(c)}/(N_A^{(c)} + N_B^{(c)})$. Sort probabilities

$\{P_A^{(c)}\}_{c=1}^C$ in ascending order. Breiman *et al.* (1984) prove that the optimal split can be obtained from the sorted list of categories. That is, if we search through all $2^{C-1} - 1$ possible splits to find the one with the maximal impurity gain, this split divides the categories in two subsets, t_L and t_R, such that the probability for class A for any category from the left subset is less or equal to the probability for class A for any category from the right subset: $\forall i \in t_L$ and $\forall j \in t_R$ must have $P_A^{(i)} \leq P_A^{(j)}$. Instead of searching through $2^{C-1} - 1$ splits, we therefore need to consider only $C - 1$.

There is no similar rigorous shortcut for three or more classes. Heuristic techniques for binary splits are described in Mehta *et al.* (1996); Chou (1991), and Coppersmith *et al.* (1999). Another heuristic technique is outlined below. Let K be the number of classes. Choose one class to be class A and group all other classes into class B. Define probability of class A in category c in the same way as for binary classification. Sort probabilities $\{P_A^{(c)}\}_{c=1}^C$ in ascending order and inspect $C - 1$ splits defined by the sorted list. Now label another class as A and put all other classes into B and repeat. After every class has been used as class A, find the best split among the $K(C - 1)$ examined partitions. We have had good experience with this technique for small K but are not aware of any paper describing it.

The difficulty of the problem increases with the number of categories. Coppersmith *et al.* (1999) investigate the performance of their algorithm for $K \leq 9$ and $C \leq 12$. At the largest values of these parameters, their technique fails to find the true best partition in 15% of simulated experiments. The failure rate goes up with the number of classes. For $C = 15$ and $K = 100$, we observe that all techniques mentioned here fail to find the optimal partition at least 50% of the time. Depending on the values of K and C and the optimized criterion (Gini, cross-entropy or twoing), any technique could be the winner. The success rate can be improved by combining two or possibly more techniques. The algorithm due to Chou (1991) performs quite poorly for large C and $K \gg C$.

These heuristic techniques are specific to the CART algorithm. C4.5 can impose multiway splits and is therefore immune to this problem.

14.7
Surrogate Splits

At every step of tree construction, a node is partitioned using the best split out of all possible splits on all variables. The chosen split (*primary split*) uses one variable (*primary split variable*) and ignores the rest, even when the other variables produce splits as powerful as the best one. This loss of information can have negative consequences.

Imagine, for instance, growing a tree on data with two identical variables. The best splits on the two variables always have equal impurity gains. Selection of the optimal split is then determined by the implementation. For example, we could always take the first variable out of two producing equal gains. The second variable would then be never split on and deemed useless. Suppose the constructed tree is used to predict class for a new observation. If the first variable is missing for this observation, the tree may not be able to produce an accurate response. If the second variable is not missing, it could be used for prediction. Unfortunately, the tree does not know that the second variable is identical to the first because it has not saved this information.

This loss of information can be mitigated by using *surrogate splits*. Surrogate splits are best splits on variables other than the primary split variable. A regular tree memorizes one split at every branch and therefore saves B splits for B branch nodes. A tree with surrogate splits memorizes the best split for each variable at every branch and saves at most BD splits for B branch nodes and D input variables. Some surrogate splits may be poor and do not deserve to be memorized. This is why the total number of saved splits can be less than BD.

The best surrogate split on a variable is not equivalent to the best split on this variable. The best split maximizes the impurity gain. The best surrogate split maximizes correlation with the primary split. Take a specific tree node. First, find the best split s^* out of all possible splits on all variables (primary split). Suppose this split is imposed on variable d^*. Then for every variable d; $d = 1, \ldots, D$; $d \neq d^*$; find the best surrogate split s_d maximizing correlation between s_d and s^*. Discard surrogate split s_d if the correlation between s^* and s_d is too small. Save the primary split s^* and up to $D-1$ surrogate splits in the tree.

Why are we interested in surrogate splits maximizing correlation with the primary split? Surrogate splits are used as substitutes for the primary split, not for independent partitioning. After the tree is grown, we commit to its unique structure. Let an observation land on a branch node in this tree. If the primary split variable for this observation is missing, the primary split cannot send this observation left or right. We can then conjecture where the primary split would send this observation (if the missing value were available) using the saved surrogate information. If the surrogate splits were obtained with disregard to the primary split, we would not know how to make such a conjecture. Effectively, we would have many trees in one. Treatment of missing data by surrogate splits is discussed further in the next section.

Correlation between two splits is called *predictive association* and computed as follows. Suppose the primary split s^* partitions a branch node in two children, t_L^* and t_R^*. Similarly, surrogate split s_d divides this branch node into $t_L^{(d)}$ and $t_R^{(d)}$. Let $P_L^{(d)} = P(t_L^* \cap t_L^{(d)})/P(t_0)$ be the probability of sending an observation left by both best split s^* and surrogate split s_d; similarly for $P_R^{(d)}$. The probabilities $P_L^{(d)}$ and $P_R^{(d)}$ are estimated relative to the branch node t_0 and sum to one if the two splits have identical effect. The predictive measure of association between splits s^* and s_d,

$$\lambda_d = \frac{\min\left(P\left(t_L^*\right), P\left(t_R^*\right)\right) - \left(1 - P_L^{(d)} - P_R^{(d)}\right)}{\min\left(P\left(t_L^*\right), P\left(t_R^*\right)\right)}, \tag{14.12}$$

ranges from $-\infty$ to $+1$. If the predictive measure of association with the best split is negative, the surrogate split can be discarded.

We demonstrate the practical usefulness of surrogate splits in the next two sections.

14.8
✂ Missing Values

Treatment of missing values has been reviewed in Section 7.2, where techniques for decision trees have been briefly mentioned. Here, we describe these techniques in detail. We focus on handling of missing values in the training data first and then discuss handling of missing values for prediction on new (test) data.

What is the best split on a variable with missing values? Consider a continuous variable. If its value is below a split threshold, send this observation left; otherwise send it right. If the value is missing, this observation cannot be sent either left or right. If such observations were included in the parent node, t_0, but not in the child nodes, t_L and t_R, the sum of the child node probabilities would be less than the probability of the parent node, $P(t_L) + P(t_R) < P(t_0)$. The splitting criterion (14.3) would then favor splits on variables with many missing values. In contrast, we should prefer splits on variables with few or no missing values because

1. They are based on more observations and are therefore more reliable.
2. They are more likely to be useful for prediction if the pattern of missingness in the test data is similar to the one in the training data.

How can we redefine the splitting criterion?

Suppose node t_0 is split three ways: some observations go left, some go right, and some are placed in a special child node called *unsplit data*. The impurity gain (14.3) can then be rewritten as

$$\Delta I_{\text{3-way}} = P(t_0)i(t_0) - P(t_U)i(t_U) - P(t_L)i(t_L) - P(t_R)i(t_R) \qquad (14.13)$$

for nodes $t \in \{t_0, t_L, t_R, t_U\}$ with probabilities $P(t)$ and impurities $i(t)$. Subscript "U" above denotes the unsplit data. If the pattern of missingness does not correlate with the class label, the expectations for $i(t_0)$ and $i(t_U)$ are equal. Setting $i(t_U) = i(t_0)$, we effectively choose to keep observations with missing values in node t_0. Noting that $P(t_0) - P(t_U) = P(t_L) + P(t_R)$, we obtain a modified impurity criterion for binary splits:

$$\Delta I_{\text{mis}} = \left[P(t_L) + P(t_R)\right] i(t_0) - P(t_L)i(t_L) - P(t_R)i(t_R). \qquad (14.14)$$

The twoing criterion (14.11) favors large child nodes and does not need to be redefined.

The new criterion finds more productive splits for data with missing values. Yet if many variables had many missing values, a typical split would produce small child nodes. The tree construction would then stop early because the tree would run out of training data. To grow a nonshallow tree, we need a way to split observations with missing values.

Breiman et al. (1984) deal with this problem by using surrogate splits. Suppose s^* is the best split found either by the corrected impurity criterion (14.14) or by twoing. This split is imposed on variable d^*. For every variable d; $d = 1, \ldots, D$; $d \neq d^*$; find the best surrogate split s_d maximizing the predictive measure of association λ_d between s_d and s^*. This predictive measure of association is computed for every s_d using observations without missing values for variable d^* and variable d. Keep surrogate splits with positive λ_d and discard the rest. Sort the computed values λ_d in descending order. Split observations with missing values for d^* using the surrogate split with the largest λ_d. If some observations remain unsplit because they have missing values both for variable d^* *and* the variable with the largest λ_d, try the surrogate split with the second largest λ_d. Continue until all observations in the branch node are sent either left or right or until you run out of surrogate splits.

Quinlan (1993) does not use surrogate splits. Instead, he sends a fraction of an observation with a missing value in d^* down to every child node. This fraction is proportional to the number of observations with known values for variable d^* sent to the respective child node.

What about prediction? Send an observation with missing values down the tree. Suppose this observation travels to a branch node in which the best split is imposed on a missing variable. If the tree has been grown without surrogate splits, we could stop here. The predicted label and class posterior probabilities for this observation would be then obtained from this branch node. Alternatively, we could send this observation down to all branches originating at this node and set the tree prediction to a weighted sum of predictions from all leaves reached by this observation. This approach is taken in Quinlan (1993). If the tree holds surrogate-split information, this observation can be sent to one of the child nodes using the surrogate split with the largest λ_d. If variable d for the largest λ_d is also missing, use the variable with the second largest λ_d, and so on. This approach is taken in Breiman et al. (1984).

The algorithms for training and prediction described above can be used for categorical variables in a similar fashion. For training, use surrogate splits to split the data. If the test data have categorical values not found in the training data, treat them as missing and split them using surrogate splits.

We have reviewed sophisticated approaches to missing values. If your tree implementation does not provide these algorithms, you can try a few simple-minded techniques. One approach is to remove any observation with at least one missing value from the training data. In this case, the tree would still need to know how to handle missing values in the test data. Another approach is to replace all missing values in the training and test data by a value outside the allowed range. Both approaches can work well if the amount of missing data is small.

14.9
Variable importance

Techniques for estimation of variable importance will be reviewed in Chapter 18. Here, we discuss a popular approach specific to decision trees.

The importance of a single decision split can be estimated using the impurity gain (14.3) or twoing criterion (14.11). It would be natural then to measure the importance of a variable by summing the importance values over all splits imposed on this variable. Divide this sum by the number of branch nodes in the tree to set the importance of a variable to the impurity gain averaged over all splits. The same logic applies to the twoing criterion.

Surrogate splits could be included in this calculation as well. If a variable is used for splitting a node, its importance increases by the value of the impurity gain. If a variable is not used for splitting this node, its importance increases by the value of the impurity gain produced by the best surrogate split on this variable.

What measure of variable importance should you prefer – the one computed with surrogate splits or the one without? The answer depends on the goal of your analysis. If you wish to find a minimal set of variables producing an accurate classifier, you should estimate the importance using best splits only. If you wish to find a set of all variables correlating with the class label, include surrogate splits.

In addition, you can use surrogate splits to estimate the degree of correlation between any two variables. Let Λ be a $D \times D$ matrix of variable associations. To compute element λ_{ij} of this matrix, sum the predictive measure of association (14.12) between splits on variables i and j over branch nodes for which variable i is the best splitting variable. Divide this sum by the number of such branch nodes. Set the diagonal elements of Λ to 1. Use the obtained asymmetric matrix to identify pairs of strongly associated variables.

We illustrate application of surrogate splits using the BaBar particle identification (PID) data. Bevan *et al.* (2013) use measurements recorded in several detector subsystems to identify four particle types: proton p, kaon K, pion π, and electron e. We focus on K/π separation for tracks above 3 GeV. Because we consider a subset of the classification problem addressed in Bevan *et al.* (2013), we effectively reduce the set of useful variables. We apply trees to about 16 000 training and 16 000 test observations with pion and kaon samples mixed roughly in equal proportion. These data have 31 variable, real or integer. About 3% of observations have a missing value for at least one variable.

We use the `ClassificationTree` class available from the Statistics Toolbox in MATLAB to grow two trees, one with and one without surrogate splits. The configurations for both trees are chosen by minimizing the cross-validated classification error. The error is then measured using an independent test set. The variable importance values estimated with and without surrogate splits are shown in Figure 14.9. The likelihood ratio of the kaon hypothesis over the electron hypothesis, `likeKvsEle`, dominates the rest of the variables. Indeed, the classification test error for the tree grown on this variable alone, 9.3%, is close to the test error for the tree grown on all variables, 8.5%. If we used the importance estimates obtained without surrogate splits, we could conclude that all other variables are barely relevant. Surprisingly, eliminating `likeKvsEle` does not lead to any loss of accuracy – the respective tree shows the same 8.5% test error. The reason becomes clear when we inspect the estimates of variable importance obtained with surrogate splits. We

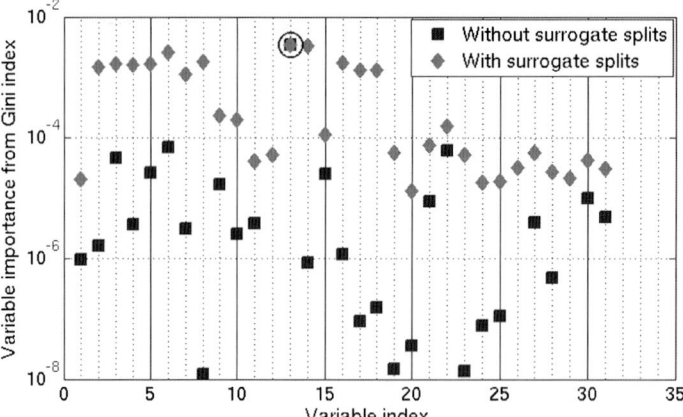

Figure 14.9 Variable importance with and without surrogate splits for the BaBar K/π particle identification data with track momenta above 3 GeV. The values are obtained by averaging the impurity gain for every variable over all branch nodes. The most important variable, likeKvsEle, is shown with a circle. The two importance estimates for this variable are equal.

observe the presence of a dozen of other powerful variables which compensate for the loss of the most powerful variable.

Using surrogate splits, we identify several pairs of strongly associated variables. We show three such pairs in Table 14.3 including the pairwise predictive measure of association for all six selected variables. The strong association for the selected pairs is not at all surprising given the physics nature of the measured variables. At least one variable in every pair with a high measure of association can likely be removed from the data without loss of accuracy.

Table 14.3 Variable associations estimated using surrogate splits for some variables in the BaBar K/π PID data. Large values are shown in boldface. The variables are: dE/dx for the kaon and proton hypotheses measured in the drift chamber (dEdxdchPullk and dEdxdchPullp), energy deposited in the electromagnetic calorimeter (ecal), energy deposited in the electromagnetic calorimeter divided by the estimated particle momentum (ecaldivp), and likelihood ratios for the proton over electron and proton over pion hypotheses (likeProvsEle and likeProvsPi).

	dEdxdchPullk	dEdxdchPullp	ecal	ecaldivp	likeProvsEle	likeProvsPi
dEdxdchPullk	1.0000	**0.8054**	0.0154	0.0205	0.0866	0.0945
dEdxdchPullp	**0.8135**	1.0000	0	0.0002	0.0145	0.0191
ecal	0.0218	0.0187	1.0000	**0.7763**	0.0268	0.0225
ecaldivp	0.0185	0.0190	**0.8854**	1.0000	0.0000	0.0062
likeProvsEle	0.1008	0.0840	0.0630	0.1932	1.0000	**0.7914**
likeProvsPi	0.1034	0.0460	0.0115	0.0115	**0.9540**	1.0000

14.10
Why Are Decision Trees Good (or Bad)?

Trees offer a number of advantages over other classification algorithms:

- Trees can easily handle data with numeric or categorical variables, or both types mixed. Other classifiers typically require preprocessing of categorical variables such as "dummifying" discussed in Section 7.1.
- Trees are resilient to the curse of dimensionality discussed in Sections 5.8 and 13.7. One-dimensional decision splits are not sensitive to the choice of the distance metric in the variable space.
- Trees can efficiently handle missing data. Simple approaches work fairly well on variables with small fractions of missing values. Advanced techniques such as surrogate splits are among the best known recipes for attacking missing data.
- A single shallow tree is highly interpretable. It can be easily visualized and understood.

Trees are not without flaws:

- A single tree can offer competitive classification accuracy only if the class boundaries are orthogonal (or close to orthogonal) to the coordinate axes. Tree implementations splitting on linear combinations of variables exist but such splits tend to be rather unstable (see below). Even if you have such an implementation, for many datasets splits on linear combinations would not add much power to orthogonal splits.
- A tree can be unstable: the best split variable in any branch can change if the tree is grown on a slightly different version of the training set. This can occur, for instance, when there are two variables with similar predictive power. Injection of a small amount of new training data or re-simulation of the training set can arbitrarily change the choice of one variable over the other.

At present, trees are used for data analysis in two very different regimes. The strengths and weaknesses mentioned above refer to a single tree. In addition, tree is a popular choice for ensemble learning.

A single tree is either grown to a deep level and then pruned to a configuration with only a few nodes or grown to a configuration with only a few nodes in one pass. The final configuration is determined by the desired balance between interpretability and accuracy. The imposed splits are then studied and justified by the analyst.

In contrast, a tree grown in an ensemble can be rather inaccurate and noninterpretable. Individual trees in an ensemble do not need to be closely inspected by the analyst. The predictive power of an ensemble is rooted in efficient collaboration of its members. Ensemble learning is reviewed in Chapter 15.

14.11
Exercises

1. Consider replacing the Gini index (14.4) with $\phi(p,q) = A(1-p^a-q^a)$, where A is a constant such that $\phi(1/2, 1/2) = 1/2$. How would increasing the value of a above 2 affect tree construction? Consider the case $a \to \infty$.

2. When the class prior probabilities are derived from the training set, $\pi_k = N_k/N$, (14.6) for prediction simplifies to
$$\hat{P}(k|t) = \frac{N_k(t)}{\sum_{i=1}^{K} N_i(t)}.$$
Consider a pure node in which only one class has a nonzero observation count. The estimated posterior probabilities for this node would be 1 for one class and 0 for all other classes. This estimate may be hard to use in some applications; for instance, it would not be possible to take an inverse logit transformation, $\log(\hat{P}(k|t)/(1 - \hat{P}(k|t)))$. This problem can be cured by the Laplace correction,
$$\hat{P}_{\text{Laplace}}(k|t) = \frac{1 + N_k(t)}{K + \sum_{i=1}^{K} N_i(t)}.$$
Modify the corrected estimate to account for arbitrary prior probabilities π_k. Then include observation weights.

3. Consider identifying four particle types, e, μ, K and π, by a decision tree. Consider the root node of the tree. Suppose the optimal split found by the Gini index would separate one particle type from the other three and the optimal split found by twoing would separate two particle types from the other two (we describe a scenario similar to the one captured in Figure 14.8). Under what circumstances would you prefer the split obtained by one criterion or the other?

4. To derive (14.14), we assume that the pattern of missingness does not correlate with the class label. Using this assumption, we leave unsplit observations t_U in the branch node t_0. Consider the case when this assumption does not hold, that is, $E[i(t_U)] \neq E[i(t_0)]$. How would you modify the split criterion (14.14)? Where would you put unsplit observations t_U if the tree is grown without surrogate splits?

5. In Section 14.9, we analyze variable importance for the BaBar PID data. A tree grown without surrogate splits finds one variable that is much more important than the rest. In contrast, a tree grown with surrogate splits finds a dozen important variables. Consider the opposite scenario for some hypothetical data. A tree grown without surrogate splits finds a dozen important variables. A tree grown with surrogate splits considers one variable out of this dozen important and the rest relatively unimportant. How would you interpret this result?

References

Bevan, A. *et al.* (ed.) (2013) *Physics of the B Factories*, Springer (to be published in 2013).

Bock, R., Chilingarian, A., Gaug, M., Hakl, F., Hengstebeck, T., Jirina, M., Klaschka, J., Kotrc, E., Savicky, P., Towers, S., Vaicilius, A., and Wittek, W. (2004) Methods for multidimensional event classification: a case study using images from a Cherenkov gamma-ray telescope. *Nucl. Instrum. Methods A*, **516**, 511–528.

Breiman, L., Friedman, J., Stone, C., and Olshen, R. (1984) *Classification and Regression Trees*, Chapman & Hall.

Chou, P. (1991) Optimal partitioning for classification and regression trees. *IEEE Trans. Pattern Anal. Mach. Intell.*, **13** (4).

Coppersmith, D., Hong, S., and Hosking, J. (1999) Partitioning nominal attributes in decision trees. *Data Min. Knowl. Discov.*, **3**, 197–217.

Kass, G. (1980) An exploratory technique for investigating large quantities of categorical data. *Appl. Stat.*, **29** (2), 119–127.

Kohavi, R. and Quinlan, J. (2002) Decision-tree discovery, in *Handbook of Data Mining and Knowledge Discovery* (eds W. Klosgen and J. Zytkow), Oxford University Press, Chapter 16.1.3, pp. 267–276.

Mehta, M., Agrawal, R., and Rissanen, J. (1996) SLIQ: A fast scalable classifier for data mining, CiteSeerX, doi:10.1.1.42.5335.

Quinlan, J. (1993) *C4.5: programs for machine learning*, Morgan Kaufmann Publishers Inc., San Francisco, CA.

Quinlan, J. (1986) Induction of decision trees. *Mach. Learn.*, **1**, 81–106.

Quinlan, J. and Rivest, R. (1989) Inferring decision trees using the minimum description length principle. *Inf. Comput.*, **80**, 227–248.

Rokach, L. and Maimon, O. (2008) *Data mining with decision trees: Theory and applications*, Machine perception and artificial intelligence, vol. 69, World Scientific.

15
Ensemble Learning

The expression *ensemble learning* refers to a broad class of classification and regression algorithms operating on many learners. Every learner in an ensemble is typically *weak*; by itself it would predict on new data quite poorly. By aggregating predictions from its weak learners, the ensemble often achieves an excellent predictive power. The number of weak learners in an ensemble usually varies from a few dozen to a few thousand. One of the most popular choices for the weak learner is decision tree. However, every classifier reviewed in this book, so far, can be used in the weak learner capacity.

If we repeatedly trained the same weak learner on the same data, all learners in the ensemble would be identical and the ensemble would be as poor as the weak learner itself. Data generation (induction) for the weak learner is crucial for ensemble construction. The weak learner can be applied to the induced data independently, without any knowledge of the previously learned weak models, or taking into account what has been learned by the ensemble participants, so far. The training data can be then modified for consecutive learners. The final step, undertaken after all weak learners have been constructed, is aggregation of predictions from these learners.

Approaches to ensemble learning are quite diverse and exercise a variety of choices for data induction and modification, weak learner construction and ensemble aggregation. The field of ensemble learning is comprised of algorithms derived from different principles and methodologies. In this chapter, we review a few well-known ensemble algorithms. By necessity we omit many others.

Ideas that laid foundation to ensemble learning can be traced to decades ago. The first practical ensemble algorithms came out in the mid 1990s. The first convincing application of an ensemble technique to particle physics is particle identification by boosted trees at MiniBOONE (Yang et al., 2005). At the time of this writing, ensemble learning has become fairly popular in the particle and astrophysics communities, although its applications have been mostly confined to AdaBoost and random forest. We aim to broaden the arsenal of the modern physicist with other algorithms.

Statistical Analysis Techniques in Particle Physics, First Edition. Ilya Narsky and Frank C. Porter.
©2014 WILEY-VCH Verlag GmbH & Co. KGaA. Published 2014 by WILEY-VCH Verlag GmbH & Co. KGaA.

15.1
Boosting

A popular class of ensemble methods is boosting. Algorithms in this class are sequential: the sampling distribution is modified for every weak learner using information from the weak learners constructed earlier. This modification typically amounts to adjusting weights for observations in the training set, allowing the next weak learner to focus on the poorly studied region of the input space.

In this section, we review three theoretical approaches to boosting, namely minimization of convex loss by stagewise additive modeling, maximization of the minimal classification margin, and minimization of nonconvex loss. The AdaBoost algorithm, which has gained some popularity in the physics community, can be explained in several ways. In one interpretation, AdaBoost works by minimizing the (convex) exponential loss shown in Table 9.1. In another interpretation, AdaBoost achieves high accuracy by maximizing the minimal margin. These two interpretations gave rise to other successful boosting algorithms, to be discussed here as well. Boosting by minimizing nonconvex loss can succeed in the regime where these two approaches fail, that is, in the presence of significant label noise. We focus on binary classification and review multiclass extensions in the concluding section.

15.1.1
Early Boosting

An algorithm introduced in Schapire (1990) may be the first boosting algorithm described in a journal publication. Assume that you have an infinite amount of data available for learning. Assume that the classes are perfectly separable, that is, an oracle can correctly label any observation without prior knowledge of its true class. Define your objective as constructing an algorithm such that, when applied to independent chunks of data drawn from the same parent distribution, learns a model with generalization error at most ϵ at least $100(1 − \delta)\%$ of the time. As usual, generalization error is measured on observations not seen at the learning (training) stage.

At the core of this algorithm, there is a simple three-step procedure which works as follows. Draw a training set. Learn the first classifier on this set. Form a new training set with observations correctly classified and misclassified by the first classifier mixed in equal proportion. Learn the second classifier on this set. Make a third training set by drawing new observations for which the first and second classifiers disagree. Learn the third classifier on this set. The three training sets must be sufficiently large to provide the desired values of ϵ and δ. To predict the label for a new observation using the learned three-step model, put this observation through the first two classifiers. If they agree, assign the predicted class label. If they do not agree, use the label predicted by the third classifier. This completes one three-step pass.

Algorithm 1 AdaBoost for two classes. Class labels y_n are drawn from the set $\{-1, +1\}$. The indicator function I equals one if its argument is true and zero otherwise.

Input: Training data $\{x_n, y_n, w_n\}_{n=1}^{N}$ and number of iterations T
1: **initialize** $w_n^{(1)} = \frac{w_n}{\sum_{n=1}^{N} w_n}$ for $n = 1, \ldots, N$
2: **for** $t = 1$ **to** T **do**
3:    Train classifier on $\{x_n, y_n, w_n^{(t)}\}_{n=1}^{N}$ to obtain hypothesis $h_t : x \mapsto \{-1, +1\}$
4:    Calculate training error $\epsilon_t = \sum_{n=1}^{N} w_n^{(t)} I(y_n \neq h_t(x_n))$
5:    **if** $\epsilon_t == 0$ **or** $\epsilon_t \geq 1/2$ **then**
6:      $T = t - 1$
7:      break loop
8:    **end if**
9:    Calculate hypothesis weight $\alpha_t = \frac{1}{2} \log \frac{1-\epsilon_t}{\epsilon_t}$
10:   Update observation weights $w_n^{(t+1)} = \frac{w_n^{(t)} \exp[-\alpha_t y_n h_t(x_n)]}{\sum_{n=1}^{N} w_n^{(t)} \exp[-\alpha_t y_n h_t(x_n)]}$
11: **end for**
Output: $f(x) = \sum_{t=1}^{T} \alpha_t h_t(x)$

Learn the classifiers in this three-step procedure recursively, using the same three-step procedure. At every level of recursion, relax the desired value of ϵ forcing it to gradually approach $1/2$. When ϵ gets sufficiently close to $1/2$, stop and return a weak learner whose generalization error is below ϵ at this recursion level (and therefore barely below $1/2$).

This algorithm is mostly of theoretical value. It relies on construction of the first and second classifier in the three-step scheme with known accuracies, but in practice these are not known. A remarkable accomplishment of Schapire's paper is the proof of concept. To construct a model with an arbitrarily small classification error, all you need is a classifier assigning correct labels to a little more than half of the data!

15.1.2
AdaBoost for Two Classes

It took a few years after Schapire's paper to develop successful practical algorithms. Proposed in Freund and Schapire (1997), *AdaBoost* (adaptive boosting) earned a reputation of a fast and robust algorithm with an excellent accuracy in high-dimensional problems. AdaBoost and similar boosting methods were named "the best available off-the-shelf classifiers" in Breiman (1998). We show this famous algorithm here with a simple modification to account for the initial assignment of observation weights (see Algorithm 1).

AdaBoost initializes observation weights to those in the input data and enters a loop with at most T iterations. At every iteration, AdaBoost learns a weak hypothesis h_t mapping the space of input variables $\mathcal{X}$ onto the class set $\{-1, +1\}$. The hypothesis is learned on weighted data, and the weights are updated at every pass; hence, every time AdaBoost learns a new weak hypothesis. After learning a new weak hypothesis, AdaBoost estimates its error by summing the weights of observations misclassified by this hypothesis. AdaBoost then computes the weight for this hypothesis using line 9 in Algorithm 1. The hypothesis weight α_t approaches $+\infty$ as the error approaches 0, and α_t approaches 0 as the error approaches $1/2$. If the error is not in the $(0, 1/2)$ interval, AdaBoost cannot compute α_t and exits. If the error is in the allowed range, AdaBoost multiplies the weights of the misclassified observations by $(1 - \epsilon_t)/\epsilon_t$ and the weights of the correctly classified observations by the inverse of this expression. Then AdaBoost renormalizes the weights to keep their sum at one. The soft score $f(x)$ returned by the final strong hypothesis is a weighted sum of predictions from the weak hypotheses. The predicted class is $+1$ if the score is positive and -1 otherwise.

15.1.2.1 Example: Boosting a Hypothetical Dataset

Let us examine step by step how AdaBoost separates two classes in the hypothetical dataset we used in Chapter 13. We show these steps in Figure 15.1. The weak learner for AdaBoost is a decision stump (a tree with two leaves) optimizing the Gini index. Before learning begins, each observation is assigned a weight of $1/7$. The first stump separates the two lower stars from the other five observations. Since the upper leaf has more crosses than stars, the two upper stars are misclassified as crosses and their weights increase. The second split separates the left star from the rest of the observations. Because the top star has a large weight, the right leaf is dominated by the stars. Both leaves predict into the star class, and the weights of the crosses increase. The third stump is identical to the second stump. Now the crosses in the right leaf have more weight than the three stars. The three stars in the right leaf are misclassified and their weights increase. The fourth stump separates the top star from the rest.

In about ten iterations, AdaBoost attains perfect class separation. The light rectangle in Figure 15.2a shows the region with a positive classification score. The region with a negative score is shown in semi-dark gray, and the region with a large negative score is shown in dark gray. The class boundary does not look as symmetric as it did in Chapter 13, but keep in mind that we operate under two severe constraints: the limited amount of statistics and the requirement to use decision stumps in the space of the original variables. From the computer point of view, this boundary is as good as the one in Figure 13.1!

Evolution of the observation weights through the first ten iterations is shown in Figure 15.2b. The three crosses always fall on the same leaf, and their weights are therefore always equal. The same is true for the two lower stars.

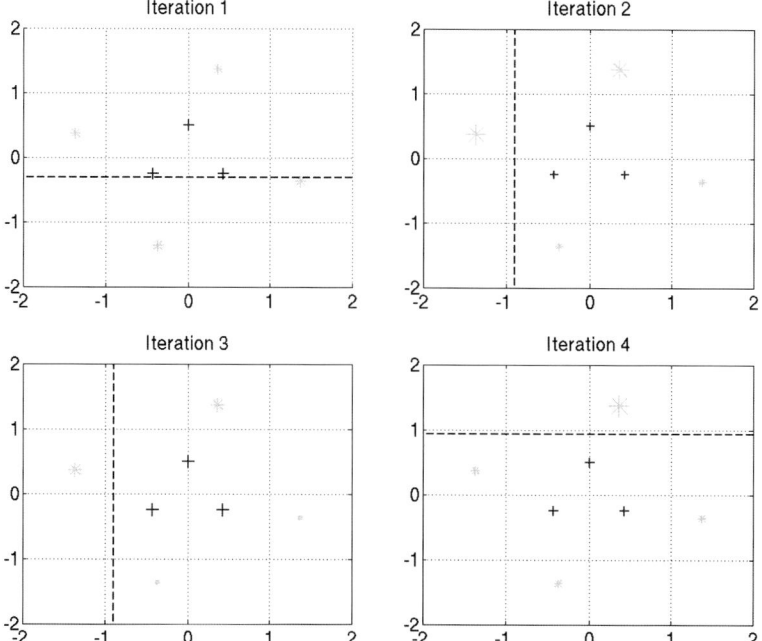

Figure 15.1 First four decision stumps imposed by AdaBoost. The size of each symbol is proportional to the observation weight at the respective iteration before the stump is applied.

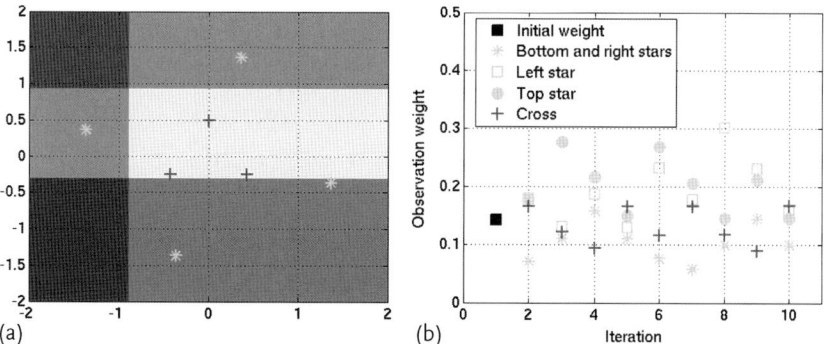

Figure 15.2 (a) Predicted classification scores, and; (b) evolution of observation weights in the first ten iterations.

15.1.2.2 Why is AdaBoost so Successful?

One way to explain AdaBoost is to point out that every weak hypothesis is learned mostly on observations misclassified by the previous weak hypothesis. The strength of AdaBoost can be then ascribed to its skillful use of hard-to-classify observations. This statement, although true in some sense, is insufficient for understanding why AdaBoost works so well on real-world data. Indeed, a few questions immediately arise for Algorithm 1:

- Why use that specific formula on line 9 to compute the hypothesis weight α_t?
- Why use that specific formula on line 10 to update the observation weights?
- What is the best way to construct h_t on line 3? If we chose a very strong learner, the error would quickly drop to zero and AdaBoost would exit. On the other hand, if we chose a very weak learner, AdaBoost would run for many iterations converging slowly. Would either regime be preferable over the other? Is there a golden compromise?
- What is the best way to choose the number of learning iterations T?
- Could there be a more efficient stopping rule than the one on line 5? In particular, does it makes sense to continue learning new weak hypotheses after the training error for the final strong hypothesis drops to zero?

These questions do not have simple answers. A decade and a half after its invention, AdaBoost remains, to some extent, a mystery. Several interpretations of this algorithm have been offered. None of them provides a unified, coherent picture of the boosting phenomenon. Fortunately, each interpretation led to discoveries of new boosting algorithms. At this time, AdaBoost may not be the best choice for the practitioner, and is most certainly not the only boosting method to try.

In the following sections, we review several perspectives on boosting and describe modern algorithms in each class. For a good summary of boosting and its open questions, we recommend Meir and Ratsch (2003); Buhlmann and Hothorn (2007), as well as two papers followed by discussion, Friedman et al. (2000) and Mease and Wyner (2008).

15.1.3
Minimizing Convex Loss by Stagewise Additive Modeling

AdaBoost can be interpreted as a search for the minimum of a convex function. Put forward by three Stanford statisticians, Friedman, Hastie and Tibshirani, in the late 1990s, this interpretation has been very popular, if not dominant, and used as a theoretical basis for new boosting algorithms. We describe this view here.

Suppose we measure the quality of a binary classifier using exponential loss, $\ell(y, f) = e^{-yf}$, for labels $y \in \{-1, +1\}$ and scalar soft score $f(x)$. Learning a good model $f(x)$ over domain $\mathcal{X}$ then amounts to minimizing the expected loss

$$E_{X,Y}[\ell(Y, f(X))] = \int_{\mathcal{X}} [e^{f(x)} P(Y = -1|x) + e^{-f(x)} P(Y = +1|x)] P(x) dx \quad (15.1)$$

for pdf $P(x)$ and conditional probability $P(y|x)$ of observing class y at x. The AdaBoost algorithm can be derived by mathematical induction. Suppose we learn $t-1$ weak hypotheses by AdaBoost to obtain an ensemble

$$f_t(x) = \begin{cases} 0 & \text{if } t = 1 \\ \sum_{i=1}^{t-1} \alpha_i h_i(x) & \text{if } t > 1 \end{cases} \quad (15.2)$$

At step t, we search for a new weak hypothesis h_t with weight α_t by minimizing the expected loss $E_{X,Y}[\ell(Y, f_t(X) + \alpha_t h_t(X))]$.

Since AdaBoost multiplies the pdf $P(y|x)$ at step i by $\exp(-y\alpha_i h_i(x))$, hypothesis t is learned on the modified joint pdf.

$$P^{(t)}(x, y) = Z^{(t)} e^{-y f_t(x)} P(y|x) P(x) \tag{15.3}$$

with a constant factor $Z^{(t)}$ for proper normalization. Define

$$P_{-1}^{(t)}(x) = e^{f_t(x)} P(Y = -1|x)$$
$$P_{+1}^{(t)}(x) = e^{-f_t(x)} P(Y = +1|x) \tag{15.4}$$

to be the conditional probabilities of observing class -1 and $+1$, respectively, at x at step t. The expected loss at step t is then

$$E[\ell(Y, f_t + \alpha_t h_t)] = \int_{\mathcal{X}} \left[e^{\alpha_t h_t(x)} P_{-1}^{(t)}(x) + e^{-\alpha_t h_t(x)} P_{+1}^{(t)}(x) \right] P(x) dx. \tag{15.5}$$

This loss is minimized when the integrated expression takes the minimal value for any x.

First, let us fix $\alpha_t > 0$ and find the optimal hypothesis $h_t(x)$. This task is easy because $h_t(x)$ can only take values -1 and $+1$. For $h_t = -1$, the expression in the square brackets under the integral (15.5) evaluates to

$$A_{-1} = e^{-\alpha_t} P_{-1}^{(t)} + e^{\alpha_t} P_{+1}^{(t)}. \tag{15.6}$$

Similarly, for $h_t = +1$ we have

$$A_{+1} = e^{\alpha_t} P_{-1}^{(t)} + e^{-\alpha_t} P_{+1}^{(t)}. \tag{15.7}$$

It is easy to see that $P_{-1}^{(t)} > P_{+1}^{(t)}$ implies $A_{-1} < A_{+1}$, and vice versa. Integral (15.5) is therefore minimized when the weak hypothesis $h_t(x)$ predicts the most probable class at x at step t.

Fortunately, $h_t(x)$ does not depend on α_t. Given $h_t(x)$, let us find the optimal value of α_t by solving

$$\frac{\partial E[\ell(y, f_t + \alpha_t h_t)]}{\partial \alpha_t} = 0.$$

Differentiating under the integral sign in (15.5) and setting the derivative to zero, we obtain

$$\alpha_t = \frac{1}{2} \log \frac{\int_{h_t=-1} P_{-1}^{(t)}(x) P(x) dx + \int_{h_t=+1} P_{+1}^{(t)}(x) P(x) dx}{\int_{h_t=-1} P_{+1}^{(t)}(x) P(x) dx + \int_{h_t=+1} P_{-1}^{(t)}(x) P(x) dx}.$$

Because $h_t(x)$ predicts into the most probable class at x, integral $\int_{h_t=-1}$ is taken over a subset of $\mathcal{X}$ on which $P_{-1}^{(t)} \geq P_{+1}^{(t)}$; similarly for $\int_{h_t=+1}$. Hence the numerator is the probability of predicting into the true class and the denominator is the probability of predicting into the false class. We obtain the formula shown on line 9 of the AdaBoost algorithm.

The outlined algorithm minimizes the expected loss over the joint distribution $P(x, y)$. In the real world, we are presented with a training set, and $P(x, y)$ is not known. To apply this algorithm in practice, we need to satisfy two conditions. First, the initial weights in the training set $\{x_n, y_n, w_n\}_{n=1}^N$ must approximate the joint probability, $w_n = P(x_n, y_n)dx$. An example of such an approximation, widely used in collider physics, is normalizing Monte Carlo background components to the luminosity observed in the data. If the weights cannot be interpreted as probabilities, they cannot be used in the manner described in the AdaBoost algorithm. Second, every weak hypothesis must use the weighted *generalization* error at step t to estimate α_t. AdaBoost computes the *training* error instead. The weak learner is therefore not allowed to overtrain and for that reason must be crude. Decision stumps or shallow trees have been a popular choice for the weak learner in the AdaBoost algorithm.

Another requirement implicit in this derivation is the optimality of the weak hypothesis. Learner h_t must achieve the Bayes error on unseen data drawn from the weighted distribution at step t. This requirement is never satisfied in practice. Indeed, if we had a learner capable of attaining the Bayes error on distribution $P(x, y)$, we would use that learner instead of the ensemble. Practical learners such as decision stumps and shallow trees optimize the class separation measured by the cross-entropy or Gini index. For convergence, such learners only need to produce error rates a little below 50%.

If $f(X)$ is the minimizer of $E_{X,Y}[e^{-Yf(X)}]$, it can be found by solving $\nabla_f E_Y[e^{-Yf(X)}|X=x]=0$. After differentiation and simple math, we obtain

$$f(x) = \frac{1}{2} \log \frac{P(Y=+1|x)}{P(Y=-1|x)}. \tag{15.8}$$

Knowing the final hypothesis $f(x)$, we can estimate the class probabilities at x.

When explained in terms of stagewise additive modeling, AdaBoost resembles a search for the minimum of a convex function. Friedman *et al.* (2000) remark that AdaBoost "builds an additive logistic regression model via Newton-like updates minimizing $E(e^{-Yf(X)})$." AdaBoost can be thought of as a line search algorithm with the search direction defined by h_t (see Appendix A).

Stepwise and stagewise algorithms have been a well-known tool in statistical analysis. Examples of stepwise fitting procedures include stepwise regression and generalized additive models (GAM). Here, we stress the distinction between the two procedure types. *Stagewise* refers to algorithms in which components fitted at earlier steps are not allowed to change at the current step. This is not the case for *stepwise* algorithms. In forward stepwise regression, for instance, one fits a linear relationship by consecutively adding each input variable to the model and keeping it in case of a significant improvement to the goodness of fit. At every step, the linear model is refitted using all included variables, and the new coefficients are obtained. In stagewise regression, the coefficients for all variables included in the model at earlier steps remain fixed.

Algorithm 2 *GentleBoost* for two classes. Class labels y_n are drawn from the set $\{-1, +1\}$.

Input: Training data $\{x_n, y_n, w_n\}_{n=1}^{N}$ and number of iterations T
1: **initialize** $w_n^{(1)} = \frac{w_n}{\sum_{n=1}^{N} w_n}$ for $n = 1, \ldots, N$
2: **for** $t = 1$ to T **do**
3:     Train regression model on $\{x_n, y_n, w_n^{(t)}\}_{n=1}^{N}$ to obtain hypothesis $h_t : x \mapsto [-1, +1]$
4:     Update observation weights $w_n^{(t+1)} = \frac{w_n^{(t)} \exp[-y_n h_t(x_n)]}{\sum_{n=1}^{N} w_n^{(t)} \exp[-y_n h_t(x_n)]}$
5: **end for**
Output: $f(x) = \sum_{t=1}^{T} h_t(x)$

After demonstrating the connection between AdaBoost and additive modeling, Friedman *et al.* (2000) derive several new boosting algorithms in the same framework. We include the derivation of a "gentle" version of AdaBoost here. Let us set weight α_t for every learner to 1 and minimize

$$E_Y \left[e^{-Y(f_t(X) + h_t(X))} \,\middle|\, X = x \right] = e^{h_t(x)} P_{-1}^{(t)}(x) + e^{-h_t(x)} P_{+1}^{(t)}(x)$$

under integral (15.5) by a Newton update (see Appendix A). The first and second derivatives with respect to h_t are given by

$$\left. \frac{\partial E_Y \left[\exp(-Y(f_t + h_t)) \right]}{\partial h_t} \right|_{h_t=0} = P_{-1}^{(t)} - P_{+1}^{(t)}$$

$$\left. \frac{\partial^2 E_Y \left[\exp(-Y(f_t + h_t)) \right]}{\partial h_t^2} \right|_{h_t=0} = P_{-1}^{(t)} + P_{+1}^{(t)} . \qquad (15.9)$$

The negative ratio of the first and second derivatives equals $E_Y^{(t)}[Y|x]$, the conditional expectation of class Y at x at step t. To estimate this conditional expectation, we could fit a classification model to the modified joint distribution (15.3) and then set $E_Y^{(t)}[Y|x] = -\hat{P}^{(t)}(Y = -1|x) + \hat{P}^{(t)}(Y = +1|x)$ for the estimated posterior probabilities $\hat{P}^{(t)}(Y = -1|x)$ and $\hat{P}^{(t)}(Y = +1|x)$. Alternatively, we could regress Y on X and approximate $E_Y^{(t)}[Y|x]$ by the response of this regression model. The latter approach gives rise to Algorithm 2.

Another boosting algorithm can be derived by maximizing the binomial log-likelihood. Class label $Y \in \{-1, +1\}$ at x can be modeled as a Bernoulli random variable with mass function

$$P(y|x) = \begin{cases} p(x) & \text{if } y = +1 \\ 1 - p(x) & \text{if } y = -1 \end{cases} .$$

Assuming (15.8), we put

$$p(x) = \frac{e^{f(x)}}{e^{f(x)} + e^{-f(x)}}$$

and obtain for the negative log-likelihood at x

$$\ell(y, f) = \log(1 + e^{-2yf}). \qquad (15.10)$$

Minimizing the expectation of $\ell(Y, f(X))$ by Newton updates, similar to GentleBoost, gives the *LogitBoost* algorithm. Details can be found in Friedman *et al.* (2000).

Friedman (2001) proposes to scale the contribution of every weak learner by a factor η, $0 < \eta \leq 1$. For example, observation weights on line 4 of GentleBoost are multiplied by $\exp[-\eta y_n h_t(x_n)]$, and the final hypothesis is set to $f(x) = \eta \sum_{t=1}^{T} h_t(x)$. This technique, called *shrinkage*, often improves the ensemble accuracy at the expense of reducing the rate of convergence and therefore increasing the number of weak learners in the ensemble. A common choice for the learning rate η is 0.1.

Interpreting AdaBoost in terms of stagewise additive modeling has proven very fruitful. This interpretation has exposed the relation between boosting and statistical learning, usually defined as modeling the posterior distributions $P(y|x)$. The proposed formalism has given rise to several new boosting methods based on well-known optimization techniques. This paradigm however is not without flaws.

First, this approach fails to provide a sensible termination condition. Convergence for Newton updates is usually tested by comparing the gradient to zero within a specified tolerance. In contrast, the proposed boosting algorithms are run for many cycles and convergence is determined by the analyst upon visual inspection of the learning curve using cross-validation or an independent test set.

Second, although shrinkage has been empirically shown to improve the accuracy, there is no convincing theoretical explanation of why it works. Since the exponential loss is a strictly convex function, Newton updates should converge to the minimum for the step size 1. The mechanism by which a shrunk ensemble converges to a better optimum is not understood.

Third, optimization by Newton-like steps can only work for convex loss functions. Convex optimization can fail in the presence of irreducible classification noise, also known as *label noise*. A rigorous proof of this proposition can be found in Long and Servedio (2010). We give an intuitive explanation. If the training set has "difficult" observations misclassified by most weak learners, the weights for these observations continue to grow with boosting iterations. Eventually, new weak learners shift their entire focus to these hard-to-classify instances and do a miserable job for the rest of the training set. The final strong classifier is then quite poor. To understand the term "label noise", imagine taking a training set with known class labels and flipping the labels for a certain fraction of observations selected at random. Long and Servedio (2010) show that a boosting algorithm trained on such data would produce generalization error 1/2 even for modest levels of noise.

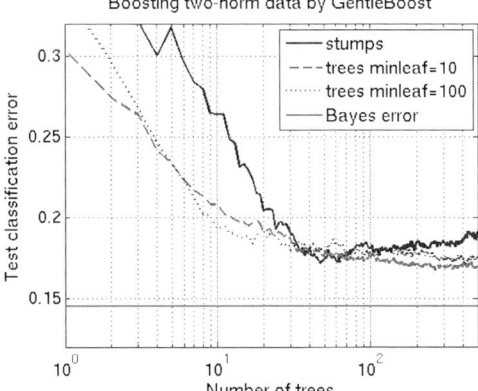

Figure 15.3 Classification error for GentleBoost ensembles with trees of various depths. The error is estimated using the test set.

15.1.3.1 Example: Binary classification in 20 Dimensions

We illustrate additive modeling using a *two-norm dataset* with $D = 20$ input variables. The two distributions are multivariate normal with identity covariance matrices. The means for all variables are $1/\sqrt{D}$ and $-1/\sqrt{D}$ for the first and second class, respectively. We set the prior probabilities for the two classes to 2/3 and 1/3 and generate 1000 observations for training and 5000 observations for testing. The Bayes error (irreducible noise) for these distributions is 14.5%.

We use the `fitensemble` function available from the Statistics Toolbox in MATLAB to grow an ensemble of decision trees by GentleBoost. We control the tree depth by varying the minimal size of a leaf node or the minimal size of a branch node. We use learning rates $\eta = 1$ and $\eta = 0.1$. By experimentation, we find that

- The lowest classification error is obtained by boosting deep trees with minimal leaf size 10.
- The lowest exponential loss is obtained by boosting semi-shallow trees with minimal leaf size 100 at the learning rate 0.1.

The learning rate does not affect the classification error for these data, but it does affect the exponential loss. The classification error and exponential loss curves for the test data are shown in Figures 15.3 and 15.4, respectively. Two observations are obvious upon inspection of these curves.

First, minimization of the classification error is not equivalent to minimization of the exponential loss. The analysis objective affects both optimal tree configuration and ensemble stopping rule. The ensemble with the smallest error, shown by the dashed line in the two plots, does a miserable job for minimizing the exponential loss. The ensemble of trees with leaf size 100, shown with the dotted line in the two plots, attains minimal exponential loss with 5 trees, but the error continues to decrease well past that mark.

15 Ensemble Learning

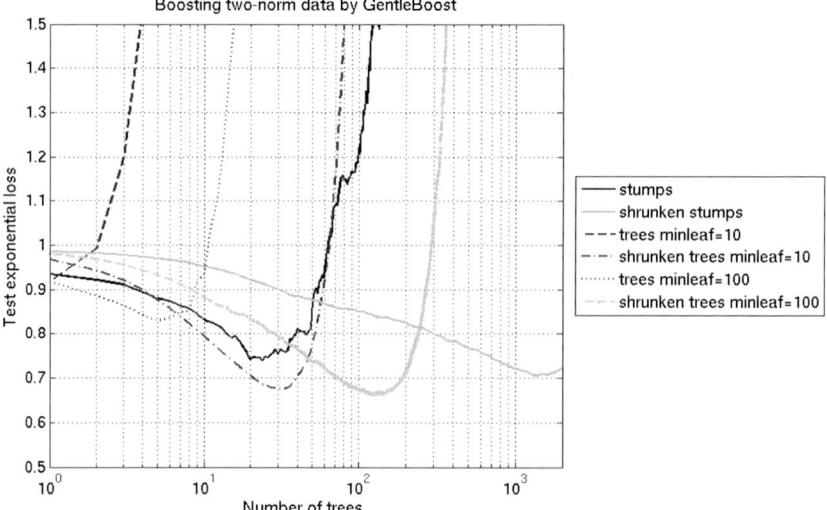

Figure 15.4 Exponential loss for GentleBoost ensembles with trees of various depths. The shrunken versions are for the learning rate $\eta = 0.1$. The loss is estimated using the test set.

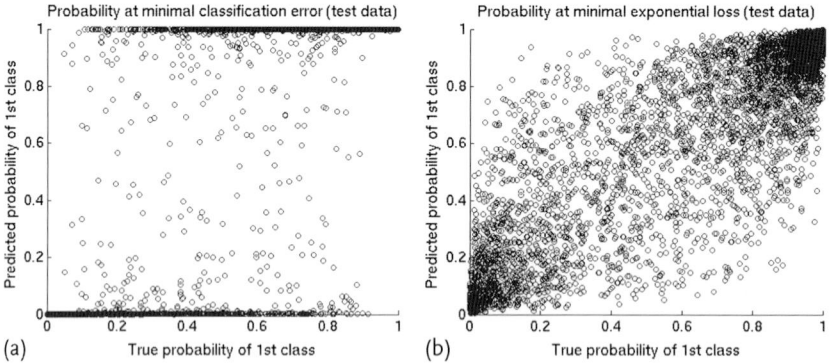

Figure 15.5 (a) Predicted probability of class 1 vs its true probability for the ensemble minimizing classification error, and; (b) the ensemble minimizing exponential loss.

Second, shrinkage is helpful for minimization of the exponential loss, especially for "aggressive" learners prone to overtraining, such as deep trees. Trying various combinations of the tree depth and learning rate proves fruitful.

We compare the true probabilities $P(y|x)$ with those predicted by GentleBoost for the test data in Figure 15.5. The ensemble obtained by minimizing the exponential loss does a much better job at prediction. The ensemble obtained by minimizing the classification error sets the predicted probability to 0 or 1 most of the time.

Above, we have shown plots for the test data. The classification error and exponential loss for the training data behave differently. The exponential loss continues to decrease as the number of weak learners grows, approaching zero in the limit. The classification error quickly drops to zero and remains there for the rest of the boosting iterations. The rate of the decrease, both for the exponential loss and classification error, depends on the strength of the weak learner and shrinkage. Regardless, the training error and exponential loss are mostly meaningless as they cannot be used by the analyst to choose the optimal leaf size, learning rate and number of trees. The 0–1 predicted probability pattern captured in Figure 15.5a is usually observed in the training data as well, even for ensembles minimizing the exponential loss in the test data. Thus, ensembles grown by boosting are most usually overtrained.

Stagewise additive modeling provides no insight into the phenomenon of overtraining. Other perspectives on boosting relate overtraining to the expected performance on new data. We describe these perspectives in the next sections.

15.1.4
Maximizing the Minimal Margin

In the previous section, we have shown how the success of AdaBoost can be explained through stagewise additive modeling. An alternative interpretation was put forward by Schapire *et al.* (1998). They explain boosting, in particular the phenomenon of overtraining, in terms of the margin theory. Similar to additive modeling, this line of research has given rise to new boosting algorithms.

In Chapter 9, we have defined statistical classification as learning the posterior distribution $P(y|x)$. We have done so for pedagogical reasons. Now we free ourselves of the burden of learning $P(y|x)$ and set to minimize the generalization error, or expected 0-1 loss, in (9.1). This task is easier as we only need to predict the most probable class at x, without knowing how probable it is.

We now introduce *classification margin*, an important concept in the learning theory. The SVM margin described in Section 13.5 has a simple geometrical interpretation. More generally, a classification margin $m(x, y; f)$ for an observation (x, y) is the difference between the soft score predicted by classifier $f(x)$ for the true class y and the maximal soft score predicted by the same classifier for one of the false classes. The margin represents the confidence of classification. The classification error (9.1) equals the probability of observing a positive margin, $\epsilon = P(m(X, Y; f) > 0)$. For binary classification $Y \in \{-1, +1\}$ and soft score f, $-1 \leq f(x) \leq 1$, the margin is simply $m = 2yf$. It is convenient to drop the factor of 2 in this formula and define the binary margin as $m = yf$. *Classification edge*,

$$\gamma(f) = \sum_{y \in \mathcal{Y}} \int_{\mathcal{X}} m(x, y; f) P(x, y) dx, \tag{15.11}$$

is the expected margin. If a binary classifier returns discrete predictions $f(x) \in \{-1, +1\}$, its error and edge are related by $\epsilon = (1 - \gamma)/2$.

As noted in Chapter 9, the generalization error is larger than the training error and must be estimated by applying the classifier to observations not seen at the training stage. We can however impose bounds on the generalization error using only information obtained in training. Consider a binary classifier with soft score $f(x) = \sum_{t=1}^{T} \alpha_t h_t(x)$ set to the weighted average of classifiers taken from domain $\mathcal{H}$. If coefficients α_t are positive with $\sum_{t=1}^{T} \alpha_t = 1$, the classifier f is said to be defined on the convex hull of space $\mathcal{H}$. Let **S** be a finite set of data drawn from the joint distribution $P(x, y)$. An important theorem proved in Schapire *et al.* (1998) and refined in Koltchinskii *et al.* (2003) states that

$$P(m(X, Y; f) \leq 0) \leq P_\mathbf{S}(m(X, Y; f) \leq \theta)$$
$$+ O\left(\frac{1}{\sqrt{N}} \sqrt{\frac{V \log^2(N/V)}{\theta^2} + \log\left(\frac{1}{\delta}\right)}\right) \quad (15.12)$$

for any $f(x)$ from the convex hull of $\mathcal{H}$ and for some θ, $0 \leq \theta < 1$, with probability at least $1 - \delta$. Above, N is the number of observations in set **S**, V is the Vapnik–Chervonenkis (VC) dimension of the classifier space $\mathcal{H}$, and $P_\mathbf{S}$ is the probability measured on the finite set **S**. The probability $P(m(X, Y; f) \leq 0)$ is the generalization error. This theorem holds in the probabilistic sense: if you took a fixed classifier $f(x)$, repeatedly drew sets of size N, and estimated the probability $P_\mathbf{S}(m(X, Y; f) \leq \theta)$ on each set, the inequality (15.12) would hold for at least $100(1 - \delta)\%$ of such estimates. We state theorem (15.12) for an infinite set $\mathcal{H}$ such as, for instance, the case of decision trees h_t learned on continuous variables X. An analogous theorem for a finite set $\mathcal{H}$ is also available; such a set would be produced, for example, by decision trees learned on discrete variables.

The VC dimension is of little practical value. The curious reader can look up its definition in Schapire *et al.* (1998) or popular textbooks such as Hastie *et al.* (2008). Its value grows with the complexity of the learner and can be estimated only for simple classifiers such as decision stumps or linear models. Neglecting the logarithmic factors and ignoring the VC dimension, we can restate inequality (15.12) in a more compact way:

$$P(m(X, Y; f) \leq 0) \leq P_{\text{train}}(m(X, Y; f) \leq \theta) + O\left(\frac{1}{\theta \sqrt{N}}\right), \quad (15.13)$$

where the probability estimate P_{train} is taken over the training set. The phenomenon of overtraining can be now understood in terms of the margin distribution. After the training error drops to zero, the ensemble continues to pull the margin distribution toward larger values. The bound on the generalization error, and the generalization error itself, then decrease.

Could we put this bound to work? When we increase θ, the first term on the right-hand side increases and the second term decreases. If we could find the optimal θ minimizing the right-hand side, we could construct a classifier with the lowest upper bound on the generalization error. Optimizing θ is not possible due to the large uncertainty on the second term. In the absence of a clear winning strategy, we can try heuristics. For instance, we could maximize the minimal margin in the training data and set θ just below this minimal margin. The P_{train} term then turns into zero and the $O(1/(\theta\sqrt{N}))$ term attains minimum over all values of θ for which the first term is zero.

For continuous variables, we could instantly accomplish this goal by putting every observation in a separate leaf of a decision tree. All training margins would then equal one and we could never come up with a better classifier. This strategy, of course, is not going to work because the VC dimension of a leafy tree approaches infinity. We need a way of optimizing the margin by simple learners only.

Suppose we have generated T weak hypotheses h_t, each mapping space $\mathcal{X}$ onto the score interval $[-1, +1]$. Define $f(x) = \sum_{t=1}^{T} \alpha_t h_t(x)$ with $\sum_{t=1}^{T} \alpha_t = 1$. We aim to find coefficients α_t maximizing the minimal margin,

$$\rho^* = \max_{\alpha} \min_{x,y} m(x, y; f).$$

We do not know the true distribution $P(x, y)$ and instead have to work with a finite training set. Let $\mathbf{M}$ be an $N \times T$ matrix of margins for N observations and T learners with elements $m_{nt} = y_n h_t(x_n)$. We then need to solve

$$\hat{\rho}^* = \max_{\alpha} \min_{n} \mathbf{m}_{n.} \cdot \boldsymbol{\alpha},$$

where $\hat{\rho}^*$ is the estimated optimal margin, and $\mathbf{m}_{n.} \cdot \boldsymbol{\alpha} = \sum_{t=1}^{T} m_{nt} \alpha_t$ is a dot product of the nth row of matrix $\mathbf{M}$ and column vector of hypothesis weights $\boldsymbol{\alpha}$. This is a linear programming (LP) problem (see Appendix A). It can be solved numerically by various software utilities such as, for instance, the `linprog` function available from the Optimization Toolbox in MATLAB.

Unfortunately, this simple-minded approach does not work. Grove and Schuurmans (1998) used the LP formulation above to optimize the weights of the weak hypotheses grown by the conventional AdaBoost in Algorithm 1. Although they could easily improve the minimal margin in the training set, the test error in their experiments increased. Maximizing the margin of an arbitrary set of hypotheses thus might prove useless. It matters how the weak hypotheses are constructed.

By duality (see Appendix A), maximizing the minimal margin is equivalent to minimizing the maximal edge,

$$\hat{\rho}^* = \max_{\alpha} \min_{n} \mathbf{m}_{n.} \cdot \boldsymbol{\alpha} = \min_{w} \max_{t} \mathbf{w}^{\mathsf{T}} \mathbf{m}_{\cdot t} = \hat{\gamma}^*, \quad (15.14)$$

over observation weights collected in a column vector $\mathbf{w}$ subject to $\sum_{n=1}^{N} w_n = 1$. Above, $\hat{\gamma}^*$ is the estimated optimal edge and $\mathbf{m}_{\cdot t}$ is the tth column of the margin matrix $\mathbf{M}$. We can then construct a boosting algorithm as follows. At step t, find the optimal weights $\mathbf{w}^{(t+1)}$ for the next step by minimizing the maximal edge of all weak hypotheses learned at previous steps,

$$\mathbf{w}^{(t+1)} = \arg\min_{\tilde{w}} \max_{i=1,\ldots,t} \tilde{\mathbf{w}}^{\mathsf{T}} \mathbf{m}_{\cdot i}. \quad (15.15)$$

Train the next weak learner using observation weights $\mathbf{w}^{(t+1)}$. Paraphrasing Demiriz et al. (2002), this strategy can be viewed as searching for a linear combination

Algorithm 3 LPBoost for two classes. Class labels y_n are drawn from the set $\{-1,+1\}$. We use $\boldsymbol{m}_{i\cdot}$ and $\boldsymbol{m}_{\cdot i}$ to specify the ith row and ith column of matrix, respectively.

Input: Training data $\{\boldsymbol{x}_n, y_n, w_n\}_{n=1}^N$, number of iterations T, and margin precision ν

1: **initialize** $w_n^{(1)} = \frac{w_n}{\sum_{n=1}^N w_n}$ for $n = 1, \ldots, N$
2: **initialize** $\mathbf{M} =$ empty matrix
3: **for** $t = 1$ **to** T **do**
4:     Train classifier on $\{\boldsymbol{x}_n, y_n, w_n^{(t)}\}_{n=1}^N$ to obtain hypothesis $h_t : \boldsymbol{x} \mapsto [-1, +1]$
5:     Compute margins $m_{nt} = y_n h_t(\boldsymbol{x}_n)$ for $n = 1, \ldots, N$
6:     Append column vector $\boldsymbol{m}_{\cdot t}$ to matrix $\mathbf{M}$
7:     Compute edge $\gamma_t = \sum_{n=1}^N w_n^{(t)} m_{nt}$
8:     Set $g_t = \min_{i=1,\ldots,t} \gamma_i$
9:     Obtain $\boldsymbol{w}^{(t+1)}$ by solving $\gamma_t^* = \min_{\tilde{\boldsymbol{w}}} \max_{i=1,\ldots,t} \tilde{\boldsymbol{w}}^\top \boldsymbol{m}_{\cdot i}$ subject to

- $\sum_{n=1}^N \tilde{w}_n = 1$
- $\forall n : \tilde{w}_n \geq 0$

10:     **if** $\gamma_t^* \geq g_t - \nu$ **then**
11:         $T = t$
12:         break loop
13:     **end if**
14: **end for**
15: Obtain optimal learner weights $\boldsymbol{\alpha} = \arg\max_{\tilde{\boldsymbol{\alpha}}} \min_{n=1,\ldots,N} \boldsymbol{m}_{n\cdot} \tilde{\boldsymbol{\alpha}}$ subject to

- $\sum_{t=1}^T \tilde{\alpha}_t = 1$
- $\forall t : \tilde{\alpha}_t \geq 0$

Output: $f(\boldsymbol{x}) = \sum_{t=1}^T \alpha_t h_t(\boldsymbol{x})$

---

of the weak hypotheses that perform best under the most pessimistic choice of observation weights. This dual formulation has led to a successful LP boosting algorithm proposed in Demiriz et al. (2002). We present *LPBoost* in Algorithm 3 with a different stopping rule, following Warmuth et al. (2006).

Let us elaborate on the stopping rule. Suppose g is the maximal *guaranteed* edge of a weak hypothesis constructed on the training set $\{\boldsymbol{x}_n, y_n\}_{n=1}^N$, where "guaranteed" is defined as follows: Varying observation weights (while keeping their sum equal to 1) and selecting a learner from the base hypothesis class $\mathcal{H}$, we always obtain a learner with edge at least g. If the weak learner attains at least edge g, the ensemble can attain the optimal edge, or equivalently the optimal margin, $\gamma^* = \rho^* \geq g$. We can use this fact to determine convergence. In particular, we could choose an accuracy parameter ν and terminate boosting as soon as the optimal margin $\hat{\rho}^*$, or equivalently the optimal edge $\hat{\gamma}^*$, in the training set exceeds $g - \nu$. In practice, we do not know g; hence we attempt to estimate it by inspecting the edges of all weak hypotheses generated so far. The value of ν is usually set small. The fitensemble function in MATLAB by default uses $\nu = 0.01$. A larger value of ν would lead to faster convergence, and a smaller value of ν would require more boosting iterations.

At every step, LPBoost finds optimal observation weights by minimizing the maximal edge of all weak hypotheses constructed so far. When this optimal edge exceeds $g - \nu$, LPBoost stops. Alternatively, we could optimize the weight distribution constraining the edges of all weak hypotheses so far constructed to be at most $g - \nu$. Convergence would be then detected by the failure to find the optimal weights satisfying this condition. This approach is taken in Warmuth et al. (2006). The *TotalBoost* algorithm minimizes mutual information, also known as the Kullback–Leibler divergence, between the current weight distribution $w^{(t+1)}$ and the initial distribution $w^{(1)}$:

$$w^{(t+1)} = \arg\min_{\tilde{w}} \sum_{n=1}^{N} \tilde{w}_n \log \frac{\tilde{w}_n}{w_n^{(1)}}. \tag{15.16}$$

In the absence of any constraints, $w^{(t+1)}$ would equal the initial distribution and all hypotheses would be identical. Requiring the edges of all past hypotheses to be below $g-\nu$, we impose t linear constraints, which force the new weight distribution to be substantially different from the one at the previous step.

Straightforward optimization of the nonlinear problem (15.16) would prove difficult. Warmuth et al. (2006) use a quadratic expansion of the Kullback–Leibler divergence in terms of $\delta = \tilde{w} - w^{(t)}$ at $\delta = 0$:

$$w^{(t+1)} = w^{(t)} + \arg\min_{\delta} \sum_{n=1}^{N} \left[\left(1 + \frac{w_n^{(t)}}{w_n^{(1)}}\right) \delta_n + \frac{1}{2w_n^{(t)}} \delta_n^2 \right]. \tag{15.17}$$

This quadratic programming (QP) problem can be solved numerically by many software packages such as, for instance, the quadprog utility available from the Optimization Toolbox in MATLAB. We give full details of TotalBoost in Algorithm 4.

TotalBoost is known to converge more reliably than LPBoost. In our studies, TotalBoost always converges in a few dozen iterations, while LPBoost sometimes constructs several hundred weak hypotheses before exceeding the prespecified maximal number of iterations.

LPBoost and TotalBoost belong to the family of *totally corrective* algorithms: every new weak hypothesis takes into account all past hypotheses. AdaBoost can be interpreted as a *corrective* algorithm: it adjusts observation weights at every step by forcing the edge of the last hypothesis to be zero. An interpretation of AdaBoost in terms of maximizing the minimal margin is offered in Ratsch and Warmuth (2005).

Algorithm 4 TotalBoost for two classes. Class labels y_n are drawn from the set $\{-1, +1\}$. We use $m_{i\cdot}$ and $m_{\cdot i}$ to specify the ith row and ith column of matrix, respectively.

Input: Training data $\{x_n, y_n, w_n\}_{n=1}^{N}$, number of iterations T, and margin precision ν

1: **initialize** $w_n^{(1)} = \frac{w_n}{\sum_{n=1}^{N} w_n}$ for $n = 1, \ldots, N$
2: **for** $t = 1$ to T **do**
3:     Train classifier on $\{x_n, y_n, w_n^{(t)}\}_{n=1}^{N}$ to obtain hypothesis $h_t : x \mapsto [-1, +1]$
4:     Compute margins $m_{nt} = y_n h_t(x_n)$ for $n = 1, \ldots, N$
5:     Compute edge $\gamma_t = \sum_{n=1}^{N} w_n^{(t)} m_{nt}$
6:     Set $g_t = \min_{i=1,\ldots,t} \gamma_i$
7:     Solve $w^{(t+1)} = \arg\min_{\tilde{w}} \sum_{n=1}^{N} \tilde{w}_n \log \frac{\tilde{w}_n}{w_n^{(1)}}$ subject to

- $\forall n : \tilde{w}_n > 0$
- $\sum_{n=1}^{N} \tilde{w}_n = 1$
- $\forall i = 1, \ldots, t : \tilde{w}^{\top} m_{\cdot i} \leq g_t - \nu$

8:     **if** $w^{(t+1)}$ cannot be found **then**
9:        $T = t$
10:        **break loop**
11:     **end if**
12: **end for**
13: Obtain optimal learner weights $\alpha = \arg\max_{\tilde{\alpha}} \min_{n=1,\ldots,N} m_{n\cdot} \tilde{\alpha}$ subject to

- $\sum_{t=1}^{T} \tilde{\alpha}_t = 1$
- $\forall t : \tilde{\alpha}_t \geq 0$

Output: $f(x) = \sum_{t=1}^{T} \alpha_t h_t(x)$

Totally corrective algorithms typically do not offer any improvement over AdaBoost either in accuracy or training speed. The time spent by LPBoost or TotalBoost on constructing one weak hypothesis exceeds this time for AdaBoost, but this increase in the training time per weak learner is compensated by the sparsity of the ensemble (see below). Totally corrective algorithms do not help address the limitation of convexity either: since by design LPBoost and TotalBoost focus on the worst-case observations, they can do a poor job for the rest of the data.

LPBoost and TotalBoost offer two important advantages:

- Well-defined convergence. The analyst does not need to guess when to stop training because the algorithm knows its stopping rule.
- Sparsity of the final hypothesis. The strong learner produced by a totally corrective algorithm is often composed of a few dozen weak hypotheses. AdaBoost in contrast often needs to be trained for a few hundred iterations to attain the maximal accuracy. Depending on the complexity of the weak learner, this sparsity can lead to a noticeable reduction in the amount of consumed memory.

15.1.4.1 Example (Continued): Binary Classification in 20 Dimensions

We apply LPBoost and TotalBoost to the dataset described in Section 15.1.3.1. To grow ensembles by LPBoost and TotalBoost, we use the default value $\nu = 0.01$ with trees of various depths.

We compare margin distributions in the training data for one of the TotalBoost ensembles and the GentleBoost ensemble attaining minimal exponential loss in Figure 15.6. For T learners, GentleBoost returns soft scores in the range $[-T, +T]$ and TotalBoost returns scores in the range $[-1, +1]$. For direct comparison, we divide GentleBoost scores by T. The training margins in the TotalBoost ensemble are all above 0.5, and the training margins for GentleBoost are stacked near 0.

Despite the dramatic improvement in the training margins, TotalBoost does not improve the test error. As seen in Figure 15.7, the TotalBoost stumps perform significantly worse than the GentleBoost stumps, and the TotalBoost trees perform comparably to the GentleBoost trees.

LPBoost and TotalBoost return comparable test errors, but TotalBoost tends to converge faster.

In Figure 15.8 we compare the training edge and test error obtained by various ensembles. Observe the general trend for LPBoost and TotalBoost: the higher the training edge, the lower the test error. This trend does not hold for GentleBoost.

Finally, we compare values of the test classification edge in Figure 15.9. TotalBoost and GentleBoost clearly have different optimization goals. A certain analogy

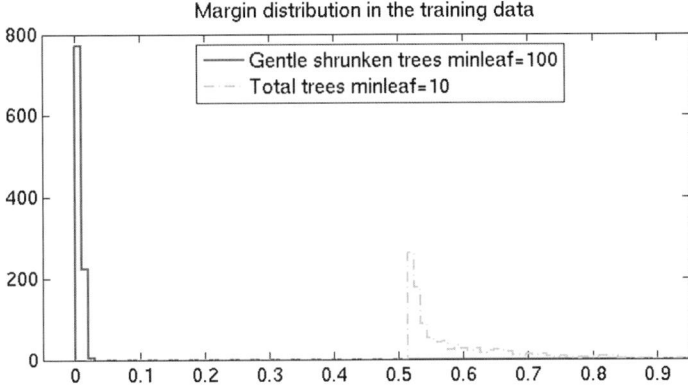

Figure 15.6 Distribution of margins in the training data for two ensembles, trees with minimal leaf size 100 boosted by GentleBoost with shrinkage, and trees with minimal leaf size 10 boosted by TotalBoost.

can be drawn here. Algorithms reviewed in Section 15.1.3 minimize the exponential loss (or a similar quantity), and algorithms reviewed in this section typically drive the classification edge to a large value. In either case, minimal exponential loss or maximal edge do not necessarily translate into minimal classification error.

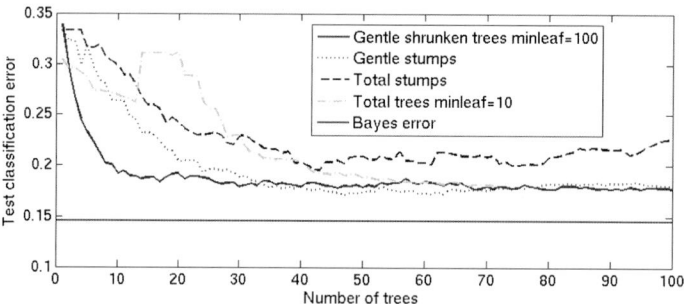

Figure 15.7 Test error for several TotalBoost and GentleBoost ensembles. The GentleBoost ensemble with shrunken trees at minimal leaf size 100 attains minimal exponential loss. TotalBoost stumps terminate at 101 iteration, at the right edge of the plot.

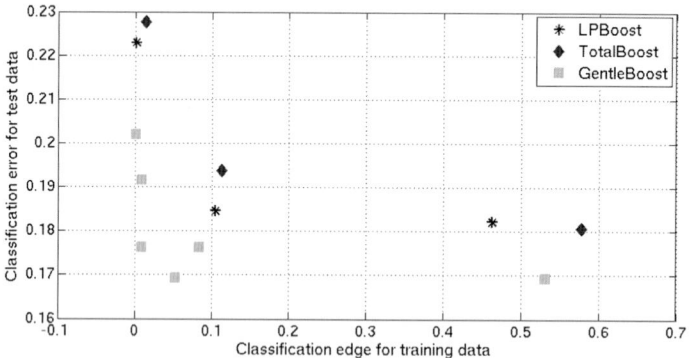

Figure 15.8 Test error versus training edge for several boosting ensembles.

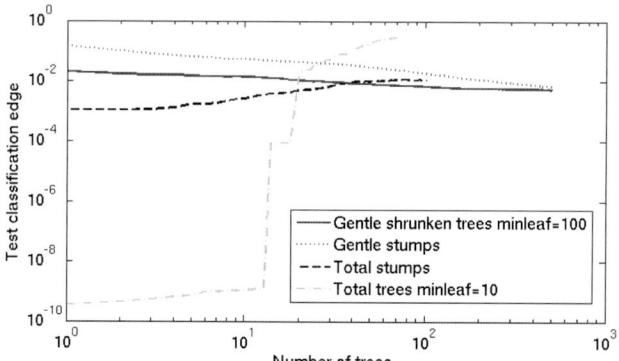

Figure 15.9 Test edge for several TotalBoost and GentleBoost ensembles.

The learning curves for TotalBoost in Figures 15.7 and 15.9 show weak hypotheses in the order in which they are generated on line 3 of Algorithm 4. After the weak hypotheses are generated, their weights are optimized on line 13. Learners with large weights can be located anywhere on the learning curve. A learner with a large weight is responsible, for instance, for the sudden error increase around iteration 15 in Figure 15.7. If you sort the weak hypotheses by their weights in descending order, the learning curve would look smoother and give an appearance of quick convergence.

15.1.5
Nonconvex Loss and Robust Boosting

In the previous two sections, we have discussed two popular classes of boosting algorithms. Stagewise additive modeling minimizes a convex loss function by incremental updates. Maximization of the minimal margin focuses on worst-case observations. Here, we turn to a different approach. Robust statistics, briefly surveyed in Section 7.3, builds a predictive model removing effects of outliers. The objective of robust fitting is thus, in a sense, opposite to that of margin optimization. Elimination of worst-case observations lets the algorithm focus on the typical case, that is, precisely the case driving the generalization error. Consequently, a substantial improvement to the classification accuracy can be at times observed.

For boosting algorithms based on the exponential loss, the weight of an observation at point x is proportional to $\exp(-y f(x))$ (see Section 15.1.3 for discussion). More generally, if a boosting algorithm minimizes loss $\ell(y, f)$, the weight at x induced by this algorithm is proportional to the magnitude of the gradient $|\partial \ell / \partial f|$. Recall that for binary classification with labels $y \in \{-1, +1\}$ the classification margin is simply $m = y f(x)$. Popular convex loss functions listed in Table 9.1 and depicted in Figure 15.10 assign large weights to observations misclassified by a large margin, moderate weights to observations near the classification boundary, and small weights to observations classified with great confidence. A boosting algorithm minimizing a convex loss then focuses on observations with large negative margins and for the most part ignores everything else. Because these observations present a real challenge, the algorithm fails to classify them anyway. At the same time, the classification accuracy for observations near the boundary suffers.

One could attempt to solve this problem by introducing a nonconvex loss or at least reducing the convexity. Indeed, several such attempts have been made. Rosset (2005) uses bounded weights for some constant $0 < p < 1$

$$\tilde{w} = \begin{cases} \dfrac{1}{1-p} & \text{if } w > \dfrac{p}{1-p} \\ \dfrac{w}{p} & \text{otherwise} \end{cases}$$

instead of the regular weights w obtained from the exponential loss. Sano et al. (2004) introduce a sigmoid loss function

$$\ell(y, f) = \frac{1}{1 + \exp(\kappa y f)}$$

for some appropriately chosen factor κ.

An interesting approach developed in Freund (1999) and Freund (2009) combines Brownian motion and the margin theory. Let us introduce a loss function fully defined by the classification margin $m = y f$,

$$\ell(m) = 1 - \operatorname{erf}\left(\frac{m - \mu}{\sigma}\right) . \tag{15.18}$$

The erf function,

$$\operatorname{erf}(a) \equiv \frac{1}{\sqrt{\pi}} \int_{-\infty}^{a} \exp(-x^2) dx , \tag{15.19}$$

looks very much like the sigmoid. The erf loss for $\mu = 0$ and $\sigma = 0.1$, multiplied by two for proper scaling, is shown in Figure 15.10. The observation weights $w = |\partial \ell / \partial m|$ then define a normal pdf with mean μ and standard deviation $\sigma/\sqrt{2}$.

Suppose we could change the shape of the loss (15.18) as the ensemble accumulates learners. To model the evolution in time, let us introduce time dependence $m = y f(x, t)$ with $\mu = \mu(t)$ and $\sigma = \sigma(t)$. Before boosting begins, we have a distribution of margins with mean $\mu_0 = \mu(0)$ and large width $\sigma_0 = \sigma(0)$. After boosting stops, we would like to have a distribution with nonnegative mean $\mu_\infty = \lim_{t \to \infty} \mu(t)$; $\mu_\infty > \mu_0$; and small width $\sigma_\infty = \lim_{t \to \infty} \sigma(t)$; $\sigma_\infty < \sigma_0$. As the margin distribution drifts toward large m, while becoming more narrow, the hard-to-classify observations at the far left move to the tail of the normal distribution and therefore get almost no weight. At the same time, observations with large

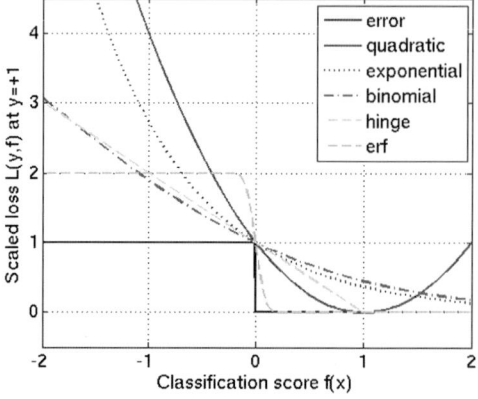

Figure 15.10 Loss functions $\ell(y, f)$ evaluated at $y = +1$ and multiplied by a constant factor to satisfy $\ell(+1, 0) = 1$. Hard-to-classify for $y = +1$ are observations with large negative scores.

positive margins move ahead of the mean to the right tail of the distribution and similarly get no weight. The boosting process gradually gives up on large-margin observations of both kinds and chooses to focus almost exclusively on uncertain observations near the classification boundary. Formally, we write

$$w(m, t) = \frac{1}{\sigma(t)\sqrt{\pi}} \exp\left[-\left(\frac{m - \mu(t)}{\sigma(t)}\right)^2\right] \tag{15.20}$$

to describe the evolution of the observation weight w versus time t.

A normal distribution with evolving width is described by Einstein's theory for Brownian motion. If the mean of the distribution evolves too, a drift term is required. Both trends are described by the Fokker–Planck equation,

$$\frac{\partial w}{\partial t} = \frac{\partial}{\partial m}[(m - \mu_\infty)w] + \frac{\sigma_\infty^2}{2}\frac{\partial^2 w}{\partial m^2}, \tag{15.21}$$

with a stationary solution at $t \to \infty$

$$w(m) = \frac{1}{\sigma_\infty \sqrt{\pi}} \exp\left[-\left(\frac{m - \mu_\infty}{\sigma_\infty}\right)^2\right]. \tag{15.22}$$

In mathematics, a random variable with density $w(m, t)$ satisfying (15.21) is called the Ornstein–Uhlenbeck process.

In the previous section, the generalization error on the left-hand side of (15.13) has been minimized by maximizing θ_0 at which $P_{\text{train}}(m \leq \theta_0)$ turns to zero. There is no fundamental reason why this heuristic approach must work better than minimizing $P_{\text{train}}(m \leq \theta)$ for some $\theta \neq \theta_0$. Varying μ_∞ and σ_∞ in the stationary solution (15.22), we can explore different values of θ. More flexibility could lead to a higher classification accuracy.

In practice, working with an infinite time horizon is inconvenient. This is why the differential equation (15.21) is rewritten to obtain the stationary solution (15.22) at the finite horizon $t = 1$. The stationary values μ_1 and σ_1, where the ∞ subscript has been replaced by 1 to indicate the new horizon, need to be chosen in advance. Freund (2009) sets σ_1 to 0.1 and varies μ_1 between 0 and 2. The value of σ_1 defines the value of σ_0. The value of μ_0 is chosen independently. Instead of setting μ_0 directly, Freund suggests finding it from

$$1 - \text{erf}\left(-\frac{\mu_0}{\sigma_0}\right) = \epsilon_0 \tag{15.23}$$

for the *error goal* ϵ_0; $0 < \epsilon_0 < 1/2$.

At every boosting step, (15.21) is solved in finite differences using the weight and margin distributions in the training data to obtain the optimal updates, $\Delta \mu$ and Δt, for the margin mean μ and evolution time t, respectively. Boosting continues until time t reaches 1 or the maximal number of boosting iterations is exceeded. We leave further details of the boosting algorithm out. The curious reader is encouraged to consult Freund (2009).

The `fitensemble` function available from the Statistics Toolbox in MATLAB provides this algorithm under the name of *RobustBoost*. The user of this RobustBoost implementation has three knobs to turn: ϵ_0, μ_1, and σ_1. In addition, the user needs to set the maximal number of boosting iterations sufficiently high to let the algorithm converge to the stationary solution. In our experiments, we keep σ_1 at the default value 0.1 and minimize the test error on a two-dimensional grid (ϵ_0, μ_1). Freund (2009) recommends setting ϵ_0 to the minimal value at which the algorithm converges within a reasonable number of boosting iterations. "Reasonable" is defined by your patience, characteristics of the data and the chosen weak learner. In our studies, we use decision trees grown on data with thousands of observations and dozens of variables. Based on this experience, values $0.05 \leq \epsilon_0 \leq 0.3$ and $0 \leq \mu_1 \leq 5$ work best. The lowest error is attained by training up to several thousand weak hypotheses.

RobustBoost and other algorithms based on nonconvex loss functions are most useful for data with label noise. The simplest way to simulate label noise is by generating two distributions according to a predefined pattern and then flipping the binary class labels at random for a fixed fraction of the data. Note that irreducible noise defined in (9.5) is in play for any pair of overlapping distributions. But if the region of the overlap is localized near the class boundary, a classifier can learn efficiently. Noise scattered over the entire space of input variables presents a more serious challenge.

In physics analysis, true class labels for simulated data are most usually known with certainty. If the training data are obtained by measurement, mislabeled observations could be an important factor. For example, a kaon sample for particle identification at high energy experiments is often collected using $D^{*+} \to D^0 \pi^+$ decays with D^0 further decaying to K^- and other particles. Decays of D^0 to K^+ are suppressed by an order of magnitude but nevertheless can be a nonnegligible background source. If the "true kaon" is defined as "the negative particle coming from a decay of D^0", label noise can be a factor.

15.1.5.1 Example: Classifying Data with Label Noise

In this example, we use a binary classification problem studied by Long and Servedio (2010) and Freund (2009). We generate 800 observations for training and 800 for testing. First, we generate the data in 21 dimension without label noise. Then we select 80 training observations at random and flip their class labels; the same is done for the test data. Without the label noise, the two classes could be perfectly separated by a plane. The best classifier for this problem therefore attains 10% generalization error.

Following Freund (2009), we set $\epsilon_0 = 0.14$, $\mu_1 = 0$ and $\sigma_1 = 0.1$. For the chosen parameters, we obtain the initial mean $\mu_0 = -1.94$ and standard deviation $\sigma_0/\sqrt{2} = 1.80$ for the margin distribution. The evolution of the margin density obtained analytically from the Fokker–Planck equation (15.21) is shown in Figure 15.11. At time $t = 1$ we end up with the target distribution with mean μ_1 and standard deviation $\sigma_1/\sqrt{2}$.

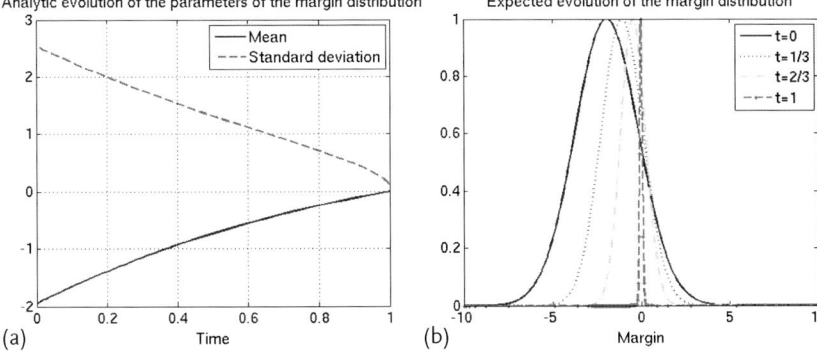

Figure 15.11 Expected evolution of the margin distribution. (a) Mean and standard deviation versus time, and; (b) margin density at several time snapshots. The distributions are normalized to unit height for easy comparison.

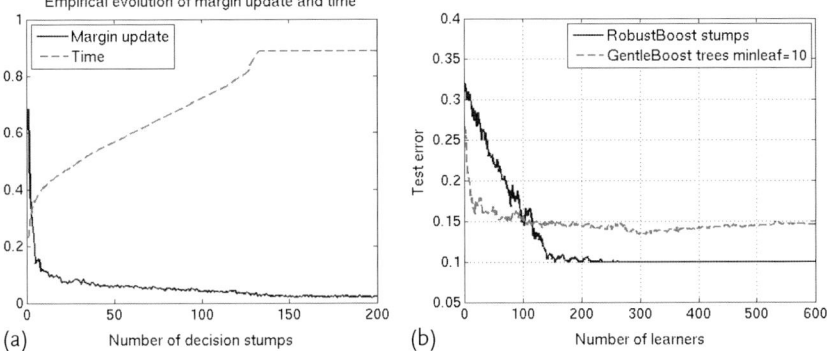

Figure 15.12 (a) Margin update and time evolution. (b) Test error for RobustBoost and the best run of GentleBoost.

We boost 1000 decision stumps by executing

```
robustStump = fitensemble(Xtrain,Ytrain,'RobustBoost',1000,'Tree',...
    'RobustErrorGoal',0.14,'RobustMaxMargin',0,'RobustMarginSigma',0.1);
```

in MATLAB for the training data Xtrain with class labels Ytrain. The margin update $\Delta\mu$ and time evolution are saved in the FitInfo property of the robustStump object. We plot them, as well as the test classification error, in Figure 15.12. Although we grow 1000 stumps, boosting could be stopped at iteration 200 or even earlier. Indeed, time stops evolving around iteration 130 and the margin updates approach zero. The MATLAB implementation does not self-terminate under these conditions, but the combination of the stalled time evolution and small margin updates is a good indicator of convergence. At this point, the test error drops to 10%.

An empirical evolution of the training margins is shown in Figure 15.13. The first stump correctly classifies a little more than half of the training observations.

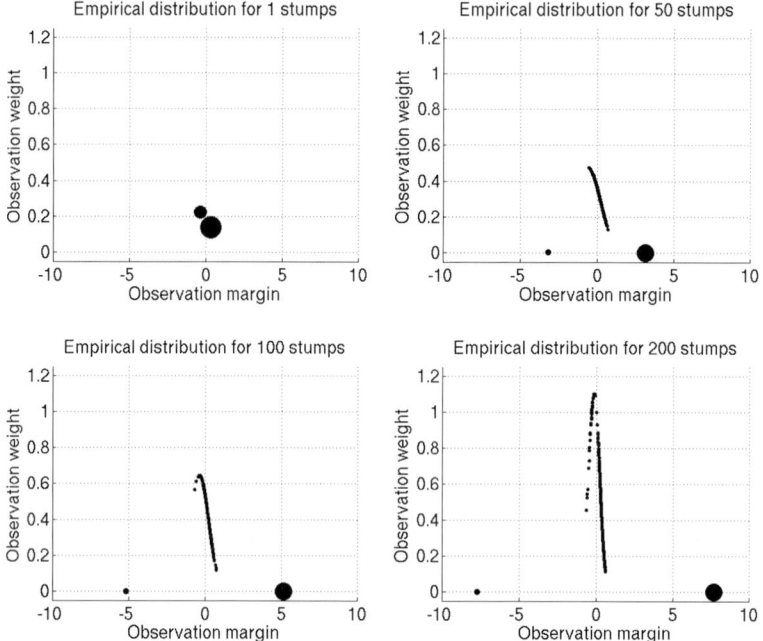

Figure 15.13 Observation weight versus margin in the training data at several time snapshots. After the first iteration, all observations have similar margins and weights. Later observations split in 3 categories. Easily classified observations (large positive margins) and badly misclassified observations (large negative margins) have negligible weights and are ignored by the boosting algorithm. Observations with uncertain classification (margins near zero) obtain large weights.

The weights for the misclassified observations in the smaller left blob are slightly larger than those for the correctly classified observations in the larger right blob. After 50 stumps are grown, every observation falls into one of the three distinct patterns. Many observations with large positive margins and a few observations with large negative margins have almost no weight, and the ensemble focuses on uncertain observations near the zero margin. The three patterns persist through the rest of the boosting cycles. The two blobs with small weight drift away from zero margin, without picking up or losing observations. The margins for the uncertain observations in the middle gradually shape into a narrow distribution with zero mean approaching the analytic distribution at $t = 1$ seen in Figure 15.11.

For the Long and Servedio problem, RobustBoost performs remarkably better than any other boosting algorithm reviewed in this chapter. The closest second is a GentleBoost ensemble of trees with at least 10 observations per leaf shown in Figure 15.12b.

15.1.6
Boosting for Multiple Classes

So far, we have discussed boosting for binary classification. Here, we briefly describe how boosting can be extended to multiple classes.

The formalism for stagewise additive modeling has been derived in Section 15.1.3 for binary classification. GentleBoost and LogitBoost construct weak regression learners, and a regression learner with scalar response can be used to separate two classes at most. A multiclass extension of LogitBoost proposed in Friedman et al. (2000) effectively fits a one-versus-all model for every class. The one-versus-all approach and other techniques for reducing a multiclass problem to a set of binary problems are discussed in Chapter 16.

AdaBoost has earned several multiclass extensions. One simple-minded approach, named AdaBoost.M1 in Freund and Schapire (1997), is to use the exact Algorithm 1. For multiple classes, this algorithm would tend to terminate early because the weak learner is required to have training error below 1/2. For two classes, a weak hypothesis with error $\epsilon < 1/2$ can be always found by assigning observations to the two classes at random and swapping the assignment if the error is above 1/2.[1] For K classes, the maximal error attainable by random assignment is $(K-1)/K$, and the requirement $\epsilon < 1/2$ would be often violated early. The number of boosting iterations could be increased by strengthening the weak learner – of course, one should then avoid the opposite danger of driving ϵ to zero. Because it can be reliably used for two classes only, AdaBoost.M1 is essentially an alternative name for the binary AdaBoost algorithm.

A better approach is to construct an error substitute bound to be less than 1/2. This substitute is called *pseudo-loss*. Instead of using one weight per observation, use K weights, one per class. Before boosting begins, initialize the training weights to

$$w_{nk}^{(1)} = \begin{cases} 0 & \text{if } y_n = k \\ \dfrac{w_n}{(K-1)\sum_{n=1}^{N} w_n} & \text{if } y_n \neq k \end{cases} \quad (15.24)$$

for K classes and N observations with class labels y_n and weights w_n. The weak hypothesis learned at iteration t generates a vector of K soft scores, $h_t(\mathbf{x}, k) \in [0, 1]$; $k = 1, \ldots, K$. For simplicity, let us assume that the soft scores are posterior probabilities and therefore $\sum_{k=1}^{K} h_t(\mathbf{x}, k) = 1$ holds for any $\mathbf{x}$. The pseudo-loss for hypothesis h_t is then set to (Drucker, 1999)

$$\epsilon_t = \frac{1}{2} \sum_{n=1}^{N} \sum_{k=1}^{K} w_{nk}^{(t)} [1 - h_t(\mathbf{x}_n, y_n) + h_t(\mathbf{x}_n, k)]. \quad (15.25)$$

For two classes and discrete scores $h_t \in \{0, 1\}$, the difference $h_t(\mathbf{x}_n, y_n) - h_t(\mathbf{x}_n, \bar{y}_n)$ between the predictions for the true class y_n and complementary class $\bar{y}_n$ is simply

---

[1] This strategy would fail if the error were exactly 1/2.

the observation margin m_{nt}. The right-hand side then turns into $(1 - \gamma_t)/2$, where γ_t is the edge of hypothesis h_t, and therefore equals the classification error. Proving that ϵ_t cannot exceed $1/2$ for an arbitrary number of classes is left as an exercise. Now compute α_t as shown on line 9 of Algorithm 1 and update the observation weights using

$$w_{nk}^{(t+1)} = \frac{w_{nk}^{(t)} \exp[\alpha_t(1 + h_t(x_n, y_n) - h_t(x_n, k))]}{\sum_{n=1}^{N} \sum_{k=1}^{K} w_{nk}^{(t)} \exp[\alpha_t(1 + h_t(x_n, y_n) - h_t(x_n, k))]}. \quad (15.26)$$

The weights for the true classes remain zero at any iteration, and only the "false" weights get boosted. This multiclass extension is named AdaBoost.M2. A slightly different version of AdaBoost.M2 allowing $\sum_{k=1}^{K} h_t(x, k)$ to exceed 1 is described in Freund and Schapire (1997).

AdaBoost.M2 makes use of weak learners with continuous soft scores ranging from 0 to 1. Algorithms of this type are often called *confidence-rated*. In contrast, the original AdaBoost algorithm is based on discrete hypotheses $h_t \in \{-1, +1\}$. A discrete multiclass extension, AdaBoost.MH, is proposed in Schapire and Singer (1999).[2] Similar to AdaBoost.M2, this algorithm uses one weight per observation per class. Similar to AdaBoost.M1, this algorithm can terminate early if the error exceeds $1/2$.

Since the definition of a classification margin naturally extends to multiple classes, LPBoost and TotalBoost can be viewed as multiclass algorithms. RobustBoost is harder to generalize, and we are not aware of its multiclass extensions.

15.2
Diversifying the Weak Learner: Bagging, Random Subspace and Random Forest

In boosting, the diversity of weak learners is achieved by reweighting the data. Although every learner in the ensemble is weak, a combination of all learners is strong. Clearly, combining weak learners is most fruitful when they differ most. Diversifying the base learner can be accomplished in many ways; reweighting is merely one of them.

In this section, we review another popular way of diversifying the weak hypotheses – by subsetting the data. Every learner can be trained on a subset of observations or a subset of variables, or both. We describe one popular algorithm in each category: bagging (bootstrap aggregation) for sampling observations and random subspace for sampling variables. Random forest can be viewed as a combination of both.

Before we plunge into analysis of the algorithms, we discuss measures of diversity.

---

[2] "H" in AdaBoost.MH stands for Hamming loss.

15.2.1
Measures of Diversity

Suppose we have two sets of labels predicted by two classifiers for the same data. If we know the true labels, we can count correct and incorrect predictions in both sets. Let us arrange observation counts, n_{ij}, in a table. We arrange the counts for the first classifier in rows and the counts for the second classifier in columns:

	correct	incorrect
correct	n_{11}	n_{10}
incorrect	n_{01}	n_{00}

A *coefficient of pairwise correlation* between the two classifiers is then

$$\hat{\rho} = \frac{n_{00}n_{11} - n_{01}n_{10}}{\sqrt{(n_{00}+n_{01})(n_{00}+n_{10})(n_{01}+n_{11})(n_{10}+n_{11})}}. \tag{15.27}$$

We use the hat notation to emphasize that this estimator is obtained using a finite set of data. Similar to the well-known linear correlation between two variables, this quantity ranges from -1 to $+1$.

The relationship between this correlation and the ensemble accuracy has been established in Breiman (2001). Suppose weak learners h_t from space $\mathcal{H}$ are trained on data drawn from pdf $P(x, y)$. Define *strength* of the weak learner prediction at x to be

$$\hat{s}(x) = \hat{P}_h(y^*|x) - \max_{y \neq y^*} \hat{P}_h(y|x), \tag{15.28}$$

where y^* is the most probable class at x according to $P(x, y)$ and $\hat{P}_h(y|x)$ is the fraction of learners that assign an observation at x to class y. This definition is similar to the one for the classification margin in Section 15.1.4, except we use probability estimates $\hat{P}_h$ in the learner space instead of probabilities (scores) estimated by a single learner $\hat{P}(y|x)$. Take $\rho = E\hat{\rho}$ to be the expectation of the pairwise correlation across all weak learners and $s = E\hat{s}$ to be the expectation of the ensemble strength over $P(x, y)$. The ensemble error then satisfies

$$\epsilon \leq \rho \frac{1-s^2}{s^2}. \tag{15.29}$$

If both ρ and s are reliably estimated on large data not used for training the ensemble, this inequality can produce a bound on the generalization error. This bound can be rather loose and thus is mostly of theoretical value.

It is instructive to see how different ensembles follow different paths to obtain a high classification accuracy. For this example, we use the MAGIC telescope data described in Bock et al. (2004). We use two thirds of the data for training and one third for testing. The best classifier for the HIACC criterion defined in the paper

Table 15.1 Mean pairwise correlation ρ and learner strength s for two ensembles trained on the MAGIC telescope data.

	ρ	s
Random forest	0.46	0.63
AdaBoost	0.000 70	0.049

is random forest with 300 trees. The trees for the random forest are grown in the default configuration with minimal leaf size 1. We also grow 500 trees with minimal leaf size 1000 by AdaBoost. For 12 742 observations in the training set, the applied tree parameters produce, on average, trees with 626 leaves for the random forest and 10 leaves for the AdaBoost. Both ensembles are constructed using the `fitensemble` function available from the Statistics Toolbox in MATLAB.

The mean pairwise correlation and strength for each ensemble are shown in Table 15.1. The trees in the random forest are strong and correlated, and the boosted trees are weak and mostly uncorrelated. The two ensembles show comparable predictive performance. For instance, the values of the HIACC criterion are 0.865 and 0.823 for the random forest and AdaBoost, respectively.

To measure the pairwise correlation, we need to know the true class labels. If they are not available, a possible measure of diversity is the *interrater agreement*, also known as the *kappa statistic*. An explanation of the effectiveness of this statistic for classification can be found in Margineantu and Dietterich (1997). Let c_{ij} be the fraction of observations assigned to class i by the first classifier and class j by the second classifier. These fractions form a coincidence matrix **C** of size $K \times K$ for K classes. The kappa statistic is defined by

$$\kappa = \frac{\sum_{k=1}^{K} c_{kk} - A}{1 - A}, \tag{15.30}$$

where

$$A = \sum_{i=1}^{K} \sum_{j=1}^{K} c_{ij} \sum_{j=1}^{K} c_{ji} \tag{15.31}$$

gives the agreement between the two classifiers by chance. Values of this statistic generally lie between 0 and 1. One indicates perfect consensus and zero attributes the agreement entirely to chance. This statistic can occasionally take negative values if the agreement is worse than that due to chance.

Similarly, we cannot estimate the strength of a learner unless we know the true labels. Define *confidence* of the learner prediction to be

$$\hat{c}(\pmb{x}) = \hat{P}_h(\hat{y}^*|\pmb{x}) - \max_{y \neq \hat{y}^*} \hat{P}_h(y|\pmb{x}), \tag{15.32}$$

where $\hat{y}^*$ is the class most frequently predicted by the weak learners at $\pmb{x}$. This confidence cannot be used as a measure of strength: it would be easy to construct an example in which a learner with low strength predicts with high confidence. For ensembles with low generalization error, these two quantities often correlate.

For the MAGIC telescope data, the mean kappa statistic is 0.57 for the random forest and 0.0027 for the AdaBoost. The respective values of the mean confidence are 0.78 and 0.058.

Other measures of diversity are discussed in Kuncheva (2004) and Rokach (2010).

15.2.2
Bagging and Random Forest

We discussed bootstrap in Sections 4.2 and 9.4.2. The key idea to this technique is estimating a statistic of interest using repeatedly drawn bootstrap replicas of the data. A bootstrap replica is obtained by sampling N out of N observations with replacement.

Here, we describe an important learning technique based on the same idea. Proposed in Breiman (1996), *bagging* (bootstrap aggregation) works as follows. Generate many bootstrap replicas of the training set, typically between a few dozen and a few hundred. Train a weak learner on each bootstrap replica. Compute the ensemble prediction by taking an (unweighted) majority vote over the weak learners.

To ensure a high degree of diversity across the weak learners, the base hypothesis must be unstable. A good candidate is a decision tree splitting the dataset to the finest level, that is, a tree with the minimal leaf size set to one. As seen in Figure 9.1, a tree with small leaves has low bias and high variance. Bagging reduces the variance of an individual tree by aggregating predictions of many trees across the ensemble. Note that bagged trees should not be pruned. Pruning them would be most usually counterproductive since it could increase their bias.

Recall from Section 9.4.2 that on average 63% of observations in a bootstrap replica are unique and the rest are their duplicates. For every weak learner, roughly 63% of observations are *in bag* (used for training this learner) and 37% are *out of bag* (not used for training this learner). The same 63-37 rule holds for observations: every observation is on average in bag for 63% of learners in the ensemble and out of bag for 37% of learners.

The out-of-bag error can be used as an unbiased estimate of the generalization error for an individual tree. The .632+ estimator (9.11) takes a weighted average of the out-of-bag and in-bag error. Since a tree split to the finest level attains zero error on continuous variables, the optimal weight ω in (9.11) can be set to 1. In addition, Breiman (2001) shows empirically that the out-of-bag error averaged over the bagged trees can be used as an unbiased estimate of the generalization error for the entire ensemble. This virtue of bagged trees can be an important CPU-saving factor. Computing out-of-bag estimates of the classification error and other quantities such as, for instance, posterior class probabilities is usually cheaper than by cross-validation.

Note that the training error of bagged trees, averaged over in-bag and out-of-bag observations, is most usually biased low for deep trees grown on continuous variables. Indeed, 63% of the trees usually predict the true class for any observation in the training data, and the rest 37% of the trees cannot flip this prediction in favor of a false class.

As you saw in Figures 15.3 and 15.4, the learning curve for an ensemble grown by boosting often reaches a minimum past which the loss begins to increase. In contrast, it is not possible to increase the generalization error by bagging more learners, just like it is not possible to obtain a less accurate Monte Carlo integral by increasing the size of the Monte Carlo set. Bagging thus avoids the problem of overtraining.

Since it is easy to choose both the optimal tree size (set the minimal leaf size to one) and the optimal ensemble size (grow as many trees as possible), bagging has a practical advantage over boosting for small datasets. Of course, bagging does not necessarily produce a more accurate model. For large datasets, growing many deep trees can consume too much memory and prove too slow. Using deep trees for large data often proves unnecessary, and bagging can be sped up by increasing the minimal leaf size. In this case, you need to find the optimal leaf size by experimentation.

Random forest, proposed in Breiman (2001), randomizes decision splits. A tree normally searches through all variables to find the best decision split. A tree in a random forest finds the best split by searching through a subset of variables. This subset is generated randomly *for every split*. For classification, Breiman (2001) recommends sampling $\sqrt{D}$ variables at random without replacement from D variables in the data.[3] A deep tree is composed of many splits and therefore makes use of significantly more than $\sqrt{D}$ variables for all its splits.

This technique diversifies trees in an ensemble and can improve the ensemble accuracy. Strictly speaking, random forest (randomization of variables) can be applied independently of bagging (randomization of observations). In practice, these two techniques are always applied together. From now on "random forest" means "an ensemble of trees grown by bagging with random selection of variables for every split".

Random forest speeds up tree construction because it evaluates splits for fewer variables. However, random forest may need more trees. Overall, random forest is faster than bagging for most real-world problems. Still it can be significantly slower than boosting decision stumps on large data.

Unlike boosting, bagging and random forest can be easily parallelized.

The original random forest algorithm does not account for class prior probabilities or observation weights. For training, this algorithm includes all observations in every bootstrap replica with equal probability. For prediction, random forest aggregates predictions of individual trees by assigning an observation to the class with the most popular vote. Chen *et al.* (2004) investigate an alternative recipe termed *weighted random forest* (WRF). For training, WRF weights every observation by the prior probability in the respective class. For prediction, WRF takes an average over posterior probabilities predicted by the individual trees and assigns an observation to the class with the largest average. The posterior probability predicted by

---

3) Round off $\sqrt{D}$ to the nearest integer.

a tree at x is computed using weights for training observations in the tree node containing x. Chen *et al.* (2004) show that this weighted version performs quite well on data with imbalanced classes (see Section 9.5 for general discussion of imbalanced data). There is no evidence suggesting that the weighted version would perform worse than the standard random forest under any circumstances. WRF, as described in Chen *et al.* (2004), assigns equal weights to all observations in the same class, but it could be easily generalized to include observations with different weights in one class.

Chen *et al.* (2004) sample observations uniformly and use their weights for tree construction. Alternatively, we can sample observations using their weights as multinomial sampling probabilities. In this case, every bootstrap replica is composed mostly of observations with large weights. A tree grown on this bootstrap replica would then ignore these observation weights because they have been taken into account in the sampling process.

In summary, there are two options available for training:

1. Unweighted sampling followed by growing a tree on weighted data.
2. Weighted sampling followed by growing a tree on unweighted data.

There is no evidence suggesting that one option favorably compares to the other. We expect that they produce similar results in most problems.

Similarly, there are two options for prediction:

1. Assign to the class with the most popular vote. Estimate $P(y|x)$ as a fraction of votes for class y by the individual trees.
2. Estimate $P(y|x)$ as an average over posterior probabilities predicted by the individual trees. Assign to the class with largest $P(y|x)$.

Again, there is no clear winner, and the results should be similar most of the time. The `fitensemble` function in the Statistics Toolbox of MATLAB uses the second option for training and prediction.

Breiman and Cutler (2004) describe several random forest extensions suitable for outlier detection and discovering structures in the data. These extensions are based on a proximity, or similarity, matrix. Take a dataset and send every observation down every tree in a random forest. When two observations land on the same leaf node, increase their proximity by one. Obtain a symmetric $N \times N$ matrix, **S**, for N observations. Divide every element in **S** by the number of trees to force the $[0, 1]$ range. Matrix $\mathbf{I}_{N \times N} - \mathbf{S}$ then holds pairwise distances. We can use this matrix to find observations that are unusually far from all others (outliers). Applying multidimensional scaling to this matrix, we can discover clusters in the data.

15.2.3
Random Subspace

Random subspace is a technique based on randomization of variables, similar in spirit to random forest. An ensemble trained by random subspace selects a subset of variables at random *for every learner*. A tree in a random forest selects d variables out of D for every split and uses more than d variables overall because it imposes many splits. A tree in a random subspace ensemble selects d variables out of D just once and uses only these d variables for all splits in this tree. Random forest was developed for trees and is applied in combination with bagging. Random subspace can be used with any base learner and is often deployed without bagging.

A random subspace ensemble selects variables for every learner by sampling at random without replacement. Variables, like observations, can be weighted. The Learn^{++}.MF algorithm proposed in Polikar *et al.* (2010) for handling missing values keeps record of how often a particular variable is selected by learners in the ensemble and reduces the sampling weight for this variable every time it is used by an individual learner. The prediction of a random subspace ensemble is computed by averaging predictions from all weak learners. We can average either labels or posterior probabilities, as discussed in Section 15.2.2.

Random subspace was introduced in the late 1990s, although the idea can be traced back to earlier publications by various authors. Ho (1995) and Ho (1998b) describe random subspace for decision trees. Ho (1998a) and Bay (1998) apply random subspace to nearest neighbor classification. Bryll *et al.* (2003) investigate the performance of random subspace with decision trees on hand-pose recognition data. Serpen and Pathical (2009) apply random subspace learning to trees, nearest neighbors, and naive Bayes on several high-dimensional datasets. Prinzie and den Poel (2007) focus on random subspace learning with naive Bayes. Skurichina and Duin (2002) apply random subspace to linear discriminant analysis.

Among all classification methods, random subspace works best, on average, with nearest neighbor learning. For trees, random forest tends to be more powerful than random subspace. Linear discriminant and naive Bayes are based on simplifying assumptions and can be rather inaccurate on real-world problems. Classification by nearest neighbors is a nonparametric method capable of attaining the Bayes error on low-dimensional data. In high dimensions, its predictive performance suffers. An ensemble of low-dimensional nearest neighbor classifiers usually offers a significant improvement over the accuracy of a single nearest neighbor classifier built on high-dimensional data.

15.2.4
Example: K/π Separation for BaBar PID

We separate kaons from pions in the BaBar particle identification data described in Bevan *et al.* (2013). We focus on the difficult part of the momentum spectrum by using tracks above 3 GeV only. We use about 16 000 training and 16 000 test observations with pion and kaon samples mixed roughly in equal proportion. These data have 31 variables, real or integer. About 3% of observations have a missing value for at least one variable.

We bag 150 trees letting every tree use all variables for every split. We also grow 150 trees for random forest; now 6 variables are selected at random for every split. For both ensembles, the minimal leaf size is set to one observation. We then train an ensemble of 150 nearest neighbor classifiers by random subspace. For nearest

Table 15.2 Classification error, mean pairwise correlation, mean strength, mean kappa statistic, and mean confidence for various ensembles trained on BaBar PID data.

	ϵ	ρ	s	κ	c
Bagging	7.5%	0.42	0.76	0.76	0.83
Random forest	7.4%	0.40	0.76	0.75	0.83
k-NN subspace	8.1%	0.34	0.67	0.60	0.74

neighbor learning, we standardize variables by subtracting the means and dividing by the standard deviations estimated from the training data. We use 3 nearest neighbors and select 8 variables for every classifier at random. The classification error and other statistics of interest are summarized in Table 15.2. All statistics are evaluated on the test set.

Although the strengths of individual trees with or without random selection of split variables are equal, random forest obtains a slightly lower error by producing trees that are more diverse. The k-NN learners are noticeably weaker than the decision trees. Yet they are more diverse, and their accuracy is not too far from the accuracy of the bagged trees. The κ and c statistics computed without knowing the true labels trail their label-aware counterparts closely.

For reference, a single decision tree with leaf size optimized by cross-validation gives 8.5% error on the test set, and a linear discriminant model errors for 8.8% of test observations. Constructing 500 trees with minimal leaf size 1000 by AdaBoost with learning rate 0.1 gives 7.6% test error.

15.3
Choosing an Ensemble for Your Analysis

Ensemble learning is well known for its superb predictive power. Although not necessarily the most accurate classifier for any given dataset, an ensemble often comes close.

The most popular weak learner is, by far, decision tree. We have reviewed the strengths and weaknesses of a single decision tree in Section 14.10. When used for ensemble learning, decision tree is grown in a different fashion. The optimal tree configuration is determined by ensemble settings. Ensembles grown by boosting typically prefer stumps (trees with two leaf nodes) or shallow trees. Ensembles grown by bagging favor trees split to the deepest level such as one observation per leaf. Pruning is not used because it could lower the ensemble accuracy by reducing the learner diversity and because the speed of training is of primary concern.

In this regime, the poor accuracy and instability of an individual tree are no longer important. The accuracy is guaranteed by the whole ensemble. Instability turns into a desirable feature ensuring sufficient diversity of individual trees. At the same time, trees grown in an ensemble retain all their attractive qualities, most notably their ability to operate in high dimensions. As pointed out in Section 14.10, trees can efficiently handle categorical variables and missing data. All these features turn ensembles of decision trees into a powerful and versatile tool.

The two most important tuning knobs for ensemble learning with trees are the stopping criterion for growing a tree, such as the minimal size of a leaf node, and the number of trees per ensemble. Random forest most usually performs quite well at the default minimal leaf settings suggested by Breiman, 1 for classification and 5 for regression, and attains the optimal accuracy with a few hundred trees. The accuracy of random forest cannot decrease when new trees are added to the ensemble. For large data, growing random forest could be sped up by increasing the minimal leaf size, often without sacrificing accuracy. For boosted trees, the optimal leaf size (or a similar stopping criterion) can be found by a simple search. The corrective algorithms reviewed in Section 15.1.4 and the robust boosting algorithm discussed in Section 15.1.5 know their stopping rules. The analyst needs to specify the number of trees large enough to let the algorithm reach convergence. Algorithms obtained by additive stagewise modeling (Section 15.1.3) can lose accuracy when the ensemble grows above a certain size. Convergence for these algorithms needs to be monitored closely. An additional tuning knob for additive stagewise boosting is the learning rate.

Random subspace is well suited for analysis of data with missing values and for reducing the data dimensionality for nearest neighbor rules.

The main weakness of ensembles is the lack of interpretability. If the optimal decision boundary between two classes had a simple functional form, an ensemble could model this boundary efficiently, but most usually the analyst would not be able to deduce this functional form from the obtained model. This problem is common to all powerful nonparametric learners including neural networks and nonlinear support vector machines (SVM).

Another difficulty with ensemble learning, in particular boosting, is the lack of conceptual understanding by the analyst. In the last few years, AdaBoost has been applied to physics analysis fairly often. Certain aspects of the boosting theory however do not sit well with the traditional mindset. Ensembles described in Sections 15.1.3 and 15.1.4 are typically much more accurate on training data than test data. Many analysts interpret this difference as a sign of overfitting. As you saw in Section 15.1.4, this "overfitting" is essential to the learning process. Maximization of the minimal margin can be justified theoretically only if the optimal margin θ in (15.13) is positive, implying the absence of misclassified observations in the training set. Note that the SVM theory (Section 13.5) is introduced in similar terms. The material in this book, we hope, could help you understand that such an "overfit" model can nevertheless be accurate on new data.

15.4
Exercises

1. As noted in Section 15.1.3, GentleBoost can be used with a weak classification or regression learner. Consider two weak learners, a linear regression model and linear discriminant, both described in Chapter 11. Compare their pros and cons as weak learners for GentleBoost.
2. Using the binomial loss (15.10), derive the LogitBoost algorithm.
3. What loss is more robust to observations misclassified by a large margin – exponential or binomial? Consult Figure 15.10.
4. Let $m = y f(x)$ be the margin for binary classification. The exponential loss is then $\ell(y, f) = \exp(-m)$. As discussed in Section 15.1.5, if a loss function takes large values for large negative margin, boosting this loss function can lead to loss of accuracy for observations near the decision boundary. Consider capping the exponential loss at $m = -1$:

$$\ell(m) = \begin{cases} \exp(-m) & \text{if } m \geq -1 \\ \exp(1) & \text{if } m < -1 \end{cases}.$$

   Derive an algorithm analogous to GentleBoost using the modified loss. What difficulties would you face if you tried to implement this algorithm?
5. Prove that the pseudo-loss (15.25) cannot exceed 1/2 for an arbitrary number of classes.
6. How would you modify the pseudo-loss (15.25) if the soft score $h_t(x, y)$ took values in the range $[-1, +1]$?
7. Prove that the pairwise classifier correlation (15.27) is bounded between -1 and $+1$.
8. Consider bagging linear discriminant described in Section 11.1. How would the strength and mean pairwise correlation for bagged discriminants compare to those for bagged decision trees? Based on this conjecture, what could you say about the generalization error for the two ensembles?

References

Bay, S. (1998) Nearest neighbor classification from multiple feature subsets. *Intell. Data Anal.*, 3, 191–209.

Bevan, A. et al. (ed.) *Physics of the B Factories*, Springer (to be published in 2013).

Bock, R., Chilingarian, A., Gaug, M., Hakl, F., Hengstebeck, T., Jirina, M., Klaschka, J., Kotrc, E., Savicky, P., Towers, S., Vaicilius, A., and Wittek, W. (2004) Methods for multidimensional event classification: a case study using images from a Cherenkov gamma-ray telescope. *Nucl. Instrum. Methods A*, 516, 511–528.

Breiman, L. (1996) Bagging predictors. *Mach. Learn.*, 24 (2).

Breiman, L. (1998) Arcing classifiers. *Ann. Stat.*, 26 (3), 801–849.

Breiman, L. (2001) Random forests. *Mach. Learn.*, 45, 5–32.

Breiman, L. and Cutler, A. (2004) Random forests, classification/clustering. http://oz.berkeley.edu/users/breiman/

RandomForests/cc_home.htm (accessed 28 July 2013).

Bryll, R., Gutierrez-Osuna, R., and Quek, F. (2003) Attribute bagging: improving accuracy of classifier ensembles by using random feature subsets. *Pattern Recognit.*, **36** (6), 1291–1302.

Buhlmann, P. and Hothorn, T. (2007) Boosting algorithms: Regularization, prediction and model fitting. *Stat. Sci.*, **22** (4).

Chen, C., Liaw, A., and Breiman, L. (2004) Using random forest to learn imbalanced data, *Statistics Tech. Rep. 666*, University of California Berkeley.

Demiriz, A., Bennett, K., and Shawe-Taylor, J. (2002) Linear programming boosting via column generation. *Mach. Learn.*, **46**, 225–254.

Drucker, H. (1999) Boosting using neural networks, in *Combining Artificial Neural Nets* (ed. A. Sharkey), Springer, Perspectives in Neural Computing, pp. 51–77.

Freund, Y. (1999) An adaptive version of the boost by majority algorithm. *Proc. 12th Annu. Conf. Comput. Learn. Theory*, ACM, COLT '99, pp. 102–113.

Freund, Y. (2009) A more robust boosting algorithm, arXiv:0905.2138v1.

Freund, Y. and Schapire, R. (1997) A decision-theoretic generalization of on-line learning and an application to boosting. *J. Comput. Syst. Sci.*, **55**, 119–139.

Friedman, J. (2001) Greedy function approximation: A gradient boosting machine. *Ann. Stat.*, **29** (5), 1189–1232.

Friedman, J., Hastie, T., and Tibshirani, R. (2000) Additive logistic regression: A statistical view of boosting. *Ann. Stat.*, **28** (2), 337–407.

Grove, A. and Schuurmans, D. (1998) Boosting in the limit: Maximizing the margin of learned ensembles. *AAAI/IAAI'98 Proc. Artif. Intell./Innov. Appl. Artif. Intell.*, pp. 692–699.

Hastie, T., Tibshirani, R., and Friedman, J. (2008) Model assessment and selection, in *The Elements of Statistical Learning*, Springer, pp. 219–260, 2nd edn.

Ho, T. (1995) Random decision forests, 3rd Int. Conf. Doc. Anal. Recognit., pp. 278–282.

Ho, T. (1998a) Nearest neighbors in random subspaces, Joint IAPR Int. Workshops Adv. Pattern Recognit.

Ho, T. (1998b) The random subspace method for constructing decision forests. *IEEE Trans. Pattern Anal. Mach. Intell.*, **20** (8), 832–844.

Koltchinskii, V., Panchenko, D., and Lozano, F. (2003) Bounding the generalization error of convex combinations of classifiers: Balancing the dimensionality and the margins. *Ann. Appl. Probab.*, **13** (1), 213–252.

Kuncheva, L. (2004) *Combining Pattern Classifiers: Methods and Algorithms*, John Wiley & Sons.

Long, P. and Servedio, R. (2010) Random classification noise defeats all convex potential boosters. *Mach. Learn.*, **78**, 287–304.

Margineantu, D. and Dietterich, T. (1997) Pruning adaptive boosting. *Proc. 14th Int. Conf. Mach. Learn.*, pp. 211–218.

Mease, D. and Wyner, A. (2008) Evidence contrary to the statistical view of boosting. *J. Mach. Learn. Res.*, **9**, 131–156.

Meir, R. and Ratsch, G. (2003) An introduction to boosting and leveraging, in *Advanced Lectures on Machine Learning*, LNCS, pp. 119–184.

Polikar, R., DePasquale, J., Mohammed, H., Brown, G., and Kuncheva, L. (2010) Learn^{++}.MF: A random subspace approach for the missing feature problem. *Pattern Recognit.*, **43** (11), 3817–3832.

Prinzie, A. and den Poel, D.V. (2007) Random multiclass classification: Generalizing random forests to random MNL and random NB, in *Database and Expert Systems Applications, Lecture Notes in Computer Science*, vol. 4653, Springer, pp. 349–358.

Ratsch, G. and Warmuth, M. (2005) Efficient margin maximizing with boosting. *J. Mach. Learn. Res.*, **6**, 2131–2152.

Rokach, L. (2010) *Pattern Classification Using Ensemble Methods, Machine Perception and Artificial Intelligence*, vol. 75, World Scientific.

Rosset, S. (2005) Robust boosting and its relation to bagging. *KDD'05*, pp. 249–255.

Sano, N., Suzuki, H., and Koda, M. (2004) A robust boosting method for mislabeled data. *J. Oper. Res.*, **47** (3), 182–196.

Schapire, R. (1990) The strength of weak learnability. *Mach. Learn.*, **5**, 197–227.

Schapire, R., Freund, Y., Bartlett, P., and Lee, W. (1998) Boosting the margin: A new explanation for the effectiveness of voting methods. *Ann. Stat.*, **26** (5), 1651–1686.

Schapire, R. and Singer, Y. (1999) Improved boosting algorithms using confidence-rated predictions. *Mach. Learn.*, **37** (3), 297–336.

Serpen, G. and Pathical, S. (2009) Classification in high-dimensional feature spaces: Random subsample ensembles. *Int. Conf. Mach. Learn. Appl.*, pp. 740–745.

Skurichina, M. and Duin, R. (2002) Bagging, boosting and the random subspace method for linear classifiers. *Pattern Anal. Appl.*, **5**, 121–135.

Warmuth, M., Liao, J., and Ratsch, G. (2006) Totally corrective boosting algorithms that maximize the margin., in *ICML*, vol. 148, ACM, vol. 148, pp. 1001–1008.

Yang, H.J., Roe, B., and Zhu, J. (2005) Studies of boosted decision trees for MiniBooNE particle identification. *Nucl. Instrum. Methods Phys. Res. A*, **555** (1/2), 370–385.

16
Reducing Multiclass to Binary

In this chapter, we describe how a classification problem with 3 or more classes can be solved by a set of binary (two-class) learners. First, we divide a multiclass problem into sub-problems, each presented as a binary classification task; this reduction is called *encoding*. Then we learn binary classifiers, one per task. Finally, to apply the learned model to a new observation, we combine these binary classifiers to predict the most likely class label and, optionally, estimate the confidence of this prediction. This last step is called *decoding*.

Most classifiers introduced in this book can be readily applied to data with multiple classes. One notable exception is SVM, for which we reviewed a few nontrivial multiclass extensions in Section 13.5.4. Given the abundance of multiclass learners, why do we need yet another technique?

First and foremost, reduction of multiclass learning to binary adds flexibility. For example, if we boost decision trees by AdaBoost.M2 described in Section 15.1.6, we can control the number of weak learners (trees) and their parameters such as the minimal leaf size. If we reduce this problem to learning a set of ensembles for binary classification by AdaBoost.M1, we in addition can control the choice of a specific reduction scheme. We can deploy one of the well-known schemes such as *complete* reduction (to be defined later). We can also convert our intimate understanding of the problem into a reduction scheme tailored to the task at hand. Suppose, for instance, that we have 4 classes: A, B, C, and D. Classes A and B are similar, and so are classes C and D, but class $A \cup B$ is very different from $C \cup D$. Instead of trying to separate all 4 classes at once, we could easily separate $A \cup B$ from $C \cup D$, and then invest more effort in separation of A from B and C from D.

Second, although multiclass learners are well known and available from various software suites, you may not have a multiclass implementation of your favorite learner at your disposal. For example, TMVA and StatPatternRecognition, two software packages recommended by Beringer *et al.* (2012), instrument decision trees capable of two-class learning only.

Third, the lack of a straightforward, well-accepted multiclass extension for SVM is no small matter, due to the huge popularity of kernel methods.

We note that popular multiclass reductions to binary such as "one against all" and "one against one" are easy to code, even for an inexperienced programmer. Advanced strategies may require more work.

Statistical Analysis Techniques in Particle Physics, First Edition. Ilya Narsky and Frank C. Porter.
©2014 WILEY-VCH Verlag GmbH & Co. KGaA. Published 2014 by WILEY-VCH Verlag GmbH & Co. KGaA.

16.1
Encoding

Techniques for reducing multiclass learning to binary have been known for at least several decades. Two simple and popular strategies are *one versus one* (OVO), also known as *all pairs*, and *one versus all* (OVA), also known as *one per class*. For K classes, OVA constructs K binary classifiers with the kth classifier separating class k from all others. OVO trains $K(K-1)/2$ binary learners, one for each pair of classes.

But why restrict our choice to these two?

The *error correcting output code* (ECOC) approach proposed in Dietterich and Bakiri (1995) generalizes the multiclass reduction scheme. Let us introduce a *binary design matrix* **M** with K rows, one per class, and L columns, one per learner. "Design matrix" is our terminology analogous to that used for multivariate regression; matrix **M** does not seem to have an agreed-upon name in the literature. Dietterich and Bakiri (1995) fill this matrix with zeros and ones. For the lth column in **M**, classes labeled with 1 form the positive class and classes labeled with 0 form the negative class. These positive and negative classes are used to train the lth binary learner. Every binary learner then predicts either 0 or 1, and for every class (every row of the design matrix) we have L binary predictions forming a *codeword*. The multiclass label is chosen by minimizing the *Hamming distance* between the predicted codeword and the respective row of the design matrix (expected codeword). The Hamming distance between two codewords is computed by counting mismatches: it is increased by 1 every time the two codewords have different values in the matching position. Formally, let $\hat{\mathbf{y}}$ be a vector of L predicted binary labels with $\hat{y}_l \in \{0, 1\}$ and let m_{kl} be an element of the matrix **M**. The Hamming distance between class k and the predicted codeword $\hat{\mathbf{y}}$ is then

$$\Delta_{\{0,1\}}(k, \hat{\mathbf{y}}) = \sum_{l=1}^{L} |m_{kl} - \hat{y}_l|. \tag{16.1}$$

This approach allows exploring reductions other than OVO and OVA in a consistent framework. One particularly useful reduction for small K is *complete*, or *exhaustive*, design. Note that a design matrix could have 2^K columns at most. Half of these 2^K columns would be redundant because swapping class labels for a binary learner would produce the same model. Besides, one column in the remaining half would be filled with zeros (or ones) only and could not be used for learning a classifier. A complete design matrix therefore has $2^{K-1} - 1$ columns. A complete matrix for 4 classes is shown in Table 16.1. This design matrix is used by the BaBar particle identification (PID) analysis described in Bevan et al. (2013) to separate particles of 4 types: electron, kaon, pion, and proton.

A natural measure of the quality of a design matrix is the minimal Hamming distance between its rows, H_K. The complete matrix above has $H_K = 4$. Suppose the codeword $\hat{\mathbf{y}}$ predicted for a new observation is 1111111. The label is then assigned to the first class (top row). If the predicted codeword has one classification error, for instance, 1111110, the label is still assigned to the first class since $\Delta_{\{0,1\}}(k, \hat{\mathbf{y}})$ is minimized at $k = 1$. If the predicted codeword has 2 errors, for example, 1111100, the multiclass prediction is ambiguous: this codeword is halfway between rows 1 and 2. In general, a binary design matrix is capable of correcting up to $\lfloor \frac{H_K-1}{2} \rfloor$ errors where $\lfloor a \rfloor$ stands for the floor of a.

At the same time, we should prefer a design matrix with large minimal distance between its columns, H_L. If two columns are similar, the respective binary learners are highly correlated. Increasing H_L would diversify the learners and likely increase the accuracy of their multiclass prediction. When we compute distance between two columns, we need to consider their complementary values. For example, the Hamming distance between the third and fourth column in the complete design matrix above is 3. But if we swap 0 and 1 in one of the columns, the new distance is 1. Swapping class labels cannot change the correlation between two binary learners. The Hamming distance between two columns should be therefore set to the minimum of their Hamming distance and their Hamming distance computed when one of the columns is replaced by its complementary set.

For the complete 4-class matrix in Table 16.1, the distance between any two rows is 4 and the distance between two columns is either 1 or 2. For the OVA matrix, the distance between any two rows or any two columns is 2. This matrix is said to be *equidistant*. The OVA matrix does not have the error-correcting property: any binary error would lead to an incorrect multiclass prediction.

Kuncheva (2005) argues that minimal distance between rows or columns is insufficient to judge the quality of **M** and proposes using average distance. The average Hamming distance between rows is defined by

$$\bar{H}_K = \frac{2}{K(K-1)} \sum_{i=1}^{K} \sum_{j>i} \sum_{l=1}^{L} |m_{il} - m_{jl}| \tag{16.2}$$

with a similar definition for the average column distance, $\bar{H}_L$. Windeatt and Ghaderi (2003) show that a design matrix with equidistant rows generally gives better multiclass accuracy compared to a nonequidistant matrix with a similar H_K value. A review of other approaches to ECOC design can be found in Rokach (2010).

The complete design cannot be used for many classes due to long training times. In their experiments, Dietterich and Bakiri (1995) use a complete design matrix when the number of classes is small, $3 \leq K \leq 7$. For $8 \leq K \leq 11$, they generate a complete design matrix and select a subset of its columns by an optimization

Table 16.1 OVA and complete design matrices for 4 classes in the binary encoding.

OVA	complete
$\begin{pmatrix} 1 & 0 & 0 & 0 \\ 0 & 1 & 0 & 0 \\ 0 & 0 & 1 & 0 \\ 0 & 0 & 0 & 1 \end{pmatrix}$	$\begin{pmatrix} 1 & 1 & 1 & 1 & 1 & 1 & 1 \\ 1 & 1 & 1 & 0 & 0 & 0 & 0 \\ 1 & 0 & 0 & 1 & 1 & 0 & 0 \\ 0 & 1 & 0 & 1 & 0 & 1 & 0 \end{pmatrix}$

procedure. For $K > 11$, they find a good design matrix by random hill climbing for an assumed value of L. Kuncheva (2005) deploys an evolutionary algorithm. As follows from the previous paragraph, the optimized criterion can vary.

The described $\{0, 1\}$ encoding has one serious limitation: it requires that every binary learner be trained on all classes. For instance, the OVO strategy cannot be represented by a binary design matrix. Allwein et al. (2000) introduce a more general framework. An element of a *ternary* design matrix can take 3 values: -1, 0, and $+1$. For every binary learner (column in the design matrix), classes labeled by $+1$ form the positive group, classes labeled by -1 form the negative group, and classes labeled by 0 are not used for training this learner. A binary learner then predicts either -1 or $+1$, and the multiclass label is assigned by minimizing a modified Hamming distance

$$\Delta_{\{-1,0,+1\}}(k, \hat{y}) = \sum_{l=1}^{L} \frac{1 - \text{sign}(m_{kl} \hat{y}_l)}{2}. \tag{16.3}$$

A class not used for training a binary learner contributes $1/2$ to the Hamming loss. Design matrices for OVO and the hypothetical problem discussed in the introduction to this chapter are shown in Table 16.2.

To convert a design matrix from the binary to ternary encoding, replace all 0 elements with -1. The opposite conversion cannot be done unless the ternary design matrix is free of zero elements.

To compute the distance between two codewords (rows), add $1/2$ for every pair in which exactly one element is zero and add 1 for every pair in which the two elements are neither zero nor equal. Take the OVO design matrix, for example. There are $K(K-1)/2$ learners and there is only one pair of nonzero matched elements in any two rows. Hence, the distance between any two rows is $(K^2 - K + 2)/4$. The complete matrix (with 0 replaced by -1) has minimal row distance 2^{K-2}. The complete design is expected to produce a more accurate multiclass model for $K > 3$.

An important issue so far ignored is the speed of training. Suppose the training time for the chosen binary learner scales as $O(N^\gamma)$ with $\gamma \geq 1$. Most classifiers conform to this assumption. For discriminant analysis and neural network with backpropagation, we can put $\gamma = 1$. The training time for decision tree typically scales as $O(N \log N)$ and would be bounded by $1 < \gamma < 2$. Kernel methods requir-

Table 16.2 OVO and customized design matrices for 4 classes in the $\{-1, 0, +1\}$ encoding.

OVO						customized		
+1	+1	+1	0	0	0	+1	+1	0
−1	0	0	+1	+1	0	+1	−1	0
0	−1	0	−1	0	+1	−1	0	+1
0	0	−1	0	−1	−1	−1	0	−1

ing computation of the full Gram matrix would scale as $O(N^2)$ or worse. The SMO algorithm for SVM described in Section 13.5.3.2 is characterized by γ between 1 and 2.2. From now on we drop the O symbol for brevity.

Any strategy based on the $\{0,1\}$ encoding gives all data to every binary learner and therefore scales as LN^γ; in particular, $L = K$ for OVA. What about OVO? Let N_l be the number of observations used to train the lth learner. Since every class is used by $K-1$ learners, we have $\sum_{l=1}^{L} N_l = (K-1)N$ with $L = K(K-1)/2$. The sum $\sum_{l=1}^{L} N_l^\gamma$ attains maximum when almost all observations are placed in one class. Set $N_1, \ldots, N_{K-1}$ to $N - K + 2$ and set $N_K, \ldots, N_L$ to 2. We immediately obtain $\sum_{l=1}^{L} N_l^\gamma \leq (K-1)N^\gamma$. The sum $\sum_{l=1}^{L} N_l^\gamma$ attains minimum when observations are divided evenly across classes. In this case, we have $N_l = 2N/K$ for every l and therefore $\sum_{l=1}^{L} N_l^\gamma \geq (K-1)(2/K)^{\gamma-1} N^\gamma$. The ratio of the training times for OVO and OVA is then

$$\left(\frac{2}{K-1}\right)^{\gamma-1} \left(1 - \frac{1}{K}\right) \leq \frac{T_{\text{OVO}}}{T_{\text{OVA}}} \leq 1 - \frac{1}{K}. \tag{16.4}$$

OVO always trains faster than OVA. This advantage could be negligible if almost all observations are concentrated in one class. Most significant speed-up can be seen at large γ and large K for observations evenly divided across classes.

16.2 Decoding

A decoding strategy has been introduced in the previous section: we can assign the multiclass label by minimizing the Hamming distance between the predicted and expected codewords. This approach faces one serious difficulty: the Hamming distance often leads to ambiguous predictions for a small number of classes. Consider, for instance, OVA for 3 classes, A, B and C. Suppose the first classifier trained to separate A from $B \cup C$ predicts A, the second classifier trained to separate B from $A \cup C$ predicts $A \cup C$, and the third classifier trained to separate C from $A \cup B$ predicts $A \cup B$. In this case, we can unambiguously assign this observation to class A. If one of the binary predictions is swapped, the dominance of class A can no longer be established. Now consider OVO for the same classes. In this case, a label can be assigned only if one class wins in both pairings, for example, A wins over B and A wins over C. If every class wins once and loses once, a clear prediction cannot be made.

The absence of a clear winner is an indication of uncertain prediction. We could break such ties at random or we could place them in the hard-to-classify category without a predicted label. In practice, physicists usually want to optimize the between-class decision threshold using a criterion specific to their analysis. For example, the PID analysis described in Bevan et al. (2013) aims at separating particles of 4 types, e, K, π, and p. For each type, they consider 6 selectors: super loose, very loose, loose, tight, very tight, and super tight. Each selector is obtained by imposing an appropriate threshold on the prediction of a multiclass model. Clearly, the Hamming distance cannot offer this kind of flexibility.

Fortunately, most classifiers are capable of producing not only hard labels but also soft scores. As discussed in Section 9.1, a soft score represents the confidence of classification. For OVA, we can immediately disambiguate predictions by assigning the multiclass label to the class with the highest individual score. In our example with 3 classes, we record the score f_A for classification into A against $B \cup C$, score f_B for classification into B against $A \cup C$, and score f_C for classification into C against $A \cup C$. We then assign the label to the class with the largest score out of $\{f_A, f_B, f_C\}$. As the scores are measured on a continuous scale, the likelihood of getting equal scores would be low. We could use the three scores to optimize the decision boundaries. Unfortunately, the same trick is going to work neither for OVO nor for many other design matrices.

Allwein et al. (2000) solve this problem by introducing continuous ternary loss functions. In their framework, loss is a function of $y f$, the product of the ternary label y and predicted score f. We show these loss functions in Table 16.3. For consistency with the Hamming distance, every loss function is set to 1/2 at $y = 0$. The exponential loss is appropriate for AdaBoost.M1 and GentleBoost discussed in Section 15.1.3. The quadratic loss is appropriate for a binary classifier predicting posterior probabilities. The multiclass label is assigned to the class (row of the design matrix) with minimal loss,

$$\Delta(k, f(x)) = \sum_{l=1}^{L} \ell(m_{kl}, f_l(x)). \qquad (16.5)$$

Above, f with elements f_l; $l = 1, \ldots, L$; is a vector of soft scores produced by the L binary learners. Other forms of loss functions can be found in Escalera et al. (2010).

Often, we wish to estimate class posterior probabilities for an observation at x. Even if all binary learners can estimate such probabilities, it is not immediately obvious how to convert them into multiclass posterior estimates. In the binary encoding, every learner uses all training data and therefore estimates sums of the posterior probabilities. For instance, the first column for the complete design matrix in Table 16.1 has 1 for classes 1–3 and 0 for class 4. The first binary learner therefore estimates the sum of the posterior probabilities for the first three classes. Let $\hat{q}$ be a column vector of L posterior probability estimates predicted by the binary

Table 16.3 Ternary loss functions for multiclass problems. Class label y is -1 for the negative class, $+1$ for the positive class and 0 if this class is not used for training this classifier.

Name	Binary loss $2\ell(y, f)$	Range of f
Hamming	$1 - \text{sign}(y f)$	$\{-1, +1\}$
Exponential	$\exp(-y f)$	$(-\infty, +\infty)$
Quadratic	$[1 - y(2f - 1)]^2$	$[0, 1]$

learners for class 1. Dietterich and Bakiri (1995) propose estimating $\hat{p}$, a column vector of K posterior probabilities for the K classes, by solving

$$\mathbf{M}^\top \hat{p} = \hat{q}. \tag{16.6}$$

For OVA, $\mathbf{M}$ is an identity matrix, and we have $\hat{p} = \hat{q}$ in accordance with our heuristic recipe for OVA disambiguation. Usually, there are more learners than classes and the system (16.6) is overdetermined. It can be solved, in the least-squares sense, by QR decomposition. MATLAB users can solve it by the backslash operator, as discussed in several places in this book. The backslash operator does not guarantee that all elements in $\hat{p}$ are positive and sum to one. For this reason, we prefer a quadratic programming (QP) formulation: Minimize $(\mathbf{M}^\top \hat{p} - \hat{q})^\top (\mathbf{M}^\top \hat{p} - \hat{q})$ subject to $\hat{p}_k \geq 0$; $k = 1, \ldots, K$; and $\sum_{k=1}^{K} \hat{p}_k = 1$. A convex QP problem with linear and bound constraints can be solved by many software suites. For example, you could use the `quadprog` utility available from the Optimization Toolbox in MATLAB.

Estimation of posterior probabilities in the ternary design is more involved. Every binary learner now predicts the posterior probability of class $+1$ relative to the total posterior probability of classes -1 and $+1$. For example, the first column for the OVO design in Table 16.2 has $+1$ in the first and -1 in the second row. The first binary learner therefore estimates $r_1 = \hat{p}_1/(\hat{p}_1 + \hat{p}_2)$. Wu et al. (2004) solve the respective system of equations for OVO by QP. We generalize their approach to an arbitrary ternary matrix. Let $r(x)$ be the computed vector of L probability ratios at x and let $\mathbf{R}$ be a $K \times L$ matrix with each row set to $r(x)$. Set $\mathbf{M}_{+1}$ to the design matrix $\mathbf{M}$ and zero out all elements in $\mathbf{M}_{+1}$ not equal to $+1$; $\mathbf{M}_{-1}$ is defined in a similar fashion. Form a $K \times L$ matrix,

$$\mathbf{Q} = \mathbf{R} \circ \mathbf{M}_{-1} + (1 - \mathbf{R}) \circ \mathbf{M}_{+1}. \tag{16.7}$$

Then find the optimal vector of posterior probability estimates $\hat{p}$ by minimizing $\hat{p}^\top \mathbf{H} \hat{p}$ for the Hessian matrix $\mathbf{H} = \mathbf{Q}\mathbf{Q}^\top$ subject to the same constraints as in the previous paragraph. If the ternary matrix consists of -1 and $+1$ only, the two prescriptions are equivalent.

Hastie and Tibshirani (1998) find $\hat{p}$ by minimizing the Kullback–Leibler divergence between the expected and observed distributions of the posterior ratios in the OVO approach. Let $\mu_{ij} = p_i/(p_i + p_j)$ be the true ratio of the posterior probabilities at x and let r_{ij} be its value measured by the binary learner trained on classes i and j. The optimal vector $\hat{p}$ then minimizes

$$\Delta_{\mathrm{KL}}(p) = \sum_{i \neq j} N_{ij} r_{ij} \log \frac{r_{ij}}{\mu_{ij}}. \tag{16.8}$$

Above, N_{ij} is the number of observations used by the learner trained on classes i and j; if the observations are weighted, N_{ij} can be replaced by their weight. Hastie and Tibshirani (1998) minimize (16.8) by a simple iterative algorithm. Zadrozny (2001) generalizes this approach to an arbitrary design matrix.

Galar et al. (2011) perform an empirical comparison of these and several other approaches to estimation of posterior probabilities for OVO and conclude that the techniques based on (16.7) and (16.8) are among the most robust decoding schemes.

To predict the class label for an observation, we can select the class with smallest loss, as discussed earlier. Alternatively, we could assign an observation to the class with largest posterior. There is no guarantee that the two choices coincide, but this is usually the case in practice. For illustration, we apply multiclass analysis to the BaBar PID data (Bevan et al., 2013). The original analysis uses ensembles of bagged trees with a complete design matrix. Instead of reproducing this analysis, we run a quick exercise. We use the complete design matrix as well and select a tree ensemble grown by AdaBoost.M1 for the binary learner. To convert the AdaBoost score f to a probability estimate, we apply the $1/(1 + \exp(-2\gamma f))$ transformation justified in Section 15.1.3. We use only 10% of the available training and test data. Our training and test sets have 92 155 observations each in 31 dimension. The electron and proton samples contribute about 1/3 each, and the kaon and pion samples contribute about 1/6 each. The minimal leaf size for the decision tree is set to 2000, and we boost 300 trees for every binary ensemble. The classification error measured on the test data is 4.4%. Out of 92 000 test observations, class labels predicted by smallest loss and largest posterior disagree in 662 cases. Most of these 662 observations are misclassified by both decoding schemes: The labels predicted by smallest loss and largest posterior are incorrect in 55 and 61% of cases, respectively.

16.3
Summary: Choosing the Right Design

Multiclass learning requires that you make three decisions: choose the binary learner or perhaps a set of binary learners for different class combinations, choose the design matrix, and choose the decoding scheme. Choosing the binary learners, of course, involves optimizing their parameters. Searching for the best combination of many components is never easy. Perhaps that is why published results on multiclass reductions are contradictory.

The ECOC framework, in particular the complete design, is undoubtedly a powerful tool. Nonetheless, OVA remains the most popular approach in practice with OVO trailing as a close second. As documented in several empirical studies, advanced strategies often, if not usually, fail to outperform OVA and OVO in classification accuracy. Rifkin and Klautau (2004), for instance, argue that it is possible to provide the ultimate classification accuracy through OVA by fine-tuning the parameters of the binary learner. At the same time, proposals for new multiclass encoding and decoding schemes often claim superiority over the simple-minded strategies. Who is right?

The answer may be as simple as you need to be diligent at every stage of the analysis. It is more likely than not that a clever design produces a more accurate multiclass model than the good old OVA strategy would. But you need to optimize the binary learner (or learners) carefully, and you might want to look into more than one decoding scheme.

References

Allwein, E., Schapire, R., and Singer, Y. (2000) Reducing multiclass to binary: A unifying approach for margin classifiers. *J. Mach. Learn. Res.*, **1**, 113–141.

Beringer, J. *et al.* (Particle Data Group), *Phys. Rev. D*, **86**, 010001 (2012). (Statistics, Chapter 36).

Bevan, A. *et al.* (ed.) *Physics of the B Factories*, Springer (to be published in 2013).

Dietterich, T. and Bakiri, G. (1995) Solving multiclass learning problems via error-correcting output codes. *J. Artif. Intell. Res.*, pp. 263–286.

Escalera, S., Pujol, O., and Radeva, P. (2010) Error-correcting ouput codes library. *J. Mach. Learn. Res.*, **11**, 661–664.

Galar, M., Fernandez, A., Barrenechea, E., Bustince, H., and Herrera, F. (2011) An overview of ensemble methods for binary classifiers in multi-class problems: Experimental study on one-vs-one and one-vs-all schemes. *Pattern Recognit.*, **44** (8), 1761–1776.

Hastie, T. and Tibshirani, R. (1998) Classification by pairwise coupling. *Ann. Stat.*, **26** (2), 451–471.

Kuncheva, L. (2005) Using diversity measures for generating error-correcting output codes in classifier ensembles. *Pattern Recognit. Lett.*, **26**, 83–90.

Rifkin, R. and Klautau, A. (2004) In defense of one-vs-all classification. *J. Mach. Learn. Res.*, **5**, 101–141.

Rokach, L. (2010) *Pattern Classification Using Ensemble Methods, Machine Perception and Artificial Intelligence*, vol. 75, World Scientific.

Windeatt, T. and Ghaderi, R. (2003) Coding and decoding strategies for multi-class learning problems. *Inf. Fusion*, **4** (1), 11–21.

Wu, T.F., Lin, C.J., and Weng, R. (2004) Probability estimates for multi-class classification by pairwise coupling. *J. Mach. Learn. Res.*, **5**, 975–1005.

Zadrozny, B. (2001) Reducing multiclass to binary by coupling probability estimates. *NIPS'01*, pp. 1041–1048.

17
How to Choose the Right Classifier for Your Analysis and Apply It Correctly

In this book, we have reviewed a number of algorithms for classification. Here, we give a few practical recipes for choosing and applying classifiers to real-world data.

17.1
Predictive Performance and Interpretability

An ideal classifier is accurate and transparent. This classifier rarely makes a mistake predicting a class label for a new observation. At the same time, it is easy to understand why the classifier predicts this particular label. Both these goals are hardly ever satisfied for multivariate data. In practice, you can either get an accurate black box or a less accurate simple model. The first question you need to answer is what is more important to you – predictive power or interpretability.

You can estimate the range of the accuracy values in a quick experiment. Often, the most powerful variables for signal-to-noise discrimination are known from a similar analysis published earlier. Train several simple classifiers on a few such variables. Good candidates are: signal box (cuts imposed on individual variables), naive Bayes, linear discriminant, logistic regression, decision tree, and linear SVM. Then add more, perhaps many, variables with potential for noise discrimination. Grow an ensemble of decision trees by random forest or boosting. Compare the accuracy values obtained by the simple and complex classifiers. Then answer two questions:

1. Is the observed difference between the accuracy values significant?
2. Is it worth losing the transparency of a simple model for the observed accuracy gain?

The first question can be answered by one of the formal tests described in Chapter 10. The answer to the second question depends on the context of your analysis, your subjective preference and the standards adopted in the field.

Statistical Analysis Techniques in Particle Physics, First Edition. Ilya Narsky and Frank C. Porter.
©2014 WILEY-VCH Verlag GmbH & Co. KGaA. Published 2014 by WILEY-VCH Verlag GmbH & Co. KGaA.

17.2
Matching Classifiers and Variables

Ensembles of decision trees are robust to irrelevant variables, outliers, and missing values. They can easily handle a mix of real, ordinal, and nominal variables. Although tree ensembles do not necessarily provide the best accuracy for any given dataset, they often come close. The two most important tuning knobs are the stopping criterion for growing a tree (for instance, the minimal size of a leaf node) and the number of trees per ensemble. Random forest most usually performs quite well at the default minimal leaf settings suggested by Breiman, 1 for classification and 5 for regression, and attains the optimal accuracy with a few hundred trees. For boosted trees, the optimal leaf size (or a similar stopping criterion) can be found by a simple search. Boosting a few hundred trees usually proves sufficient, but sometimes a few thousand are needed. An additional tuning knob for some types of boosting is the learning rate described in Chapter 15.

Tree ensembles are a good default for data with more observations than variables (tall data). Data with more variables than observations (wide data) are atypical for physics analysis. Classification of wide data is most usually combined with variable selection. Linear methods such as SVM described in Section 13.5 and discriminant with regularization described in Section 11.1.5 often achieve good accuracy and provide semi-automated variable selection tools. Categorical variables can be "dummified" following the prescription in Section 7.1, and missing values can be imputed as discussed in Section 7.2.4. If wide data have many categorical variables or missing values, tree ensembles can be a competitive choice.

For any data, the power of a classifier can be improved by making use of known relations among the variables and the response. For example, in particle physics the mass of a resonance, m, is measured by reconstructing the energy E and momentum p and putting $m = \sqrt{E^2 - p^2}$. If all three variables can be good discriminators, pass all three to the classifier. If you pass only E and p, a powerful nonparametric learner could still find a good decision boundary between background and signal. Yet a classifier would learn more efficiently if it did not have to discover the $m = \sqrt{E^2 - p^2}$ transformation without your help.

17.3
Using Classifier Predictions

The next important question is how you wish to process the classifier prediction. Are you looking for a simple decision boundary or do you need a reliable estimate of the classification confidence at any point? As discussed in Chapter 9, most classifiers predict not only hard class labels but also soft scores. In physics, classification analyses generally fall in two categories. In the first case, the analyst searches for the optimal threshold on the predicted soft score by inspecting, for example, the associated ROC curve (see Section 10.2). Observations on one side of this threshold are then treated as signal candidates, and those on the other side are assumed to be background (for simplicity, we discuss binary classification only). The physics result in this case is based on the counts of the signal and background candidates seen in the data. This approach is called *event-counting analysis*. In the second case, the analyst passes the soft score, as well as other variables, to a likelihood fit used to estimate the signal size. We refer to this approach as *likelihood-fit analysis*.

Estimation of the statistical and systematic uncertainties for the signal magnitude is easier in the event-counting approach. The likelihood-fit analysis requires that the shapes of the soft score distributions for signal and background be accurately modeled. Verifying the accuracy of the soft score distribution may not be trivial. Monte Carlo simulation can differ from the real data in many ways. Comparing simulated and observed univariate distributions is not hard, but multivariate comparison can be challenging. If the soft score is used in likelihood fitting, the multivariate distribution for the variables passed to the classifier must be well modeled. A useful trick for verifying the quality of multivariate modeling is *slicing*, that is, inspecting the distribution of one variable while keeping all other variables fixed. Since the effort invested in such verification rapidly grows with dimensionality, using more than a few variables in this case does not seem like a good idea. In contrast, an event-counting analysis could still allow classification in dozens or hundreds of dimensions followed by conservative estimation of the uncertainties on the signal size.

17.4
Optimizing Accuracy

If your ultimate goal is accuracy, you should search for the optimal classifier. The amount of effort you need to invest in this search generally increases with data size and dimensionality. Almost any powerful, nonlinear classifier can do well for few observations in few dimensions. Even if some classifiers appear to have larger accuracy, this superiority cannot be convincingly demonstrated on small datasets. Many classifiers reviewed in this book can be applied with great success to tall data with many observations in up to a few dozen dimensions. You have especially many options if all variables are continuous, with smooth distributions, without missing values, and without outliers. Neural networks, SVM and kernel ridge regression with polynomial and Gaussian kernels, locally weighted regression, nearest neighbor rules, and, of course, decision tree ensembles are viable choices. Search for the optimal classifier not only across learning algorithms but also across parameters for each algorithm.

17.5
CPU and Memory Requirements

Finally, evaluate the CPU and memory requirements for your application, especially if this application must run continuously in an automated framework. The execution time and memory footprint scale differently for different algorithms. This applies to both training and prediction. The dependence of the time and memory on the data size and dimensionality is set by the underlying theory as well as the software implementation. In particular, kernel methods can quickly run out of memory because they typically require computing the full $N \times N$ Gram matrix for N observations. Clever tricks, most notably SMO and ISDA algorithms for SVM training described in Section 13.5.3, can ease the memory burden.

18
Methods for Variable Ranking and Selection

In particle physics, data are often analyzed using only thoroughly selected and well understood variables. The analysis chain, from preliminary crude selection requirements to last-stage refined techniques, consists of sequential modules, each operating on a few, hardly more than ten, variables. Optimal variables for a specific analysis are usually known in advance and often re-used from one version of the analysis to another. Analysts who are interested in this traditional approach may not find this chapter useful.

At the same time, studies of particle and astrophysics data with dozens or even hundreds of variables are not unknown. For instance, Yang *et al.* (2005) use more than 50 variables in their study of particle identification at MiniBOONE. In such a study, the analyst starts with many variables, a good fraction of which may not provide any background-signal separation power, and attempts to gain insight into their relative importance. Such insight leads to a better understanding of the data and sometimes new ideas for its analysis. This view prevails in other communities, for example biostatistics, where motivating variable selection by some fundamental principles is less common.

Estimation of importance can be followed by elimination of some variables. This step can be carried out to:

- reduce the CPU time and memory requirements of the learning algorithm;
- improve the classification accuracy for an algorithm sensitive to noise;
- eliminate potential sources of systematic error, and;
- make the analysis more understandable for your colleagues.

There is vast literature on variable selection. After giving basic definitions, we focus on practical, comprehensible algorithms. We then illustrate various algorithms on physics datasets.

Adopting the vocabulary of machine learning, we often say "feature" instead of "variable". "Attribute" and "predictor" are other popular names found in the literature.

18.1
Definitions

In this section, we define the terms "feature ranking" and "feature selection". Then we discuss two forms of feature relevance, weak and strong, and describe the two popular problems in feature selection: Minimal-Optimal and All Relevant.

18.1.1
Variable Ranking and Selection

Ranking is assigning a measure of importance to each variable. This measure is usually nonnegative, with a large value indicating an important variable. Selection is identifying a subset of all variables satisfying some optimality criterion.

Selection requires more effort than ranking. Any ranking procedure can be turned into a selection procedure by introducing a threshold. Variables with importance values below this threshold are discarded, and variables with importance values above this threshold are kept *in the model*. The choice of this threshold can be subjective or motivated by a rigorous statistical test. Often, a trade-off between the model accuracy and the number of in-model variables is considered.

The rigorous statistical test mentioned above computes a *p*-value for the null hypothesis "there are no useful features in this subset" against its alternative. A subset can be formed of one or more features. The decision, to keep or discard this subset, is made by comparing the computed *p*-value with some prespecified level. In a typical feature selection scheme, such a test may be carried out many times on different subsets. The prespecified level in this case needs to be adjusted for multiple tests. We discuss this problem in Section 18.3.2.

18.1.2
Strong and Weak Relevance

We would like to identify useful features in the observed data. But what does "useful" mean? This question can be answered in a number of ways. John *et al.* (1994) give several definitions of "usefulness", or *relevance*. We use two of them, also adopted by Nilsson *et al.* (2007) and Yu (2008), to explain various concepts in this chapter.

Suppose we have D features. Let X_d be a random variable for the dth feature, and let x_d be a specific value of X_d. Let X be a set of all measured features. Let $S \subset X$ be a subset of X. Similarly, s denotes a specific value of the random set S. Let $\bar{X}_d$ be a subset complementary to X_d; "complementary" means that this subset includes all features but X_d. Let Y be a random variable for the response; for classification, Y represents class. We are now ready to give formal definitions of feature relevance.

We say that variable Y is conditionally independent of variable X_d on subset S if the two conditional probabilities,

$$P(y|x_d, s) = P(y|s), \tag{18.1}$$

are equal. Borrowing notation from Nilsson *et al.* (2007), we denote this by

$$Y \perp X_d | S . \tag{18.2}$$

If Y is not conditionally independent of X_d on S, we write $Y \not\perp X_d | S$. Feature X_d is *strongly relevant* if the response is not conditionally independent of this feature on its complementary set,

$$Y \not\perp X_d | \bar{X}_d . \tag{18.3}$$

Feature X_d is *weakly relevant* if it is not strongly relevant and we can find a subset on which the response conditionally depends on this feature,

$$Y \perp X_d | \bar{X}_d \quad \cap \quad \exists S \subset \bar{X}_d \; : \; Y \not\perp X_d | S . \tag{18.4}$$

If a feature is neither strongly nor weakly relevant, we call it *irrelevant*.

Set $M(Y)$, $M \subseteq X$, is called a *Markov blanket*, or *Markov boundary*, for the response Y if the response is conditionally independent on M of all features not in M,

$$Y \perp \bar{M}(Y) | M(Y) . \tag{18.5}$$

The Markov boundary for the response Y therefore contains at least all strongly relevant features in X. To define a Markov boundary, $M(X_d)$, for variable X_d, we could re-use the definition above with one modification. Obviously, a variable must be excluded both from its Markov boundary and the complementary set. We postulate that $X_d \notin M(X_d)$ and subtract the variable from the complementary set explicitly. Set $M(X_d)$ is a Markov boundary for variable X_d if

$$X_d \perp (\bar{M}(X_d) \setminus X_d) | M(X_d) . \tag{18.6}$$

The above definitions are given for data (X, Y) irrespective to how these data are analyzed. In practice, we learn a classifier $f(x)$ mapping domain $\mathcal{X}$ onto domain $\mathcal{Y}$. Assume that this classifier predicts a vector of posterior class probabilities; we emphasize this by putting f in bold. A feature X_d is *relevant to the classifier* $f(x)$ if the classifier prediction depends on it:

$$\exists \; x'_d, x''_d, x'_d \neq x''_d, \bar{x}_d \; : \; f(x'_d, \bar{x}_d) \neq f(x''_d, \bar{x}_d) . \tag{18.7}$$

(We state this definition loosely; refer to Nilsson *et al.* (2007) for rigor.) To be able to detect, for instance, strongly relevant features in the sense of definition (18.3), we must use a classifier for which (18.3) implies (18.7) and vice versa. This requirement would be satisfied by the *perfect classifier*, that is, one predicting the true posterior probabilities for all classes at any x. In practice, a classifier can err on either side. If the classifier is crude, its prediction may not show any sensitivity to strongly relevant features. If the classifier is noisy, its prediction may depend on irrelevant features.

Using these definitions, Nilsson et al. (2007) identify two classes of feature selection algorithms:

- *Minimal-Optimal*: Find the Markov boundary for the response variable.
- *All Relevant*: Find all relevant (strongly or weakly) features.

Most publications on feature selection attempt to solve the Minimal-Optimal problem, sometimes without stating it explicitly. In our experience, most physics analyses making use of feature selection techniques also aim at solving the Minimal-Optimal problem. Nilsson et al. (2007) prove that a theoretical algorithm with time complexity $O(D^2)$ for the Minimal-Optimal problem exists. An example of such an algorithm is sequential backward elimination reviewed in Section 18.2.2. Note however that Nilsson et al. (2007) limit this proof to the case of strictly positive distributions $P(x, y)$, which may not hold in practice. The All-Relevant problem is computationally harder. In principle, this problem amounts to an exhaustive search through all 2^D feature subsets.

Earlier we stated that the Markov boundary for the response must include at least strongly relevant features. A subset of weakly relevant features may need to be included as well. Suppose the feature set contains two identical variables significantly correlating with the class label and independent of all other variables. By definitions (18.3) and (18.4), both are weakly relevant and neither is strongly relevant. Clearly, the minimal-optimal set must include one of them.

Yu (2008) formalizes the problem by introducing redundant variables. Feature $X_d \in S$ is *redundant with respect to set* S if it is weakly relevant and has a nonempty Markov boundary $M(X_d)$ in S. The minimal-optimal set S^* then contains all strongly relevant features found in the full set X and weakly relevant features in X that are not redundant with respect to S^*. This optimal set can be constructed by sequentially eliminating one irrelevant or redundant feature at a time and re-evaluating redundancy for the remaining features. Take the two identical features from the example above. These are redundant on the full feature set X. If one of these two features is removed, the other one is no longer redundant. Since the remaining feature is independent of all others, it will not be recognized as redundant in the consequent steps and therefore never eliminated.

Nilsson et al. (2007) prove that for a strictly positive distribution $P(x, y)$ the Markov boundary for the response Y, or equivalently the minimal-optimal set, is equal to the set of all strongly relevant features. Weakly relevant features are therefore redundant on any set including the minimal-optimal set as a subset. To appreciate this point, consider a binary classification problem shown in Figure 18.1. The empirical class distributions shown in Figure 18.1a are perfectly separated in each variable. Both variables appear weakly relevant and neither seems strongly relevant, implying that one of them can be discarded. Now fit the sample for each class to a normal pdf and mix the two distributions in the proportion set by the sample sizes. A contour plot of the resulting joint pdf $P(x, y)$ shown in Figure 18.1b indicates that the optimal decision boundary is given by the diagonal line connecting the upper left and lower right corners. Both variables are strongly relevant, and neither can be discarded. This example illustrates a common dilemma faced by practition-

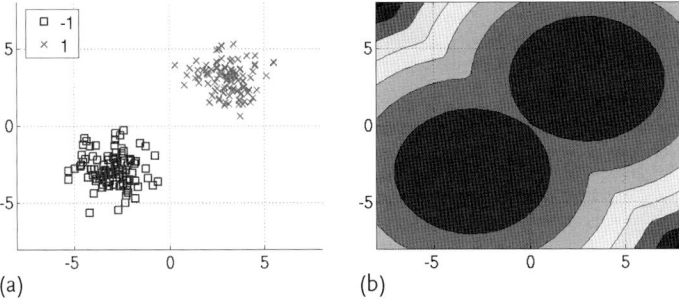

Figure 18.1 Using two features for binary classification. (a) Finite sample from a bivariate pdf $P(x, y)$. (b) Contour plot of log $P(x, y)$.

ers. While the true joint pdf $P(x, y)$ is most usually strictly positive in the entire domain of definition, the analyst may not be able to estimate this pdf reliably in a sub-domain. In practice, this lack of an estimate is often accounted for by zeroing the pdf in the respective sub-domain. The optimal decision boundary then cannot be unambiguously determined, and weakly relevant variables may be included in the minimal-optimal set. The notion of redundancy thus helps solve the problem with insufficient information.

18.2 Variable Ranking

As noted in Section 18.1.1, ranking amounts to assigning a measure of importance to each variable. Algorithms for computing such measures fall in three broad categories: filters, wrappers, and embedded methods.

John *et al.* (1994) are credited for introducing the terms "filter" and "wrapper". A *filter* performs feature ranking without learning a classification model. Filters often use simple measures of dependence between a feature and the class label, and among features. Examples of such measures are the Pearson correlation and mutual information. A *wrapper* ranks features by computing their effect on the predictive power of a chosen classifier. A wrapper must be sufficiently generic, that is, it can be used with most, if not all, classifiers. For example, sequential backward elimination (SBE) at every step examines all features and discards the one with the smallest effect on the classification accuracy. SBE proceeds recursively until only one feature is retained in the model.

In addition to wrappers and filters, we recognize embedded algorithms. An *embedded algorithm* measures the importance of a feature using tools specific to the chosen classifier. For instance, the importance of a variable for decision tree induction can be estimated by summing gains in the Gini index (or a similar criterion) over all splits imposed on this variable.

We describe filters, wrappers, and embedded methods for variable ranking, but the same classification applies to variable selection. As noted in Section 18.1.1, any ranking procedure can be turned into a selection procedure by introducing a threshold on the feature rank. More complex selection algorithms fit in these three categories as well.

The distinction between filters, wrappers, and embedded methods can be fuzzy. For instance, ReliefF reviewed in Section 18.2.3 is usually described as a filter method, but in fact it is intimately tied to the nearest neighbor search. We can expect that features ranked important by ReliefF be powerful for nearest neighbor rules and other local learners. A decision tree, on the other hand, may focus on a different feature subset.

Based on the discussion in Section 18.1.2, you may be tempted to conclude that filters are used for selecting strongly relevant features, while wrappers and embedded methods are best suited for finding features relevant to the chosen classifier. In practice, a wrapper or an embedded method coupled with a powerful classifier can identify strongly relevant features more efficiently than a simple-minded filter. Wrappers are often favored for accuracy, and filters are often preferred for speed. The speed and accuracy for embedded methods vary. For example, the gain in the Gini index is obtained at no extra cost because it needs to be computed for tree induction anyway; this measure however is inaccurate for reasons to be discussed later. The ACE algorithm reviewed in Section 18.3.3 grows ensembles of decision trees with surrogate splits, and tends to be accurate and slow. We will discuss options at both sides of the spectrum.

18.2.1
Filters: Correlation and Mutual Information

Two simple measures, correlation and mutual information, can be used to estimate feature relevance. Their main advantage is fast computation. The Pearson correlation is well known to physicists. Two other forms of correlation, the Kendall tau and Spearman rank correlation, can be used for ordinal variables and continuous variables with outliers. Mutual information is a more reliable measure of dependence, although its use in physics analysis at this point is rather limited.

18.2.1.1 Measures of Correlation

One approach to identifying important variables is by computing their pairwise correlations with the response. Similarly, to identify irrelevant and redundant variables, we can compute pairwise correlations for all variables in the data.

Three types of correlation are encountered in statistics: Pearson correlation, Kendall tau, and Spearman rank correlation. The Pearson correlation is more popular in physics analysis than the other two. This correlation can be computed only if both variables are defined on interval scales. The other two correlation statistics require that both variables be defined on ordinal scales. The interval and ordinal scales were defined in Section 7.1. Any statistic defined on an ordinal scale could be applied to an interval scale, but not the other way around. When applied to continuous variables, the Kendall tau and Spearman rank correlation statistics show less sensitivity to outliers compared to the Pearson statistic.

We give definitions only for estimators of the respective statistics, using the hat notation. For two observed paired vectors $\mathbf{x} = \{x_n\}_{n=1}^N$ and $\mathbf{y} = \{y_n\}_{n=1}^N$, the *Pearson correlation* is given by

$$\hat{r} = \frac{\sum_{n=1}^N (x_n - \bar{x})(y_n - \bar{y})}{\sqrt{\sum_{n=1}^N (x_n - \bar{x})^2 \sum_{n=1}^N (y_n - \bar{y})^2}}. \tag{18.8}$$

Above, $\bar{x}$ and $\bar{y}$ are the observed means. The *Kendall tau* is defined by

$$\hat{\tau} = \frac{\sum_{i=1}^N \sum_{j=1}^N I(x_i > x_j \cap y_i > y_j) - \sum_{i=1}^N \sum_{j=1}^N I(x_i > x_j \cap y_i < y_j)}{N(N-1)/2}. \tag{18.9}$$

The indicator function I evaluates to 1 when its argument is true and 0 otherwise. The tau statistic has several versions designed to work with ties, $x_i = x_j$ or $y_i = y_j$, described in textbooks. To compute the *Spearman rank correlation*, replace elements in $\mathbf{x}$ and $\mathbf{y}$ by their observed ranks: sort $\mathbf{x}$ in ascending order and replace every x_n with its position in the sorted vector; similarly for $\mathbf{y}$. If all values in a vector are unique, the computed ranks are integers; in case of ties, a rank has a $1/T$ remainder for T observations tied for this rank. After replacing original values with their ranks, compute the Spearman correlation $\hat{\rho}$ using (18.8). All three correlation statistics range from -1 to $+1$. A value close to zero indicates that the two variables are independent, and a value with magnitude close to 1 indicates strong dependence.

As you may recall from Section 7.1, class labels are nominal and therefore have no provision for rank comparison. Correlation between a continuous variable and a class label can nevertheless be computed for binary classification by representing each class with an integer. In case of three or more classes, none of the discussed correlation statistics applies.

Correlation is an unreliable measure of dependence. If its magnitude is close to 1, we can stipulate strong dependence. The inverse is not true: if the correlation is close to zero, the two variables can nevertheless exhibit strong dependence. Take a uniform random variable X with support $[-1, 1]$, and let $Y = |X|$. The correlation between X and Y is zero; yet X gives complete information about Y.

18.2.1.2 Mutual Information

The three correlation measures reviewed in the previous section can fail to detect a strong dependence between two variables in fairly simple cases. A more robust measure of dependence is mutual information introduced in Section 8.4, also known as the Kullback–Leibler divergence. Mutual information between continuous variables X and Y is defined by

$$J(X, Y) = \int_X \int_Y P_{X,Y}(x, y) \log \frac{P_{X,Y}(x, y)}{P_X(x) P_Y(y)} dx\, dy \tag{18.10}$$

for the joint density $P_{X,Y}(x,y)$ and marginal densities $P_X(x)$ and $P_Y(y)$. If one of the variables is discrete, replace the respective integral by a sum. The integral (or sum) in (18.10) can be computed for any domains $\mathcal{X}$ and $\mathcal{Y}$ with nonzero Lebesgue measure, including nominal variables. As shown in Section 8.4, mutual information must be nonnegative.

We use log above to denote the natural logarithm. Sometimes $\log_2$ is used in the definition (18.10) instead. This choice is of no practical consequence since the absolute scale for mutual information never matters.

Mutual information is less intuitive than correlation. It is not bounded from above. A feature selection procedure based on the definition (18.10) would favor continuous variables over discrete, and discrete variables with many states over discrete variables with few states. This selection bias toward multistate features can be cured by normalization. An alternative definition of the mutual information,

$$J(X, Y) = H(X) - H(X|Y), \qquad (18.11)$$

in which X and Y can be swapped, leads to a *symmetric uncertainty* measure (Yu and Liu, 2004),

$$\tilde{J}(X, Y) = \frac{2 J(X, Y)}{H(X) + H(Y)}. \qquad (18.12)$$

For discrete variables, the measure (18.12) lies between 0 and 1. Zero indicates independence of X and Y, and one implies that X is completely explained by Y (and vice versa). The entropy $H(X) = -\sum_{\mathcal{X}} P(x) \log P(x)$ for discrete X is called the *Shannon entropy*. Unfortunately, the *differential entropy*, $H(X) = -\int_{\mathcal{X}} P(x) \log P(x) dx$, for continuous X does not approach the Shannon entropy as the integration step approaches zero. The two entropy definitions differ by an infinite constant. The symmetric uncertainty (18.12) can be consistently applied to discrete variables only. Continuous features need to be discretized, and the value of $\tilde{J}(X, Y)$ would depend on the arbitrarily chosen discretization step. Note that the unnormalized version (18.10) is not exposed to this fundamental arbitrariness because it is based on the ratio of the probability density (or mass) functions.

18.2.1.3 Computing Mutual Information

Although more robust than pairwise correlation, mutual information may not be trivial to compute. Computing mutual information for two categorical variables is easy. Computing mutual information for two continuous or one continuous and one categorical variable takes some effort.

For two continuous variables, mutual information can be computed by nearest neighbors. For every two-dimensional point (x_n, y_n) in a set with N points, Kraskov et al. (2004) and Kraskov et al. (2011) find the Chebyshev, or infinite-norm, distance to its Kth nearest neighbor, $d_K^{(n)}$. To find the Kth nearest neighbor, compute $|x_i - x_n|$ and $|y_i - y_n|$; $i = 1, \ldots, N$; $i \neq n$; and set $d_i = \max(|x_n - x_i|, |y_n - y_i|)$. Sort the d_i values in ascending order and set $d_K^{(n)}$ to the Kth element in the sorted list. Kraskov et al. (2004) count points within $d_K^{(n)}$ of the nth point in x and y; let these counts be $N_X^{(n)}$ and $N_Y^{(n)}$, respectively. The mutual information is approximated by

$$\hat{J}(X, Y) = \psi(K) - \frac{1}{N} \sum_{n=1}^{N} \left[N_X^{(n)} + N_Y^{(n)} \right] + \psi(N), \quad (18.13)$$

where $\psi(v) = -d \log \Gamma(v)/dv$ is the digamma function. Kraskov et al. (2004) recommend $K = 3$ as the rule-of-thumb choice.

Leonenko et al. (2008) find three Kth nearest neighbors to the nth point. The first nearest neighbor is found using x, the second one is found using y, and the third one is found using (x, y). All three neighbors are found using the Euclidean distance. Let $d_X^{(n)}$, $d_Y^{(n)}$ and $d_{XY}^{(n)}$ be the respective distances. The mutual information is then estimated by putting

$$\hat{J}(X, Y) = \log(N-1) - \psi(K) + 2\log 2 - \log \pi + \frac{1}{N} \sum_{n=1}^{N} \left[d_X^{(n)} + d_Y^{(n)} - 2 d_{XY}^{(n)} \right]. \quad (18.14)$$

They recommend $K = 7$ as a good computational choice.

The two nearest-neighbor methods give good approximations for smooth bivariate distributions. Before computing $\hat{J}(X, Y)$, it is best to standardize variables and, if necessary, apply univariate transformations to smoothen their distributions. kd-trees discussed in Section 13.6.3 can speed up the nearest neighbor search in two dimensions.

Computing mutual information for one categorical and one continuous variable requires discretization of the continuous variable. We are not aware of any papers discussing such discretization. The choice of optimal binning for estimation of univariate entropy is discussed in Hall and Morton (1993). One might expect that the bin width optimized for the univariate case should be close to optimal in the bivariate case. Hall and Morton (1993) propose choosing the optimal bin width h^* by minimizing a penalized log-likelihood,

$$h^* = \arg\min_{h} \left[\frac{1}{N} \sum_{b=1}^{B} I(N_b > 0) + \log(Nh) - \frac{1}{N} \sum_{b=1}^{B} N_b \log N_b \right], \quad (18.15)$$

where the summation is taken over B bins with N_b counts per bin. The first term on the right-hand side is the penalty due to the number of bins with nonzero counts. This empirical rule is derived for densities with tails decaying at the rate of $\beta |x|^{-\alpha}$ with $\alpha > 1$. The optimal bin width is expected to scale as $h^* \propto N^{-\gamma}$ with $\gamma = (\alpha - 1)/(3\alpha - 1)$.

Semi-continuous variables present the hardest case. Estimates of the mutual information obtained by the nearest-neighbor methods can be very inaccurate, and the optimal bin width found by (18.15) typically takes an impractical low value. For the lack of a better prescription, we bin such variables using the normal reference rule described in Chapter 5. An example of such a variable is taken from

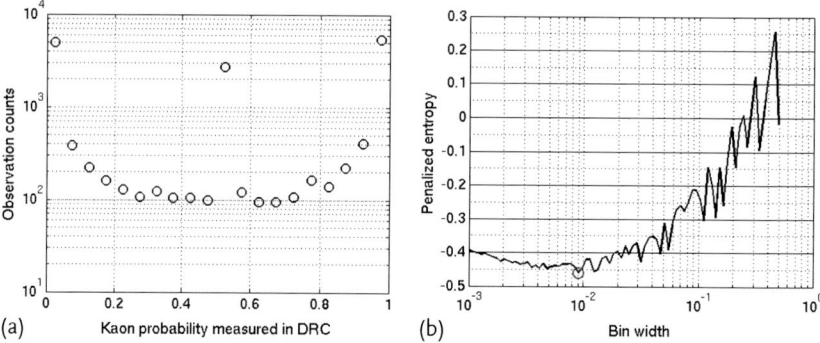

Figure 18.2 (a) Probability of a kaon track as measured by the DIRC detector in BaBar on the logarithmic scale. (b) Penalized log-likelihood of the entropy versus bin width. The optimal bin width is shown with a circle.

the K/π identification analysis described in Bevan et al. (2013). The kaon probability measured by the detector of internally reflected Cherenkov light (DIRC) for tracks above 3 GeV is histogrammed in Figure 18.2a. 15 841 observations are used to make this histogram, of which 2628 are exactly at 0.5 and 504 are exactly at 1. An attempt to find the optimal bin width using the empirical rule (18.15) predictably fails: the found value is unacceptably small, 1.1×10^{-4}. The mutual information between this variable and the class label (kaon or pion) estimated at this bin width is 0.498. The bin width computed by the normal reference rule is 5.9×10^{-2} with the respective mutual information value 0.385. To verify this computation, we remove observations below 0.01, at 0.5 and above 0.99. As shown in Figure 18.2b, the empirical rule (18.15) puts the optimal bin width for the remaining nonpeaking distribution at 9.1×10^{-3}. Using this width to discretize the full distribution, we estimate the mutual information to be 0.390. The small difference between this estimate and the one obtained by the normal reference rule is tolerable.

18.2.2
Wrappers: Sequential Forward Selection (SFS), Sequential Backward Elimination (SBE), and Feature-based Sensitivity of Posterior Probabilities (FSPP)

In the previous section, we have described simple measures of variable importance computed without learning a classifier. Here, we describe algorithms for ranking features by evaluating their effect on the classification accuracy or a similar measure of the predictive power. These algorithms can be used in combination with any classifier included in this book.

Sequential selection is a popular and powerful tool for feature ranking in low-dimensional data. It comes in two flavors: *sequential forward selection* (SFS) and *sequential backward elimination* (SBE). Prior to executing either algorithm, you need to define a criterion, $Q(S, Y)$, for assessing the quality of an arbitrary feature subset S for class labels Y. A common choice is classification error or some other measure of the predictive power for a classifier applied to S and Y. By convention, a lower value of Q indicates a better feature subset. An obvious choice for Q is the expected loss $E_{S,Y}\ell(Y, f(S))$ introduced in (9.1).

Let S be the current set of selected features and $\bar{S}$ be the complementary set of all features in X not included in S. SFS starts with a null set, $S = \emptyset$. At every pass, SFS loops over features in $\bar{S}$. At every step of this loop, SFS takes feature $X_d \in \bar{S}$ and evaluates the criterion on the selected set extended with this feature, $Q(S \cup X_d, Y)$. After the loop is completed, SFS finds the feature X_d^* with the smallest criterion and moves this feature from the complementary set to the selected set, $S = S \cup X_d^*$ and $\bar{S} = \bar{S} \setminus X_d^*$, where $\setminus$ is a set subtraction operator. SFS then proceeds to the next pass. This procedure is repeated until there are no features left in the complementary set, that is, $\bar{S}$ equals $\emptyset$.

SBE starts with the full feature set, $S = X$, and works in the opposite direction. At every pass, SBE loops over features in S. At every step of the loop, SBE evaluates the criterion on the selected set with one feature removed, $Q(S \setminus X_d, Y)$, for some $X_d \in S$. After the loop is completed, SBE finds the feature X_d^* with the largest criterion and moves this feature from the selected set to the complementary set, $S = S \setminus X_d^*$ and $\bar{S} = \bar{S} \cup X_d^*$. This procedure is repeated until S is empty.

SFS and SBE are usually described as variable *selection* methods. In fact, SFS and SBE are merely search strategies. The order in which features are added or removed defines their ranks. In addition, the analyst can impose a stopping rule. Without such a rule, SFS and SBE can be used for variable ranking only. The stopping rule can be as simple as "stop when the selected set has a prespecified number of features" or "stop when the criterion Q at this pass is worse than the one at the previous pass". Stopping rules based on statistical tests are discussed in Section 18.3.2.

Shen *et al.* (2008) propose ranking features by their influence on the posterior probability estimates in binary classification. Let $Y \in \{-1, +1\}$. To estimate the importance of feature d, the *feature-based sensitivity of posterior probabilities* (FSPP) algorithm compares posterior class probabilities $P(y|x)$ obtained on the full set X to those obtained on a subset with the dth feature removed, $X^{(-d)}$:

$$\hat{i}_d = \frac{1}{N} \sum_{n=1}^{N} \left| \hat{P}(Y = +1|x_n) - \hat{P}\left(Y = +1|x_n^{(-d)}\right) \right| . \quad (18.16)$$

The posterior probabilities need to be estimated by a certain classifier. If the classifier tends to overtrain, as ensembles do, the value of $\hat{i}_d$ should be computed using the posterior probabilities estimated either on cross-validated or independent test data. Shen *et al.* (2008) use SVM with a sigmoid transformation applied to the SVM scores for conversion to probability estimates. Yang *et al.* (2009) extend this analysis to neural networks. Arnosti and Danyluk (2012) generalize this prescription to any classifier.

A slightly different version of the FSPP algorithm permutes values across the dth variable instead of removing the feature. Shen *et al.* (2008) prove that the two versions of FSPP give identical answers in the limit of infinite statistics.

Although this method is designed for posterior probabilities, we see no fundamental reason for not using raw classification scores, assuming that removing or permuting a feature does not change their scale.

It seems counterintuitive to assign the same importance to a variable increasing and reducing the posterior probability for the true class at x by the same amount. Arnosti and Danyluk (2012) propose replacing (18.16) with

$$\hat{i}_d = \frac{1}{N} \sum_{n=1}^{N} \left[\hat{P}(y_n|x_n) - \hat{P}\left(y_n|x_n^{(-d)}\right) \right]. \qquad (18.17)$$

For two classes, there is a monotone mapping from the estimated posterior probability to the classification margin, and we can replace $\hat{P}(y|x) - \hat{P}(y|x^{(-d)})$ with the difference of the two margins, $m(x, y; f) - m(x^{(-d)}, y; f^{(-d)})$, for classifier f trained on the full feature set and classifier $f^{(-d)}$ trained on the set with the dth feature removed. As defined in Section 15.1.4, the margin $m(x, y; f)$ is the difference between the score for class y and the score for the complementary class $\bar{y}$. We can use the margin-based definition for data with more than two classes as well. In that case, the margin is the difference between the score for class y and the largest score for the other classes.

Alternatively, we could estimate the importance of the dth feature by comparing the observed margin distributions $m(x_n, y_n; f)$ and $m(x_n^{(-d)}, y_n; f^{(-d)})$. Since the margins do not need to be normally distributed, we apply the right-tailed Wilcoxon signed-rank paired test to compute a p-value for the null hypothesis "the median of the difference $m(x_n, y_n; f) - m(x_n^{(-d)}, y_n; f^{(-d)})$ does not exceed zero". This test assumes that the observed pairs, $\{m(x_i, y_i; f), m(x_i^{(-d)}, y_i; f^{(-d)})\}$, and $\{m(x_j, y_j; f), m(x_j^{(-d)}, y_j; f^{(-d)})\}$, are independent for any i and j; $i \neq j$. If the p-value is small, feature d is likely to be relevant for classifier f.

18.2.2.1 Example: SBE and FSPP for the MAGIC Telescope Data

Using random forest for the underlying classifier, we apply SBE and FSPP to the MAGIC telescope data described in Bock et al. (2004). Variable ranks obtained by SBE and estimates of variable importance by FSPP are shown in Figure 18.3. We also compute mutual information between each variable and the class label. To compute the mutual information, we optimize the bin width following the prescription (18.15) for the first seven variables. This prescription fails for the last three variables, and we bin them using the normal reference rule. Variable importance estimates computed by ReliefF, discussed in Section 18.2.3, are also shown. The ranks of the ten variables in the MAGIC telescope data obtained by the three techniques are shown in Table 18.1. The four methods agree that fAlpha is the most important variable. Predictions for the other variables differ.

As seen from this example, different ranking methods can disagree about the relative importance of the analyzed variables. Interpretation of such results is inevitably subjective.

We include a MATLAB example below.

18.2 Variable Ranking

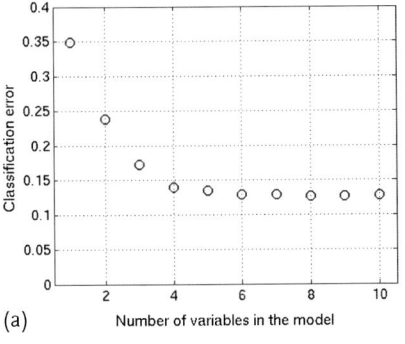
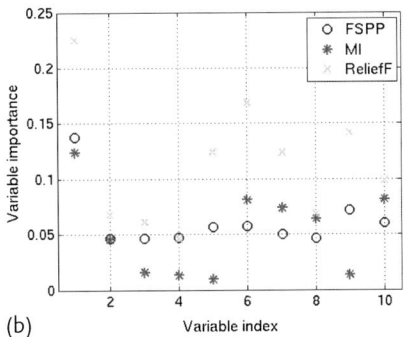

Figure 18.3 (a) Ranking variables in MAGIC telescope data by SBE. (b) Variable importance computed by FSPP, mutual information and ReliefF. The SBE curve shows classification error versus the number of variables in the model. Random forest with 100 trees is used for SBE and FSPP. Importance estimates for ReliefF are multiplied by 10 to put them on the same scale.

Table 18.1 Ranks for variables in the MAGIC telescope data computed by SBE, FSPP, mutual information, and ReliefF.

	SBE	FSPP	MI	ReliefF
fAlpha	1	1	1	1
fAsym	8	9	10	6
fConc	5	10	6	9
fConc1	10	6	7	5
fDist	6	5	8	7
fLength	4	7	2	10
fM3Long	7	4	3	8
fM3Trans	9	3	9	2
fSize	2	8	4	3
fWidth	3	2	5	4

Contents
- Load data
- Prepare a pool of parallel workers
- Run sequential feature selection
- Obtain feature ranks and criterion values
- Estimate feature importance by FSPP.

Load data
We use the MAGIC telescope data from the UCI repository http://archive.ics.uci.edu/ml/datasets/MAGIC+Gamma+Telescope. Earlier we split these data in training and test sets in the proportion 2 : 1. In this exercise, we use only the training

data and estimate the model accuracy by cross-validation. These data have about 12 742 observations and 10 features.

```
load MagicTelescope
size(Xtrain)

ans =
    12742   10
```

Prepare a pool of parallel workers

To speed up computation, we use the Parallel Computing Toolbox. In this example, parallel computation is used twice. First, we estimate the classifier accuracy for SBE using 5-fold cross-validation with each fold processed in parallel. Second, we rank variables by the FSPP method: we remove one variable at a time and evaluate the classifier accuracy on the feature set with this variable eliminated. Evaluation of the reduced feature sets is carried out in parallel.

To prepare for parallel computation, we open a pool with 3 workers. The computer used in this example has 4 cores, and we could open at most 4 workers.

```
matlabpool open 3
```

Run sequential feature selection

The `sequentialfs` function in the Statistics Toolbox provides SFS and SBE. To use this function, we specify the optimized criterion, Q. We set Q to an anonymous function handle with 4 input arguments: xtrain (training data), ytrain (class labels for the training data), xtest (test data), and ytest (class labels for the test data). We pass the training data to the `fitensemble` function to grow an ensemble of 100 trees by bagging. Then we use the `loss` method of the trained ensemble to estimate the classification error for the test data. Function `sequentialfs` assumes that the criterion is computed by taking a *sum* over all observations in the test set, and `loss` returns the *mean* classification error. To compensate for this discrepancy, we multiply the output of `loss` by the number of observations in the test set.

```
Q = @(xtrain,ytrain,xtest,ytest) ...
    size(xtest,1)*loss( fitensemble(xtrain,ytrain,...
    'Bag',100,'Tree','type','classification'), xtest, ytest);
```

Having set up Q, we define options for `sequentialfs`. By setting UseParallel to true, we tell `sequentialfs` to execute cross-validation in the parallel mode. We use cross-validation with 5 folds specified by the CV argument. We set the search direction to "backward" for SBE. By setting NFeatures to zero, we tell `sequentialfs` to continue eliminating features until none are left. If we neglected to do so, `sequentialfs` would only remove features as long as Q would not not increase.

The `sequentialfs` function returns two output arguments, a vector of in-model feature indices and a search history. The search history shows criterion values and in-model features at every step. Here, the vector of in-model features would be empty because we set the minimal number of in-model features to 0. Since we do not need it, we replace the first output with a tilde.

```
options = statset;
options.UseParallel = true;
[~,history] = sequentialfs(Q,Xtrain,Ytrain,'CV',5,...
    'Direction','backward','NFeatures',0,'Options',options);
```

Obtain feature ranks and criterion values

The `history` output is a logical array of size D-by-D for D features. Every row of this array shows features included in the model at the respective step. Features with large ranks (unimportant) are eliminated first, and features with low ranks (important) are eliminated last. We save the feature ranks in variable `ranked`.

```
D = size(Xtrain,2);
ranked = 1 + D - sum(history.In,1)

ranked =
     1     8     6    10     7     4     5     9     2     3
```

Estimate feature importance by FSPP

First, we bag 100 trees on the full feature set and obtain `PfitFull`, posterior class probability estimates, by 5-fold validation. Since this is binary classification, we only need posterior probability estimates for one class. We retain one column of the N-by-2 array of probabilities for the "Gamma" class.

```
cvens = fitensemble(Xtrain,Ytrain,'Bag',100,'Tree',...
        'type','classification','kfold',5);
[~,PfitFull] = kfoldPredict(cvens);
PfitFull = PfitFull(:,strcmp(cvens.ClassNames,'Gamma'));
```

We eliminate one feature at a time and estimate `Pfit`, the posterior class probabilities without this feature. The mean magnitude of the change in the posterior probability estimates for the "Gamma" class is used to compute the importance of this feature. We save these importance values in the `fsppImp` variable.

The loop over D features is a `parfor` loop, where `parfor` stands for "parallel for". One pass of the loop runs independently on one of the pool workers. Instead of looping through the features consecutively, we process up to 3 features at a time.

```
fsppImp = zeros(1,D);
parfor d=1:D
    X = Xtrain;
    X(:,d) = [];
    cvens = fitensemble(X,Ytrain,'Bag',100,'Tree',...
            'type','classification','kfold',5);
    [~,Pfit] = kfoldPredict(cvens);
    Pfit = Pfit(:,strcmp(cvens.ClassNames,'Gamma'));
    fsppImp(d) = mean(abs(PfitFull-Pfit));
end
```

18.2.3
Embedded Methods: Estimation of Variable Importance by Decision Trees, Neural Networks, Nearest Neighbors, and Linear Models

Unlike filters and wrappers reviewed in the earlier sections, embedded algorithms for feature ranking are tied to unique properties of the respective classifiers. Here, we provide a short summary of embedded techniques for popular classification algorithms. The reader is encouraged to study the references for detail.

Measures of variable importance for decision trees are reviewed in Section 14.9. Random forest and boosted trees can use the same measures by averaging estimates obtained for individual trees in an ensemble. Draminski *et al.* (2008) grow an ensemble of trees by random subspace and estimate the importance of a feature by averaging impurity gains due to splits on this feature. Instead of using simple averaging, they weight each impurity gain by the accuracy of the respective tree measured on an independent test set.

A good review of methods for neural nets can be found in Gevrey *et al.* (2003) with quantitative comparison studies performed on small low-dimensional datasets in Olden *et al.* (2004).

If continuous features are standardized (transformed to zero mean and unit variance), their importance estimates can be derived from magnitudes of the respective coefficients in a linear model. It is not unusual to select variables for linear discriminant analysis (LDA) by thresholding on the coefficient magnitudes. For K classes, LDA constructs $K(K-1)/2$ separating hyperplanes, and every feature is thus associated with $K(K-1)/2$ with coefficients. You can threshold on the coefficient with the largest magnitude for this feature. A sequential feature selection procedure based on the coefficient magnitude is described in Guyon *et al.* (2002) for binary linear SVM.

ReliefF, introduced in Robnik-Sikonja and Kononenko (2003), is often described as a filter method, but has strong ties to nearest neighbor classification. Let x_r be a randomly selected observation with class label y_r. ReliefF finds a set $\mathcal{H}$ of nearest *hits* (observations of the same class) and a set $\mathcal{M}_k$ of nearest *misses* (observations of a different class) for each class $k = 1, \ldots, K$; $k \neq y_r$. The sizes of the K sets, $|\mathcal{H}|$ and $|\mathcal{M}_k|$, are typically set equal. The importance of the dth feature for the rth observation, $\hat{i}_d^{(r)}$, is then estimated by

$$\hat{i}_d^{(r)} = -\frac{1}{|\mathcal{H}|} \sum_{x_i \in \mathcal{H}} |x_{rd} - x_{id}| + \frac{1}{|\mathcal{M}|} \sum_{k \neq y_r} \frac{\pi_k}{1 - \pi_{y_r}} \sum_{x_i \in \mathcal{M}_k} |x_{rd} - x_{id}|, \quad (18.18)$$

where x_{rd} is the value of the dth feature for the rth observation, π_k is the prior class probability for class k, either chosen by the analyst or estimated empirically from the observed data, and $\mathcal{M} = \mathcal{M}_1 \cup \ldots \cup \mathcal{M}_K$; $k \neq y_r$; is the set of all misses. Random sampling is repeated R times and the importance of the dth feature, i_d, is estimated by averaging, $\hat{i}_d = \sum_{r=1}^{R} \hat{i}_d^{(r)}/R$. If a feature is a good discriminator for nearest neighbor rules, distances $|x_{rd} - x_{id}|$ are expected to be small for hits and large for misses, yielding a large value for $\hat{i}_d$. The feature scale is important. Just

18.3
Variable Selection

In the previous section, we have reviewed techniques for estimation of variable importance. Ranking variables by their importance does not tell us (yet) what variables can be eliminated. To reduce the data dimensionality, we need to do a bit more work.

As discussed in Section 18.1.1, any ranking method can be turned into a selection algorithm by imposing a threshold on feature importance values. This simple-minded approach tends to be quite poor for identifying strongly relevant and eliminating redundant features. Indeed, with the exception of SFS and SBE, all ranking methods discussed above can assign equal ranks to identical features.[1] In this section, we review better algorithms for solving the Minimal-Optimal problem.

18.3.1
Optimal-Set Search Strategies

We have reviewed two sequential ranking strategies, SFS and SBE, in Section 18.2.2. To use them for feature selection, we could apply a simple stopping rule: terminate the search as soon as it fails to improve the value of the criterion Q.

In the limit of infinite statistics, this stopping rule works perfectly well for SBE. If propositions $Q(S \cup X_d, Y) < Q(S, Y)$ and $Y \not\perp X_d | S$ are equivalent, the optimal set S^* found by SBE converges in probability to the Markov boundary of Y: for any $0 < \delta < 1$, we can find a number of observations N_0 such that for any sample with more than N_0 observations we have $P(M(Y) = S^*) \geq 1 - \delta$. In other words, SBE is consistent, as proven in Nilsson et al. (2007).

In contrast, SFS is not consistent. A simple example is the XOR problem. Let $Y = (X_1 \neq X_2)$ for class label Y and independent binary features X_1 and X_2. Add a few irrelevant variables. Generate the two classes in equal numbers. Start SFS and let it add at most one feature at a time. In the infinite statistics limit, every feature by itself shows no correlation with the class label and SFS fails to find any relevant features. In a finite sample, an irrelevant feature may correlate with the class label by chance and be picked by SFS in the first pass. Since SFS does not remove features from the optimal set, this irrelevant feature will never be excluded.

In practice, SFS can be nevertheless preferred over SBE for two reasons:

---

[1] A decision tree typically finds the optimal split in a deterministic manner, for instance, by selecting the variable with a smaller index in the input data out of two identical features. Linear SVM can find any solution $\beta_1 x_1 + \beta_2 x_2$ from the infinite set $\beta_1 + \beta_2 = $ const thus breaking the symmetry between identical features x_1 and x_2 arbitrarily, subject to the choice of the underlying optimization utility and starting point.

1. Speed. If SFS and SBE are used for ranking, they require the same execution time. If they are used for selection, SFS often finds a few powerful features past which there is no improvement in the criterion. Many analysts are comfortable stopping there. It would take much longer for SBE to get to that optimal subset if the number of features is large.
2. Sensitivity to strongly relevant features in small datasets. SFS is the recommended choice if there are a few strongly relevant features mixed with many redundant variables. SBE can accidentally lose a strongly relevant feature in a small sample because its effect is polluted by many redundant features.

Both SFS and SBE are *greedy* search strategies. At every step, they select the feature that improves Q most. It is well known that greedy search algorithms tend to converge to the global optimum slowly. Compare, for instance, the steepest descent and Newton's method for quadratic optimization. Since the new gradient at the minimum of a line search is orthogonal to the direction of this line search, the steepest descent turns at the right angle moving to the minimum in many short perpendicular steps. The steepest descent thus converges to the optimum linearly, while Newton's method converges quadratically. By analogy, SFS adds features to the optimal set in small increments and can miss a good combination of features capable of improving Q by a large margin in one pass. Furthermore, SFS is quite rigid because it is not allowed to remove a feature even if this feature becomes useless when others are added subsequently.

Instead of adding one feature at a time and never removing any features, we can add n best and remove $r < n$ worst features at every step of a sequential algorithm. The *add n remove r* extension is typically more successful than vanilla SFS at obtaining feature sets with lower classification error. It is more CPU-intensive as well since more subsets need to be evaluated at every selection step. The *add 2 remove 1* version performs slightly better than regular SFS in a study of feature selection algorithms by Palombo and Narsky (2009). A similar extension, *remove r add n* with $n < r$, exists for SBE.

It is also known that deterministic algorithms get stuck at local optima more often than randomized procedures. Randomization techniques, summarized in Stracuzzi (2008), include the Las Vegas filter algorithm, random mutation hill climbing, simulated annealing, and genetic algorithms. These algorithms are CPU-intensive on most datasets.

Stracuzzi and Utgoff (2004) propose *random variable elimination* (RVE), most useful when the number of strongly relevant features, R, is small relative to the total number of features D. RVE recursively removes B features until only R features remain in the model. The value of B changes at every level of recursion. An attempt to remove B features from the model is considered successful if no significant loss of accuracy is observed. Since the removed subset can contain strongly relevant features, several attempts may be needed. Using the current dimensionality of the data and the assumed value of R, RVE computes the expected number of removal attempts before one successful reduction occurs. Given $t(F, D)$, the time needed to train classifier F on D features, RVE estimates the expected cost of removing B

redundant features as a function of B, R, and the number of in-model features before the removal attempt. RVE then selects the value of B minimizing this cost. Stracuzzi and Utgoff (2004) show that the average execution time of RVE scales as $O(R \cdot t(F, D))$. For example, the training time for decision trees should scale linearly versus dimensionality, and the complexity of RVE should be $O(R D)$. If the value of R is not known in advance, Stracuzzi and Utgoff (2004) propose estimating it by a binary search.

18.3.2
Multiple Testing: Backward Elimination by Change in Margin (BECM)

So far, we have been discussing feature selection without tying it to statistical significance. For instance, SFS adds features to the optimal set until an improvement for the optimized criterion Q obtained at the current step drops below an arbitrarily chosen threshold. In the limit of infinite statistics, an improvement in the value of Q represents a real effect, and the threshold can be set to zero. For a finite sample, the observed improvement may not be statistically significant, and choosing this threshold may not be straightforward. Moreover, the change ΔQ for a feature at any step of SFS is a random variable, and these random variables can have different distributions for different features. Therefore selecting the best candidate for addition by comparing the values of ΔQ directly for a set of features could not be sensible.

Following the classical statistics, we solve this problem by significance testing. To compare the effects of two features on the model accuracy, we form two respective null hypotheses and compare their p-values. In addition, we frame the stopping criterion for SFS and SBE as a decision based on a test of an appropriately stated sequence of hypotheses.

Before we roll out the full formalism, let us recap what we know about hypothesis testing. First, we state a null hypothesis, H_0, and its alternative, H_1. Then we design a test with a desired level, α, to guarantee that the Type I error, the probability of rejecting the null hypothesis when it is true, does not exceed α. This test is based on a statistic, T, deemed optimal for this problem. The distribution of this statistic under the null hypothesis is known: it is either approximated analytically or modeled nonparametrically by Monte Carlo and resampling techniques. To apply this test, we obtain the value of the chosen statistic for the observed data, t_0, and compute a p-value, the probability of observing equally or less likely values of T under the null. The meaning of "equally or less likely" depends on the choice of T. Typically we need to compute either one-sided, $P(T \leq t_0)$ or $P(T \geq t_0)$, or two-sided tail probability, $P(|T| \geq |t_0|)$. The null hypothesis is rejected if the p-value does not exceed α and not rejected otherwise.

Sometimes we also need to choose among several test procedures. In that case, we fix the test level α and compute the Type II error, the probability of not rejecting the null hypothesis when the alternative is true, for several likely alternatives. The test power, defined as one minus the Type II error, represents the probability of rejecting the null hypothesis when the alternative is true. The best test procedure is chosen by maximizing the power.

This classical testing setup can be applied in a straightforward manner to one null hypothesis and its alternative. The case of feature selection is more complex; since we have more than one feature, we need multiple null hypotheses and consequently multiple tests. For an illustration, consider data with D independent features. Suppose we wish to select features correlating with the class label. To do so, we perform a series of individual tests inspecting one feature at a time. For every test, we state the null hypothesis as "the correlation between this feature and the class label is zero". The statistic of interest, R_d, is the correlation between the dth feature and the class label, and r_d is its value observed in the data. The p-value for the dth test is then $P(|R_d| \geq |r_d|)$. For simplicity, assume that the distribution of the correlation between each feature and the class label is known and the p-value can be computed analytically. Set the level for each individual test at 5%. The probability of rejecting the dth null hypothesis (the probability of recognizing the dth feature as having a significant correlation with the class label) then equals 5%. But the probability of rejecting at least one individual null (the probability of finding at least one feature correlating with the class label) is $1-(1-0.05)^D$. For $D=100$, for instance, this probability exceeds 99%. Even if all our 100 features exhibit no correlation with the class label, in 99% of experiments we conclude by sheer chance that at least one of them correlates with the class label. This appears unsatisfactory.

Our procedure is flawed because we asked the wrong question! Instead of asking "how often we make a Type I error for an individual test", we should ask "how many Type I errors we expect to commit in all individual tests" or perhaps "how likely it is to commit at least one Type I error in all individual tests". We introduce notation in Table 18.2 and summarize error-rate measures for multiple tests in Table 18.3. Techniques for controlling FWER (defined in Table 18.3) and other error rates are reviewed in Dudoit *et al.* (2003). Various measures of power are available as well. We mention only one: the probability of rejecting a false hypothesis, $E\,N_{11}/N_1$, or *average power*.

Following the most popular traditional approach, from now on we focus on FWER. Our goal is thus to design a procedure such that $P(N_{01} > 0)$ does not exceed the desired level α. We make no assumptions about the number of true and false individual null hypotheses, N_0 and N_1, and our procedure must guarantee $P(N_{01} > 0) \leq \alpha$ for any pair of their values. In other words, we wish to have *strong*

Table 18.2 Notation used for discussion of multiple testing. Although we set all numbers in uppercase, the numbers of rejected and not rejected hypotheses are random variables, while the numbers of true and false null hypotheses are fixed unknown parameters.

	Not rejected	Rejected	Row total
True null	N_{00}	N_{01}	N_0
False null	N_{10}	N_{11}	N_1

Table 18.3 Error rates controlled in multiple testing procedures. E is expectation and $N = N_0 + N_1$ is the total number of tested hypotheses.

Abbreviation	Name	Formula
PCER	Per-comparison error rate	$E\,N_{01}/N$
PFER	Per-family error rate	$E\,N_{01}$
FWER	Family-wise error rate	$P(N_{01} > 0)$
FDR	False discovery rate	$E[N_{01}/(N_{01} + N_{11})]$

control of FWER. If our procedure guaranteed $P(N_{01} > 0) \leq \alpha$ for $N_1 = 0$ only, that is, under the *complete null*, we would have *weak control* of FWER.

A common solution is to ensure the strong control of FWER at the overall level α by carrying out a set of individual tests at levels less than α. We have reviewed two such techniques in Section 10.4 in the context of comparing multiple classifiers. Both techniques have disadvantages. Let us return to the example described earlier in this section. Suppose we have D features and wish to select those correlating with the class label. The Bonferroni procedure dictates that we perform D individual tests, each at the level α/D. The Bonferroni correction is conservative: it rejects a true individual null hypothesis less often than any other procedure reviewed here. The Sidak correction would set the level for each individual test at $1 - (1-\alpha)^{1/D}$, reducing the amount of conservatism. This correction however was derived for independent individual hypotheses and would only provide weak control.

A procedure by Westfall and Young (1993) can deal with dependent features by making use of the order statistic for the individual p-values. Let P_d be a random variable for the p-value of the dth test and let $P_{\min} = \min_{d=1,\ldots,D} P_d$ be a random variable representing the minimal p-value in a set of D individual tests. The procedure rejects null hypotheses with $P(P_{\min} \leq p_d) \leq \alpha$, where p_d is the observed p-value for the dth test. Although an improvement over the Sidak correction, this procedure suffers from two shortcomings. The distribution of $P_{\min}$ must be learned by simulation or resampling the observed data; in particular, the tails of this distribution must be accurately estimated for the popular test levels, 5% and 10%. Simulating enough experiments to provide the required tail accuracy can be expensive. Besides, this procedure can guarantee weak control only.

Holm (1979) proposes a sequentially rejective modification to the Bonferroni procedure. Sort the observed p-values for the N individual tests in ascending order: $p_{(1)}, \ldots, p_{(N)}$. Find the *smallest* index n^* in the sorted list such that $p_{(n^*)} > \alpha/(N - n^* + 1)$. Reject all hypotheses with indices $n < n^*$ in the sorted list and do not reject the rest. If n^* cannot be found, reject all hypotheses. Shaffer (1986) describes a generalization of the Holm procedure: instead of comparing $p_{(n)}$ with $\alpha/(N - n + 1)$, compare with $\alpha/N_0(n)$, where $N_0(n)$ is the maximal number of true hypotheses remaining after the previous $H_{(1)}, \ldots, H_{(n-1)}$ hypotheses are rejected. The value of $N_0(n)$ may not be equal to $N - n + 1$ if the falsity of some individual hypotheses rejected earlier logically implies the falsity of some individual hypotheses to be tested later. The procedures by Holm and Shaffer are less conservative than Bonferroni's and provide strong control.

Simes (1986) introduces a modified Bonferroni procedure for testing the complete null. Building on his work, Hommel (1988) devises a sequentially rejective test, further simplified by Hochberg (1988). The Hochberg procedure is a mirror reflection of Holm's. Again, sort the observed *p*-values for the N individual tests in ascending order: $p_{(1)}, \ldots, p_{(N)}$. Find the *largest* index n^* in the sorted list such that $p_{(n^*)} \leq \alpha/(N - n^* + 1)$. Reject all hypotheses with indices $n \leq n^*$ in the sorted list. If n^* cannot be found, do not reject any hypotheses. The Hochberg procedure rejects at least as many hypotheses as the one by Holm (1979).

If the sorted *p*-values increase rapidly enough, the procedures by Holm and Hochberg give identical answers. An example in which they produce different results is shown in Figure 18.4.

The procedures by Holm and Hochberg sort the individual hypotheses by their observed *p*-values. In practice, the order in which the hypotheses are analyzed may be set by the feature selection algorithm. Suppose D features are ranked by SFS or SBE. This ranking defines D feature subsets: the zeroth subset is empty (no features), the first subset has one feature ranked first (either added first by SFS or eliminated last by SBE), the second subset has two features ranked first and second, and so on. Form D individual hypotheses, $H_{(0)}, \ldots, H_{(D-1)}$, one per subset. State the null hypothesis $H_{(d)}$ as: A classifier learned on this subset is at least as accurate as the one learned on all features. Assume that a classifier learned on S' is at least as accurate as the same classifier learned on S if S' includes S as a subset, $S \subset S'$. Then if hypothesis $H_{(d)}$ is true, all hypotheses with higher indices $H_{(d+1)}, \ldots, H_{(D-1)}$ must be also true. Compute *p*-values $p_{(0)}, \ldots, p_{(D-1)}$ for these hypotheses. Apply the Holm procedure to the specified hypothesis list, that is, find the smallest d^* such that $p_{(d^*)} > \alpha/(D - d^*)$ and reject all hypotheses $H_{(d)}$ with

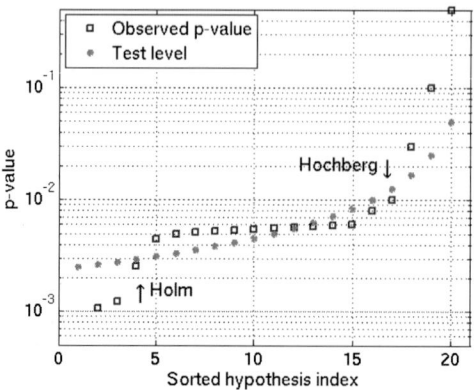

Figure 18.4 Illustration of Holm's and Hochberg's procedures. Holm's procedure rejects 4 hypotheses, and Hochberg's procedure rejects 17 hypotheses. The stars represent the individual test level $\alpha/(N - n + 1)$ for $\alpha = 0.05$ and $N = 20$.

$d < d^*$. Use the subset defined by d^* as the minimal feature set needed to attain the maximal accuracy for the chosen classifier.

Following the proof in Holm (1979), we show that this procedure provides strong control of FWER. Suppose there are D_1 false, $H_{(0)}, \ldots, H_{(D_1-1)}$, and D_0 true, $H_{(D_1)}, \ldots, H_{(D-1)}$, hypotheses; $D_0 + D_1 = D$. To reject one true hypothesis, the procedure must reject the first true hypothesis with index D_1 in this list. Let $P_{(d)}$ be the random variable for the dth p-value in the formed hypothesis list. The probability of rejecting the first true hypothesis is

$$P\left(P_{(D_1)} \leq \frac{\alpha}{D_0}\right) \leq \sum_{d=D_1}^{D-1} P\left(P_{(d)} \leq \frac{\alpha}{D_0}\right) \leq D_0 \frac{\alpha}{D_0} = \alpha. \qquad (18.19)$$

The Type I error therefore does not exceed α no matter what D_0 and D_1 are.

Above, we have assumed that a classifier learned on S is at least as accurate as the same classifier learned on a subset of S. This assumption can be violated for many learners. Ensembles of decision trees usually conform to this assumption for sufficiently large statistics, thanks to their resilience in the presence of noisy and redundant features.

The Holm procedure is best suited for SFS, provided we first estimate Q using all features. Indeed, as soon as we find the first d^* such that $p_{(d^*)} > \alpha/(D - d^*)$, we can stop the selection and return the current subset. If we prefer SBE, the Hochberg procedure is computationally more efficient.

From now on we focus on SBE. We describe the BECM procedure (*backward elimination by change in margin*) summarized in Algorithm 5. This procedure performs two consecutive tasks (two stages) at every SBE step:

1. Selecting the least relevant feature, and;
2. Evaluating the predictive equivalence of the full and reduced feature sets.

A procedure similar to the one described below can be trivially designed for SFS.

As required for SBE, we start with the full set of features, $S = X$. At stage 1, we select the least relevant feature in the current set S. We use the margin-based modification of the FSPP statistic (18.17) and consider two random variables: the classification margin predicted by a classifier trained on the set S and the margin predicted by the same classifier trained on the set S with one feature eliminated, $S' = S \setminus X_d$. Our classifier of choice is an ensemble of decision trees grown either by random forest or boosting. We compute a p-value for every feature X_d in S under the null hypothesis "the median of the difference between the two random variables does not exceed zero". Since removing a strongly relevant feature must lead to a decrease in the classification margin, the alternative hypothesis is "the median is greater than zero". We apply the two classifiers to the same set of data obtained either by cross-validation or by using out-of-bag observations. The p-value is then computed for the two sets of classification margins using a right-tailed version of the Wilcoxon signed-rank paired test. After the p-values for all $|S|$ features in S are computed, we select the feature with the largest p-value, X_{d^*}, and form a new subset, $S^* = S \setminus X_{d^*}$.

Algorithm 5 BECM algorithm for feature selection in classification. OOB stands for out-of-bag, S_i stands for ith feature in set S, and $m(\mathbf{X}, \mathbf{y}; f)$ returns a vector of classification margins for model f applied to data with class labels. Here, we compute OOB margins, but cross-validation can be used instead as well.

Input: Learnable model F
Input: Training data $\{\mathbf{X}_{\text{train}}, \mathbf{y}_{\text{train}}\}$ in space (X, Y) with dimensionality $D = |X|$
Input: Test data $\{\mathbf{X}_{\text{test}}, \mathbf{y}_{\text{test}}\}$ in the same space as $\{\mathbf{X}_{\text{train}}, \mathbf{y}_{\text{train}}\}$
Input: Test level for comparing with the full model α

1: Learn $f_X = F(\mathbf{X}_{\text{train}}, \mathbf{y}_{\text{train}})$
2: Compute predictions $\hat{\mathbf{y}}_X = f_X(\mathbf{X}_{\text{test}})$
3: Compute margins for OOB data $\mathbf{m}_X = m(\mathbf{X}_{\text{OOB}}, \mathbf{y}_{\text{train}}; f_X)$
4: **initialize** $S = X$
5: **initialize** $f_S = f_X$
6: **initialize** $\mathbf{m}_S = \mathbf{m}_X$
7: **for** $d = 1$ to D **do**
8:     **initialize** $\mathbf{p}_S = $ a vector of length $|S|$ filled with zeros
9:     **for** $i = 1$ to $|S|$ **do**
10:         $S' = S \setminus S_i$
11:         Learn $f_{S'} = F(\mathbf{S}'_{\text{train}}, \mathbf{y}_{\text{train}})$
12:         Compute margins for OOB data $\mathbf{m}_{S'} = m(\mathbf{S}'_{\text{OOB}}, \mathbf{y}_{\text{train}}; f_{S'})$
13:         Compare $\mathbf{m}_S$ and $\mathbf{m}_{S'}$ using a right-tailed version of the Wilcoxon signed-rank paired test. Assign the p-value to the ith element of $\mathbf{p}_S$.
14:     **end for**
15:     Find X_{d*}, the feature corresponding to the largest value in $\mathbf{p}_S$
16:     $S^* = S \setminus X_{d*}$
17:     Learn $f_{S^*} = F(\mathbf{S}^*_{\text{train}}, \mathbf{y}_{\text{train}})$
18:     Compute predictions $\hat{\mathbf{y}}_{S^*} = f_{S^*}(\mathbf{S}^*_{\text{test}})$
19:     Run the binomial test on $\mathbf{y}_{\text{test}}$, $\hat{\mathbf{y}}_X$, and $\hat{\mathbf{y}}_{S^*}$. Assign the p-value to p.
20:     **if** $p \leq \alpha/d$ **then**
21:         break loop
22:     **end if**
23:     Compute margins for OOB data $\mathbf{m}_{S^*} = m(\mathbf{S}^*_{\text{OOB}}, \mathbf{y}_{\text{train}}; f_{S^*})$
24:     $S = S^*$
25:     $f_S = f_{S^*}$
26:     $\mathbf{m}_S = \mathbf{m}_{S^*}$
27: **end for**
Output: S

At stage 2, we train a new ensemble on the set S^* and compare its accuracy with that for the ensemble trained on all features. We call the former ensemble a *reduced* *model* and the latter ensemble a *full model*. To perform this comparison, we execute the binomial test described in Section 10.3. We count observations misclassified by the full model and correctly classified by the reduced model, n_{01}, and observations correctly classified by the full model and misclassified by the reduced model, n_{10}. The *p*-value is given by (10.7). If the computed *p*-value does not exceed $\alpha/(D-|S^*|)$, where $|S^*|$ is the number of features in set S^*, we stop the procedure and return the set S. Otherwise we update the current set, $S = S^*$, and go back to stage 1. Comparing the *p*-value to $\alpha/(D - |S^*|)$ amounts to performing the Hochberg procedure.

We apply the BECM algorithm to the MAGIC telescope data with 10 features described in Bock *et al.* (2004) and BaBar K/π particle identification data for particle momenta above 3 GeV with 31 features described in Section 14.9. We choose random forest with 100 trees for the learning model and set the level for the binomial test at $\alpha = 0.05$. Due to the nondeterministic nature of random forest, the selected feature set can vary from one application to another. We run the feature selection procedure 5 times on each dataset and record the number of times a feature is selected in Tables 18.4 and 18.5. BECM selects 6–7 features out of 10 in the MAGIC telescope data. When applied to the K/π data, BECM selects between 6 and 8 features 4 times out of 5 and produces one unusual run selecting 25 features. We do not show this unusual run in Table 18.5.

The BECM algorithm is best suited for analyzing data with not too many variables and not too large fractions of redundant variables. If the data are polluted with many redundant features, it may take a long time for BECM to eliminate them all due to the sequential nature of this algorithm.

Table 18.4 Number of times a feature is selected in 5 runs of the BECM algorithm on the MAGIC telescope data. All 10 features are included.

Variable	BECM
fAlpha	5
fAsym	0
fConc	4
fConc1	1
fDist	5
fLength	5
fM3Long	2
fM3Trans	0
fSize	5
fWidth	5

Table 18.5 Number of times a feature is selected in 4 runs of the BECM algorithm and 5 runs of the ACE algorithm on the K/π data. Features not included in this table have not been selected once by either algorithm. DCH is drift chamber, DIRC is detector of internally reflected Cherenkov light, and EMC is electromagnetic calorimeter.

Variable	Explanation	BECM	ACE
charge	charge	0	5
dEdxdchPulle	dE/dx pull in DCH for electron hypothesis	0	5
dEdxdchPullk	dE/dx pull in DCH for kaon hypothesis	3	5
dEdxdchPullp	dE/dx pull in DCH for proton hypothesis	0	5
dEdxdchPullpi	dE/dx pull in DCH for pion hypothesis	4	5
drcprobK	Kaon probability measured by DIRC	4	5
drcprobP	Proton probability measured by DIRC	4	2
drcprobPi	Pion probability measured by DIRC	0	1
ecal	Energy deposit in EMC	3	1
ecaldivp	Energy deposit in EMC over track momentum	1	0
likeKvsPi	Likelihood of kaon vs pion	4	0
likeKvsPro	Likelihood of kaon vs proton	0	1
likeProvsEle	Likelihood of proton vs electron	0	4
likeProvsPi	Likelihood of proton vs pion	1	0
nbgphot	Number of background photons in DIRC	0	5
ncry	Number of crystals in EMC cluster	0	1
ndch	Last layer with a hit in DCH	4	4
nphot	Number of signal photons in DIRC	0	5
pIPpi	Momentum at interaction point under pion hypothesis	0	4
s1s9	EMC s1s9 ratio	0	1
s9s25	EMC s9s25 ratio	0	3
secmom	EMC second moment in θ and ϕ	0	1
theta	Polar angle	0	5
zmom20	EMC Zernike moment (2,0)	0	4
zmom42	EMC Zernike moment (4,2)	0	1

18.3.3
Estimation of the Reference Distribution by Permutations: Artificial Contrasts with Ensembles (ACE) Algorithm

As discussed in Chapter 4, the equality of two distributions can be evaluated by a permutation test. Suppose we have two samples, one labeled A and one labeled B. Let T be the two-sample statistic of interest and t_0 be its value observed for these two samples. Pool observations from both samples together and permute their labels at random. Record the value of T observed in the permuted data. Repeat many times. Shuffle the labels and record the value of T to form a *reference distribution*. Under the null hypothesis "the two distributions are equal", the value of t_0 should be consistent with the recorded values of T. Set the *p*-value to the probability of observing a value more extreme than t_0 under the reference distribution.

We can use this approach to find strongly relevant features in a classification setting. Let **X** be an $N \times D$ matrix for N observations and D features and let **y** be the associated vector of N class labels. Let T_d be a measure of importance for the *d*th feature, X_d, such as, for instance, the mean gain in the Gini index due to decision splits on this feature. Note that T_d may depend on other features as well; for example, a decision tree may choose to impose less powerful splits on this variable in the presence of preferred alternative split candidates. Shuffle observed values for this feature across the dataset, that is, copy the dataset **X** into **X̃** and then set the *d*th column in **X̃** to a random permutation of the *d*th column in **X**. Measure T_d on the shuffled data. Repeat many times to get a reference distribution for T_d. Shuffling obliterates any dependence between the *d*th feature and the class label and therefore mimics the distribution of T_d for an irrelevant feature with as many states (unique values) as the real variable.

This technique lays foundation to several measures of feature importance. First, we describe a few such measures for bagged trees due to Breiman (2002). These measures are used for feature ranking; however, applying these measures to feature selection would be also possible. We then describe the ACE algorithm due to Tuv *et al.* (2009) built on bagging decision trees with surrogate splits. Surrogate splits, discussed in Section 14.7, are used to estimate the feature importance and predictive association among the features.

Breiman (2002) proposes three measures of feature importance for random forest making use of out-of-bag information. The first measure is based on classification error. Grow one decision tree on a bootstrap replica of the training data and compute its classification error for out-of-bag observations (observations not used for growing this tree). Randomly permute the *d*th feature across these out-of-bag observations and measure the classification error for the same tree again. Record the difference between the error for the permuted data and error for the original data. Continue growing trees until a desired number is reached, for instance, 100 trees. For the *d*th feature, we obtain 100 recorded values of this difference. Divide the mean of this sample by its standard deviation to obtain a standardized importance estimate. Permuting a strongly relevant feature would lead to an increase in the classification error, and a large value of this estimate would therefore indicate an important feature. The standardization allows uniform comparison across features with few and many states. Two other measures proposed in Breiman (2002) are computed in the same way. Classification edge replaces classification error for the second measure, and the count of lowered classification margins in the out-of-bag data is used for the third measure.

As you recall from Section 14.9, feature importance for a decision tree can be estimated by summing changes in the split criterion such as the Gini index over all splits imposed on a given feature in the tree. This measure can be naturally generalized for an ensemble by averaging the importance estimates over individual learners. Trees grown by bagging are usually deep and these importance estimates are computed using many splits. Since the focus is on feature selection rather than classification accuracy, it would make sense to consider all variables for every split. In other words, random selection of variables for every split by random forest should not be used. If the trees are grown with surrogate information, the number of splits used to measure the importance for any given feature increases even further.

The importance estimates based on the summed changes in the split criterion are not terribly useful by themselves. To evaluate their significance, we need to generate reference distributions. Tuv *et al.* (2009) permute every column (feature) in the input data **X** and append these permuted columns, called *artificial contrasts*, to the end of **X**. The new data **X̌** have twice as many features (columns) as the original data **X**, with first D columns taken by the original features and last D columns taken by their permuted replicas. Then Tuv *et al.* (2009) grow many ensembles on **X̌** recording $2D$ importance estimates, one per feature, for every ensemble. The first D estimates are used to obtain the distributions for the real features, and the last D estimates provide the reference distributions. In our experiments, we typically grow 50 ensembles with 50 trees per ensemble.

The ACE algorithm described in Tuv *et al.* (2009) then proceeds in two stages. First, ACE eliminates irrelevant features by comparing the distributions for the real features to the reference distributions for the artificial contrasts. For each real feature, Tuv *et al.* (2009) compute a *p*-value for the *t*-test of the null hypothesis "the mean of the importance distribution for this real feature does not exceed the mean of the importance distribution for the respective artificial contrast" against the obvious alternative. Unlike Tuv *et al.* (2009), we prefer the Wilcoxon rank sum test for the null hypothesis "the importance measure for this real feature is stochastically not greater than the importance measure for the respective artificial contrast" against the equally obvious alternative. The *t*-test assumes that the two samples are drawn from normal distributions, and the rank sum test is nonparametric. A continuous random variable A is said to be stochastically not greater than a continuous random variable B if $P(A \leq z) \geq P(B \leq z)$ for any z; $-\infty < z < \infty$. Tuv *et al.* (2009) apply the Bonferroni correction to reject the null hypotheses for features with *p*-values below α_1/D. Instead, we apply the Hochberg procedure. The hypotheses rejected at the first stage form a set of relevant features. The list of relevant features sorted by their *p*-values in ascending order is passed to the second stage of the algorithm.

At the second stage, ACE eliminates redundant variables using distributions of the feature associations estimated by surrogate splits. This elimination, again, is based on comparison of the distributions for the real features and their artificial contrasts. Let the current size of the relevant feature set (composed of real features only) be D_R. Start with the first feature in the sorted set, the one with the lowest *p*-value. Run $D_R - 1$ tests comparing the distributions of the association between this feature and every other real feature in the current set to the distributions of the association between this feature and the artificial contrast for each real feature in the current set. Tuv *et al.* (2009) use the *t*-test for this comparison. As earlier, we deviate from their prescription to apply the Wilcoxon rank sum test in combination with the Hochberg procedure: find the last feature in the current set with a *p*-value *p* such that $p \leq \alpha_2/(N_R - d^* + 1)$ and remove all features with indices not exceeding d^*. Here, $N_R = D_R(D_R - 1)/2$ is the maximal number of true hypotheses for the set of size D_R. After the current set has been reduced, continue to the next feature in the current set. If there are no features left, exit the second stage.

After the two stages are executed, the selected set contains at least one feature. ACE then removes the effect of the selected features and repeats the two-stage search. The effect of the selected features can be removed in several ways. Tuv *et al.* (2009) train an ensemble using the selected features only and compute residuals by subtracting the predicted response from the observed response. They substitute these residuals for the observed response, remove the selected features from the data and search for relevant features in the remaining set. For regression, computing the residuals is straightforward. In case of classification, Tuv *et al.* (2009) compute residuals using a multiclass logistic regression model described in Friedman *et al.* (2000). In either case, Tuv *et al.* (2009) compute the predicted response for the training data. In contrast, we compute the predicted response for out-of-bag observations. For regression, we compute the residuals by simple subtraction. For classification, we remove observations with correctly predicted labels. We then remove the selected features from the data and repeat the two-stage search on the reduced set of features.

Tuv *et al.* (2009) continue searching for relevant features until no such features can be found. If a large independent dataset is available, we use an additional termination condition. Every time we add new features to the current set, we learn an ensemble on all features selected so far and run the binomial test (10.7) to compare the classification accuracy of the ensemble learned on the current feature set with the accuracy of the ensemble learned on all features. The binomial test is based on this independent dataset. For regression, the binomial test is replaced by a test comparing the distributions of the residuals. If the *p*-value returned by the binomial test is above $\alpha_3/(D - D_R)$, we stop the search and return the current set.

Our version of the ACE algorithm thus requires three test levels α_i; $i = 1, 2, 3$. In our experiments, we set them all to 0.05.

ACE tends to select more features than the BECM algorithm. We run ACE 5 times on the MAGIC telescope and BaBar K/π data. ACE selects all 10 features in the MAGIC telescope data in every run and selects on average twice as many features as BECM does in the K/π data (see Table 18.5). These datasets do not demonstrate the computational advantage of the ACE algorithm. ACE can be much faster than backward elimination for data dominated by irrelevant or redundant features, thanks to its ability to remove many features at once.

18.4
Exercises

1. Let $Z_i \sim N(0, 1)$; $i = 1, \ldots, 5$; be independent normal random variables with zero mean and unit variance. Let $Y = Z_1 + Z_2 + Z_3$ be the response variable. Consider two feature sets:

a) $X_1 = Z_1$
    $X_2 = Z_2$
    $X_3 = Z_1 - Z_2$
    $X_4 = Z_4 + Z_5$
    $X_5 = Z_3 + Z_5$

b) Same as above except $X_3 = Z_1 + Z_2$.

Identify strongly relevant, weakly relevant, and irrelevant features in both sets. Construct Minimal-Optimal and All Relevant feature sets for each problem. Are the Minimal-Optimal sets unique?

2. Compute the Pearson correlation between each feature and the response variable for the problem above. Using these correlation coefficients, rank the features in both sets.
3. Using the ranks computed in Problem 2, apply SFS and SBE to the two feature sets described in Problem 1. For SFS, stop before you add the first irrelevant or redundant feature. For SBE, stop before you eliminate the first nonredundant feature. Compare the sets obtained by SFS and SBE.
4. Come up with an example of two variables for which the Pearson correlation is zero and the mutual information is not zero.

References

Arnosti, N. and Danyluk, A. (2012) Feature selection via probabilistic outputs, 29th Int. Conf. Mach. Learn.

Bevan, A. et al. (ed.) (2013) *Physics of the B Factories*, Springer, to be published.

Bock, R., Chilingarian, A., Gaug, M., Hakl, F., Hengstebeck, T., Jirina, M., Klaschka, J., Kotrc, E., Savicky, P., Towers, S., Vaicilius, A., and Wittek, W. (2004) Methods for multidimensional event classification: a case study using images from a Cherenkov gamma-ray telescope. *Nucl. Instrum. Methods A*, **516**, 511–528.

Breiman, L. (2002) Looking inside the black box, 2nd WALD lecture presented at the 277th meeting of the Institute of Mathematical Statistics, Banff, Alberta, Canada.

Draminski, M., Rada-Iglesias, A., Enroth, S., Wadelius, C., Koronacki, J., and Komorowski, J. (2008) Monte Carlo feature selection for supervised classification. *Bioinformatics*, **24** (1), 110–117.

Dudoit, S., Shaffer, J., and Boldrick, J. (2003) Multiple hypothesis testing in microarray experiments. *Stat. Sci.*, **18** (1), 71–103.

Friedman, J., Hastie, T., and Tibshirani, R. (2000) Additive logistic regression: A statistical view of boosting. *Ann. Stat.*, **28** (2), 337–407.

Gevrey, M., Dimopoulos, I., and Lek, S. (2003) Review and comparison of methods to study the contribution of variables in artificial neural network models. *Ecol. Model.*, **160**, 249–264.

Guyon, I., Weston, J., Barnhill, S., and Vapnik, V. (2002) Gene selection for cancer classification using support vector machines. *Mach. Learn.*, **46**, 389–422.

Hall, P. and Morton, S. (1993) On the estimation of entropy. *Ann. Inst. Stat. Math.*, **45** (1), 69–88.

Hochberg, Y. (1988) A sharper Bonferroni procedure for multiple tests of significance. *Biometrika*, **75** (4), 800–802.

Holm, S. (1979) A simple sequentially rejective multiple test procedure. *Scand. J. Stat.*, **6**, 65–70.

Hommel, G. (1988) A stagewise rejective multiple test procedure based on a modified Bonferroni test. *Biometrika*, **75** (2), 383–386.

John, G., Kohavi, R., and Pfleger, K. (1994) Irrelevant features and the subset selection problem. *Proc. 11th Int. Conf. Mach. Learn.*, pp. 121–129.

Kraskov, A., Stogbauer, H., and Grassberger, P. (2004) Estimating mutual information. *Phys. Rev. E*, **69**, 066138.

Kraskov, A., Stogbauer, H., and Grassberger, P. (2011) Erratum: Estimating mutual information [Phys. Rev. E 69, 066138 (2004)]. *Phys. Rev. E*, **83**, 019903.

Leonenko, N., Pronzato, L., and Savani, V. (2008) A class of Renyi information estimators for multidimensional densities. *Ann. Stat.*, **36** (5), 2153–2182.

Nilsson, R., Pena, J., Bjorkegren, J., and Tegner, J. (2007) Consistent feature selection for pattern recognition in polynomial time. *J. Mach. Learn. Res.*, **8**, 589–612.

Olden, J., Joy, M., and Death, R. (2004) An accurate comparison of methods for quantifying variable importance in artificial neural networks using simulated data. *Ecol. Model.*, **178**, 389–397.

Palombo, G. and Narsky, I. (2009) A numeric comparison of variable selection algorithms for supervised learning. *Nucl. Instrum. Methods Phys. Res. A*, **612** (1), 187–195.

Robnik-Sikonja, M. and Kononenko, I. (2003) Theoretical and empirical analysis of ReliefF and RReliefF. *Mach. Learn.*, **53**, 23–69.

Shaffer, J. (1986) Modified sequentially rejective multiple test procedures. *J. Am. Stat. Assoc.*, **81** (395), 826–831.

Shen, K.Q., Ong, C.J., Li, X.P., and Wilder-Smith, E. (2008) Feature selection via sensitivity analysis of SVM probabilistic outputs. *Mach. Learn.*, **70**, 1–20.

Simes, R. (1986) An improved Bonferroni procedure for multiple tests of significance. *Biometrika*, **73** (3), 751–754.

Stracuzzi, D. and Utgoff, P. (2004) Randomized variable elimination. *J. Mach. Learn. Res.*, **5**, 1331–1362.

Stracuzzi, J. (2008) Randomized feature selection, in *Computational Methods of Feature Selection* (eds H. Liu and H. Motoda), Chapman & Hall, pp. 41–62.

Tuv, E., Borisov, A., Runger, G., and Torkkola, K. (2009) Feature selection with ensembles, artificial variables, and redundancy elimination. *J. Mach. Learn. Res.*, **10**, 1341–1366.

Westfall, P. and Young, S. (1993) *Resampling-Based Multiple Testing: Examples and Methods for p-Value Adjustment*, Wiley.

Yang, H.J., Roe, B., and Zhu, J. (2005) Studies of boosted decision trees for MiniBooNE particle identification. *Nucl. Instrum. Methods Phys. Res. A*, **555** (1/2), 370–385.

Yang, J.B., Shen, K.Q., Ong, C.J., and Li, X.P. (2009) Feature selection for MLP neural network: The use of random permutation of probabilistic outputs. *IEEE Trans. Neural Netw.*, **20** (12), 1911–1922.

Yu, L. (2008) Feature selection for genomic data analysis, in *Computational Methods of Feature Selection* (eds H. Liu and H. Motoda), Chapman & Hall, pp. 337–353.

Yu, L. and Liu, H. (2004) Efficient feature selection via analysis of relevance and redundancy. *J. Mach. Learn. Res.*, **5**, 1205–1224.

19
Bump Hunting in Multivariate Data

In supervised learning, signal and background distributions can be simulated. Classification aims at finding the optimal decision boundary between the two distributions and, optionally, estimating the probability of observing signal at any point in space.

In unsupervised learning (not discussed in this book), no prior knowledge about data is available. The analysis goal is discovering structures in observed data.

In bump hunting, only the background distribution is known. The goal is locating regions in space where observed data deviate from the expected distribution due to background. Such deviations may be a signature of an unknown process, from now on referred to as *signal*.

In a modified version of this problem, the analyst may have some prior knowledge of the sought signal. For example, we may be looking for a Gaussian-like peak on top of uniform background. We may know the peak width but not the location or vice versa. We may have some knowledge of both location and width, within loose bounds.

In this chapter, we focus on problems with little prior information about signal. We review nonparametric algorithms suitable for signal search in multivariate low-dimensional data. We limit the discussion to the statistical problem defined above. In practice, such discoveries require verification. These verification procedures would depend on the physics of the experiment and be designed on a case-by-case basis.

Bump hunting can be viewed as an extension to goodness-of-fit techniques discussed in Chapter 3. In either case, the probability of observing the data given the expected distribution is computed. Goodness-of-fit analysis stops there. If this probability is small, a search for a new signal needs to identify a set of observations most responsible for this small value.

The statistical significance of the found signal is set to the probability of obtaining this or larger signal from the background process alone. An important factor in calculation of this probability is the *look elsewhere effect*. In essence, the probability of finding a signal in a specific region must account for the fact that many regions are inspected by the search. Take a histogram with 100 bins and a Poisson mean of 5 counts in each bin. Suppose we record 15 counts in one bin. The Poisson probability of observing 15 or more counts with mean 5 is 6.9×10^{-5}, and the equivalent significance is 4.0σ. Before taking the measurement, we did not know that such a large excess would be seen in this specific bin. Therefore we need to estimate the probability of observing 15 counts in *any* bin. This probability is, roughly, 100 times the probability of recording 15 counts in one bin, and the respective significance is only 2.7σ. Quote the latter number as an estimate of the signal significance.

We use words "bump" and "excess" throughout this chapter. Techniques discussed here could be equally well applied to detect dips and deficit. The most common type of analysis in particle physics is searching for a signal excess above the known background level.

19.1
Voronoi Tessellation and SLEUTH Algorithm

One technique for bump hunting is the SLEUTH algorithm described in DØ Collaboration (2000). This algorithm is based on Voronoi tessellation.

Let $\mathbf{X} = \{x_n\}_{n=1}^{N}$ be a set of distinct points in a multivariate domain $\mathcal{X}$. A *Voronoi cell*, $V(x_n)$, around point n is a set of all points in the domain $\mathcal{X}$ closer to x_n than to any other point in $\mathbf{X}$. Formally, we put $x \in V(x_n)$ iff $\forall i \neq n: \|x - x_n\| < \|x - x_i\|$. SLEUTH uses the Euclidean distance $\|x - y\|$ between points x and y.

Construction of Voronoi cells for points in a dataset is called *Voronoi tessellation*. The computational cost for this procedure grows exponentially with data dimensionality. On a modern desktop, Voronoi tessellation can be performed in up to a few dimensions.

If the expected distribution is uniform, the expected number of observations in a Voronoi cell is proportional to the cell volume. In addition to background, observed data may contain signal events. The observed excess above the expected level can be quantified using a Poisson probability, $P_n = 1 - \exp(-\lambda_n)$, for the Poisson mean $\lambda_n = B|V(x_n)|/|\mathcal{X}|$. Here, B is the expected background in the entire observable region with volume $|\mathcal{X}|$, and $|V(x_n)|$ is the volume of the nth cell. A small value of P_n indicates a significant effect. Neighbor cells can be combined into larger regions for which the Poisson probability can be computed in a similar manner.

Because computing probabilities for a uniform distribution is simple, SLEUTH transforms the background pdf to uniform within a hypercube. As discussed in Section 3.4.2, this transformation is uniquely defined when the variables are independent. A distribution of D correlated variables can be transformed to uniform in $D!$ ways. Results of the signal search can be sensitive to the transformation choice.

Algorithmically, SLEUTH can be represented by two loops. The outer loop varies the number of observations per region, M, from 1 to N. The inner loop inspects possible divisions of the observed set $\mathbf{X}$ into regions with M observations each.

The number of observations per region is a useful constraint. It simplifies the optimization problem and often leads to better convergence. Without this constraint, the bump hunting algorithm would search for the optimal region across all possible divisions of the observed set into subsets of any size. Tackling such a big problem at once could prove difficult.

Thus, the outer loop is not specific to the SLEUTH algorithm. Sometimes the size of the sought signal is known, at least roughly, a priori. The outer loop can be then shortened or avoided entirely. All other algorithms reviewed in this chapter operate in a similar way.

The most challenging part of SLEUTH is dividing a set with N observations into subsets with M observations each. For simplicity, put $N = MR$, where R is the number of regions after division. A set with N observations can be split into R subsets of equal size in $N!/((M!)^R R!)$ ways. In practice, fewer divisions are allowed because each region must be composed of neighbor cells. Yet for $M = \sqrt{N}$ the number of possible divisions can be impractically high. SLEUTH uses heuristic rules to select preferable regions and deploys a simplex-like algorithm to traverse a set of promising divisions. For every value of M, a region with the smallest Poisson probability, $P_{\min}(M)$, is found.

To find the region with the most significant excess, SLEUTH applies the same search procedure many times to samples drawn from the background distribution. Let $P_{\min}^{(0)}(M)$ be the smallest Poisson probability observed in the data at this value of M. SLEUTH draws B observations from the background distribution and searches for signal at fixed M. This search on background-only samples is repeated many times to obtain a distribution of $P_{\min}(M)$. Let $\tilde{P}_{\min}(M)$ be the fraction of these experiments with $P_{\min}(M)$ below $P_{\min}^{(0)}(M)$, or equivalently a p-value for testing the hypothesis of no excess at M. Let $\tilde{P}_{\min}^{(0)}$ be the minimum of $\tilde{P}_{\min}(M)$ over all values of M, and let M_0 be the respective value of M. The region with the smallest Poisson probability $P_{\min}^{(0)}(M_0)$ is where the most significant excess is seen.

To estimate the significance of the observed signal, DØ Collaboration (2000) applies (without mentioning explicitly) the procedure by Westfall and Young described in Section 18.3.2. Note that random variables $P_{\min}(M_1)$ and $P_{\min}(M_2)$ are not independent: the same observations can contribute to the excess observed at M_1 and M_2. This lack of independence calls for the procedure by Westfall and Young; otherwise a simpler recipe could be used. To apply this procedure, one must estimate the distribution of $\min_M \tilde{P}_{\min}(M)$. This can be done as follows. Draw a sample from the background distribution. Search over $M = 1, \ldots, N$ and record the smallest Poisson probability, $P_{\min}^{(B)}(M)$, for every value of M. Re-draw many background-only samples to obtain a distribution of $P_{\min}(M)$. Let $\tilde{P}_{\min}(M)$ be the fraction of these experiments with $P_{\min}(M)$ below $P_{\min}^{(B)}(M)$, and let $\tilde{P}_{\min}^{(B)} = \min_M \tilde{P}_{\min}(M)$ be the minimum of $\tilde{P}_{\min}(M)$ over all values of M. Repeat this two-loop search many times to obtain a distribution of $\tilde{P}_{\min}^{(B)}$. Set the signal significance to the fraction of $\tilde{P}_{\min}^{(B)}$ values below $\tilde{P}_{\min}^{(0)}$.

This two-stage procedure appropriately includes the look elsewhere effect. Since we do not know a priori what region produces the most significant excess at fixed M, we account for the possibility of observing a similar effect in other regions. Since we do not know a priori what value of M produces the region with the most significant excess, we account for the possibility of observing a similar effect for different values of M.

The SLEUTH algorithm has been designed for continuous variables. If continuous and categorical variables are mixed, distance between two points is not trivially defined. Without such a definition, Voronoi cells cannot be constructed.

19.2
Identifying Box Regions by PRIM and Other Algorithms

Physicists often search for bumps with simple geometrical shapes, in particular, multidimensional boxes. Such bumps have an important advantage: they can be easily visualized and interpreted.

Suppose two datasets are recorded, $\mathbf{X}_{bgr}$ and $\mathbf{X}_{obs}$. The expected data $\mathbf{X}_{bgr}$ are drawn, either by simulation or measurement, from the background distribution. The experimentally observed data $\mathbf{X}_{obs}$ may include an unknown signal. A typical objective is to construct a box $\mathcal{B}$ minimizing the cumulative Poisson probability $P(N_\mathcal{B} \geq N_{obs}; \hat{\lambda}_\mathcal{B})$. Here, N_{obs} and $\hat{\lambda}_\mathcal{B}$ are, respectively, the observed number of observations and expected number of background observations in this box. The mean estimate $\hat{\lambda}_\mathcal{B}$ is determined by counting observations from the set $\mathbf{X}_{bgr}$ in $\mathcal{B}$ and multiplying this count by an appropriate scale factor.

Instead of working with a finite set $\mathbf{X}_{bgr}$, we could model the continuous background distribution from which this dataset is drawn. This approach, taken by the SLEUTH algorithm, can reduce the uncertainty of the background estimate $\hat{\lambda}_\mathcal{B}$ by eliminating its discreteness. Modeling the background distribution can be challenging, even in few dimensions. An inaccurate model of the background distribution can lead to vastly inaccurate estimates $\hat{\lambda}_\mathcal{B}$. In many dimensions, estimates obtained from a finite set $\mathbf{X}_{bgr}$ are generally safer than those based on kernel density estimation.

As always, it is important to obtain an unbiased estimate for the significance of the found bump. One strategy, discussed in Section 19.1, is calibrating the probability of observing an excess of this size on many datasets $\mathbf{X}_{bgr}^{(t)}$; $t = 1, \ldots, T$; drawn from the background distribution. An alternative approach is to follow the usual recipe for supervised learning described in Section 9.3. Split both $\mathbf{X}_{bgr}$ and $\mathbf{X}_{obs}$ into three subsets for training, validating, and testing. Use the training sets to search for the optimal box. Use the validation sets to optimize parameters of the search algorithm, that is, choose the parameter set with best results obtained from the validation data. Use the test sets to obtain an unbiased estimate of the optimized objective such as $P(N_\mathcal{B} \geq N_{obs}; \hat{\lambda}_\mathcal{B})$.

What strategy is best? The answer depends on the anticipated signal size. If the new phenomenon is likely to produce just a few observations, dividing the observed set $\mathbf{X}_{obs}$ could lead to unacceptable loss of sensitivity.

The difficultly in searching for a multivariate box arises from discreteness of the empirical distribution. The observed count N_{obs} is discrete. If the expected count $\lambda_\mathcal{B}$ is estimated using a finite set $\mathbf{X}_{bgr}$, it is discrete as well. The search algorithm must use derivative-free optimization. A popular class of such algorithms is *direct search* methods.

An example of a direct search method is *coordinate descent*. A box in a D-dimensional space is defined by D lower and D upper bounds, $\{l_d\}_{d=1}^D$ and $\{u_d\}_{d=1}^D$. A coordinate descent search traverses these $2D$ variables sequentially updating one variable at a time. Every update changes one lower or one upper bound to decrease the objective. These updates must respect the obvious constraints $l_d < u_d$ and an optional constraint on the minimal box size (in this chapter, "box size" means "number of observations in the box"). Note that we have not specified how each update is found. The term "coordinate descent" describes a number of algorithms with various heuristic steps.

A more flexible algorithm in this class is *pattern search* described, for example, in Nocedal and Wright (2006). Let v be a current estimate of the optimal point in the $2D$-dimensional space and let $\{p_k\}_{k=1}^K$ be a search pattern, that is, a set of search directions. Form a mesh of K points by moving away from the current estimate in each direction, $\{v + \alpha p_k\}_{k=1}^K$, where α is the step size. Poll the points on this mesh by computing their objective function values. If a point with a lower value of the objective is found, move the current estimate v to that point and double the step size α for the next iteration. Otherwise halve the step size α, construct a new mesh, and poll again. Coordinate descent can be viewed as a special case of pattern search in which vectors p_k are aligned with coordinate axes. In this case, we have $K = 4D$ and the mesh is formed by adding $+\alpha$ and $-\alpha$ to every element in v. Note however that pattern search is more greedy: the complete poll strategy moves the best guess in the direction of the largest decrease on the mesh. Coordinate descent updates each element in turn even if updating a different element would produce a larger decrease.

If the objective can have several local minima with size above the minimal bump size, randomized search methods, in particular genetic algorithms and simulated annealing, may succeed where direct search fails.

The *patient rule induction method* (PRIM) due to Friedman and Fisher (1999a) bears some resemblance to pattern search. Initially, the algorithm draws a box around the entire data. Then it proceeds in two stages, *peeling* and *pasting*. Let $\mathcal{B}$ be the current box estimate with $N_\mathcal{B}$ observations.

At the peeling stage, PRIM can reduce the box by increasing l_d or decreasing u_d in dimension d. A reduction is considered if two conditions are satisfied:

1. Not more than $\alpha N_\mathcal{B}$ observations are peeled away from the box.
2. The box size after the reduction does not fall below the minimal allowed size.

Among all possible peels in all dimensions, PRIM chooses the one reducing the objective most. PRIM then shrinks the box and searches for the new optimal reduction. Peeling continues as long as the objective can be decreased by an allowed reduction.

At the pasting stage, PRIM relaxes the bounds of the box found at the peeling stage. Similar to peeling, an expansion is allowed if not more than $\alpha N_\mathcal{B}$ observations are pasted back into the box. Among all possible expansions in all dimensions, PRIM chooses the one reducing the objective most. PRIM then inflates the box and searches for the new optimal expansion. Pasting continues as long as the objective can be decreased by an allowed expansion.

The peel parameter α controls the degree of patience. If α is set close to 1, PRIM can fail to find the optimal box due to excessive greediness. If α is set close to 0, PRIM searches slowly and updates the objective in small decrements with large variance. Friedman and Fisher (1999a) recommend a value in the range $0.05 - 0.1$. For nominal variables, Friedman and Fisher (1999a) propose peeling or pasting at most one level at a time irrespective of the number of observations per level.

An in-depth discussion of PRIM properties can be found in Friedman and Fisher (1999b). An implementation of the PRIM algorithm in StatPatternRecognition reviewed in Chapter 20 has been applied to several analyses at BaBar such as the one described in Aubert *et al.* (2009).

All algorithms discussed in this section can search for more than one box. Find the first box and remove observations inside this box from X_{obs}. If the expected background is estimated using a finite set X_{bgr}, remove observations inside this box from X_{bgr} too. If the expected data are modeled by a continuous background distribution, set the background pdf in this box to zero. Then repeat the search. This strategy, described in Friedman and Fisher (1999a), can apply to pattern search and coordinate descent as well.

19.3
Bump Hunting Through Supervised Learning

In principle, any supervised learning technique described in this book can be used for bump hunting.

As in the previous section, assume that two datasets are recorded, X_{bgr} and X_{obs}. The first set is drawn from the background distribution, and the second set, observed experimentally, may include signal. Label observations in the first set by -1 and observations in the second set by $+1$. Train a binary classifier on these data. Compute classification scores by cross-validation or using out-of-bag data if available. Consider observations with large scores for the $+1$ class as signal candidates.

If we know the size of the sought signal, s, we can focus on s observations with largest scores for the positive class. To measure the bump quality, define an appropriate statistic such as, for instance, the average of these s largest scores.

To estimate the significance of the found bump, apply the bump search procedure to background-only data. Draw a dataset from the background distribution with the same size as X_{obs}. Label X_{bgr} by -1 and the new dataset by $+1$. Train a binary classifier and obtain predictions by cross-validation or using out-of-bag data. Record the average over s largest scores for the positive class. Repeat many times to obtain the empirical distribution of the chosen statistic. Set the p-value of the observed bump to the fraction of these experiments in which the average over s largest scores exceeds the value obtained for X_{obs}. This procedure assumes that statistical fluctuations in X_{bgr} can be neglected because X_{bgr} is much larger than X_{obs}.

If the size of the sought signal is not known, loop over possible values of s to find the one with the smallest p-value.

The learning parameters of the classifier would need to be found by minimizing the p-value as well. This calls for a two-loop setup, one loop for searching for the optimal bump size and another loop for searching for the optimal learning parameters. A similar two-loop approach is used by the SLEUTH algorithm.

A bump found by a powerful learner such as a neural net or boosted trees may not have a simple geometrical structure. Observations with large positive-class scores may be far from each other in the space of input variables. Bumps found by such powerful learners need to pass additional "sanity" checks. Of course, the lack of geometrical constraints on the bump shape can be to the analyst's advantage too.

References

Aubert, B. *et al.* (BaBar Collaboration) (2009) B meson decays to charmless meson pairs containing η or η' mesons. *Phys. Rev. D*, **80**, 112002–112012.

DØ Collaboration (2000) Search for new physics in $e\mu X$ data at DØ using SLEUTH: A quasi-model-independent search strategy for new physics. *Phys. Rev. D*, **62**, 092004.

Friedman, J. and Fisher, N. (1999a) Bump hunting in high dimensional data. *Stat. Comput.*, **9** (2), 123–143.

Friedman, J. and Fisher, N. (1999b) Rejoinder: Discussion of bump hunting in high dimensional data, *CiteSeerX*, doi:10.1.1.29.9582.

Nocedal, J. and Wright, S. (2006) *Numerical Optimization*, Springer Series in Operations Research, Springer, 2nd edn.

20
Software Packages for Machine Learning

In this book, we have reviewed many algorithms. Some of these algorithms are easy to code. Most however could not be used unless a software implementation was available.

Likelihood fits reviewed in Chapter 2 are well supported by software developed in the physics community. MINUIT (2013) is a set of utilities for function minimization developed by Fred James, a CERN physicist, in the 1970s. ROOT (2013), in particular its two extensions, RooFit and RooStats, provides high-level interface for likelihood and chi-square fitting using MINUIT as the underlying engine. Overall, physicists have a lot of expertise coding, executing, and interpreting likelihood fits to parametric distributions. A researcher in particle or astrophysics is unlikely to turn to software suites produced outside of the community.

The situation with respect to machine learning algorithms is different. Whether you wish to estimate the goodness of fit, construct a nonparametric density estimator, apply a variable transformation, or fit a supervised learning model, you should consider non-ROOT software suites. Here, we focus on tools for supervised learning. We refrain from giving accurate snapshots of functionality for various suites and merely describe their general strengths. All reviewed suites are open source, or at least mostly open source; the meaning of "mostly" is explained below.

20.1
Tools Developed in HEP

Beringer *et al.* (2012) mention two machine learning suites: TMVA (2013) and SPR (2013). Both were made available to the broad public in 2006. The two projects are aimed primarily at supervised learning, coded in C++ and hosted at Sourceforge. Both packages provide boosting and bagging, decision trees, neural networks, and linear discriminant analysis. Detailed documentation can be found at the respective links.

TMVA was developed as an extension to ROOT and at present is included in the ROOT distribution. Although we have never run TMVA on non-Linux platforms, we believe that the cross-platform compatibility is supported for all ROOT extensions included in the distribution. TMVA is likely the most popular machine learning package in particle and astrophysics today. This package offers more classification algorithms than SPR, in particular nearest neighbor rules and SVM reviewed in Chapter 13, RuleFit (Friedman and Popescu, 2008), and naive Bayes described in Section 5.10 (named "projective likelihood estimator" in the documentation).

SPR was primarily developed by Narsky with contributions from several other physicists. The package was under active development between 2005 and 2008. This development work ceased in 2008 when Narsky left HEP. At the time of writing this book (SPR version 3.3.2), all implementations appear functional.

The distinctive features of SPR are the framework for multiclass reduction to binary described in Chapter 16, the PRIM algorithm described in Chapter 19, and the sequential "add n remove r" feature selection algorithm described in Chapter 18. A study in Palombo (2012) shows that ensembles of decision trees in SPR are faster and consume less memory than those in TMVA. This advantage is especially significant for random forest.

20.2
R

R (2013) is a computing environment popular among statisticians. This environment implements the S language designed in the 1980s and described in Becker *et al.* (1988). Unlike strongly-typed languages such as C++ or Java, R has dynamic typing and a user-friendly design. It is well suited for interactive data analysis and can be learned in little time. Newcomers may find some syntax cumbersome. For example, to compute $y = x^\mathsf{T} A x$ in R, you would type

```
y <- x%*%A%*%x
```

for a scalar y, vector x, and square matrix A. Elementwise multiplication $C = A \circ B$ however looks much cleaner:

```
C <- A*B
```

Originally developed by Ihaka and Gentleman at the University of Auckland, New Zealand, R gained popularity in the 1990s. At present, the R core team is responsible for maintaining and improving the base R functionality. Additional extension packages can be uploaded by individual contributors to the Comprehensive R Archive Network (CRAN). These extension packages play a vital role. None of the machine learning algorithms described in this book are found in core R, but many are available from extension packages. R is supported on Linux, Windows, and Mac.

Due to dynamic typing, R code is slow. CPU-intensive utilities are typically implemented in C/C++ or Fortran with a thin layer of R code on top. The speed of such utilities can be comparable to those coded in strongly-typed languages. For example, the `randomForest` R package based on the original Fortran code by

Breiman and Cutler is faster than bagged trees in SPR on some datasets and slower on others. Of course, such code does not have the same transparency as pure R.

The core R functionality is well documented. Documentation for extension packages follows an established format. Users can post questions to the R-help mailing list read by many contributors.

Thanks to the large community of users and contributors, R stays on top of major trends in software development. Support for parallel computing is included in core R. RStudio (2013) provides a desktop environment integrating a command window, history, workspace, file editor, graphics, and other tools in one bundle. The Big Data project for R undertaken by Revolution Analytics instruments a number of popular algorithms for data with size exceeding the memory capacity.

Perhaps the most attractive feature of R is access to most up-to-date statistical algorithms. If you are looking for an implementation of a technique described in a recent publication, there is a good chance you will find it at CRAN.

Although R provides solid support for functional programming, its object-oriented framework is rather primitive. Debugging and profiling R code at present is difficult because the call stack is not matched to line numbers in the code.

The most annoying feature of R is insufficient documentation and the lack of consistency across its contributed packages. Installation and application of the core R is a breeze. Installing and using extension packages can be tricky. The quality of support provided by their developers varies. These shortcomings are typical for most, if not all, software produced by a large community without an organized structure.

20.3
MATLAB

MATLAB (2013) (MATrix LABoratory) is a programming environment for algorithm development, data analysis, visualization, and numeric computation, with an estimated one million users. Produced and sold by MathWorks, MATLAB was originally developed as a set of utilities for matrix algebra. Modern versions of these utilities are included in core MATLAB. A lot of functionality is provided in add-on packages, or *toolboxes* in MATLAB lingo. Many algorithms described in this book are included in the Statistics Toolbox; some of these need an Optimization Toolbox license. Others can be coded using utilities provided in the Statistics, Optimization and Global Optimization Toolboxes. Neural network with backpropagation and radial basis function modeling are included in the Neural Network Toolbox. Other toolboxes useful for data analysis in particle and astrophysics are Curve Fitting and Symbolic Math. MATLAB is supported on Windows, Linux and Mac.

Intuitive interface and excellent documentation are the two traditional strengths of MATLAB software. The learning curve for a novice is quick. For example, here is how the two matrix operations shown above would look:

```
y = x'*A*x;
C = A.*B;
```

Unlike all other software suites reviewed here, MATLAB is not free. Since MATLAB is purchased by many universities and research labs worldwide, an individual researcher often can use it with no out-of-pocket expense. If you are a student, you can buy MATLAB at a deep discount.

Comparing MATLAB with R is a bit misguided: R is an environment aimed at statistical data analysis, and MATLAB provides many other tools. Nevertheless we point out some common and distinct features in the two environments.

Just like R, MATLAB has dynamic typing and therefore tends to be slower than strongly-typed languages. Source code for utilities implemented in the MATLAB language is usually included in the distribution. CPU-extensive utilities such as neural networks and decision trees are coded in C or C++ and wrapped in MATLAB interface. The source code for such utilities is usually not distributed to customers. Unlike R, the MATLAB Coder can convert MATLAB code to C/C++ source or libraries.

The MATLAB integrated desktop environment, offered years before Rstudio, provides a similar bundle. Distributed computing is supported by the Parallel Computing Toolbox and the MATLAB Distributed Computing Server. Unlike R, MATLAB comes with a first-class profiler and debugger.

The `comp.soft-sys.matlab` news forum and the MATLAB Answers site hosted at MathWorks support communication among MATLAB users and developers. Neither forum is meant to replace the MathWorks technical support available via phone or online. The MATLAB File Exchange hosted at MathWorks is a searchable repository of user-rated code submissions available for free download. These submissions are maintained by individual developers, either employed by MathWorks or not.

MATLAB supports a full-fledged object-oriented (OO) system. All design patterns developed for strongly-typed languages can be used in MATLAB code as well. Many design problems can be solved in a MATLAB-specific way, thanks to dynamic typing and virtues of the OO system not found in other languages.

20.4
Tools for Java and Python

We have described two C++ machine learning suites developed by physicists and two popular data analysis environments with dynamic typing. Here, we turn our attention to other languages, Java and Python.

WEKA (2013) (Waikato Environment for Knowledge Analysis) is a collection of machine learning algorithms for data mining developed primarily by the Machine Learning Group at University of Waikato, New Zealand, with contributions from others. Suites for data mining, or business analytics, typically focus on tools for data flow such as database communication, data cleaning, categorization, visualization and trend discovery. To that extent, physicists inevitably develop tools of their own, in frameworks supported by individual experiments and often heavily based on ROOT. Although data mining suites provide a great deal of machine learning algorithms, they hide these algorithms under the high-level interface. WEKA, for instance, delivers several graphical user interface (GUI) bundles. Explorer, the most flexible of these, can build graphical models for complete data analysis chains. The Explorer documentation explains how to construct and execute such models. Machine learning algorithms are not mentioned in the Explorer guide; their parameters are briefly described in the documentation for the formal Java class interface. RapidMiner (2013), the successor of WEKA, although providing a greater variety of data flow patterns and machine learning algorithms, follows the same trend.

A Python module for machine learning is provided as scikit-learn (2013). Started in 2007, the toolkit was released for broad consumption in 2010. At present, this project is partially funded by INRIA and Google. The toolkit includes more machine-learning algorithms than TMVA or SPR. Its functionality is comparable to, if not more extensive than, the machine-learning coverage in the Statistics Toolbox of MATLAB. Unlike MATLAB, scikit-learn does not provide algorithms for statistical analysis not attributed to machine learning such as, for instance, distribution fitting and random number generation from a specified distribution. The User Guide gives many examples of usage, and the built-in Python help is of good quality.

20.5
What Software Tool Is Right for You?

We have described a few toolkits for machine learning. There are many others. Which one is best for you?

Ideally, one would choose software by reviewing available options and making an informed choice. In practice, this choice often follows the path of least resistance: my colleagues use this package, and so should I. This approach can serve you fine. Indeed, you would not need to convince others that these algorithms work because your colleagues are familiar with them already, you could use add-on utilities developed by your colleagues for this software, and if you had a question, you could easily find someone to ask. Of course, if physicists always chose tools well known to their colleagues, most algorithms in this book would never be used.

If you are in a position to (re-)evaluate the choice of a machine learning suite for your needs, here are a few things to consider.

If your machine learning application must be intimately tied to a C++ analysis framework, favor a C++ toolkit or a toolkit capable of producing C/C++ libraries. In the absence of such a requirement, any suite reviewed here, as well as many others, could work.

If you anticipate applying machine learning algorithms again in the future, consider learning a powerful suite such as MATLAB or R. It may take a bit longer to finish your current project, but you will be rewarded in the long term by the access to a great variety of algorithms. This is especially true if one day you leave physics and apply your machine-learning skills elsewhere.

If you plan to work with big data and run CPU-intensive algorithms, evaluate your toolkit on data of the appropriate size and complexity. Monitor both execution time and memory consumption. A common mistake is comparing the speed of two packages or two algorithms within one package on a small dataset and assuming that the obtained conclusion would hold for larger volumes of data. Execution times for different algorithms and different implementations of the same algorithm scale differently with the number of observations and variables.

References

Becker, R., Chambers, J., and Wilks, A. (1988) *The New S Language*, Chapman & Hall, New York.

Beringer, J. *et al.* (Particle Data Group), *Phys. Rev. D*, **86**, 010001 (2012). (Statistics, Chapter 36).

Friedman, J. and Popescu, B. (2008) Predictive learning via rule ensembles. *Ann. Appl. Stat.*, **2** (3), 916–954.

MATLAB (2013) MATLAB, The Language of Technical Computing. http://www.mathworks.com/products/matlab/ (accessed 29 July 2013).

MINUIT (2013) MINUIT: Function Minimization and Error Analysis. http://lcgapp.cern.ch/project/cls/work-packages/mathlibs/minuit/ (accessed 29 July 2013).

Palombo, G. (2012) Comparison of the CPU and memory performance of StatPattern-Recognitions (SPR) and Toolkit for Multi-Variate Analysis (TMVA). *Nucl. Instrum. Methods Phys. Res. A*, **672** (21), 6–12.

R (2013) R: A Language and Environment for Statistical Computing. http://www.R-project.org (accessed 29 July 2013).

RapidMiner (2013) Report the Future. http://rapid-i.com/ (accessed 29 July 2013).

ROOT (2013) ROOT. http://root.cern.ch/ (accessed 29 July 2013).

RStudio (2013) Integrated Development Environment for R. http://rstudio.org/ (accessed 29 July 2013).

scikit-learn (2013) scikit-learn: machine learning in Python. http://scikit-learn.org/stable/ (accessed 29 July 2013).

SPR (2013) StatPatternRecognition. http://sourceforge.net/projects/statpatrec/ (accessed 29 July 2013).

TMVA (2013) Toolkit for Multivariate Data Analysis with ROOT. http://tmva.sourceforge.net/ (accessed 29 July 2013).

WEKA (2013) Weka 3: Data Mining Software in Java. http://www.cs.waikato.ac.nz/ml/weka/ (accessed 29 July 2013).

Appendix A: Optimization Algorithms

A.1
Line Search

To minimize a function $f(x)$ over x, we construct a search sequence x_k with

$$x_{k+1} = x_k + \alpha_k p_k, \tag{A1}$$

where p_k is the direction of descent at point x_k: $(\nabla f_k)^T p_k < 0$ for $f_k \equiv f(x_k)$. The descent direction is often found using

$$p_k = -\mathbf{B}_k^{-1} \nabla f_k \tag{A2}$$

for a symmetric nonsingular matrix $\mathbf{B}_k$. For instance, $\mathbf{B}_k = \mathbf{I}$ for steepest descent, and $\mathbf{B}_k = \nabla^2 f_k$ in Newton's method. If $f(x)$ is a scalar function of a scalar argument, the Newton update at $\alpha = 1$ is simply

$$\Delta x = -\frac{df/dx}{d^2 f/dx^2}. \tag{A3}$$

The search (A1) is *globally convergent* if

$$\lim_{k \to \infty} \|\nabla f_k\| = 0. \tag{A4}$$

The search sequence converges to a minimum in the sense (A4) if every step satisfies the two Wolfe conditions:

$$\text{sufficient decrease condition} \quad f_{k+1} \leq f_k + c_1 \alpha_k (\nabla f_k)^T p_k \tag{A5}$$

$$\text{curvature condition} \quad (\nabla f_{k+1})^T p_k \geq c_2 (\nabla f_k)^T p_k \tag{A6}$$

for $0 < c_1 < c_2 < 1$.

An acceptable value of α_k can be found by *backtracking*. Choose an initial estimate α_k. In particular, set the initial α_k to 1 for Newton's method. Choose parameters ρ and c_1 such that $0 < \rho < 1$ and $0 < c_1 < 1$. Continue reducing α_k by a factor of ρ until the sufficient decrease condition (A5) is satisfied. The curvature condition (A6) is then not needed to ensure convergence. The backtracking strategy is popular for Newton's method.

A.2
Linear Programming (LP)

The following *primal* problem

$$\min_{x} c^T x \tag{A7}$$

$$\text{subject to} \quad \mathbf{A}x = b \tag{A8}$$

$$x \geq 0 \tag{A9}$$

can be solved by Lagrange multipliers. Let

$$L(x, \lambda, s) = c^T x - \lambda^T(\mathbf{A}x - b) - s^T x \tag{A10}$$

be the Lagrangian function. Define

$$q(\lambda, s) = \inf_{x} L(x, \lambda, s). \tag{A11}$$

The *dual* problem is then

$$\max_{\lambda, s} q(\lambda, s) \tag{A12}$$

$$\text{subject to} \quad s \geq 0. \tag{A13}$$

The dual has a solution only if

$$\mathbf{A}^T \lambda + s = c. \tag{A14}$$

Otherwise we could construct a sequence x_k such that $\lim_{k \to \infty} L(x_k, \lambda, s) = -\infty$ implying $q(\lambda, s) = -\infty$. If (A14) holds, (A11) simplifies to $q(\lambda, s) = b^T \lambda$, and the dual LP problem takes its conventional form,

$$\max_{\lambda} b^T \lambda \tag{A15}$$

$$\text{subject to} \quad \mathbf{A}^T \lambda + s = c \tag{A16}$$

$$s \geq 0. \tag{A17}$$

Elements of the vector s are often called *slack variables*.

Using (A8) and (A14), it is trivial to show that

$$c^T x - b^T \lambda = s^T x. \tag{A18}$$

Since all elements in x and s must be nonnegative, the left-hand side of (A18) must be nonnegative as well. The difference between the primal and dual objective $c^T x - b^T \lambda$ is called *feasibility gap*.

A theorem known as *first-order necessary optimality conditions* states that if x solves the primal problem (A7)–(A9), we can find solutions to the dual problem (A15)–(A17), λ and s, satisfying the Karush–Kuhn–Tucker (KKT) conditions

$$\mathbf{A}^\top \lambda + s = c \tag{A19}$$

$$\mathbf{A} x = b \tag{A20}$$

$$x \geq 0 \tag{A21}$$

$$s \geq 0 \tag{A22}$$

$$s^\top x = 0. \tag{A23}$$

The *complementarity condition* (A23) implies that the feasibility gap at the solution is zero. Slack variables s_i can be strictly positive only for active constraints $x_i = 0$.

The primal problem is *infeasible* if it is not possible to find x satisfying conditions (A8) and (A9). The primal problem is *unbounded* if we can find a sequence x_k satisfying conditions (A8) and (A9) such that $\lim_{k \to \infty} c^\top x_k = -\infty$. Similar definitions exist for the dual problem.

The *strong duality* theorem proves two propositions for the primal (A7)–(A9) and dual (A15)–(A17) problems:

1. If either problem has a finite solution, so does the other, and the two objective values are equal.
2. If either problem is unbounded, the other problem is infeasible.

The formulation (A7)–(A9) can be used to solve the SVM problem in Section 13.5.

The LP problem discussed in Section 15.1.4 replaces the equality constraint (A8) with an inequality constraint:

$$\max \rho \tag{A24}$$

$$\text{subject to} \quad \rho \mathbf{1} \leq \mathbf{M} \alpha \tag{A25}$$

$$\alpha \geq 0 \tag{A26}$$

$$\alpha^\top \mathbf{1} = 1. \tag{A27}$$

Above, $\mathbf{M}$ is an $N \times T$ matrix of classification margins for N observations and T classifiers, and α is a vector of T learner weights. The dual formulation can be derived by introducing a vector w of N Lagrange multipliers for the term $\mathbf{M}\alpha - \rho \mathbf{1}$. Both primal and dual problems are solved by optimizing $w^\top \mathbf{M} \alpha$ subject to appropriate constraints.

Index

Symbols
0-1 loss, 167, 343
F
  – 5 × 2 cross-validation paired test, 213
  – distribution, 213
k-means, 280, 281
p-value, 21, 40, 57, 141, 145, 153, 233, 386, 396, 403, 419, 422
χ^2, 42, 46, 59
  – distribution, 141, 146, 195, 212, 213, 224, 225, 233, 234
  – test, 41

A
acceptance region, 19, 44
AdaBoost, 185, 333–338, 345, 365, 366, 371, 376
  – AdaBoost.M1, 357, 371, 376
  – AdaBoost.M2, 185, 358, 371
  – pseudo-loss, 357
Anderson–Darling test, 50
arbiter tree, 126
artificial contrasts with ensembles (ACE), 412
asymptotic, 13
asymptotic mean integrated squared error (AMISE), 294
asymptotic significance level, 20

B
backward elimination by change in margin (BECM), 407
bagging, 126, 127, 171, 361–363
  – decision trees, 361, 364, 365, 382
  – in bag, 361
  – out of bag, 361, 411, 413
balanced box decomposition tree, 300
Bartlett test, 151, 225
batch learning, 124, 125, 258, 301
Bayes error, 167, 169, 338

Bayes neural network, 260
Bayesian statistics, 5, 10, 16
BC_a confidence interval, 76, 77
bias, 7, 13
binned data, 39
binomial
  – confidence interval (Clopper–Pearson), 175, 206
  – distribution, 9, 21, 97, 131, 141, 174, 179, 181, 205, 208, 212, 215, 316, 318
  – test, 212, 409, 413
binomial loss, 167
Bonferroni correction, 216
boosting, 126, 210, 366
  – confidence-rated, 358
  – corrective, 347
  – decision trees, 197, 334, 338, 341, 344, 345, 349, 354, 360, 365, 382, 400
  – learning rate, 340
  – shrinkage, 340
  – totally corrective, 347
bootstrap, 65, 100, 179, 180, 207, 361
  – .632+ estimator, 180
  – bias estimation, 67
  – confidence intervals, 68–70
  – leave-one-out, 179
  – parametric, 70
  – replica, 65
  – smoothed, 70
Brownian motion, 352
bump hunting, 121, 417–423

C
canonical correlation analysis (CCA), 238
categorical variable, 129–132, 324
  – dummy variables, 327
  – nominal, 129, 320
  – ordinal, 129, 320

Cauchy distribution, 10, 66
  – median, 66
centering, 145, 146, 228, 229, 237, 238, 269, 270
Chernoff–Lehmann theorem, 44
class label
  – known, 165, 252
  – predicted, 165
  – true, 165
class posterior odds, 132, 222, 223, 231, 233, 234
class posterior probability, 140, 166–168, 183, 191, 202, 222, 224, 233, 241, 247, 260, 265, 288, 294, 309, 310, 312, 324, 339, 340, 343, 357, 361, 362, 364, 376, 377, 387, 395, 396
class prior probability, 166, 182, 183, 184, 188, 191, 199, 222, 223, 286, 294, 312, 362, 400
classification, 165
  – bias, 170, 361
  – binary, 165, 269
  – coefficient of pairwise correlation, 359
  – confidence, 360
  – hard label, 166, 167, 376, 382
  – irreducible noise, 170, 354
  – multiclass, 126, 131, 165, 166, 185, 223, 225, 230, 234, 235, 236, 293, 296, 357, 358, 371–378, 413
  – ordinal, 131
  – prediction strength, 359
  – soft score, 166, 196, 200–202, 205, 224, 265, 283, 284, 334, 336, 343, 357, 358, 376, 382, 396, 422
  – statistical, 340, 343
  – variance, 170, 361
classification cost, 190
classification edge, 343, 345, 349, 358
classification error, 167, 171, 173–177, 178, 181, 182, 195, 196, 198, 309, 313, 316, 341, 343, 349, 354–356
classification margin, 283, 343, 349, 351, 352, 358, 359, 366, 396, 407, 433
classifier diversity, 358–365, 373
clustering, 121, 184, 280
  – k-means, 280
condition number, 150, 151, 282
conditional likelihood, 21
conditional Monte Carlo, 64
confidence interval, 10, 14, 17
  – confidence set, 10
  – hypothesis test, 45
  – pivotal quantity, 45
confidence level, 11
confusion matrix, 106

consistent estimator, 42, 97, 99
correlation, 391
covariance matrix, 66, 147–158, 160, 221, 222, 223–231, 233, 238, 269, 274
Cramér–von Mises test, 50
critical region, 11, 28
cross-entropy, 257, 261, 310, 338
cross-validation, 58, 78–82, 101, 126, 177–179, 315
  – $\mathcal{K}$-fold, 81
  – folds, 177
  – leave-one-out, 79, 140, 178
  – repeated, 178, 214
curse of dimensionality, 102, 302, 303, 327

D

datasets
  – BaBar PID data, 325, 364, 365, 378, 394
  – ionosphere data, 152, 154–158, 197, 198
  – MAGIC telescope data, 239–246, 275–278, 280–282, 295, 297, 299, 359, 396–399, 409, 413
  – two-norm data, 331, 341
decision tree, 127, 130, 138, 139, 171, 173, 183, 365, 371, 374, 381
  – branch node, 309
  – C4.5, 138, 139, 321
  – CART, 307, 321
  – CHAID, 307
  – impurity gain, 138, 309, 312, 319, 321, 323, 325
  – leaf node, 166, 171, 173, 176, 187, 307, 309, 312, 313, 316, 334, 360, 361
  – node impurity, 307, 309, 310, 314, 319
  – optimal pruning level, 315
  – optimal pruning sequence, 314
  – predictive association, 139, 322, 325, 326, 411, 412
  – probabilistic splits, 139
  – pruning, 313–318
  – risk, 313
  – surrogate splits, 139, 321–323, 324, 390, 411, 412
  – terminal node, 309
deconvolution, 112
deflation, 161, 237
degrees of freedom (DOF), 59, 214, 224, 233
density estimation, 89, 120, 121, 294, 295, 296, 299, 301
  – empirical (epdf), 90
  – error, 95
  – histogram, 90
  – kernel, 56

– Monte Carlo, 111
– nonparametric, 93
– optimal, 94
– orthogonal series, 108
– parametric, 89, 93
deviance, 310
dimensionality reduction, 146, 147, 230, 236, 279
discriminant analysis, 183, 221–231, 374
– linear, 166, 206, 210, 222, 232, 236, 364, 400
– pseudolinear, 197, 202
– quadratic, 222

E

efficient estimator, 7, 8, 13
eigenvalue decomposition (EVD), 148–150, 230, 266
– generalized, 230
elbow plot, 152
entropy, 118, 160
– cross-entropy, 257
– differential, 159, 392
– Shannon, 159, 392
error correcting output code (ECOC), 126
– complete design, 372, 374
– exhaustive design, 372
error function, 252, 257
expectation-maximization algorithm, 135
expected prediction error (EPE), 78
exponential family, 9
exponential loss, 167

F

false positive rate (FPR), 196, 227
feature irrelevance, 387, 412
feature ranking, 389–401
– embedded algorithm, 389
– filter, 389
– wrapper, 389
feature redundancy, 388, 412
feature selection, 386
– embedded algorithm, 389
– filter, 389
– wrapper, 389
feature strong relevance, 387, 411
feature weak relevance, 387
feature-based sensitivity of posterior probabilities (FSPP), 395
Feldman–Cousins (FC), 36
Fisher discriminant, 221, 229
Fisher information, 6, 47
– matrix, 6
Fokker–Planck equation, 353

frequentist statistics, 5, 10, 14
– coverage, 22–25
fuzzy rules, 301
– membership function, 301

G

Gauss–Markov theorem, 37
Gauss–Seidel method, 291
generalization error, 169, 178, 180, 312, 338, 340, 344, 351, 361
genetic algorithms, 262
genetic crossover, 262
GentleBoost, 187, 338, 339, 341, 349, 356, 357, 376
Gini diversity index, 127, 310, 334, 338, 389, 411
goodness of fit (GOF), 39
Gram matrix, 266, 268, 270, 272, 273, 279, 288, 296, 299, 375
Gram–Schmidt orthogonalization, 161, 237
Green's function, 272

H

Hamming loss, 372
Hessian, 6, 232, 233, 261, 377
heterogeneous value difference metric (HVDM), 302
heteroscedastic, 75, 85
hinge loss, 167
histogram, 22, 39, 90, 91
– binning, 97
Hosmer–Lemeshow test, 234, 244
Hotelling transform, 147
Householder reflections, 269
hypothesis, 11
– composite, 11, 19, 47
– simple, 59
hypothesis test, 11
– p-value, 211
– alternative hypothesis, 11, 211, 403, 407
– confidence interval, 45
– critical region, 11
– decision rule, 44
– level, 403
– likelihood ratio, 46
– multiple, 216
– null hypothesis, 11, 211, 212, 217, 233, 386, 396, 403, 405–407, 410, 412
– one-sample, 39
– power, 12, 28, 40, 211, 403
– replicability, 214
– α-level test, 211
– α-size test, 211
– score, 46

– simple, 28
– two-sample, 39
– Type I error, 11, 211, 216, 403
– Type II error, 12, 211, 403
– uniformly most powerful (UMP), 12, 21, 29, 59, 215
– Wald, 46

I

IB3 algorithm, 301
ideogram, 93
 – Gaussian, 93
independent component analysis (ICA), 146, 149, 158–162, 236
independent identically distributed (i.i.d.), XV
indicator function, 55, 91
influence function, 83
integrated squared error (ISE), 89, 96
interrater agreement, 360
irrelevant feature, 387
iterative single data algorithm (ISDA), 291, 292, 302, 384

J

jackknife, 70–77, 101
 – delete-d, 74
 – generalized, 74

K

kappa statistic, 360
Karhunen–Loeve transform, 147
kd-tree, 300
kernel density estimation, 56, 92, 201, 202, 205, 207
 – adaptive, 103
 – fuzzy rules, 301
 – multivariate, 92, 106
 – standard deconvolution, 116
kernel dot product, 267, 268, 286
kernel function, 92, 266
kernel regression, 269, 270
kernel ridge regression, 274–278, 279, 295, 296, 302, 383
kernel trick, 268, 274, 286, 288
Kolmogorov–Smirnov (KS) test, 49
Kullback–Leibler divergence, 159
kurtosis, 162
 – excess, 162
 – sample, 225

L

label noise, 340, 354
labeled data, 165
Laplace approximation, 261
learning curve, 81
learning rate, 187, 258, 270, 340, 341
least squares estimation, 13, 22, 60, 201, 232, 236, 269, 377
likelihood
 – equation, 12
 – equation roots, 12
 – extended, 22
 – function, 5
 – maximum likelihood estimator (MLE), 12
 – ratio, 14, 46
 – with missing data, 134, 135
linear correlation, 7
 – Kendall tau, 391
 – Pearson, 389, 391
 – Spearman rank, 391
linear discriminant analysis (LDA), 304, 381
linear regression, 235, 236, 274
 – intercept, 232
 – multiple, 237
 – multivariate, 236
 – regression of an indicator matrix, 236
 – ridge, 274
linear statistic, 72, 75
linearly separable, 252, 283
local density tests, 55
location parameter, 17
log odds, 254
logistic regression, 231–235, 338, 381
logit function, 231, 254
LogitBoost, 339, 340, 357
loss
 – binomial, 167, 339
 – exponential, 167, 341, 376
 – Hamming, 130, 374–376
 – hinge, 167, 283
 – quadratic, 167, 372, 376
loss function, 165, 257, 260
LPBoost, 346, 348, 349, 358

M

machine learning, 121
Mahalanobis distance, 41, 140, 225, 243
maintained hypothesis, 43, 47
Mann–Whitney U test, 200
marginal likelihood, 16
Markov blanket, 387
Markov boundary, 387–389, 401
maximum likelihood, 12
McNemar's test, 211, 215
mean integrated squared error (MISE), 97, 202
mean squared error (MSE), 6, 95, 127, 168, 236
median, 73

Mercer's Theorem, 286
meta learning, 126
minimal covariance determinant, 140
minimal volume ellipsoid, 140
missing data, 132–139, 323, 324
    – augmentation, 137
    – casewise deletion, 135
    – hot deck imputation, 138
    – ignorable missingness, 133
    – imputation, 137
    – imputation by regression, 138
    – lazy evaluation, 136
    – missing at random (MAR), 132
    – missing completely at random (MCAR), 132, 136, 138
    – missing not at random (MNAR), 132
    – reduced model, 136
    – testing MCAR, 133
Monte Carlo
    – bootstrap, 66
    – density estimation, 111
    – permutation test, 64
multiclass learning
    – error correcting output code (ECOC), 372
    – one versus all (OVA), 372
    – one versus one (OVO), 372
multinomial distribution, 22, 48, 141, 185, 234, 363
multiple hypothesis test, 216, 404
    – Bonferroni correction, 216, 405, 412
    – complete null, 405
    – Hochberg procedure, 406, 407, 409, 412
    – Holm procedure, 405, 407
    – Sidak correction, 216, 405
    – strong control, 405
    – weak control, 386, 405
multivariate regression, 236
mutual information, 159, 347, 377, 389, 391
    – symmetric uncertainty, 392

N

Nadaraya–Watson estimator, 294, 295, 299
naive Bayes, 105, 210, 364, 381
nearest neighbor rules, 184, 186, 268, 364, 365, 383
    – approximate neighbor, 300
    – IB3 algorithm, 301
neural network, 280, 374
    – feed-forward, 254–260
    – prior distribution, 260
Neyman modified chi-square, 42
Neyman smooth test, 51
nominal variable, 129, 165, 302, 320

nonsmooth statistics, 73
normal distribution, 14
nuisance parameter, 19, 43

O

one versus all (OVA), 236, 357, 372, 375–377
one versus one (OVO), 372, 374–377
one-sided sampling, 184
online learning, 124, 125, 258, 301
ordinal variable, 129, 320
outliers, 139–141
    – masking, 140
overfitting, 366
overtraining, 173, 366

P

pairwise similarity, 265
parallel learning, 125–127, 362, 399
partial least squares, 232, 236–239
    – loadings, 237
    – scores, 237
    – weights, 237
patient rule induction method (PRIM), 421
Pearson chi-square, 41
perceptron,
    – multi-layer, 254
    – perceptron criterion, 252
permutation
    – sampling, 63–65
    – test, 64
permutation sampling, 152, 180, 395, 410–413
pivotal quantity, 17, 45
point spread function, 112
Poisson distribution, 21, 22
pooled data, 57
posterior distribution, 260
power, 12
power divergence family, 48
principal component analysis (PCA), 146, 147–158, 230, 237
    – correlation PCA, 149
    – covariance PCA, 149
    – loadings, 149
    – nontrivial component, 152
    – principal components, 149
    – scores, 149
prior distribution, 260
profile likelihood, 20
proportional odds model, 132, 233
pseudo-inverse, 228, 238, 268

Q

QR decomposition, 228, 269, 377
quadratic loss, 167

quadratic programming (QP), 289, 347, 377
quantile
- confidence bounds, 207
- QQ plot, 225
queue, 292

R

radial basis functions (RBF), 273, 302
- RBF network, 267, 280
random forest, 361–363, 362, 364, 366, 400, 411
- balanced, 185
- weighted, 362
random subspace, 126, 127, 136, 363, 364, 366, 400
random variable elimination (RVE), 402
Rao–Cramér–Frechet (RCF) bound, 7–10, 35
receiver operating characteristic (ROC), 179, 196–210, 227, 382
- threshold averaging, 205
- vertical averaging, 205
reduced model, 363
redundant feature, 388
reflection, 145
regression, 127, 128, 138, 251, 294, 413
- bias, 169
- irreducible noise, 169
- linear, 235
- locally weighted, 295–297, 383
- multiple, 127, 131, 168, 265, 269, 296, 339
- multivariate, 127, 296
- stepwise, 338
- variance, 169
regularization, 114, 260, 270–278
- Tikhonov, 117
ReliefF, 390, 400
Representer Theorem, 272
resubstitution error, 173, 313, 344
ridge regression, 274
RobustBoost, 354, 355, 358
Rosenblatt's theorem, 96
runs, 51
RUSBoost, 185, 187

S

sample size, 5
scale
- interval, 129, 390
- nominal, 129
- ordinal, 129, 390
scaling, 145, 146, 160, 237, 238, 269, 270, 298
score, 6, 48
- score statistic, 48
- test, 46

scree plot, 152
semi-supervised learning, 122
sequential backward elimination (SBE), 388, 389, 395, 413
- backward elimination by change in margin (BECM), 407
- remove r add n, 402
sequential forward selection (SFS), 395
- add n remove r, 402
sequential minimal optimization (SMO), 290, 291, 302, 375, 384
Sidak correction, 216
sigmoid function, 253, 278, 352, 395
signed-rank test, 396, 407
significance level, 11, 12, 40, 44
- asymptotic, 20
signum, 252
simple test, 47
singular value decomposition (SVD), 150, 228, 238
- thin, 150
smoothed bootstrap, 70
stacked generalization, 126
stagewise modeling, 338
standardization, 145
stationary sequence, 75
stepwise modeling, 338
stratified sampling, 181, 182
strongly relevant feature, 387
Student t, 19
- 10-fold cross-validation test, 214
- 5 × 2 cross-validation paired test, 213
- 10 × 10 cross-validation test with calibrated degrees of freedom, 214
- distribution, 64, 212, 213
- test, 412
subsampling, 75
substitution method, 8
sufficient statistic, 13, 18, 21
supervised learning, 122
support vector machines (SVM), 166, 232, 265, 295, 302, 371, 375, 400
- bias term, 284
- box constraint, 267, 286
- dual problem, 284, 287
- linear, 283–285, 381
- nonlinear, 285, 286, 383
- primal problem, 284, 287
- support vectors, 285
- working set algorithm, 289
synthetic minority oversampling technique (SMOTE), 165, 186
systematic error, 112, 210, 383, 385

T

tables, 39
tall data, 123, 247, 382
target coding, 252
test power, 211
test replicability, 214
Tikhonov regularization, 117, 260
Tomek links, 184, 186
TotalBoost, 347–349, 358
training error, 173
transductive learning, 122
true label, 252
true positive rate (TPR), 196, 227
twoing criterion, 319, 323
Type I error, 211
Type II error, 211

U

unbiased estimator, 7
unbinned data, 39
unfolding, 112–120
uniformly most powerful (UMP) test, 12
unlabeled data, 165
unsupervised learning, 121

V

value difference metric (VDM), 302
Vapnik–Chervonenkis dimension, 344
variable ranking, 386, 389–401
 – embedded algorithm, 389
 – filter, 389
 – wrapper, 389
variable selection, 386
 – embedded algorithm, 389
 – filter, 389
 – wrapper, 389

W

Wald statistic, 47
Wald test, 46
Watson test, 51
weakly relevant feature, 387
weight decay, 260
whitening, 161
wide data, 123, 247, 382
Wilcoxon rank sum test, 200, 412